T0266875

High-Voltage
Engineering

ELECTRICAL AND COMPUTER ENGINEERING
A Series of Reference Books and Textbooks

FOUNDING EDITOR

Marlin O. Thurston
Department of Electrical Engineering
The Ohio State University
Columbus, Ohio

1. Rational Fault Analysis, *edited by Richard Saeks and S. R. Liberty*
2. Nonparametric Methods in Communications, *edited by P. Papantoni-Kazakos and Dimitri Kazakos*
3. Interactive Pattern Recognition, *Yi-tzuu Chien*
4. Solid-State Electronics, *Lawrence E. Murr*
5. Electronic, Magnetic, and Thermal Properties of Solid Materials, *Klaus Schröder*
6. Magnetic-Bubble Memory Technology, *Hsu Chang*
7. Transformer and Inductor Design Handbook, *Colonel Wm. T. McLyman*
8. Electromagnetics: Classical and Modern Theory and Applications, *Samuel Seely and Alexander D. Poularikas*
9. One-Dimensional Digital Signal Processing, *Chi-Tsong Chen*
10. Interconnected Dynamical Systems, *Raymond A. DeCarlo and Richard Saeks*
11. Modern Digital Control Systems, *Raymond G. Jacquot*
12. Hybrid Circuit Design and Manufacture, *Roydn D. Jones*
13. Magnetic Core Selection for Transformers and Inductors: A User's Guide to Practice and Specification, *Colonel Wm. T. McLyman*
14. Static and Rotating Electromagnetic Devices, *Richard H. Engelmann*
15. Energy-Efficient Electric Motors: Selection and Application, *John C. Andreas*
16. Electromagnetic Compossibility, *Heinz M. Schlicke*
17. Electronics: Models, Analysis, and Systems, *James G. Gottling*
18. Digital Filter Design Handbook, *Fred J. Taylor*
19. Multivariable Control: An Introduction, *P. K. Sinha*
20. Flexible Circuits: Design and Applications, *Steve Gurley, with contributions by Carl A. Edstrom, Jr., Ray D. Greenway, and William P. Kelly*

High-Voltage Engineering

Theory and Practice

Second Edition, Revised and Expanded

Mazen Abdel-Salam
Assiut University
Assiut, Egypt

Hussein Anis
Cairo University
Giza, Egypt

Ahdab El-Morshedy
Cairo University
Cairo, Egypt

Roshdy Radwan
Cairo University
Giza, Egypt

CRC Press
Taylor & Francis Group
Boca Raton London New York

CRC Press is an imprint of the
Taylor & Francis Group, an **informa** business

First edition published as *High-Voltage Engineering: Theory and Practice*, edited by Magdi E. Khalifa.

CRC Press
Taylor & Francis Group
6000 Broken Sound Parkway NW, Suite 300
Boca Raton, FL 33487-2742

First issued in paperback 2019

© 2000 by Taylor & Francis Group, LLC
CRC Press is an imprint of Taylor & Francis Group, an Informa business

No claim to original U.S. Government works

ISBN-13: 978-0-8247-0402-5 (hbk)
ISBN-13: 978-0-367-39819-4 (pbk)

This book contains information obtained from authentic and highly regarded sources. Reasonable efforts have been made to publish reliable data and information, but the author and publisher cannot assume responsibility for the validity of all materials or the consequences of their use. The authors and publishers have attempted to trace the copyright holders of all material reproduced in this publication and apologize to copyright holders if permission to publish in this form has not been obtained. If any copyright material has not been acknowledged please write and let us know so we may rectify in any future reprint.

Except as permitted under U.S. Copyright Law, no part of this book may be reprinted, reproduced, transmitted, or utilized in any form by any electronic, mechanical, or other means, now known or hereafter invented, including photocopying, microfilming, and recording, or in any information storage or retrieval system, without written permission from the publishers.

For permission to photocopy or use material electronically from this work, please access www.copyright.com (http://www.copyright.com/) or contact the Copyright Clearance Center, Inc. (CCC), 222 Rosewood Drive, Danvers, MA 01923, 978-750-8400. CCC is a not-for-profit organization that provides licenses and registration for a variety of users. For organizations that have been granted a photocopy license by the CCC, a separate system of payment has been arranged.

Trademark Notice: Product or corporate names may be trademarks or registered trademarks, and are used only for identification and explanation without intent to infringe.

Visit the Taylor & Francis Web site at
http://www.taylorandfrancis.com

and the CRC Press Web site at
http://www.crcpress.com

" . . . and say 'O my Lord increase me in knowledge..''

The Sublime Quran (20:114)

Preface to the Second Edition

Not long after the first edition of this book saw light, Professor Mohamed Khalifa—editor of that edition—passed away after a long and gallant battle with cancer. Professor Khalifa assumed the leadership of high-voltage engineering research in Egypt and the region for nearly three decades. He rose to a position of respect among the international and national scientific communities, where he was known to be an extremely devoted researcher and a gifted teacher. The authors of this book will always remember him as a distinguished scholar, a great mentor, and a beloved colleague.

The need for electrical and electronics engineers to be well trained in high-voltage techniques and to be familiar with related basic theories and concepts is still being felt more keenly today, world wide, than ever before. This book deals with a variety of topics which largely fall under the heading of high-voltage engineering. Such topics as electrical insulation and its breakdown under various stresses have direct applications in industries where electrical and electronic equipment is manufactured. Knowledge of the techniques for generating high-voltage DC, impulse, and AC low and high frequencies is essential for engineers in charge of the design and manufacture of high-voltage power supplies. Such equipment has a wide range of applications, extending from aerospace electronics, radio-transmitting stations, and particle accelerators used in the medical and nuclear physics fields, to television receivers. The subject of gas discharges is another example, as they are encountered not only as noisy corona on HV and EHV overhead power lines and as troublesome arcs in switches, but also in electrostatic precipitators, ozone production plants, gas-discharge lamps, arc furnaces, plasma torches, and ion-implantation equipment, to give just a few examples.

Selected areas of this vast field of high-voltage engineering represent essential ingredients in the information and training that should be acquired by all electrical and electronics engineers. Included are engineers responsible for the design and operation of power transmission and distribution systems at all voltage levels, those responsible for the design and manufacture of electrical and electronic equipment, those in charge of the design and installation of industrial plants and radio transmitting stations, and engineers and scientists involved in the design and installation of particle accelerators in medical and nuclear research. Not to be overlooked are the staff of high-voltage research centers and university faculties active in research, education, and training in fields of high-voltage engineering.

The book requires no prerequisites other than the physics and mathematics courses normally taught to undergraduates in electrical and electronics engineering.

The first edition of this book was published over a decade ago. Over this relatively short period of time high-voltage engineering and related areas have witnessed marked development, manifested by a rapid growth in manufactured high-voltage apparatus combined with the vast accumulation of diverse knowledge and skills. This development assures for high-voltage engineering a brilliant future, in which its topics are slowly becoming a favored subject for research by graduate students. It also points to the need for a well-revised version of this book. The second edition of the book, in endeavoring to describe the present state of the art, is substantially enlarged and deals with more diverse areas.

This edition of the book has undergone many changes and additions. The material included in the first edition has been thoroughly updated by citing numerous new references. New material has been added, while older material has been carefully revised. To further enhance the role of the book as a textbook, all chapters now include a set of exercises problems or solved examples. The book now caters further to the needs of practicing engineers by introducing two new chapters on industrial applications of high-voltage engineering and related issues of hazards and safety.

The large amount of information now available on high-voltage engineering and the time required to organize it led to the decision to divide the book into three parts. The first part, High-Voltage Engineering Theory, outlines the theories and models of insulating materials and dielectric phenomena. It comprises Chapters 2 through 8. The second part, High-Voltage Engineering Practice, on the other hand, addresses numerous practical issues and deals with common high-voltage equipment, some important design problems, and high-voltage testing techniques, which are indispensable to the practicing engineer. It comprises Chapters 9 through 18. The third part, High-Voltage Engineering Applications, comprising Chapters 19 and 20, and is devoted to timely applications of high-voltage engineering not only in the electric power utility but in industry at large.

Chapter 2 is a rigorous but clear discussion of methods of field computation and mapping. New in this edition is the versatile boundary-element method of field computation. In Chapters 3 and 4 the physical phenomena acting in different types of gas discharge are reviewed. A practical account of long air-gap characteristics is added in this edition. In Chapters 5 and 6 corona and arc discharges are described in some detail. Liquid and solid insulating materials are surveyed briefly from the point of view of electrical engineers in Chapters 7 and 8. In this edition more comprehensive accounts of streamers in oils and tracking and treeing in solids are offered.

In the first chapter of Part II (Chapter 9), the principles of design of high-voltage busbars are discussed, together with their insulation and ampacity. Metal-clad gas-insulated switchgear, which is now widely used at the HV and EHV levels, is described in Chapter 10, which covers modern diagnostic practices. The various types of circuit breakers and cables are discussed in Chapters 11 and 12, including an account of solid-state breakers and superconducting cables. In this edition basic circuit interruptions are elaborated, and controlled switching is added.

Chapters 13 through 15 are assigned to a treatment of power system grounding, external and internal overvoltages imposed on system insulation, and techniques adopted for insulation coordination. This edition describes modern computer-based methods of overvoltage evaluation as well as modern distribution grounding systems. The role of insulator contamination in insulation coordination is discussed in this edition at some length.

Chapters 16 through 18 focus on the area of insulation testing, covering the topics of high-voltage generation, measurements, and standard specifications. The remarkable development in modern measurement tools over the past decade is reflected in this new edition.

Two new chapters are introduced in this edition of the book, namely Chapters 19 and 20, which are devoted to present-day applications. Whereas Chapter 19 describes some of the industrial applications of high-voltage engineering outside the area of electric utilities, Chapter 20 deals with the timely issues of electrostatic hazards and safety in some industries.

It is hoped that electrical and electronics engineers reading this book will find it rewarding and stimulating, and that it will enrich their knowledge of several segments of the field of high-voltage engineering, which continues to grow in importance and scope.

Many persons have contributed to this book in various ways, and their efforts are most sincerely appreciated. We—the authors—are indebted to our colleagues for their keen interest, helpful suggestions, constructive criticism, and wholehearted cooperation. We appreciate the valuable help we have received from our numerous graduate students over the years. We also appreciate the courtesy of permission, freely granted by firms and organizations, to reproduce many illustrations included in this book, as well as the enthusiastic and sincere cooperation from the team at Marcel Dekker, Inc.

Mazen Abdel-Salam
Hussein Anis
Ahdab El-Morshedy
Roshdy Radwan

Contents

Part II

Part III

1

Introduction

M. KHALIFA and H. ANIS *Cairo University, Giza, Egypt*

High voltage is present around us in a variety of forms. Lightning discharges are the only known "natural" form of high voltage. In the global system, lightning discharges transfer positive charge upward to restore the system's dynamic balance. The regular current flow between the positively charged ionosphere and the negatively charged earth is thus controlled by global thunderstorm activity.

On the other hand, all other forms of high voltage are man-made or "synthetic" to fulfill specific goals. The presence of man-made high voltage falls into three main purpose-oriented modes in which the interdisciplinary nature of high-voltage technology is fully apparent from the numerous and varied industrial processes in which it appears. The first—and, by far, the most well known—mode is the use of high voltage in electric power transmission to avoid excessive line currents which would render the transmission system uneconomical. For a given power, a higher voltage requires smaller current and, therefore, smaller conductor cross-sectional area and lower conductor costs. Moreover, the current-dependent copper losses are reduced.

The second mode in which high voltage is utilized is based on the fact that bodies charged under high voltage develop an electrostatic force. Applications of this property may be found in cathode-ray tubes, particle accelerators, xerography, spray painting, and electrostatic precipitators which remove vast quantities of dust from the flue gases of factories and power stations which would otherwise pollute the atmosphere.

The third mode of high-voltage presence makes use of the ability of high voltage to initiate ionization in dielectric materials where energy is subsequently released in controlled quantities. To this mode several applications belong, e.g., ignition in internal combustion engines, gas-discharge lamps, and ozone generators that eliminate the unpleasant smells from urban waste as it passes through sewage-treatment plants.

Other areas of high-voltage engineering include the design of high-voltage power supplies for various aerospace and ground applications; high-voltage, high-frequency sources for radio transmission; design, operation, and testing problems for HV, EHV, and UHF overhead lines, cables, and other equipment, and their insulation coordination; various industrial, medical, and other applications of gas discharge; and many other high-voltage phenomena. The field of high-voltage engineering covers all these areas.

The extension of high-voltage technology from energy transmission to encompass many other industrial applications has involved various areas of electrical engineering and science, e.g., circuit analysis, measurement techniques, and transient recording. Physics and chemistry also play prominent roles in high-voltage technology, in which the physical and chemical phenomena governing the electrical properties of insulating materials are of fundamental significance to high-voltage engineers.

1.1   HIGH-VOLTAGE POWER SYSTEMS

Electric power is expected to be the fastest-growing source of end-use energy supply throughout the world over the next two decades. Demand for electricity is projected to grow to 19 trillion kilowatt-hours in 2015, nearly doubling present electricity demand (Fig. 1.1) (US Department of Energy, 1997). In terms of fuel, the world global energy demand is to increase from 9.5 BTOE (billion tons oil equivalent) in 1996 to 13.6 BTOE in 2020, which represents an average annual growth rate of about 1.4% (Hammons et al., 1999).

To meet the ever-increasing demand for electricity, larger power stations are being built for efficient utilization of water power, conventional fuel, and nuclear fuel. It is now quite common to find 1000 MW being generated in a single power station (Fig. 1.2) (Chrobak and Pollak, 1986).

1.1.1   History of High Voltage

Around the year 1910 voltages up to 100 kV were used, and from about 1920 several higher-voltage systems were developed in various parts of the

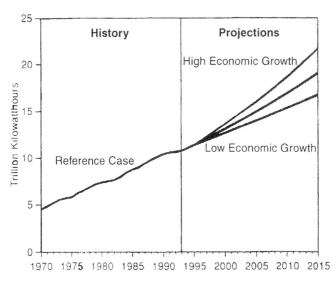

Figure 1.1   World electricity consumption.

world. Powers of 50 MW or so were transmitted over distances of approximately 50 km. During the period from 1930 to 1950 many hydroelectric systems were constructed throughout the world which could deliver 250 MW for 400 km at voltages between 200 kV and 300 kV. In 1954 a voltage of 380 kV was adopted as an international standard and during the 1960s powers up to 1000 MW were transmitted over distances of 1000 km, first at a voltage of 550 kV and then at 765 kV or 800 kV. An ultra-high-voltage (UHV) transmission is currently operating in Russia at 1150 kV with a power capacity of nearly 10,000 MW. Figure 1.3 shows the historical development of high voltage levels.

Some of the earliest electricity supply systems used direct current (DC) for the transmission and distribution of electric power. With the advent of the transformer, however, DC systems were quickly replaced by AC systems in which transmission at much higher voltages was economically more attractive. However, in recent years high-voltage direct current (HVDC) transmission has regained some of its earlier popularity. This is attributed to the fact that HVDC has no transmission distance limitation, no stability or reactive power problems, and produces many economical advantages when used in combination with AC transmission. The high reliability of modern solid-state converters, along with other technical and economic advantages, makes DC transmission sometimes preferable to AC. The disadvantages of HVDC are basically the difficulty to tap off power at points

Figure 1.2 Nuclear power station. (Courtesy of Asea Brown Boveri AG, Switzerland.)

along the line, and the excessive station costs. As a result, there is a critical distance at which DC transmission becomes more economical to adopt than AC. The highest DC transmission voltages are now found in Brazil at ±600 kV and in Russia at ±750 kV.

Rather than having every power station or small group of power stations look after its own loads, tie lines connect power stations into an integrated network over an entire country or a group of neighboring countries. These interconnections greatly improve the reliability and economy of the system and make available the benefits of diversity among the peaks of the component loads. The extra-high-voltage (EHV) grid covering Western Europe operates at 400 kV AC and is interconnected with that of Eastern Europe at substations in Austria (Putz, 1987). The United States and Canada are interconnected at voltage levels of about 750 kV AC and

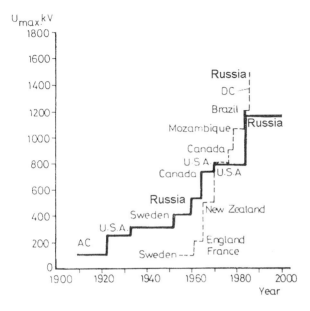

Figure 1.3 Historical development of transmission AC and DC voltages.

±400 kV DC. The partial schematic of a typical power network in Figure 1.4 shows the main components of an outdoor 500 kV substation; Figure 1.5 shows a partial view (Bethmann and Boehle, 1985).

1.2 CLASSIFICATION OF VOLTAGE

The International Electrotechnical Commission (IEC Standards, 1983) defines high voltage as being greater than 1000 V AC, or greater than 1200 V DC. Table 1.1 lists the classes and the most common AC voltages in each class. The standard levels in a country are decided in view of the country's operating limits. Every national transmission system has a minimum and maximum voltage with, usually, a permissible difference of about 200% between these extremes.

1.3 INSULATION OF ELECTRICAL EQUIPMENT

The weak link in the chain of reliability of equipment is still the insulation, and the science of materials has contributed greatly to the development of improved insulation systems for high-voltage apparatus. Equipment insulation is stressed both by continuous normal voltage and by relatively slow

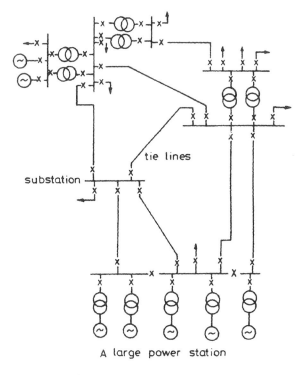

tie lines

substation

A large power station

Figure 1.4 Schematic of a power network showing busbars, transformers, generators, breakers, and feeders.

Figure 1.5 A 500 kV outdoor substation. (Courtesy of South Wales Switchgear.)

Table 1.1 Standard Operating Voltages in Europe and the USA

Voltage class	Normal line voltage	
	Europe (50 Hz)	USA (60 Hz)
Low (LV)	220/240	120 (1-ph)
	380/415	208
	650	600
	1000	—
	kV	kV
Medium (MV)	—	2.4
	5	6.9
	11	12.47
	22	23
	33	34.5
	66	69.0
High (HV)	110	115
	132	138
	156	161
	220	230
Extra high (EHV)	275	287
	380	345
	400	500
	800	765
Ultra high (UHV)	1150	

dynamic and fast transient over-voltages. For insulation to withstand the various stresses, it has to be carefully designed, coordinated, and properly tested. The test voltages have to exceed the normal rated voltages by some factor of safety. The lightning-impulse test voltage is about 10 times the rated AC voltage for low-voltage equipment. The corresponding ratio is only about 3 for UHV equipment (IEC Standards, 1976), for the sake of system economy.

To achieve an economically and technically sound design for equipment insulation, adequate information needs to be acquired. This applies to equipment in power as well as electronics systems, regardless of the size. It applies equally to components and to devices. We should be able to calculate and measure system over-voltages. We should also be able to translate these voltage magnitudes and durations to electrical stresses across the vulnerable parts of equipment insulation and, accordingly, to dimension them. This involves computing and, in some cases, measuring electric fields. Additional necessary information includes the physics of breakdown of solid, liquid, and gaseous insulating materials, and the design, service performance, and testing of insulating components.

REFERENCES

Bethmann J, Boehle B. Brown Boveri Rev, 72:464–471, 1985.

Chrobak E, Pollak M. Brown Boveri Rev 73:224–230, 1986.

Hammons TJ, Richardson B, Abaza M, Browne J, Lay KL, Lindahl G. IEEE Power Eng Rev 19(5): 3–8, 1999.

IEC Standards. Geneva: International Electrotechnical Commission (IEC), Insulation Coordination, Publication 71, 1976.

IEC Standards. Geneva: International Electrotechnical Commission (IEC), Voltage Standards, Publication 38, 1983.

Putz W. Electra 114 (October):21–24, 1987.

US Department of Energy (DOE). IEEE Power Engineering Rev 17(11):16–18, 1997.

2

Electric Fields

M. ABDEL-SALAM *Assiut University, Assiut, Egypt*

2.1 INTRODUCTION

Knowledge of electric fields is necessary in numerous applications in the design and operation of electrical and electronic equipment. To name just a few of these, it is necessary:

1. For the design of insulation and for assessing electrical stresses in high-voltage sources, machine windings, and cables as well as within electronic components
2. In the study of gas discharges
3. In the design of UHV substations and the effects of electric fields in their vicinity
4. In industrial applications such as electrostatic filters and xerography

In this chapter we survey how the electric field is determined for various gap geometries. For space-charge-free fields, we seek solutions to Laplace's equation that satisfy the boundary conditions. In the presence of a space charge, a solution to Poisson's equation is obtained. In some gap geometries, the electric fields can simply be expressed analytically in a closed-form solution; in others, the electric field problem is complex because of the sophisticated boundary conditions, including media with different permittivities and conductivities. In such cases we must resort to experimental modeling or numerical techniques.

2.2 ANALYTICAL CALCULATION OF SPACE-CHARGE-FREE FIELDS

The general governing equation for space-charge-free fields is Laplace's equation

$$\nabla^2 \phi = 0$$

where ϕ is the electrical potential. Laplace's equation has a closed-form solution for geometries with great symmetry and with one-dimensional fields, as explained in detail in books on electromagnetic fields (Shen and Kong, 1983).

2.2.1 Simple Geometries

Configurations consisting of spheres or cylinders occur frequently in high-voltage equipment and the electric field E can easily be calculated for many of these configurations. Table 2.1 contains formulas for the potential ϕ, field E, maximum field E_{max}, and field efficiency factor η:

$$\eta = \frac{E_{av}}{E_{max}} = \frac{V}{dE_{max}} \tag{2.1}$$

where $E_{av} = V/d$ (d is the electrode spacing and V is the applied voltage).

For other cases with electrode surfaces that can be expressed by analytical relations, the field calculation can be simplified by the well-known transformation of coordinates. For example, with two confocal paraboloidal electrodes, the field at any point space x from the tip of the inner paraboloid along its axis is given by

$$E = \frac{V}{\ln(F/f)} \frac{1}{x + f} \tag{2.2}$$

where f and F are the focal lengths of the inner and outer paraboloids, respectively. For a needle–plane gap with needle-tip radius $r \ll h$, the gap length

$$E(x) = \frac{2V}{\ln(4h/r)} \frac{h}{h(2x + r) - (x)^2} \tag{2.3}$$

Table 2.1 Potential, Field, Maximum Field, and Ratio η of Average to Maximum Fields

Configuration	Potential	E	E_{max}	$\eta = E_{av}/E_{max}$	Field of application
Concentric spheres	$\phi(r) = \dfrac{a}{(b-a)}\dfrac{V}{r}(b-r)$	$E(r) = \dfrac{Vba}{r^2(b-a)}$	$E_{max} = \dfrac{Vb}{a(b-a)}$	$\dfrac{a}{b}$	Spherical capacitors, capacitance representation of the dome of a Van de Graaff generator and the structure of the room
Coaxial cylinders	$\phi(r) = \dfrac{V\ln(b/r)}{\ln(b/a)}$	$E(r) = \dfrac{V}{r\ln(b/a)}$	$E_{max} = \dfrac{V}{a\ln(b/a)}$	$\dfrac{a\ln(b/a)}{b-a}$	Cable bushing and GIS
Separated equal spheres	Two-dimensional field	Two-dimensional field	$E_{max} = \dfrac{V}{2R}$ if $d \gg R$	$\dfrac{2R}{d}$ if $\dfrac{d}{R} \gg 1$	Sphere gap for HV measurements, etc.
Equal parallel cylinders	Two-dimensional field	Two-dimensional field	$E_{max} = \dfrac{V}{2R\ln[(d+2R)/R]}$	$\dfrac{2R\ln(d/R)}{d}$ if $\dfrac{d}{R} \gg 4$	Overhead transmission-line arrangements

2.2.2 Transmission-line Conductors to Ground

Consider a multiconductor transmission line of very long conductors with
potentials V_i and charges q_i per unit length ($i = 1, \ldots, \text{n}$) strung parallel to
the ground plane (Fig. 2.1). The effect of ground is usually simulated by
mirror images of the charges q_i. This guarantees that the potential at every
point on the plane is zero.

The potential at an arbitrary point $M(x, y)$ due to conductor 1 and its
image (Fig. 2.1) is given by

$$\phi_{M1} = \frac{q_1}{2\pi\varepsilon_0} \ln\left(\frac{b_{1M}}{a_{1M}}\right) \tag{2.4}$$

where a_{1M} and b_{1M} are the distances shown in Figure 2.1. Where the conductor separations and heights above ground are much larger than their

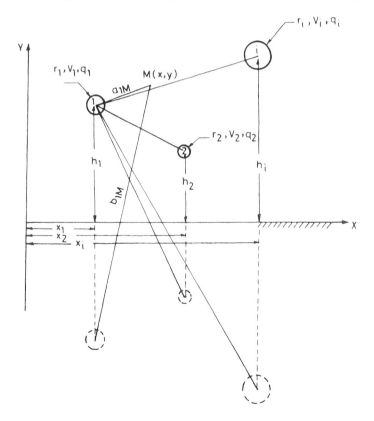

Figure 2.1 Transmission-line conductors over the ground plane. h_i and r_i are the
height above ground and radius of the ith conductor.

diameters, a useful approximation is to consider that the line charge is located at the axis of each conductor.

The field at point M due to conductor 1 and its image is the gradient of ϕ_{M1}, that is,

$$
\begin{aligned}
\bar{E}_{M1} &= -\frac{\partial \phi_{M1}}{\partial x}\bar{u}_x - \frac{\partial \phi_{M1}}{\partial y}\bar{u}_y \\
&= \frac{q_1(x - x_1)}{2\pi\varepsilon_0}\left(\frac{1}{a_{1M}^2} - \frac{1}{b_{1M}^2}\right)\bar{u}_x \\
&\quad + \frac{q_1}{2\pi\varepsilon_0}\left(\frac{y - h_1}{a_{1M}^2} - \frac{y + h_1}{b_{1M}^2}\right)\bar{u}_y
\end{aligned}
\tag{2.5}
$$

Thus the resultant potential at M due to all the transmission-line conductors is given by

$$
\phi_M = \sum_{i=1}^{n} \frac{q_i}{2\pi\varepsilon_0}\ln\frac{b_{iM}}{a_{iM}}
\tag{2.6}
$$

Subsequently, the resultant field at M has the components

$$
(E_x)_M = \sum_{i=1}^{n} \frac{q_i(x - x_i)}{2\pi\varepsilon_0}\left[\frac{1}{(x - x_i)^2 + (y - h_i)^2} - \frac{1}{(x - x_i)^2 + (y + h_i)^2}\right]
\tag{2.7}
$$

$$
(E_y)_M = \sum_{i=1}^{n} \frac{q_i}{2\pi\varepsilon_0}\left[\frac{y - h_i}{(x - x_i)^2 + (y - h_i)^2} - \frac{y + h_i}{(x - x_i)^2 + (y + h_i)^2}\right]
\tag{2.8}
$$

If the point M is placed on the first conductor, then $\phi_M = V_1$, $b_{1M} = 2h_1$, $a_{1M} = r_1$, and equation (2.6) takes the form

$$
V_1 = \frac{q_1}{2\pi\varepsilon_0}\ln\frac{2h_1}{r_1} + \frac{q_2}{2\pi\varepsilon_0}\ln\frac{b_{12}}{a_{12}} + \frac{q_3}{2\pi\varepsilon_0}\ln\frac{b_{13}}{a_{13}} + \ldots + \frac{q_n}{2\pi\varepsilon_0}\ln\frac{b_{1n}}{a_{1n}}
\tag{2.9}
$$

For brevity, equation (2.9) may be written in the form

$$
\left.\begin{aligned}
V_1 &= q_1 P_{11} + q_2 P_{21} + q_3 P_{31} + \ldots \\
V_2 &= q_1 P_{12} + q_2 P_{22} + q_3 P_{32} + \ldots \\
&\ \ . \\
&\ \ . \\
&\ \ . \\
V_i &= q_1 P_{1i} + q_2 P_{2i} + q_3 P_{3i} + \ldots
\end{aligned}\right\}
\tag{2.10}
$$

or

$$
[V] = [P][q]
$$

where $P_{ij} = (1/2\pi\varepsilon_0)\ln(b_{ij}/a_{ij})$ is the potential coefficient in meters per farad. It is evident that $P_{ij} = P_{ji}$ and that they are positive quantities, independent of the sign of charge and potential. Equations (2.10) make up the first set of Maxwell's equations.

2.2.2.1 Transmission Line with Single Conductor

For a monopolar DC line with a single conductor, equation (2.9) reduces to

$$V = qP = \frac{q}{2\pi\varepsilon_0}\ln\frac{2h}{r} \tag{2.11}$$

It can easily be shown that the equipotentials around the conductor take the form of circles. The field at the conductor surface reaches its maximum value at the point facing the ground, as expected, and can be easily evaluated.

$$E_{max} = \frac{V}{r\ln(2h/r)} \tag{2.12}$$

The field strength at the conductor surface is decided primarily by the conductor radius and whether the conductors are bundled. For bipolar DC lines and for three-phase AC lines (Fig. 2.2) the spatial field variations can be calculated using equations (2.7) and (2.8). Knowledge of the maximum surface field for different conductor bundles is necessary for estimating the corona discharge phenomena on HV and EHV transmission lines.

2.2.2.2 Transmission Line with Bundled Conductors

In practical designs, the line conductor radius r, the separation d between the subconductors of a bundle, the separation D between phases, and their height h above ground would comply with the following relation:

$$r < d \ll D \text{ and } h \tag{2.13}$$

The effects of the image charges can safely be neglected when evaluating the fields at the conductor surfaces. The magnitude and direction of the maximum conductor surface field can be evaluated using the line conductor charges and potential coefficients. These can in turn, be evaluated in terms of the line voltage and geometry. For example, in case of a monopolar bundle 2, DC line of voltage V, equation (2.10) reduces to

$$V = P_{11}q_1 + P_{12}q_2 = P_{21}q_1 + P_{22}q_2 \tag{2.14}$$

The potential coefficients P_{11}, P_{12}, and P_{22} depend on the conductor arrangement in the bundle (Table 2.2). The maximum field strength E_{max} at either subconductor of the bundle is expressed as

$$E_{max} = \frac{V}{\ln(4h^2/rd)}\left(\frac{1}{r} + \frac{1}{d}\right) \tag{2.15}$$

Table 2.2 Potential Coefficients of Equation (2.14) for a Bundle of Two Subconductors when Arranged Horizontally and Vertically

Horizontal bundle — 2 ($\leftarrow \circ \quad \circ \rightarrow$)	Vertical bundle — 2 $\left(\begin{smallmatrix}\uparrow\\\circ\\\circ\\\downarrow\end{smallmatrix}\right)$
$P_{11} = \dfrac{1}{2\pi\varepsilon_0}\ln\dfrac{2h}{r} = P_{22}$	$P_{11} = \dfrac{1}{2\pi\varepsilon_0}\ln\dfrac{2(h+d/2)}{r}$
$P_{12} = \dfrac{1}{2\pi\varepsilon_0}\ln\dfrac{\sqrt{4h^2+d^2}}{d} = P_{21}$	$P_{12} = \dfrac{1}{2\pi\varepsilon_0}\ln\dfrac{2h}{d} = P_{21}$ if $h \gg d$
$\quad = \dfrac{1}{2\pi\varepsilon_0}\ln\dfrac{2h}{d}$ if $h \gg d$	$P_{22} = \dfrac{1}{2\pi\varepsilon_0}\ln\dfrac{2(h-d/2)}{r}$
$q_1 = \dfrac{V}{P_{11}+P_{12}} = \dfrac{2\pi\varepsilon_0 V}{\ln(2h/r)+\ln(2h/d)}$	$q_1 = \dfrac{(P_{22}-P_{12})V}{P_{11}P_{22}-P_{12}^2}$
$\quad = q_2$	$q_2 = \dfrac{(P_{11}-P_{12})V}{P_{11}P_{22}-P_{12}^2}$

For three-phase AC or bipolar DC lines, the maximum surface field can be similarly calculated for bundles of two, three, and four subconductors (El-Arabaty et al., 1977).

2.2.2.3 Approximations

For quick estimation of the maximum surface field for overhead line conductors, some expressions were proposed by previous workers (e.g., Timascheff, 1975; Comber and Zaffanella, 1974). They suggested the follow-

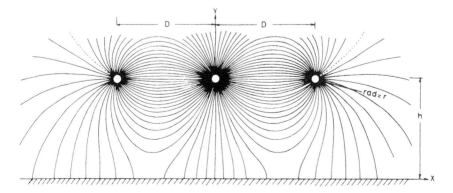

Figure 2.2 Three-phase transmission line with a sketch of field lines at the instant when the center phase voltage is maximum.

ing relation for the angular variation of the surface field E around the subconductor of a multiconductor bundle:

$$E_\theta = E_a\left[1 + \frac{2r}{C}(n-1)\cos\theta\right] \tag{2.16}$$

where E_a is the average field obtained from the well-known relation

$$E_a = \frac{q}{2\pi\varepsilon_0 r} \tag{2.17}$$

with the parameters defined as given in Figure 2.3.

2.2.3 Fields in Multidielectric Media

Usually a low-voltage or a high-voltage insulation system (e.g., a high-voltage bushing and a machine insulation) comprises more than one insulating material. The electric field distribution in a multidielectric insulation system can easily be explained as a parallel-plate capacitor with two layers of different permittivities ε_1 and ε_2. The electric displacement is the same in both layers; hence the electric field intensities E_1 and E_2 are related as $E_1/E_2 = \varepsilon_2/\varepsilon_1$. This means that partial replacement of the gas insulation with a solid material does not improve the dielectric strength of the whole. On the contrary, it becomes more heavily stressed, the higher the ε value of the inserted solid dielectric slab.

    The constancy of the electric flux density at the interface of a multi-dielectric system makes it possible to increase the field uniformity by an appropriate choice of dielectric permittivities. A typical example is the grading of insulation in coaxial cables (see Section 12.12).

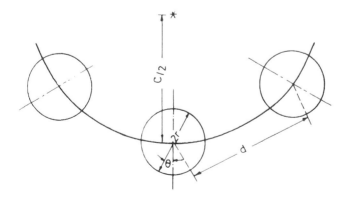

Figure 2.3   Phase geometry of a bundle of n subconductors.

2.2.3.1 Refraction Law of Electric Field

For AC voltage applications, free charges are absent at the interface between dielectrics and the polarization charges define the boundary conditions. Then the angles of incidence (θ_1) and refraction (θ_2) are related as follows:

$$\frac{\tan \theta_1}{\tan \theta_2} = \frac{E_{t1}/E_{n1}}{E_{t2}/E_{n2}} = \frac{E_{n2}}{E_{n1}} = \frac{\varepsilon_1}{\varepsilon_2} \tag{2.18}$$

On the other hand, in DC voltage applications, accumulation of free charges at the interface takes place because of the differing conductivities of the materials (interfacial polarization). More is said about such cases in Section 2.5.

Equation (2.18) points out that the electric displacement flux lines penetrating from a dielectric of a high ε into one of a much lower ε are forced to leave the material nearly perpendicular to its surface. This means that the equipotential surfaces in the lower-permittivity dielectric are forced to be nearly parallel to the interface, and the dielectric of the much higher ε behaves almost like a conductor as $\varepsilon \to \infty$.

2.3 EXPERIMENTAL ANALOGS FOR SPACE-CHARGE-FREE FIELDS

The infinitely long, smooth cylindrical conductors parallel to each other and to ground with separations much larger than their diameters represent overhead transmission lines. They are only special simple cases compared to such conductor arrangements as in multicore cables, gas-insulated switchgear, transformers, and machines. There, conductors take various shapes, their sizes are comparable to their separations, and more than one dielectric medium is involved.

For assessing the electric field distributions in such complex arrangements, sometimes in three dimensions, analytical methods are quite unsuitable. Two other approaches are in use, experimental analogs and numerical techniques. The former is the subject of this section and the latter is covered in Section 2.4.

The potential distribution in conductive media in current equilibrium conditions satisfies Laplace's equation, the same as the electric fields in space-charge-free regions. This fact makes it possible to obtain solutions to many difficult electrostatic field problems by constructing an analogous potential distribution in a conductive medium where the potential and field distributions can be measured directly. The conductors and insulation arrangements can be represented using an electrolytic tank, a sheet of semiconducting paper, or by a mesh of resistors.

2.3.1 Electrolytic Tank

This method has been widely used for decades. Equipotential boundaries are represented in the tank by specially formed sheets of metal. For example, a single dielectric problem such as a three-core cable may be represented using a flat tank as shown in Figure 2.4. Different permittivities are represented by electrolytes of different conductivities separated by special partitions. Otherwise, the tank base can be specially shaped. The conductance of the entire model is a scale model of the capacitance of the system being represented, care being taken to minimize the errors.

2.3.2 Semiconducting Paper Analog

A somewhat less accurate but attractively simple alternative to the electrolyte is semiconductive paper. Its surface resistivity is of the order of 1 kΩ per square (Fig. 2.5). Errors in this method result, among other things, from the nonhomogeneity of the paper resistivity, its dependence on the ambient humidity, and the contact resistance to the electrodes.

2.3.3 Resistive-Mesh Analog

In this analog, the continuous field is replaced by a discrete set of points as depicted by a mesh of resistors (Fig. 2.6). Replacement of the continuous field by discrete resistors does, of course, introduce an error arising from the finite mesh size. This error may be reduced by reducing the mesh size.

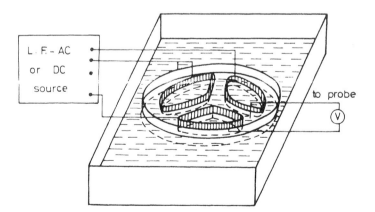

Figure 2.4 Electrolytic tank model of a three-core cable represented at the instant when one core is at zero voltage, the same as the sheath.

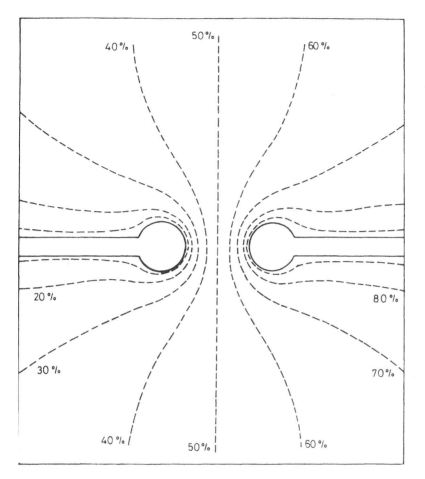

Figure 2.5 Field plot between two spheres with skanks, as plotted by a semiconducting-paper model.

For the mesh point 0 of Figure 2.6, the continuity equation of current demands that

$$\frac{V_1 - V_0}{R} + \frac{V_2 - V_0}{R} + \frac{V_3 - V_0}{R} + \frac{V_4 - V_0}{R} = I_0 \qquad (2.19)$$

or

$$V_1 + V_2 + V_3 + V_4 - 4V_0 = I_0 R$$

Thus the resistive network analog can approximately represent Laplacian fields ($I_0 = 0$) as well as Poissonian fields ($I_0 \neq 0$). For $I_0 = 0$,

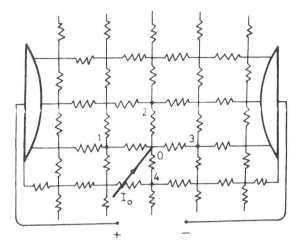

Figure 2.6 Resistive mesh analog of the field pattern between two electrodes.

$$V = \frac{V_1 + V_2 + V_3 + V_4}{4} \tag{2.20}$$

Thus the potential at each mesh point is determined by the adjacent point potentials, which are in turn determined by others. Each potential value represents, in fact, an interpolation between those one step away from it, the same as in numerical calculations using finite differences (Section 2.4.4).

Due to discretization, simulation of electrostatic fields in the vicinity of curved surfaces of the electrodes is bound to be of reduced accuracy. Nevertheless, the analog can be used in three-dimensional field problems by extending the mesh grid in three dimensions. Time dependence may also be represented using impedance networks containing capacitors and inductors.

2.4   NUMERICAL COMPUTATION OF SPACE-CHARGE-FREE FIELDS

Several numerical techniques have recently been reported in the literature for solving Laplace's and Poisson's equations for the fields between complex electrode arrangements. Each numerical technique has advantages and drawbacks. Some techniques may complement others. They are discussed briefly below.

2.4.1 Successive Imaging Technique

The successive imaging technique is based on the concept of imaginary point or line charges located outside the region of field evaluation, but chosen such that their field within this region is identical to that of the induced charges on the electrode boundaries of the region. In two cases only is it possible to implement the successive imaging technique (Khalifa and Abdel-Salam, 1974a) when the region of field evaluation is a single-permittivity medium. The first case is the field between a long, thin wire with a charge q (C/m) and a parallel long, sizable, conducting cylinder. The usual practice is to represent the cylinder by a line charge as an image within the cylinder, parallel to its axis and displaced from it by a distance

$$\delta = \frac{r^2}{D} \tag{2.21}$$

The two charges on the wire and cylinder are equal and opposite. The field outside the cylinder can be computed in terms of the charges, which in turn are determined by the voltage applied to the cylinder.

The second case is that of a field between a point charge Q and a charged spherical conductor. The sphere may be represented by a point charge as an image equal to $-Qr/D$, located a distance $\delta = r^2/D$ away from its center. The imaging technique cannot be extended to multidielectric media or to three-dimensional fields.

2.4.1.1 Successive Imaging among Cylinders

A system of parallel cylindrical conductors, as in gas-insulated switchgear (GIS), is used as an example to explain how the image charges are determined. For the system of conducting cylinders shown in Figure 2.7a, their respective potentials are V_1, V_2, ..., V_4 and their surface charges are q_1, q_2, ..., q_4. Of course, if they are placed near the ground or its equivalent, each conductor will also have its image (see Fig. 2.1).

Now, as discussed above, each cylinder presents an equal and opposite image for each of the charges existing outside it (Fig. 2.7b). The image locations are calculated according to equation (2.21). It is noticed that each of the n cylindrical conductors has $(n-1)$ images. If the ground is considered, the number of images inside each cylinder will be $(2n-1)$. These line charges are used to compute the field in the space *outside* the cylinders. The computations would not yield an equipotential surface exactly coinciding on each cylinder with its respective voltage. Therefore, to reduce errors in the results, a next-imaging process is performed.

At any step in the imaging process, each of the images in a conductor obtained by the preceding step is resolved into $(2n-1)$ new images. The

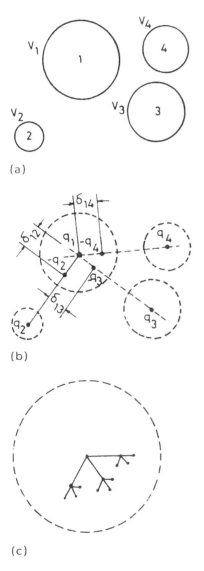

Figure 2.7 (a) Group of conductors to be modeled; (b) first-order images (i.e., of charges q_2, q_3, and q_4 inside cylinder 1); (c) locating higher-order images.

largest distance of each of these $(2n - 1)$ new images from the image it replaces can be used as a criterion to determine whether the successive imaging process should be continued. A large distance of this type indicates that further imaging is necessary to improve accuracy.

If the imaging process is continued uniformly for all the conductors, the number of images in any conductor after the mth imaging process will be $(2n - 1)^m$. The system of images starts to form a branching network in each conductor, as suggested in Figure 2.7c.

Let N line charges represent a system of N conductors $(N = 2n)$ at various potentials. Suppose that the number of line charges found to represent the system after termination of the imaging process is M. Choosing N points, one on each conductor surface, to be the N points of known potentials, the following equation needs to be solved for the unknown charge density vector $[q']$:

$$[V] = [P'][q'] \qquad (2.22)$$

where $[V]$ is an $N \times 1$ known vector of potentials of the N conductors, $[P']$ is an $N \times M$ known vector of the potential-coefficient matrix, and $[q']$ is an $M \times 1$ vector of all line charges used in the system. However, the M line charges representing the system are linearly related to the original N line charges by

$$[q'] = [G][q] \qquad (2.23)$$

where $[G]$ is an $M \times N$ matrix evaluated during the imaging process.

The electric potential and field may be calculated at any point outside the conductor boundaries as

$$\phi(a) = \sum_{i=1}^{M} p_i' q_i' \qquad (2.24)$$

$$E(a) = \frac{1}{2\pi\varepsilon_0} \sum_{i=1}^{M} \frac{q_i'}{r_{ia}} \bar{u}_{ia} \qquad (2.25)$$

where r_{ia} is the distance from q_i' to the point a and $\bar{u}_{ia}$ is a unit vector along the direction from q_i' to a. As a practical application, the successive imaging technique has been used to calculate equipotential contours for an overhead transmission line with two circuits and two ground wires (Fig. 2.8).

2.4.1.2 Successive Imaging between Spheres

Given two spheres located near, and isolated from, each other, with potentials V_1 and V_2, the same imaging process as that described above for parallel cylinders can easily be adopted here with $n = 2$, but only if the spheres are far from ground.

The advantages characterizing the successive imaging technique are the well-defined coordinates of the image charges, and the accuracy, which can be improved by increasing the number of successive images.

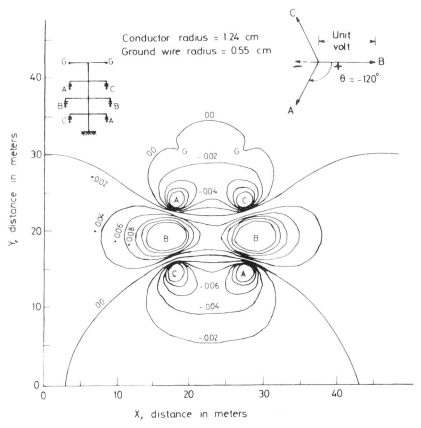

Figure 2.8 Equipotential contours plotted for a double-circuit overhead test line using a successive imaging technique.

Recently, successive imaging has been combined with the charge-simulation technique to compute the electric field of monopolar, homopolar, and bipolar conductor bundles (Abdel-Salam and El-Mohandes, 1987). Details are given in Section 2.4.3.

2.4.2 The Dipole Method

The dipole method was developed by Thanassoulis and Comsa in 1971 to calculate the electric field in the vicinity of bundled conductors. It is an extension of the successive imaging technique described above (Abdel-Salam and El-Mohandes, 1989).

To establish this model for a bundle of n subconductors, each of them with $(2n - 1)$ image charges, the effect of earth (ground) is included.

According to equation (2.21), the n images inside each subconductor caused by the effect of ground can safely be assumed concentrated at the subconductor axis. They are of the same polarity as the subconductor itself. The other images, corresponding to the other members of the bundle, will have the opposite polarity, and their displacements from the subconductor axis are significant.

A dipole is composed of each of the negative charges together with one of the positive charges at the subconductor axis. The dipoles with the remaining axial charge can be used in computing the field distribution around the bundle. The accuracy of the method is satisfactory for practical separations exceeding 10 times the subconductor radius.

2.4.3   Charge-Simulation Technique

The charge-simulation technique can be employed successfully for the computation of an electric field between electrodes in a medium where one or more dielectrics are involved. The technique is described for single- and two-dielectric arrangements. The technique is discretization of the integral-equations technique explained in Section 2.4.9.

2.4.3.1   Single-Dielectric Arrangements

In this case the distributed charges on the surfaces of the stressed electrodes are replaced by a number (n) of fictitious discrete charges arranged inside the electrodes, that is, outside the space in which the field is to be computed (Singer et al., 1974). The number of charges and their coordinates are arbitrary. However, the larger the number of charges, the higher the possible accuracy of computation. The magnitudes of these discrete charges are to be such that, under their integrated effect, each electrode has its assigned potential as checked at every contour point on its surface, that is,

$$\sum_{j=1}^{n} P_j q_j = V \tag{2.26}$$

where q_j is the discrete charge and P_j is the associated potential coefficient.

Checking could be at a number m of contour points selected equal to or exceeding the number of discrete charges (n) and used for setting a group of simultaneous equations (Fig. 2.9):

$$[P][q] = [V] \tag{2.10}$$

Thus, the unknown charges q_j can be evaluated.

The charges simulating a given electrode can be chosen as point charges, ring charges, or as finite, semi-infinite, or infinite line charges. This choice should suit the shape of the electrode being simulated. For

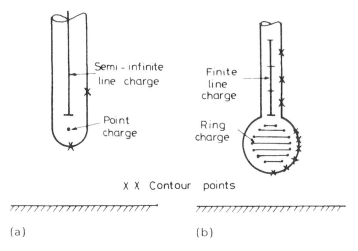

Figure 2.9 Arrangement of simulation charges and contour points: (a) hemispherically capped rod; (b) sphere with cylindrical shank.

example, spherical electrodes can easily be simulated by point charges. For fields with axial symmetry, ring charges centered on the axis are an effective method of simulation. Finite and infinite line charges have commonly been used to simulate the charge on the surfaces of cylindrical conductors. Other charge shapes, such as oblate spheroid, prolate spheroid, axihyperbola, and elliptic cylinder, have been proposed. Shell and annular plate charges have also been suggested (Mukherjee and Roy, 1983).

In terms of the calculated charges q_j, the electric field at any point in space around the electrodes can be determined:

$$\bar{E} = E_x \bar{u}_x + E_y \bar{u}_y = \sum_{j=1}^{n} f_j q_j \tag{2.27}$$

The potential and field coefficients for the different discrete charges are reported elsewhere (Singer et al., 1974).

For sufficient accuracy, effort should be devoted to choosing the proper number and location of simulation charges. The calculated potential cannot easily equal V at every point on the electrode surface under the effect of the simulation charges. Therefore, the magnitudes of the charges q_j are chosen such that the calculated potentials at a large number of points $m > n$ deviate only slightly from the actual potential V. The deviations are minimized by the least-squares techniques (Abdel-Salam and Ibrahim, 1977).

This was found to improve considerably the stability of point matching. Indeed, the choice $m > n$ represents a step forward in improving the conventional charge-simulation technique. The improved technique showed better accuracy than the conventional one, even with a smaller number of simulation charges (Khalifa et al., 1975).

In another version of the charge-simulation technique, instead of the electrode charges, surface field values on the electrodes are first evaluated from the known voltages (V) of these electrodes by solving a set of simultaneous equations analogous to (2.10). The values of charges on the electrode surface are calculated from the field values there. The method has been applied (Foo and King, 1976) successfully to symmetrical and asymmetrical overhead bundled transmission lines. This version of the charge-simulation technique has the advantage of yielding surface fields directly with simple computations.

2.4.3.2 Two-Dielectric Arrangements

In this case dipoles are aligned by the electric field and compensate for each other in the bulk of each dielectric, leaving a net surface charge on the boundary between the dielectrics. This surface charge can be simulated by discrete charges on both sides of the dielectric boundary.

Figure 2.10 shows, for example, the cross-section of a metal electrode with an applied voltage (V) meeting dielectrics I and II. At the electrode, there are n_e charges and contour points, with n_1 points on the interface between the electrode and dielectric I and $(n_e - n_1)$ points on the interface between the electrode and dielectric II. The n_e charges contribute to the potential and field in both dielectrics. At the dielectric boundary, there are n contour points with n discrete charges inside the surface of dielectric I, contributing to the potential and field in dielectric II, and vice versa. In total, there are $(n_e + n)$ contour points and $(n_e + 2n)$ discrete charges (Fig. 2.10).

The system of simultaneous equations required for determining the unknown discrete charges is formulated according to the following boundary conditions:

1. The potential at the n_1 contour points on the electrode/dielectric I interface must be equal to the applied voltage V:

$$\sum_{j=1}^{n_e} P_j q_j + \sum_{j=1}^{n} P_{\mathrm{II}j} q_{\mathrm{II}j} = V \tag{2.28}$$

Similarly, the potential at the $(n_e - n_1)$ contour points on the electrode/dielectric II interface must be equal to V.

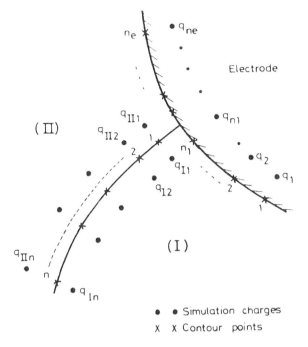

Figure 2.10 Discrete simulation charges at an electrode and at a dielectric boundary.

2. The potential at each of the n contour points on the interface between the two dielectrics is the same whether seen from either side; that is,

$$\sum_{j=1}^{n} P_{Ij}q_{Ij} = \sum_{j=1}^{n} P_{IIj}q_{IIj} \tag{2.29}$$

3. The displacement at the interface between the two dielectrics is continuous. Let the coefficient f_j' be defined as the contribution of the charge q_j to the component of the electric field normal to the dielectric interface. Then the continuity is expressed as

$$\varepsilon_I\left(\sum_{j=1}^{n_e} f_j'q_j + \sum_{j=1}^{n} f_{IIj}'q_{IIj}\right) = \varepsilon_{II}\left(\sum_{j=1}^{n_e} f_j'q_j + \sum_{j=1}^{n} f_{Ij}'q_{Ij}\right) \tag{2.30}$$

where ε_I and ε_{II} are the permittivities of dielectrics I and II, respectively.

Thus $(n_e + 2n)$ linear equations are formulated for the calculation of the same number of unknown discrete charges. In a two-dielectric arrangement, the contour points were chosen large compared with the number of discrete charges to achieve better matching of the boundary conditions (Abdel-Salam and Stanek, 1987b).

The charge-simulation technique has been applied successfully for calculating the electric field in and around practical suspension insulators (Khan and Alexander, 1982), and around overhead transmission lines (El-Arabaty et al., 1977). The simulation technique was also used for field calculation in arrangements including free potential electrodes, such as grading foils in condenser bushings and in pot heads of HV cables or suspended particles in gas-insulated systems (Abdel-Salam, 1987). The simulation technique was also attempted for computing electric fields, including surface and volume resistance of the insulation, in such as pollution layers on insulation surfaces (Takuma et al., 1981b; Abdel-Salam and Stanek, 1987a; Singer, 1981).

With an increased number of dielectrics, solution of field problems using the charge-simulation technique becomes more and more complex. Therefore, the technique is limited to field calculations in arrangements with only one or two dielectrics. On the other hand, three-dimensional fields with and without symmetry can be handled without excessive computation (Fig. 2.11) (Chen and Pearmain, 1983). Another major advantage is the simple simulation of curved interfaces between dielectrics and conductors. The values of unknown charges satisfy the boundary conditions when their number and locations are chosen judiciously. Here experience plays an important role.

Abdel-Salam and El-Mohandes (1987) have proposed an efficient charge-simulation technique (ECST) for calculating electric fields around conductor bundles of EHV transmission lines. In the ECST, the number n_e of simulation charges and their coordinates are no longer arbitrary but, rather, are dependent on the number of subconductors of the bundle and how they are arranged in space. The ECST was applied for bundles 2, 3, and 4 monopolar, homopolar, and bipolar DC transmission lines and showed high accuracy for the potential and field values calculated compared with results using the conventional technique. Moreover, the number of simulation charges involved with the ECST could thus be much reduced, with a considerable saving of the computational time. As an example, for a bundle 2 of a monopolar DC line, using the ECST, the number of charges could be reduced from 16 to 3 while accuracy was improved from 0.1% to 10^{-5}% (Abdel-Salam and El-Mohandes, 1987).

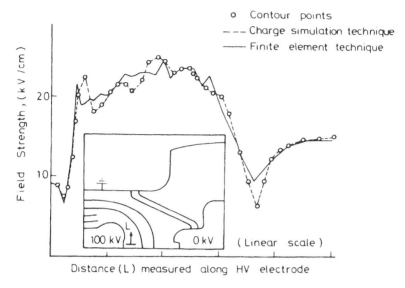

Figure 2.11 Surface field calculations along the HV electrode and the dielectric spacer within a gas-insulated switch gear using the charge-simulation and finite-element techniques. [From Okubo and Metz (1978).]

2.4.4 Finite-Difference Technique

The basis of this technique is the replacement of a continuous domain representing the entire space surrounding the high-voltage electrodes with a rectangular or polar grid of discrete "nodes" at which the value of unknown potential is to be computed. Thus we replace the derivatives describing Laplace's equation with "divided-difference" approximations obtained as functions of the nodal values.

The finite-difference technique is applicable for three-dimensional fields. However, a two-dimensional field is treated here for simplicity. Figure 2.12 shows a square grid with its sides parallel to the X or Y axis. All the elements have the same depth in the Z direction. Among the nodes of the grid, only nodes 0 to 4 will be of immediate interest, where nodes 1 to 4 surround node 0. Suppose that the potentials at these nodes are ϕ_1, ϕ_2, ϕ_3, ϕ_4, and ϕ_0. As the potential within the field region is continuous, it is possible to expand the potential at any point, such as node 0, using the well-known Taylor's series. If the series is terminated by neglecting the terms containing third and higher derivatives of the potential, then

$$\left.\begin{array}{l}
\phi_1 = \phi_0 - \left(\dfrac{\partial\phi}{\partial x}\right)\Delta x + \dfrac{1}{2}\left(\dfrac{\partial^2\phi}{\partial x^2}\right)(\Delta x)^2 \\[2ex]
\phi_2 = \phi_0 - \left(\dfrac{\partial\phi}{\partial y}\right)\Delta y + \dfrac{1}{2}\left(\dfrac{\partial^2\phi}{\partial y^2}\right)(\Delta y)^2 \\[2ex]
\phi_3 = \phi_0 + \left(\dfrac{\partial\phi}{\partial x}\right)\Delta x + \dfrac{1}{2}\left(\dfrac{\partial^2\phi}{\partial x^2}\right)(\Delta x)^2 \\[2ex]
\phi_4 = \phi_0 + \left(\dfrac{\partial\phi}{\partial y}\right)\Delta y + \dfrac{1}{2}\left(\dfrac{\partial^2\phi}{\partial y^2}\right)(\Delta y)^2
\end{array}\right\} \tag{2.31}$$

As Laplace's equation in Cartesian coordinates states that

$$\frac{\partial^2\phi}{\partial x^2} + \frac{\partial^2\phi}{\partial y^2} = 0$$

for a square grid, equation (2.31) yields

$$\phi_0 = \frac{1}{4}(\phi_1 + \phi_2 + \phi_3 + \phi_4) \tag{2.32}$$

Equation (2.32) is the finite-difference form of Laplace's equation for a two-dimensional field.

The next step is to apply equation (2.32) in order to evaluate the potential at every node in the space between the electrodes. Of course, at the nodes located on the high-voltage electrode surface, $\phi = V$. This completes the set of equations relating the node potentials $[\phi]$ to the boundary coefficients $[B]$:

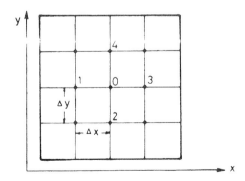

Figure 2.12 Regular grid for finite-difference technique, indicating the node numbers.

$$[\phi][P] = [B] \tag{2.33}$$

where $[P]$ is an $n \times n$ matrix of the coefficients for n-unknown node potentials, $[\phi]$ is an $n \times 1$ column matrix of the unknown potentials, and $[B]$ is an $n \times 1$ column matrix of the sum of known potentials of the bounding electrodes. Equation (2.33) can be solved by the Gaussian elimination method for problems with a relatively small number of nodes (James et al., 1985). For complex cases, iterative schemes are most efficient in combination with successive overrelaxation methods.

There is no reason why this technique cannot be used for computing the field in multidielectric arrangements. Referring to Figure 2.12, assume the nodes 2, 0, and 4 to be at the interface between two dielectrics of permittivities ε_1 (containing node 1) and ε_2 (containing node 3). At the interface, the potential and displacement are continuous. Therefore, the resultant field may be composed by superposition of two Laplacian fields, and equation (2.32) takes the form

$$\phi_0 = \frac{1}{4}\left(\phi_2 + \phi_4 + \frac{2\varepsilon_1}{\varepsilon_1 + \varepsilon_2}\phi_1 + \frac{2\varepsilon_2}{\varepsilon_1 + \varepsilon_2}\phi_3\right) \tag{2.34}$$

The finite-difference technique has thus been applied successfully for field calculation in switchgear design and development (Ryan et al., 1983). Figure 2.13 shows the computed field within a gas-insulated switchgear. The areas of field concentrations received detailed computations with finer grids. Potential values between the nodes are obtained by interpolation. The electric field components at any node are easily expressed in terms of the potential values at the nodes surrounding it.

Some difficulties are encountered in solving many problems using the finite-difference technique, and therefore its efficiency is considerably limited. A regular grid is not suitable for curved boundaries or interfaces because they intersect grid lines at points other than nodes. Therefore, the grid must be denser near the curved boundaries. The technique is well suited for use in field regions that are finite in space; otherwise, the number of nodes becomes very large and solution of equation (2.33) may be difficult even for machine computation.

2.4.5 Combined Charge-Simulation and Finite-Difference Technique

As noted in Section 2.4.3, the charge-simulation technique (CS) is well suited for computing fields in the vicinity of curved electrodes placed in open (unbounded) space with one or two dielectrics. On the other hand, the finite-difference technique (FD) is applicable to bounded space with

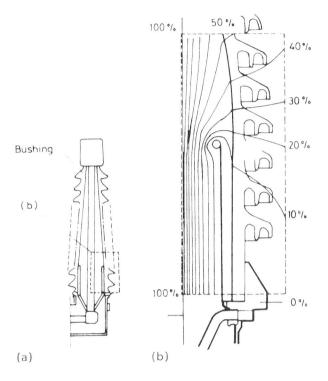

Figure 2.13 Field computed by the finite-difference technique for a 300 kV gas-insulated system: (a) part of single-phase layout, showing area enlarged in (b); (b) plot of equipotentials at lower end of bushing. [From Scott et al. (1974).]

multidielectrics. A combination of CS and FD can be utilized in some cases to obtain the advantages of both methods (Abdel-Salam and El-Mohandes, 1989).

To apply the combined method, the entire field is divided into two domains where each technique is used separately (Fig. 2.14). In the CS domain, the surface charges on the electrodes are represented by n_c unknown discrete charges with a set of contour points on the electrode surfaces. The FD domain is replaced by a rectangular grid whose n_g nodes are of unknown potentials. Naturally, equation (2.10) applies to the CS domain and equation (2.33) applies to the FD domain. The coupling between the two domains is based on the condition of continuity of potential and displacement at the n_b coupling points along the CS–FD boundary.

While analyzing the CS domain, the FD domain is represented by a number n_b of discrete charges located inside it. Also, each of the n_b coupling points on the boundary between the two domains has its own potential to be

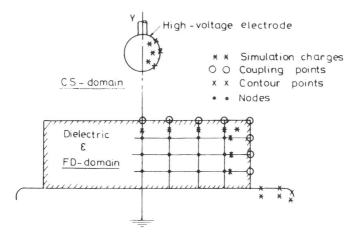

Figure 2.14 Entire field in air and a dielectric divided into CS and FD domains. Being symmetrical, only one half is studied.

evaluated in addition to the potentials of the n_g grid nodes. Therefore, the total number of unknowns in the entire space is $(n_c + 2n_b + n_g)$, to be evaluated by simultaneous solution of n_c equations (2.10) for the CS domain, n_g equations (2.33) for the FD domain, and $2n_b$ equations for the equality of potentials and continuity of displacements at the n_b coupling points.

2.4.6 Finite-Element Technique

According to this technique, the space in which the electric field is to be calculated is divided up into triangular elements as shown in Figure 2.15. All the elements have the same depth in the Z direction, as in the FD technique (Zienkiewicz, 1977, Segerlind, 1984).

Solution of electric field problems by the finite-element method is based on the fact, known from variational calculus, that Laplace's equation is satisfied when the total energy functional is minimal. If the permittivity of the medium is ε, the electrostatic energy functional F for a flat two-dimensional field is

$$F = \int\int_A \frac{1}{2}\varepsilon(\text{grad } \phi)^2 dA \tag{2.35}$$

where A represents the area scanned by the triangular elements.

The potentials ϕ at the different nodes are unknown variables and, for minimum energy functionals,

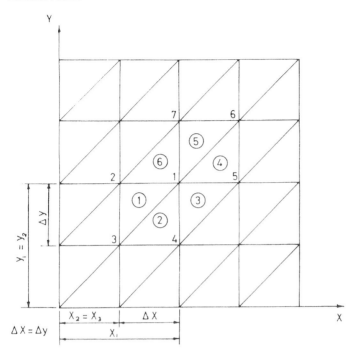

Figure 2.15 Grid for the finite-element technique and arrangement of the elements.

$$\frac{\partial F}{\partial \phi} = 0 \qquad (2.36)$$

for all the potentials at the nodes (Silvester and Ferrari, 1983). It is obvious that, when ϕ_1 is varied while all other potentials remain constant, only the energy in the adjoining triangles is affected. Therefore, in the partial derivative $\partial F / \partial \phi_1 = 0$, only the energies in the six triangles number 1 to 6 in Figure 2.15 need to be considered. For the energy functional in the six triangles,

$$F = F_1 + F_2 + F_3 + F_4 + F_5 + F_6 \qquad (2.37)$$

In triangle 1,

$$(\text{grad } \phi)^2 = (\text{grad } \phi)_x^2 + (\text{grad } \phi)_y^2 \qquad (2.38)$$

Assuming a linear variation of the potential ϕ as a function of x and y within the triangle,

$$(\text{grad } \phi)^2 = \left(\frac{\phi_1 - \phi_2}{\Delta x}\right)^2 + \left(\frac{\phi_2 - \phi_3}{\Delta y}\right)^2 \tag{2.38a}$$

and the functional F_1 becomes, according to equation (2.35),

$$F_1 = \frac{1}{2}\int\int_A \varepsilon(\text{grad } \phi)^2 dA = \frac{1}{4}\varepsilon\left[\left(\frac{\phi_1 - \phi_2}{\Delta x}\right)^2 + \left(\frac{\phi_2 - \phi_3}{\Delta y}\right)^2\right]\Delta x \Delta y$$

and the partial derivative takes the form

$$\frac{\partial F_1}{\partial \phi_1} = \frac{1}{2}\varepsilon\frac{\phi_1 - \phi_2}{(\Delta x)^2}\Delta x \Delta y \tag{2.39}$$

For the other triangles, similar equations can be obtained. For $\Delta x = \Delta y$, the total derivative becomes (Fig. 2.15)

$$\frac{\partial F}{\partial \phi_1} = \frac{1}{2}\varepsilon[8\phi_1 - 2(\phi_2 + \phi_4 + \phi_5 + \phi_7)] \tag{2.40}$$

Thus

$$\phi_1 = \frac{1}{4}(\phi_2 + \phi_4 + \phi_5 + \phi_7)$$

which is the same as the finite-difference solution (Section 2.4.4).

With a typical triangular element (Fig. 2.15), the potential is approximated by the polynomial

$$\phi = a + bx + cy + a_1x^2 + b_1xy + c_1y^2 + \cdots \tag{2.41}$$

Any increase in accuracy by including higher-order terms is outweighed by the increase in computation time and storage requirements. Most of the computations are satisfactorily based on a first-order approximation. Thus the actual continuous potential distribution over the XY plane is replaced by a piecewise-planar approximation.

The coefficients a, b, and c in equation (2.41) may be found from the three independent simultaneous equations obtained by constraining the potential to assume vertex values ϕ_1, ϕ_2, and ϕ_3 at the three vertices of coordinates shown in Figure 2.15:

$$\begin{bmatrix} \phi_1 \\ \phi_2 \\ \phi_3 \end{bmatrix} = \begin{bmatrix} 1 & x_1 & y_1 \\ 1 & x_2 & y_2 \\ 1 & x_3 & y_3 \end{bmatrix}\begin{bmatrix} a \\ b \\ c \end{bmatrix} \tag{2.42}$$

That is,

$$\phi = \sum_{m=1}^{3}\phi_m\alpha_m(x, y) \tag{2.43}$$

where α_m is a linear function of x and y, easily evaluated. It can easily be verified that the variables α_m are interpolatory on the three vertices of the triangle; that is,

$$\alpha_m(x_n, y_n) = \begin{cases} 0 & m \neq n \\ 1 & m = n \end{cases}$$

where (x_n, y_n) are the x and y coordinates of the vertices. The potential gradient may be found from equation (2.43) as

$$\text{grad } \phi = \sum_{m=1}^{3} \phi_m \nabla \alpha_m \tag{2.44}$$

So, according to equation (2.35), the energy of this element is

$$F^e = \frac{1}{2} \sum_{m=1}^{3} \sum_{n=1}^{3} \phi_m \int \int_A \varepsilon \nabla \alpha_m \nabla \alpha_n \, \mathrm{d}A \, \phi_n \tag{2.45}$$

where the dielectric within the element is assumed isotropic and the permittivity ε is constant. However, a solution is also possible if ε changes from element to element. For brevity, define the elements

$$S_{mn}^e = \int \int_A \varepsilon \nabla \alpha_m \nabla \alpha_n \, \mathrm{d}A$$

of the asymmetrical matrix $[S]^e$ for the triangular element (Fig. 2.16)

$$[S]^e = \begin{bmatrix} S_{11}^e & S_{12}^e & S_{13}^e \\ S_{21}^e & S_{22}^e & S_{23}^e \\ S_{31}^e & S_{32}^e & S_{33}^e \end{bmatrix} \tag{2.46}$$

The matrix $[S]^e$ is the well-known "stiffness matrix" of the element under consideration, as it contains the sensitivity of the energy functional with respect to the potentials. The elements of $[S]^e$ are determined as

$$S_{ij}^e = \frac{1}{4A} (\Delta Y_i \Delta Y_j + \Delta X_i \Delta X_j) \tag{2.46a}$$

where ΔY_i and ΔX_i (where $i = 1, 2, 3$ are the local node numbers which must be counterclockwise, as indicated by the arrow in Fig. 2.16) are expressed in terms of the coordinates of element nodes

$$\left. \begin{array}{lll} \Delta Y_1 = (y_2 - y_3), & \Delta Y_2 = (y_3 - y_1), & \Delta Y_3 = (y_1 - y_2) \\ \Delta X_1 = (x_3 - x_2), & \Delta X_2 = (x_1 - x_3), & \Delta X_3 = (x_2 - x_1) \end{array} \right\} \tag{2.46b}$$

$$A = \frac{1}{2}(\Delta Y_2 \Delta X_3 - \Delta Y_3 \Delta X_2) \tag{2.46c}$$

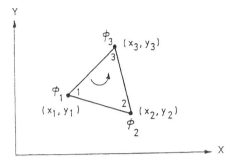

Figure 2.16 Typical triangular element; the local node numbering 1–2–3 must be counterclockwise, as indicated by the arrow.

Note that

$$\Delta Y_1 + \Delta Y_2 + \Delta Y_3 = 0 = \Delta X_1 + \Delta X_2 + \Delta X_3$$

hence

$$\sum_{i=1}^{3} S_{ij}^{e} = 0 = \sum_{j=1}^{3} S_{ij}^{e}$$

Equations (2.45) and (2.46) yield the energy of the element, expressed as

$$F^{e} = \frac{1}{2}\varepsilon[\phi]^{tr}[S]^{e}[\phi] \tag{2.47}$$

where $[\phi]$ is the column vector of vertex potential values of the element and "tr" denotes transposition.

The minimization of the energy functional for the specific element is expressed as

$$\frac{\partial F^{e}}{\partial[\phi]} = [S]^{e}[\phi] = 0 \tag{2.48}$$

Any composite assembly of triangular elements may be built up one triangle at a time. It is therefore sufficient to consider how continuity is enforced when one triangular element is added to an already existing assembly. This ends up with the formation of the global stiffness matrix $[S]$. Example (11) explains how the individual element stiffness matrices are assembled to obtain the global stiffness matrix.

For minimum energy functionals, and with properly arranged nodes, equations (2.35) and (2.48) yield

$$[[S]_{ff}[S]_{fb}] \begin{bmatrix} [\phi]_f \\ [\phi]_b \end{bmatrix} = 0 \qquad (2.49)$$

where $[\phi]_b$ are the bound (known) potentials of the nodes lying on the electrode surfaces, and $[\phi]_f$ are the free (unknown) potentials of the other nodes. This equation can be rearranged so that the potential at every point in the space between the electrodes can be evaluated in terms of the electrode voltages. Thus

$$[\phi]_f = -[S]_{ff}^{-1}[S]_{fb}[\phi]_b = [T]^{-1}[\phi]_b \qquad (2.50)$$

The solution of the Laplace equation takes the form of a set of nodal potential values as obtained from equation (2.50).

By proper numbering of the nodes, the matrix $[T]$ takes the form of a narrow diagonal band of symmetrical structure. The order of $[T]$ can reach several thousands to cover an entire two-dimensional field. However, its bandwidth is about 5–15. Fortunately, techniques have been developed for efficiently solving such sparse matrix equations, which can significantly decrease the computation time (Jennings, 1977).

Figure 2.11 shows a comparison of the finite-element (FE) and charge-simulation (CS) techniques for calculating the electric field at the contacts within a gas-insulated switchgear. The small deviation between the field values obtained by the two techniques is attributed to two reasons. The first is that with the CS technique the electrode surface is not exactly equipotential. The second reason is the discretization of the field space into finite triangular elements when using the FE technique. We should add that the FE method needs more computer storage and time than does the CS technique. On the other hand, the accuracy of calculation is improved with finer elements.

As in the finite-difference (FD) technique, the electric field components at any node can be expressed in terms of the potential values at the nodes surrounding it. The precision of field calculation using the finite-element method increases significantly with the number of elements.

Similarly to the FD technique, the FE technique is suitable for bounded regions. Many physical problems of interest, however, are only partly bounded (i.e., a portion of the boundary is at infinity). This difficulty is usually surmounted by assuming an artificial boundary, with an assumed potential, to be located far enough from the region of interest as to minimize the resulting errors. The price paid for doing this is that Laplace's equation must be solved for the entire larger region. This adds to the computation time. Some alternatives to this technique have recently been reported (Cendes and Hamann, 1981).

The flexibility of the FE technique is its greatest advantage with respect to the traditional FD technique. Elements can have various shapes and can easily be adapted to any shape of boundary and interface geometries. Further, its set of equations usually has a symmetric positive-definite matrix of coefficients. This property is not ensured in the FD technique. A third advantage of the FE technique is claimed to be simpler programming because of the easier introduction of boundary conditions. In general, it is more powerful for a wide range of problems with complicated geometries. On the other hand, it involves tedious preparation of the input data, most of which is a description of the element grid topology. For large problems, the time for grid preparation and editing becomes prohibitive. Some of these difficulties have recently been alleviated by the development of grid generation algorithms, adaptive grid refinement algorithms, and grid optimization (Shephard and Gallagher, 1979).

2.4.7 Combined Charge-Simulation and Finite-Element Technique

A comparison of the finite-element (FE) and charge-simulation (CS) techniques shows that each has advantages and disadvantages (Okubo et al., 1982; Tan and Steinbigler, 1985; Steinbigler, 1979). A combination of FE and CS can yield the advantages of both techniques, similar to the case of coupling CS and FD (Section 2.4.5) (Abdel-Salam and El-Mohandes, 1989). To apply the combined method, the space is divided into two domains—the FE and CS domains—which are treated separately. The coupling of the FE and CS domains is based on the fact that the potential and field strength must be continuous along the boundary between the domains (i.e., at all the boundary points). As an application of the combined technique, the axisymmetric three-dimensional field of a high-voltage bushing was computed (Fig. 2.17). The figure shows the configuration of the bushing and the computed equipotential surfaces, which speaks for itself.

2.4.8 Boundary-Element Method

The boundary-element method has recently been developed for the analysis of two- and three-dimensional electric fields between electrodes in multidi-electric media. The fundamental idea of the boundary-element method is to divide the electrode or insulator surfaces into a large number of surface elements. Plane triangular elements have been used (McWhirter and Oravec, 1979; Okon, 1985). For curved surfaces, curved surface triangles in parametric form were attempted (Misaki et al., 1982; Andjelic, 1984) to achieve realistic field simulation by a variable charge distribution on these elements.

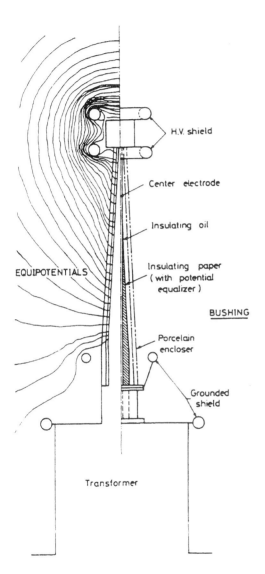

Transformer

Figure 2.17 Calculation of three-dimensional axisymmetrical field of a high-volume bushing using a combined charge-simulation/finite-element technique. [From Okubo et al. (1982).]

2.4.8.1 Plane Rectangular Elements

An electrode or insulator surface in the $X-Y$ plane at $z = z_1$ is divided into N_e elements. The charge density distribution in each is expressed as

$$\sigma = K_0 + K_1 x + K_2 y + K_3 xy \tag{2.51}$$

where K_0, K_1, K_2, and K_3 are coefficients related to the charge density values at the nodes of the element (Abdel-Salam et al., 1997).

The potential ϕ_e at any point $P(x_p, y_p, z_p)$ resulting from the charge distribution within one of the elements of Figure 2.18 is

$$\phi_e(x, y, z) = \frac{1}{4\pi\varepsilon} \int_{x_1}^{x_2} \int_{y_1}^{y_2} \frac{K_0 + K_1 x + K_2 y + K_3 xy}{\sqrt{(x - x_p)^2 + (y - y_p)^2 + (z_1 - z_p)^2}} dy\, dx$$

The potential ϕ at the point P due to all the elements is

$$
\begin{aligned}
\phi = \frac{1}{4\pi\varepsilon_0 (x_2 - x_1)(y_2 - y_1)} \sum_{k=1}^{N_e} [&\sigma_1(x_2 y_2 I_c - y_2 I_x - x_2 I_y + I_{xy}) \\
 &- \sigma_2(x_1 y_2 I_c - y_2 I_x - x_1 I_y + I_{xy}) \\
 &+ \sigma_3(x_1 y_1 I_c - y_1 I_x - x_1 I_y + I_{xy}) \\
 &- \sigma_4(x_2 y_1 I_c - y_1 I_x - x_2 I_y + I_{xy})]
\end{aligned}
\tag{2.52}
$$

where k stands for the element number; (x_1, y_1), (x_2, y_2), (x_3, y_3), and (x_4, y_4) are the $x-y$ coordinates of the nodes belonging to the kth element; and σ_1 to σ_4 are the charge density values at the nodes of the kth element. I_c, I_x, I_y, I_{xy} are integral values expressed in terms of the coordinates of the nodes of the kth element and the coordinates (x_p, y_p, z_p) of the point P.

The field components (E_x, E_y, E_z) due to the boundary-element charges are obtained by differentiating the potential expressed by equation (2.52):

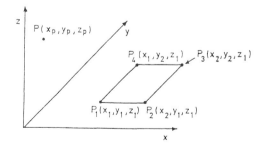

Figure 2.18 Plane rectangular element.

$$E_x(x_p, y_p, z_p) = \frac{-1}{4\pi\varepsilon(x_2 - x_1)(y_2 - y_1)} \sum_{k=1}^{N_e}$$

$$[\sigma_1(x_2 y_2 I_{cx} - y_2 I_{xx} - x_2 I_{yx} + I_{xyx})$$
$$- \sigma_2(x_1 y_2 I_{cx} - y_2 I_{xx} - x_1 I_{yx} + I_{xyx})$$
$$+ \sigma_3(x_1 y_1 I_{cx} - y_1 I_{xx} - x_1 I_{yx} + I_{xyx})$$
$$- \sigma_4(x_2 y_1 I_{cx} - y_1 I_{xx} - x_2 I_{yx} + I_{xyx})]$$

$$E_y(x_p, y_p, z_p) = \frac{-1}{4\pi\varepsilon(x_2 - x_1)(y_2 - y_1)} \sum_{k=1}^{N_e}$$

$$[\sigma_1(x_2 y_2 I_{cy} - y_2 I_{xy} - x_2 I_{yy} + I_{xyy})$$
$$- \sigma_2(x_1 y_2 I_{cy} - y_2 I_{xy} - x_1 I_{yy} + I_{xyy})$$
$$+ \sigma_3(x_1 y_1 I_{cy} - y_1 I_{xy} - x_1 I_{yy} + I_{xyy})$$
$$- \sigma_4(x_2 y_1 I_{cy} - y_1 I_{xy} - x_2 I_{yy} + I_{xyy})]$$

$$E_z(x_p, y_p, z_p) = \frac{-1}{4\pi\varepsilon(x_2 - x_1)(y_2 - y_1)} \sum_{k=1}^{N_e}$$

$$[\sigma_1(x_2 y_2 I_{cz} - y_2 I_{xz} - x_2 I_{xz} + I_{xyz})$$
$$- \sigma_2(x_1 y_2 I_{cz} - y_2 I_{xz} - x_1 I_{yz} + I_{xyz})$$
$$+ \sigma_3(x_1 y_1 I_{cz} - y_1 I_{xz} - x_1 I_{yz} + I_{xyz})$$
$$- \sigma_4(x_2 y_1 I_{cz} - y_1 I_{xz} - x_2 I_{yz} + I_{xyz})]$$

$$(2.53)$$

where

$I_{cx}, I_{cy},$ and I_{cz} are respectively $\dfrac{\partial I_c}{\partial x}, \dfrac{\partial I_c}{\partial y},$ and $\dfrac{\partial I_c}{\partial z}$

$I_{xx}, I_{xy},$ and I_{xz} are respectively $\dfrac{\partial I_x}{\partial x}, \dfrac{\partial I_x}{\partial y},$ and $\dfrac{\partial I_x}{\partial z}$

$I_{yx}, I_{yy},$ and I_{yz} are respectively $\dfrac{\partial I_y}{\partial x}, \dfrac{\partial I_y}{\partial y},$ and $\dfrac{\partial I_y}{\partial z}$

$I_{xyx}, I_{xyy},$ and I_{xyz} are respectively $\dfrac{\partial I_{xy}}{\partial x}, \dfrac{\partial I_{xy}}{\partial y},$ and $\dfrac{\partial I_{xy}}{\partial z}$

The potential and field equations for other elements such as plane triangular, cylindrical, spherical, conical, and torodial elements are reported in the literature (Gutfleisch, 1989).

2.4.8.2   Boundary Conditions

The boundary conditions are the Dirichlet condition at the electrode surface and the Neumann condition at the insulator surface. The calculated potential is equal to the applied voltage V at the electrode surface

$$\phi = V \tag{2.54a}$$

At the surface of the insulator, the continuity of the normal electric flux density (in the Z direction) is satisfied, i.e.,

$$\varepsilon_r E_{zd} = E_{za} \tag{2.54b}$$

where E_{zd} and E_{za} are the z components of the electric field calculated at any point on the insulator surface when seen from the insulator and surrounding-gas sides, respectively.

$$E_{zd} = E_z - \frac{\sigma}{2\varepsilon_0}$$

$$E_{za} = E_z + \frac{\sigma}{2\varepsilon_0}$$

where σ is the value of the surface charge density at the point and is evaluated by equation (2.51).

2.4.8.3   Boundary Points

To satisfy the boundary conditions, boundary points are chosen on the electrode/insulator surfaces equal in number to the unknown charge density values σ at the element nodes.

2.4.8.4   Determination of the Unknowns

With the aid of equations (2.52) and (2.53) for the potential and field, the boundary conditions expressed by equations (2.54a) and (2.54b) are satisfied at the boundary points chosen on the electrode and insulator respectively. This results in a set of simultaneous equations whose solution determines the values of the unknowns.

2.4.8.5   Application Example

The example deals with a disc insulator of a three-phase GIS and is shown in Figure 2.19a. A rib is placed around each HV conductor in order to prolong the creepage path. The insulator material has a relative permittivity of 3.8. The HV conductors, the grounded outer tube, and the rotationally symmetric insulator ribs were simulated by ring-shaped surface charges according to the charge simulation technique, and the plane part of the disc insulator was represented by quadrangular and triangular boundary elements (Gutfleisch et al., 1994). Symmetry with respect to the plane $z = 0$ was taken into account by image charges. Details of the different components and the discretization are illustrated in Figure 2.19b.

Figure 2.19c shows equipotential contours plotted in the plane $z = 10$ mm. In this Figure 2.19d equipotential and field lines in a sectional plane through the points A and B are shown. The influence of ribs appears in an increase of the field strength, especially at the conductor with the potential 1.0 kV.

2.4.9   Integral-Equations Technique

The integral-equations technique is applicable for evaluating the electric potential and field in a space composed of conducting and dielectric regions (Fig. 2.20). All the surfaces of the problem under study are subdivided into a number of segments. The charge density is assumed uniform on each of these segments, but of an unknown magnitude. The charge density $\sigma(\bar{\Omega})$ is assumed σ_{12} $(\bar{\Omega})$, σ_{13} $(\bar{\Omega})$, and σ_{23} $(\bar{\Omega})$ on the subsurfaces A_{12}, A_{13}, and A_{23}, respectively. $\bar{\Omega}$ represents the location of the segmental charge. The unit vectors u_{ni} are inward normal to the boundary surface A_i of the ith region $(A_i = A_{ij} + A_{ik})$.

By using Green's theorem, Laplace's equation can be expressed in surface integral form subject to the boundary conditions

$$\phi_1 = \phi_2 = V \text{ on the interface } A_{12}$$

$$\phi_1 = \phi_3 = V \text{ on } A_{13}$$

$$\phi_2 = \phi_3 \text{ on } A_{23}$$

$$\phi_3 = 0 \text{ on the infinite boundary of the third region}$$

$$\frac{\partial \phi_1}{\partial u_{n1}} = 0 \text{ on the boundary surface } A_1 \text{ (conducting surface)}$$

$$\varepsilon_2 \frac{\partial \phi_2}{\partial u_{n2}} = \neg \varepsilon_3 \frac{\partial \phi_3}{\partial u_{n3}} \text{ on } A_{23}$$

The solution of Laplace's equation is obtained by formulating the following two equations. One equation results from the condition that the potential must equal V on the conductor boundaries. The potential at any contour point on A_1 is equal to the sum of potential contributed by all surface charges; that is, by polarization charge on the interface between dielectrics and by both free and polarization charges on the conductor–dielectric interfaces. As each dielectric is assumed homogeneous, there is no volume polarization charge. Hence

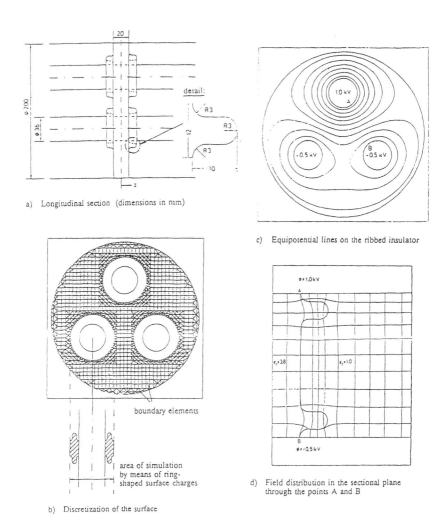

a) Longitudinal section (dimensions in mm)

c) Equipotential lines on the ribbed insulator

boundary elements

area of simulation
by means of ring-
shaped surface charges

d) Field distribution in the sectional plane
through the points A and B

b) Discretization of the surface

Figure 2.19 Disc insulator of a gas-insulated switchgear (dimensions in mm): (a) longitudinal section; (b) discretization of the surface; (c) equipotential lines on the ribbed insulator; (d) field distribution in the sectional plane through points A and B.

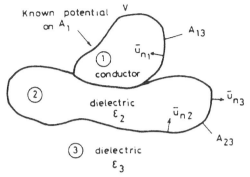

Figure 2.20 System consisting of a conductor and two dielectrics.

$$\int\int_A \sigma(\bar{\Omega})G(\bar{P}, \bar{\Omega}) \, dA = V \qquad \bar{P} \text{ on } A_1 \qquad (2.55)$$

where A is the collection of subsurfaces A_{12}, A_{13}, and A_{23}. $\bar{P}$ is the location of the contour point. $G(\bar{P}, \bar{\Omega}) = 1/4\pi(\bar{P} - \bar{\Omega})$ is a generating function that satisfies $-\nabla_x^2 G(\bar{P}, \bar{\Omega}) = \delta(\bar{P} - \bar{\Omega})$, where δ is the Dirac delta.

The second equation results from the condition that the displacement is continuous at the interface A_{23} between dielectrics; that is,

$$\frac{\varepsilon_2 + \varepsilon_3}{2}\sigma(\bar{P}) = (\varepsilon_3 - \varepsilon_2)\int\int_A \sigma(\bar{\Omega})\frac{\partial G(\bar{P}, \bar{\Omega})}{\partial u_{n3}} dA \qquad \bar{P} \text{ on } A_{23} \qquad (2.56)$$

Therefore, each integral in equations (2.52) and (2.56) is reduced to a sum of terms, each being a constant depending on the system geometry multiplied by an unknown charge density $\sigma(\bar{\Omega})$. The constants for any problem can be evaluated using analytical and numerical integration schemes (Singer, 1984). By applying equations (2.55) and (2.56) at several contour points, a set of linear equations are obtained that can easily be solved by standard techniques. Once the charge densities are known, the potential and field at any point can be determined (Daffe and Olsen, 1979).

Equations (2.55) and (2.56) can be applied to regions composed of several conductors and several dielectrics. In this case A_1 becomes the collection of all conductor–dielectric interfaces, A_{23} becomes the collection of all dielectric–dielectric interfaces (Singer, 1984).

The integral-equations technique is appealing because the potential must be computed only at the required points. Another attractive quality is that it involves fewer unknown quantities than do the FD and FE techniques, thus requiring considerably less computer time. It may be argued that the equations describing the charge-simulation technique could be obtained by discretizing the integral equations explained above. That is true, but it would result in a larger number of linear equations to solve than the integral-equations solution, in which the sources are treated as surface charges.

2.4.10 Monte Carlo Technique

Laplace's equation for the electric potential is solved here by numerically simulating random walks of a fictitious particle beginning from a point chosen in the space for which the potential distribution is to be evaluated. To illustrate the application of this technique, a conductor at a potential of $+V$ centered in a rectangular duct at zero potential will be considered (Fig. 2.21). Let the space between the conductor and the duct be ruled off in a Cartesian grid, as for the finite-difference technique. Choose at random either of the letters x or y and either of the signs $+$ or $-$, and move the particle one step from the chosen point P in the direction thus identified (say, $+x$). Again, choose at random one of the four possible directions and move the particle ahead one more step. The path traced out by this sequence of random directions is called a "random walk" (Fig. 2.18). Eventually, this walk would reach either the conductor or the duct wall and would thus be terminated. This random walk of the particle is repeated a fairly large number of times (n). Then the potential at P is defined as

$$V_{\mathrm{P}} = \lim_{n \to \infty} \frac{\Sigma V_n}{n} \tag{2.57}$$

Each V_n will equal V or 0, depending on whether the corresponding walk terminates on the conductor or on the duct wall (Fig. 2.21). The rate of convergence of V_{P} to the correct value is proportional to $\sqrt{n}$, so many random walks are necessary if a fine grid is chosen and reasonable accuracy

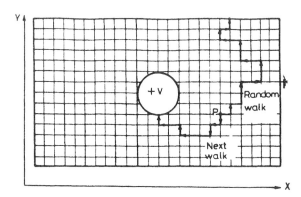

Figure 2.21 Random walks in the Monte Carlo technique.

is required. The sequence of instructions that is used to govern the motion of the fictitious particle is derived from a sequence of random numbers. Methods for generating these numbers are described elsewhere.

As was true for the FD and FE techniques, it is assumed that the region between boundaries is circumscribed. Again, however, there are techniques that can be used to apply the Monte Carlo technique to unbounded regions. The most interesting characteristic of Monte Carlo solutions is that the potential can be computed for each point at a time. Neither does a large array of potentials need to be stored, nor do a large number of simultaneous equations need to be solved. Despite this advantage, Monte Carlo techniques have been of only limited use for one- or two-dimensional problems because the time needed to calculate the potential at each point is not small. However, they are very appealing for solving three-dimensional problems, where they tend to supersede other techniques (Anis and Abd-Allah, 1989).

2.5 ANALYTICAL CALCULATIONS OF FIELDS WITH SPACE CHARGES

In high-voltage engineering, space charge is mainly developed by corona discharges at highly stressed electrodes. At alternating voltages, the space charge is constrained in the vicinity of the stressed electrodes and oscillates with periodic reversal of the electric field. This increases the difficulty of field calculation (Abdel-Salam et al., 1984; Abdel-Salam and Shamloul, 1988; Abdel-Salam and Abdel-Aziz, 1994). On the other hand, the space charges created by DC corona fill the entire inter-electrode space. Analytical calculation of electric fields with space charges created by DC corona is possible for simple configurations such as concentric spheres and coaxial cylinders.

2.6 NUMERICAL COMPUTATION OF FIELDS WITH SPACE CHARGES

In more complex electrode geometries such as wire–plane gaps, analytical calulcation of the electric field is extremely difficult because of the nonlinear nature of space-charge generation by corona. A relatively simple case is that of unipolar space charge (e.g., unipolar DC lines). The equations in this case are as follows:

$$\nabla \cdot \bar{E} = \frac{\rho}{\varepsilon} \qquad \bar{J} = k\rho\bar{E} \qquad \nabla \cdot \bar{J} = 0 \qquad \bar{E} = -\nabla\Phi \tag{2.58}$$

where $\bar{E}$ and $\bar{J}$ are the electric field and current density vectors at any point in space. The first is Poisson's equation, the second defines the relationship betwee $\bar{J}$ and $\bar{E}$, the third is the continuity equation for the ions, and the fourth is the electric field in terms of the potential Φ. ρ is the ion charge density and k is the ion mobility.

Similarly, bipolar DC ionized fields are described by the following equations:

$$\left.\begin{array}{c} \nabla \cdot \bar{E} = \dfrac{\rho_+ - \rho_-}{\varepsilon} \qquad \bar{J}_+ = k_+ \rho_+ \bar{E} \qquad \bar{J}_- = k_- \rho_- \bar{E} \qquad \bar{J} = \bar{J}_+ + \bar{J}_- \\[2mm] \nabla \cdot \bar{J}_+ = -\dfrac{\xi \rho_+ \rho_-}{e} \qquad \nabla \cdot \bar{J}_- = \dfrac{\xi \rho_+ \rho_-}{e} \qquad \bar{E} = -\nabla \Phi \end{array}\right\} \qquad (2.59)$$

where the subscripts $+$ and $-$ correspond to positive and negative ions. ξ denotes the recombination coefficient for ions and e is the electronic charge.

Most attempts in the literature resort to simplifying assumptions (Khalifa and Abdel-Salam, 1974b; Abdel-Salam et al., 1982). The two most important such assumptions are:

1. The space charge affects only the magnitude, not the direction of the electric field. This assumption was originally suggested by Deutsch in 1933.
2. The surface field at the coronating electrode remains constant at its value corresponding to corona onset.

Both assumptions compensate for the lack of adequate boundary conditions, and thus equations (2.58) and (2.59) can be solved along a number of flux lines of the imposed electric field. Iterative methods have been used for the conditions to be closely fulfilled at the electrodes.

Deutsch's assumption continued to be a basis until Abdel-Salam and Khalifa (1974) showed that its use was not a necessity. They based their computation of the resultant local field on a physical understanding of the problem as a superposition of the applied electric field and the space charge. Also, the surface field of conductors in corona could be calculated (Khalifa and Abdel-Salam, 1973) and measured. It was found to decrease significantly with corona intensity. Computation may follow the finite-element or charge-simulation technique either separately or combined with the method of characteristics or the method of residues, as shown below.

2.6.1 Finite-Element Technique

By using some additional assumptions, Takuma et al. (1981a) succeeded in applying the finite-element method to a field problem with space charge in the case of wire–plane gaps. Instead of assumption 2 above, they postulated a constant space-charge density around the wire periphery, at a value to be determined experimentally. This added some empiricism to the analysis.

More recently, Abdel-Salam et al. (1983, 1992, 1993, 1994, 1995a) applied the finite-element method for analyzing the electric field with monopolar space charge in wire–plane gaps. They were able to eliminate both Deutsch's assumption (1) and that of constant surface field intensity (2). They thus could compute the field in the interelectrode space without empiricism. From equation (2.58), we get

$$- \nabla \cdot (\varepsilon \nabla \Phi) = \rho \tag{2.60}$$

$$- \nabla \cdot (k\rho \nabla \Phi) = 0 \tag{2.61}$$

Abdel-Salam et al. (1983) considered the potential Φ as the sum of two components, the applied electrostatic value ϕ_s and the elemental potential caused by the space charge ϕ_{sc}, that is,

$$\Phi = \phi_s + \phi_{sc} \tag{2.62}$$

The elemental potential ϕ_{sc} is bound to be zero at both electrodes, where $\Phi = \phi_s$. The potential ϕ_s was computed for the contour points by the charge-simulation (CS) technique as explained in Section 2.4.3. Now, to find ϕ_{sc}, substitute equation (2.62) into equations (2.60) and (2.61) to get

$$\nabla(\varepsilon \nabla \phi_{sc}) = -\rho \tag{2.60a}$$

$$\nabla(k\rho \nabla \phi_{sc}) = 0 \tag{2.61a}$$

or, in a general form,

$$\nabla(\gamma \nabla \phi_{sc}) = \beta \tag{2.63}$$

where γ is equal to ε and $k\rho$ in equations (2.60a) and (2.61a), respectively, and β is equal to $-\rho$ and zero in equations (2.60a) and (2.61a), respectively.

In wire–plane gaps, and according to the finite-element technique, the inter-electrode space is an unbounded region and is usually defined by an artificial boundary as explained in Section 2.4.5. The inter-electrode space is divided into triangular elements, thus forming a grid, and the potential ϕ_{sc} within each element was approximated as a linear function of space coordinates.

To solve equations (2.60a) and (2.61a) in their general form (2.63), an energy functional given in equation (2.64) is used instead of equation (2.35) of the space-charge-free fields.

$$F = \int \int_A \left\{ \frac{\gamma}{2} \left[\left(\frac{\partial \phi_{sc}}{\partial x} \right)^2 + \left(\frac{\partial \phi_{sc}}{\partial y} \right)^2 \right] - \beta \phi_{sc} \right\} dA \tag{2.64}$$

Representing the variation of ϕ_{sc} over each element in terms of the nodal values of ϕ_{sc} and imposing the minimization condition $\partial F / \partial \phi_{sc} = 0$ for the entire set of elements, we get a set of simultaneous equations for ϕ_{sc} at the nodes for a given ρ distribution. Solution of this set of equations gives the nodal values of ϕ_{sc}: $[\phi_{sc1}]$ and $[\phi_{sc2}]$, corresponding to equations (2.60a) and (2.61a), respectively. Iterations are used until the values of both ϕ_{sc1} and ϕ_{sc2} agree within an acceptable tolerance (Abdel-Salam et al., 1983). The relation for correcting the charge density ρ_i to ρ_n is

$$\rho_n = \rho_i \left(1 + \frac{f_{ac}}{2} \frac{\phi_{sc1} - \phi_{sc2}}{\phi_{sc1} + \phi_{sc2}} \right) \tag{2.65}$$

where f_{ac} is an acceleration factor chosen equal to 0.5.

The method of analysis was extended to bipolar ionized fields (Abdel-Salam and Al-Hamouz, 1995b)

2.6.2 Finite-Element Technique Combined with the Method of Characteristics

A self-consistent description of the charge density ρ and potential Φ structures of a monopolar DC ionized field was proposed by Davis and Hoburg (1986) on the basis of simultaneous computation of both ρ and Φ. The technique employs iterative use of the finite-element method to compute potential and electric field distributions, and the method of characteristics to compute the charge density distribution.

Use of the finite-element method to determine the distribution of Φ for a known charge density ρ in a region with permittivity ε is based on minimizing the energy functional

$$F = \int \int_A \left[\frac{1}{2} \varepsilon (\text{grad } \Phi)^2 - \rho \Phi \right] dA \tag{2.66}$$

to obtain the solution of Poisson's equation ($\nabla^2 \Phi = -\rho/\varepsilon$). It is noted that the energy functional given by equation (2.35) is obtained from the general equation (2.66) by setting $\rho = 0$ for space-charge-free fields. Once the Φ distribution is determined, the electric field $\bar{E}$ distribution is obtained simply.

According to the method of characteristics proposed by Davis and Hoburg (1986), the partial differential equations (2.58) that describe the effect of a known field pattern on the spatial distribution of ρ can be converted into an ordinary differential equation for ρ along specific space–time trajectories. Combination of equations (2.58) leads to a nonlinear partial differential equation governing the evolution of ρ:

$$\bar{E} \cdot \nabla \rho = -\frac{\rho^2}{\varepsilon} \tag{2.67}$$

Along the "characteristic lines" defined by

$$\frac{d\bar{r}}{dt} = k_i \bar{E} \tag{2.68}$$

equation (2.67) becomes

$$\frac{d\rho}{dt} = -\frac{k_i \rho^2}{\varepsilon} \tag{2.69}$$

and yields the solution

$$\rho = \frac{1}{(1/\rho_0) + (k_i t/\varepsilon)} \tag{2.70}$$

where ρ_0 is the charge density at the starting point of the characteristic line.

For a given starting point at the surface of the coronating electrode, the characteristic line is traced out by numerically integrating equation (2.68). At intersections with triangle boundaries, time t is determined, then used in equation (2.70) to compute ρ along the characteristic line.

2.6.3 Charge-Simulation Technique

In the application of the CS technique to space-charge-free fields, approximate solutions to Laplace's equation are obtained by placing fictitious discrete charges outside the region in which the solution is sought (Section 2.4.3). This technique was adapted by Hornstein (1980, 1984) to include corona space charges. He concentrated them at some discrete points inside the region under study. Both sets of discrete charges have to be chosen to satisfy the boundary conditions of Poisson's equation (2.58). These boundary conditions are given partially in terms of known electrode potentials and are obtained partially by assuming that the magnitude of the electric field at the coronating electrodes remains constant at the onset value or varies in the manner computed by Khalifa and Abdel-Salam (1973). The contribution of these charges to the total electric field should satisfy the condition of current continuity.

To solve a given unipolar DC ionized field, first a space-charge-free field solution is found. This field represents conditions just prior to corona onset. Then, for a given value of corona current i_c, equipotential charge shells enclosing the wire and represented by discrete line charges are added to the volume, as depicted schematically in Figure 2.22 for a wire–plane geometry. Since these equipotential shells do not contribute to the surface field of the wire, the field remains at its onset level.

The equipotential charge shell nearest the wire is first located and simulated by M line charges uniformly distributed around its contour. Their images with respect to ground are also set. The M selected contour points are also chosen spread about the equipotential contour as shown in

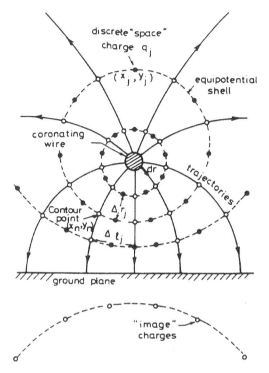

Figure 2.22 Schematic representation of trajectories, discrete space charge, and equipotential contours for a wire–plane gap (not drawn to scale).

Figure 2.22 and located midway between the line charges. Under the constant levels of the equipotential contours, M equations are formulated as

$$\sum_{j=1}^{M} \frac{q_j}{2\pi\varepsilon_0} \ln \frac{1}{\bar{r}_n - \bar{r}_j} = \text{constant at each } \bar{r}_n \tag{2.71}$$

where $\bar{r}_n$ and $\bar{r}_j$ are, respectively, the location of the contour point and line charge q_j.

The values of q_j should also satisfy the discretized form of the current continuity equation

$$\sum_{j=1}^{M} \frac{q_j}{\Delta r_j \Delta l_j} k E \Delta l_j = i_c \tag{2.72}$$

where Δr_j and Δl_j are distances defining an elemental volume with unit axial length around q_j, as shown in Figure 2.22. The quantity $(q_j/\Delta rj\Delta l_j)$ is the equivalent local space charge density ρ, and i_c is the corona current per unit length of the wire with the applied voltage V. Equations (2.71) and (2.72) formulate M equations with M unknowns, to be solved simultaneously.

After the first equipotential charge shell and its "image" charges are found, their electric field contributions are included in integrating outward to the next shell, and so on until the entire space is filled with shells of known discrete simulation charges. This defines the charge-density distribution over the space surrounding the coronating wire.

The applied voltage necessary to sustain i_c is equal to the onset voltage augmented by the potential contribution of each shell plus the potential of the images calculated at the wire surface. The $I–V$ characteristics of corona computed by the method proposed by Hornstein (1980, 1984) agreed satisfactorily with those measured by experiment.

The method was extended to the ionized fields around monopolar bundle conductors (Abdel-Salam and Abdel-Sattar, 1989; Abdel-Salam and Mufti, 1998).

Sometimes the space charge, such as a cloud of ions developed by an avalanche, is of known magnitude and travels in an inter-electrode space. The ion cloud may be approximated by point or ring charges (Zeitoun et al., 1976). As the space charges are known in magnitude, it is only necessary to calculate the charges simulating the electrodes. The potentials of contour points selected on the electrode surface are expressed as

$$[P][q] + [P_s][q_s] = [V] \tag{2.10}$$

where $[P_s]$ is the potential-coefficient matrix of the $[q_s]$ space charges and $[V]$ is the applied voltage. The simulation charges $[q]$ of the electrodes are simply expressed as

$$[q] = [P]^{-1}\{[V] - [P_s][q_s]\} \tag{2.73}$$

The potential at any point in space is then calculated as

$$\Phi = \sum_{j=1}^{n} P_j q_j + \sum_{j=1}^{n_s} P_{sj} q_{sj} \tag{2.74}$$

where n and n_s are, respectively, the number of simulation charges of the electrode and the number of discrete space charges. The electric field is calculated similarly.

The concept above was used for simulating electron avalanches growing in the ionization zone around stressed electrodes, such as in point–plane gaps. This was the basis for calculating the development of corona Trichel pulses in negative corona and burst pulses and onset-streamer pulses in positive corona (Ibrahim and Singer, 1982, 1984). Recently, the concept was used to calculate the corona pulses from positive surges propagating along overhead transmission lines (Abdel-Salam and Stanek, 1987a).

2.6.4 Charge-Simulation Technique Combined with the Method of Residues

The charge-simulation technique was extended to the calculation of the electric field vector $\bar{E}$ directly from Poisson's equation. The space charge was represented by discrete values at the nodes of a grid that was superposed on the inter-electrode space. This method is more accurate in the calculation of $\bar{E}$ than would be possible from an FE solution of Poisson's equation for space potential Φ, where $\bar{E}$ is obtained by numerical differentiation of Φ. The updated estimates of the space-charge values are obtained from the continuity equations for current density, using a weighted residual method (Oin et al., 1986). These charge densities are utilized for the calculation of $\bar{E}$ in the next iteration cycle.

The region of interest being defined by an artificial surface is subdivided into small triangular elements. The applied voltages are defined at ground and wire surfaces in wire–plane gaps. In the presence of corona, the electric field E consists of two components, the applied electrostatic component $\bar{E}_s$ and the component $\bar{E}_{sc}$ due to the space charge. The component $\bar{E}_s$ is computed by the CS technique, as explained in Section 2.4.3. The component E_{sc} can be easily computed from the point form of Gauss's law.

To complete the iteration cycle, the space charges ρ at the grid nodes are computed knowing $\bar{E}$. For this purpose, the residual error R_{es} of the current continuity at each node is determined as

$$R_{es} = \nabla \cdot k\rho E \qquad (2.75)$$

Setting the integral of R_{es} over the region of interest to zero leads to a set of algebraic equations that can be solved for nodal values of ρ. A weighting function for each element can be included in the integration of R_{es}.

The values of the electric field at the wire surface in each iteration cycle are compared with the onset value, and the deviation is used to increase or decrease ρ appropriately to speed up the convergence. The CS technique combined with this method of weighted residuals has been used successfully in solving monopolar and bipolar DC ionized fields (Oin et al., 1986).

2.7 ELECTRIC STRESS CONTROL AND OPTIMIZATION

Field stresses are controlled in much high-voltage equipment such as cable terminations, high-voltage bushings, and potential transformers. Measures have been suggested and employed with the aim of optimizing the stress throughout every part of the insulation of high-voltage equipment in order to achieve the most economical designs.

2.7.1 Electric Stress Control

For example, the stress-controlled capacitor bushing will be discussed briefly to shed some light on the problems involved. A capacitor bushing is used to bring a high-voltage conductor through the ground case, for example, a tank of a high-voltage transformer, without excessive electric fields between the conductor and the edge of the hole in the tank. This is achieved in practice by the introduction of floating screens separating dielectric layers that are interleaved at equal distances between the conductor and the grounded case, thus creating a uniform radial field. The bushing of a uniform dielectric should have a hyperbolic surface profile.

Another example is a high-voltage line gripped for suspension through bushing without floating screens. Consider a cylindrical conductor of radius a with a cylindrical bushing of length l and thickness t (Fig. 2.23). Usually, the clamp holding the bushing is grounded; i.e., the voltage at $r = a$ is V and at $r = a + t$ it is zero. The electric stress in the bushing at radius r is

$$E = \frac{V}{r \ln[(t + a)/a]} \tag{2.76}$$

Clearly, the highest stress would be at the lowest r, i.e., at $r = a$. If E_{bd} is the breakdown strength of the bushing material, then the required thickness of the bushing is obtained as

$$\ln \frac{t + a}{a} = \frac{V}{aE_{bd}}$$

or

$$t = a\left(e^{V/aE_{bd}} - 1\right) = \frac{V}{E_{bd}} + \frac{1}{2!}\frac{V^2}{aE_{bd}^2} + \frac{1}{3!}\frac{V^3}{a^2 E_{bd}^3} + \cdots \tag{2.77}$$

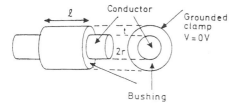

Figure 2.23 High-voltage line gripped for suspension through bushing— schematic view and cross-section.

which is much larger than the thickness of a flat bushing whose thickness is V/E_{bd}. Not only is the cylindrical bushing costly, but also the stress is not uniform within the bushing materials. The highest stress is close to the conductor and the bushing is not stressed to the limit elsewhere.

In graded bushing, the stress is uniformly distributed, the bushing material is properly utilized and the bushing thickness is equal to that of the flat bushing. This is achieved by shaping the profile of the bushing material so that the length of the bushing is $2Z$ at radius r (Fig. 2.24). The bushing is considered as a series of parallel-plate capacitors, each of thickness Δr. The concept is to make the stress across each capacitor the same so that the bushing material is properly utilized. As the capacitors are connected in series, they all carry the same charge. Also, for the same stress across each capacitor, the voltage difference is the same for all bushing capacitors, which in turn implies the same capacitance. The capacitance in each section of thickness Δr is

$$C = 2\frac{\varepsilon 2\pi rz}{\Delta r} = \text{constant} \tag{2.78}$$

This implies that the bushing length must satisfy the curve for the profile (Hoole and Hoole, 1996)

$$rz = \text{constant} \tag{2.79}$$

Thus the bushing of a uniform dielectric should have a hyperbolic surface profile (Fig. 2.24), the same as for bushings with floating screens.

2.7.2 Electric Stress Optimization

High-voltage insulators usually serve as supports and spacers of HV electrodes with respect to earthed frames, as in gas-insulated systems or the ground plane in air (see Chapters 9 and 10). Flashover takes place along the insulator surface if the potential field is high enough to sustain a discharge. Therefore, the tangential field should be kept below the critical level

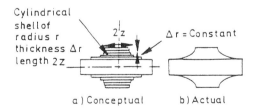

Cylindrical
shell of
radius r
thickness Δr
length 2z

$2z$

$\Delta r = \text{Constant}$

a) Conceptual   b) Actual

Figure 2.24 Model of graded bushing (a) as thin rings of insulation; (b) smoothed model.

everywhere along the insulator surface while keeping the overall length of the insulator at the minimum for a given voltage rating, to minimize the cost of material and installation.

Optimization of an HV insulator design is achieved by correcting its profile in elemental steps. At any point on the insulator surface the tangential field is derived from the potential difference between two test points close to the point under consideration. If the tangential field exceeds the set level, either the potential difference is decreased or the distance between the test points is increased (Gronwald, 1983). Recently, a method developed to calculate the tangential field component was applied to mathematical expressions of the profile to be corrected. The method was based on a modified charge-simulation technique whereby the magnitudes of the simulation charges were determined by minimizing a novel error function formulated over the HV electrode and insulator surface (Fig. 2.25) (Abdel-Salam and Stanek, 1987b).

Shaping of high-voltage electrodes within insulator systems is aimed at bringing the surface field stress down to the minimum possible, thus raising the withstand voltage of a given gap (e.g., a limited space inside a high-voltage laboratory). Such electrode optimization is based on an interactive process of charge simulation in which the contour points are shifted after each computation of field values (Fig. 2.26). Other optimization techniques for electrodes are based on an "optimized" simulation charge obtained by minimizing an objective function (Moller and Yousef, 1984).

2.8 SOLVED EXAMPLES

(1) A point charge q is located at a distance D from the center of a conducting sphere of radius R at zero potential. Locate the image of the charge q inside the sphere and determine its value.

Solution: Let the image charge q' be located at a distance b from the sphere center to satisfy the zero potential at $r = R$ (Fig. 2.27). The potential Φ at any point P outside the sphere is

$$\Phi = \frac{1}{4\pi\varepsilon_0}\left(\frac{q}{s} + \frac{q'}{s'}\right) \qquad (2.80)$$

where

$$s = [r^2 + D^2 - 2r\,D\cos\theta]^{\frac{1}{2}}$$
$$s' = [b^2 + r^2 - 2rb\cos\theta]^{\frac{1}{2}}$$

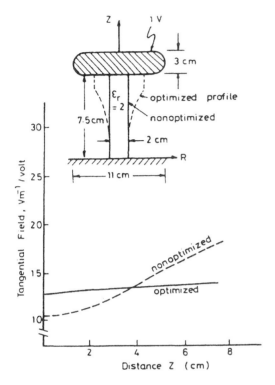

Figure 2.25 Tangential-field distribution along a profile for optimized and non-optimized insulators. [From Abdel-Salam and Stanek (1987b).]

At $r = R$, the potential Φ must be zero, so that q and q' must be of opposite polarity.

$$\frac{q}{s} + \frac{q'}{s'} = 0, \qquad \text{i.e.,} \quad \left(\frac{q}{s}\right)^2 = \left(\frac{q'}{s'}\right)^2_{r=R}$$

$$q^2[b^2 + R^2 - 2Rb\cos\theta] = q'^2[R^2 + D^2 - 2RD\cos\theta] \qquad (2.81)$$

Since equation (2.81) must be true for all values of θ,

$$\left.\begin{aligned} q^2(b^2 + R^2) &= q'^2(D^2 + R^2) \\ q^2 2Rb\cos\theta &= q'^2 2RD\cos\theta, \qquad \text{i.e., } q^2b = q'^2D \end{aligned}\right\} \qquad (2.82)$$

Eliminating q and q' yields a quadratic equation in b:

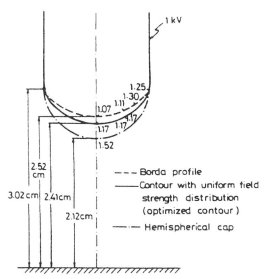

Figure 2.26 Optimization of a rod–plane gap. Borda profile: field values = 1.07–1.25 kV/cm; optimized contour: field value = 1.17 kV/cm; hemispherical cap: field values = 1.25–1.52 kV/cm.

$$b^2 - bD\left[1 + \left(\frac{R}{D}\right)^2\right] + R^2 = 0 \tag{2.83}$$

whose solution is

$$b = \frac{D}{2}\left\{\left[1 + \left(\frac{R}{D}\right)^2\right] \pm \left[1 - \left(\frac{R}{D}\right)^2\right]\right\} \tag{2.84}$$

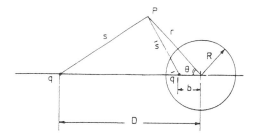

Figure 2.27 Image of a point charge q outside a conducting sphere at zero potential.

Take the lowest negative root so that the image charge is inside the sphere:

$$b = R^2/D \tag{2.85}$$

Substitute in equation (3):

$$q' = -qR/D \tag{2.86}$$

(2) A line charge λ is located at a distance a from the axis of a conducting cylinder of radius R at a given potential. Locate the image of the charge λ inside the cylinder and determine its value.

Solution: The image charge $-\lambda$ is located inside the cylinder (Zahn, 1979) at a distance $2a$ from the charge λ (Fig. 2.28). At a point $P(x, y)$, the potential Φ due to the charge λ and its image

$$
\begin{aligned}
\Phi &= \frac{\lambda}{2\pi\varepsilon_0} \ln \frac{R_2}{R_1} \\
&= \frac{\lambda}{2\pi\varepsilon_0} \ln \frac{(x+a)^2 + y^2}{(x+a)^2 + y^2}
\end{aligned} \tag{2.87}
$$

For the cylinder to be equipotential

$$
\begin{aligned}
\frac{(x+a)^2 + y^2}{(x-a)^2 + y^2} &= K_1 \\
\left(x - a\frac{K_1+1}{K_1-1} \right)^2 + y^2 &= \left(\frac{2a\sqrt{K_1}}{K_1-1} \right)^2
\end{aligned} \tag{2.88}
$$

which is the equation of a cylinder whose radius is equal to R

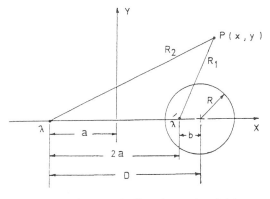

Figure 2.28 Image of a line charge λ outside a conducting cylinder.

$$R = \frac{2a\sqrt{K_1}}{(K_1 - 1)} \tag{2.89}$$

with center at

$$x(= D - a) = a\frac{K_1 + 1}{K_1 - 1}$$

i.e.,

$$a + \frac{a(K_1 + 1)}{K_1 - 1} = D = \frac{2aK_1}{K_1 - 1} \tag{2.90}$$

Solve equations (2.89) and (2.90) simultaneously to obtain

$$a = \frac{D^2 - R^2}{2D} \tag{2.91}$$

The image charge is spaced a distance b from the center

$$b = D - 2a = \frac{R^2}{D} \tag{2.92}$$

(3) Calculate the electric field at a point along the axis of a planar ring charge of radius b and charge density ρ_l.

Solution: Let the ring be located in the $X–Y$ plane (Fig. 2.29). The potential Φ at an arbitrary point P along the Z axis is

$$\Phi = \frac{1}{4\pi\varepsilon_0}\oint\frac{\rho_l}{|l|}\,dl$$

$$= \frac{\rho_l}{4\pi\varepsilon_0}\int_0^{2\pi}\frac{1}{\sqrt{b^2 + z^2}}\,bd\Psi = \frac{b\rho_l}{2\varepsilon_0\sqrt{b^2 + z^2}} \tag{2.93}$$

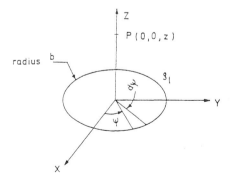

Figure 2.29   Electric field of a planar ring charge on the $X–Y$ plane.

The electric field generated by the ring is related to the potential Φ by

$$\bar{E} = -\nabla\Phi \tag{2.94}$$

Equation (2.94) implies that $\bar{E}$ may have up to three components, but one can surmise that the components other than the z component of $\bar{E}$ are zero at all points on the Z axis, since this charge distribution is symmetric about that axis. Thus, the z component of $\bar{E}$ is

$$E_z = \frac{-\partial\Phi}{\partial z} = \frac{bz\rho_l}{2\varepsilon_0[b^2 + z^2]^{3/2}}$$
$$= \frac{zq_r}{4\pi\varepsilon_0[z^2 + b^2]^{3/2}} \tag{2.95}$$

where q_r is the value of the ring charge $(= 2\pi b\rho_l)$.

(4) What is the electric field caused by a charged shell (Faraday's cage) of radius a and charge q at a distance D from its center?

Solution: Slice the shell into a ring at an angle θ and of thickness $a\,d\theta$ (Fig. 2.30). The radius of the ring is $a \sin\theta$ and its surface area is $2\pi a \sin\theta \cdot a\,d\theta = 2\pi a^2 \sin\theta\,d\theta$. The ring charge is $(q/4\pi a^2)2\pi a^2 \sin\theta\,d\theta$ $(= q\sin\theta\,d\theta/2)$, assuming that the charge q is uniformly distributed over the surface of the sphere.

At the point P(0, 0, D) (Fig. 2.30), the field due to the ring charge has one component along the Z axis [see example (3)].

$$dE_z = \frac{\frac{q}{2}\sin\theta\,d\theta(D - a\cos\theta)}{4\pi\varepsilon_0(D^2 - 2aD\cos\theta + a^2)^{3/2}} \tag{2.96}$$

$$E_z = \int_0^\pi dE_z = \frac{1}{4\pi\varepsilon_0}\frac{q}{2}\frac{1}{D^2}\left[\frac{a+D}{[(a+D)^2]^{1/2}} - \frac{a-D}{[(a-D)^2]^{1/2}}\right] \tag{2.97}$$

Take the positive root only.

$$\left.\begin{aligned}
E_z &= \frac{1}{4\pi\varepsilon_0}\frac{q}{2}\frac{1}{D^2}\left[\frac{a+D}{a+D} - \frac{a-D}{D-a}\right] = \frac{q}{4\pi\varepsilon_0 D^2} & \text{for } D > a \\
&= \frac{1}{4\pi\varepsilon_0}\frac{q}{2}\frac{1}{D^2}\left[\frac{a+D}{a+D} - \frac{a-D}{a-D}\right] = 0 & \text{for } D < a
\end{aligned}\right\} \tag{2.98}$$

The latter equation is a very important result with practical applications. It means that there is no electric field in the shell (Faraday's cage). This is utilized in the power industry to repair high-voltage lines without ever switching off the line. A conductive cage suspended from pulley-like rollers travels down line using the latter as rails, and workers repair and service the line from inside. Although they are dealing with high voltages, the fact that they are in a conductive cage prevents a voltage difference (i.e., an electric

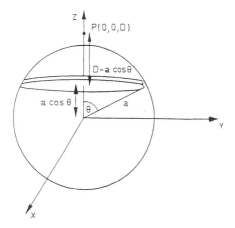

Figure 2.30 Building up rings into a charged shell.

field) from developing across their bodies with the associated health hazards.

The same result explains why people in a car are unaffected when a bolt of lightning hits the car. However, the charge deposited on the conducting body of the car raises it to a high voltage, so that it is dangerous for someone inside to step out to the ground, since now each foot of the passenger would be at a different potential.

(5) A sphere gap consists of two spheres with $R = 0.25\,\text{m}$ each. The gap between their surfaces is $0.5\,\text{m}$. Calculate the charges and their locations to make the sphere potentials 1 and 0. Calculate the maximum field at the sphere surface. Distance between spheres centers is 1 m.

Solution:
The charge of the stressed sphere (sphere 1) $Q_1 = 4\pi\varepsilon_o RV = \pi\varepsilon_0$ (Fig. 2.31, Table 2.3).

Field at surface E (i.e., at point P)

$$= \frac{Q_1}{2\pi\varepsilon_0}\left[\frac{1}{(0.25)^2} + \frac{0.067}{(0.25-0.067)^2} + \frac{0.0048}{(0.25-0.067)^2} + \cdots\right.$$

$$+ \frac{Q_1}{2\pi\varepsilon_0}\left[\frac{0.25}{(0.75-0.065)^2} + \frac{0.01795}{(0.75-0.067)^2} + \frac{0.00128}{(0.75-0.067)^2} + \cdots\right]$$

$$= 18.718\frac{Q_1}{4\pi\varepsilon_0} = 4.6795\text{V/m per volt}$$

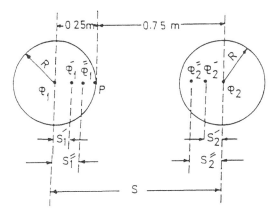

Figure 2.31 Successive imaging between two spheres; one is stressed by unit voltage and the other is grounded.

(6) The dimensions of a $\pm400\,\text{kV}$ bipolar DC line are shown in Figure 2.32. Calculate:

(a) the charge on each bundle and on each subconductor;

(b) the maximum, minimum, and average surface fields on the subconductors
   (i) omitting the charges of the second pole and image subconductors,
   (ii) considering the charges of the second pole but omitting the charge on the image subconductor;

(c) the average maximum surface voltage gradient of the bundle under (b)(ii).

Solution:

(a) Replace each bundle by an equivalent conductor of radius

$$r_{\text{eq}} = \sqrt{0.0175 \times 0.45} = 0.088\,74\,\text{m}$$

As the applied voltage is $400\,\text{kV}$

$$400 \times 10^3 = \frac{Q}{2\pi\varepsilon_0}\ln\frac{2H}{r_{\text{eq}}} - \frac{Q}{2\pi\varepsilon_0}\ln\frac{\sqrt{(2H)^2 + d^2}}{d}$$

where H = bundle height and d = pole-to-pole spacing.

$$400 \times 10^3 = \frac{Q}{2\pi\varepsilon_0}\left[\ln\frac{24}{0.088\,74} - \ln\frac{25.63}{9}\right]$$

Table 2.3 Charges in Example 5

Charges inside sphere 1		Charges inside sphere 2	
Magnitude	Distance from sphere center	Magnitude	Distance from sphere center
Q_1	$S_1 = 0.0$	$Q_2 = Q_1 R/s$ $= -0.25Q_1$	$\dfrac{R^2}{S} = (0.25)^2/1$ $= 0.0625$
$Q_1' = \dfrac{0.25Q_1}{1 - 0.0625}$ $= 0.067Q_1$	$S_1' = \dfrac{(0.25)^2}{1 - 0.0625}$ $= 0.067$	$Q_2' = \dfrac{-0.067Q_1 \times 0.25}{1 - 0.67}$ $= -0.017\ 95\ Q_1$	$S_2' = \dfrac{(0.25)^2}{1 - 0.067}$ $= 0.067$
$Q_1'' = \dfrac{0.017\ 95\ Q_1 \times 0.25}{1 - 0.067}$ $= 0.0048\ Q_1$	$S_1'' = \dfrac{(0.25)^2}{1 - 0.067}$ $= 0.067$	$Q_2'' = \dfrac{-0.0048\ Q_1 \times 0.25}{1 - 0.067}$ $= -0.001\ 28\ Q_1$	$S_2'' = \dfrac{(0.25)^2}{1 - 0.067}$ $= 0.067$
$Q_1''' = \dfrac{0.001\ 28\ Q_1 \times 0.25}{1 - 0.67}$ $= 0.000\ 34\ Q_1$	$S_1''' = \dfrac{(0.25)^2}{1 - 0.067}$ $= 0.067$		

$\therefore$ charge per bundle $\quad Q = 4.88\ \mu C/m$

charge per subconductor $q = Q/2$

$$= 2.44\ \mu C/m$$

(b) (i) Maximum surface field $= \dfrac{q}{2\pi\varepsilon_0}\left[\dfrac{1}{r} + \dfrac{1}{R}\right]$

where r = subconductor radius and R = subconductor-to-subconductor spacing.

$$\therefore \text{Maximum field} = 43.925 \times 10^3 \left(\dfrac{1}{0.0175} + \dfrac{1}{0.45}\right)$$

$$= 2607\ kV/m$$

Maximum surface field $= \dfrac{q}{2\pi\varepsilon_0}\left[\dfrac{1}{r} - \dfrac{1}{R}\right]$

$$= 2412 \quad kV/m$$

Average surface field $= \dfrac{q}{2\pi\varepsilon_0}\dfrac{1}{r} = 2510\ kV/m$

(ii) Consider the two subconductors on the left (Fig. 2.32).

E_{01} (field at outer point of subconductor #1)

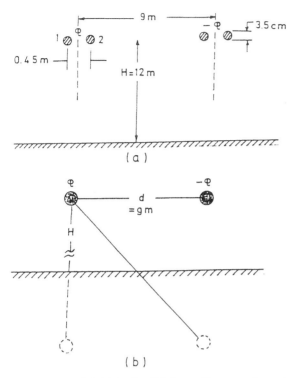

Figure 2.32   (a) A bipolar DC line with two subconductors per pole; (b) the line after replacing each pole by an equivalent conductor.

$$= \frac{q}{2\pi\varepsilon_0}\left[\frac{1}{r} + \frac{1}{R}\right] - \frac{q}{2\pi\varepsilon_0}\left(\frac{1}{d+r} + \frac{1}{d+R+r}\right)$$
$$= 2598\,\text{kV/m}$$

E_{02} (field at outer point of subconductor #2)

$$= \frac{q}{2\pi\varepsilon_0}\left[\frac{1}{r} + \frac{1}{R}\right] - \frac{q}{2\pi\varepsilon_0}\left(\frac{1}{d-R-r} + \frac{1}{d-r}\right)$$
$$= 2617.6\,\text{kV/m}$$

E_{I1} (field at inner point of subconductor #1)

$$= \frac{q}{2\pi\varepsilon_0}\left[\frac{1}{r} - \frac{1}{R}\right] - \frac{q}{2\pi\varepsilon_0}\left(\frac{1}{d-r} + \frac{1}{d+R-r}\right)$$
$$= 2422.43\,\text{kV/m}$$

E_{I2} (field at inner point of subconductor #2)

$$= \frac{q}{2\pi\varepsilon_0}\left[\frac{1}{r}-\frac{1}{R}\right] - \frac{q}{2\pi\varepsilon_0}\left(\frac{1}{d-R+r}+\frac{1}{d+r}\right)$$
$$= 2422.39 \, \text{kV/m}$$

The average maximum gradient is defined as the arithmetic average of the two maxima

$$= (2598 + 2617)/2 = 2607.8 \, \text{kV/m}$$

This is almost equal to the average gradient obtained by omitting the charges of the other pole.

For a three-phase AC line, the effect of charges on other phases can usually be ignored because, when the charge on the conductor of one phase is at peak value, the charges on the other phases are passing 50% of their peak values, but are of opposite polarity. This has an even smaller effect than what has been shown for the bipolar DC line, where the charge of the second pole is equal and opposite to the charge of the conductor under consideration.

(7) Two infinite conducting sheets lie on the planes $x = -1$ and $x = 1$. A line charge of $1C/m$ is placed at $x = 0$. Determine the electric field induced at $x = 1$.

Solution: The line charge $q(= 1C/m)$ has an image of $-1C/m$ in the right sheet at $x = 2$. This in turn has an image of $1 \, C/m$ in the left sheet of $-1C/m$ at $x = 6$. In general, it results in an infinite sequence of images, as shown in Figure 2.33.

By superposition, the total electric field is

$$E = \frac{q}{2\pi\varepsilon_0} - \frac{q}{2\pi\varepsilon_0}\left[\frac{1}{3}+\frac{1}{7}+\cdots\right] + \frac{q}{2\pi\varepsilon_0}\left[1+\frac{1}{5}+\frac{1}{9}\cdots\right]$$
$$+ \frac{q}{2\pi\varepsilon_0}\left[\frac{1}{5}+\frac{1}{9}+\cdots\right] - \frac{q}{2\pi\varepsilon_0}\left[\frac{1}{3}+\frac{1}{7}+\cdots\right]$$
$$= 27.18 \times 10^9 \text{V/m}$$

This example serves for field assessment in a wire-duct electrostatic precipitator whose wire has a charge q per unit length. The duct is represented by the two conducting sheets.

(8) Design a graded cylindrical bushing that, for mechanical reasons, has to be 20 cm long outside. The conductor inside the bushing has a radius of 2 cm and is stressed by an AC voltage of 150 kV. The strength of the bushing material is 50 kV/cm. Calculate the volume of insulator required for the graded design and compare it with the volume for regular design of the bushing.

Solution: The thickness of graded design
$$= \text{thickness of flat bushing}$$
$$= V/E_{bd} = 150\sqrt{2}/50 = 4.24 \, \text{cm}$$

The bushing profile follows the curve (Fig. 2.24)
$$zr = 10(4.24 + 2) = 62.4 \, \text{cm}^2$$

The volume of graded design V_1
$$= \int_2^{6.24} 2(2\pi r)z \, dr = 4\pi \times 62.4(62.4 - 2)$$
$$= 3324.76 \, \text{cm}^2$$

The thickness of regular design, as obtained form equation (2.77)
$$t = 2\left[e^{150\sqrt{2}/2 \times 50} - 1\right] = 14.68 \, \text{cm}$$

Thus, the volume of regular design V_2
$$= \pi[(2 + 14.68)^2 - (2)^2]$$
$$= 8614.95$$

Therefore, the volume of the graded design is about 38% of that of the regular design. This saving in insulating material is added to the proper utilization of the material where the stress is uniformly distributed.

(9) Find the potential at the grid nodes of the rectangular cable of Figure 2.34a. The inner core is at a potential of 100 V and the outer sheath, being grounded, is at zero potential. The inner core and the sheath are $2 \times 2 \, \text{cm}^2$ and $6 \times 6 \, \text{cm}^2$ in cross-section, respectively.

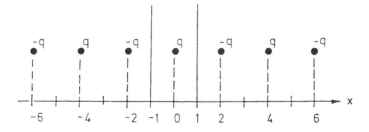

Figure 2.33 Images of the line charge causing successive images.

Solution: A symmetric quarter of the cable is considered and a square grid is mapped. The grid points are numbered column by column from top to bottom.

For the grid nodes (6–12) located at the surface of the grounded sheath, the potential is specified and equal to zero.

For the grid nodes (13–15) located at the core surface, the potential is specified and equal to 100 V.

For nodes like (1) on the boundary normal to which the potential does not vary, i.e., the Y direction, an equation is necessary since the potential is unknown; i.e., $\partial \Phi / \partial y|_{\text{node } 1} = 0$

Such nodes are surrounded by only three others (nodes 2, 6, 13). A fictitious node 2′ is introduced as shown in Fig. 2.34b;

$$\frac{\phi_2 - \phi_2'}{2\Delta Y} = 0$$
$$\therefore \phi_2 = \phi_2'$$

Therefore, equation (2.32) is reduced to

$$\phi_1 = \frac{1}{4}(\phi_6 + 2\phi_2 + \phi_{13}) \tag{2.99}$$

Similarly, for nodes like (5), on the boundary normal to which the potential does not vary, i.e., in the X direction

$$\partial \Phi / \partial x|_{\text{node } 5} = 0$$

Such nodes are surrounded by only three others (nodes 4, 12, 15). A fictitious node 4′ is introduced as shown in Fig. 2.34c;

$$\phi_4 = \phi_4'$$

and equation (2.32) takes the form

$$\phi_5 = \frac{1}{4}(\phi_{12} + 2\phi_4 + \phi_{15}) \tag{2.100}$$

For the other nodes (2,3,4), equation (2.32) takes the form

$$\phi_2 = \frac{1}{4}(\phi_7 + \phi_{14} + \phi_1 + \phi_3) \tag{2.101}$$

$$\phi_3 = \frac{1}{4}(\phi_8 + \phi_4 + \phi_2 + \phi_{10}) \tag{2.102}$$

$$\phi_4 = \frac{1}{4}(\phi_{11} + \phi_{14} + \phi_3 + \phi_5) \tag{2.103}$$

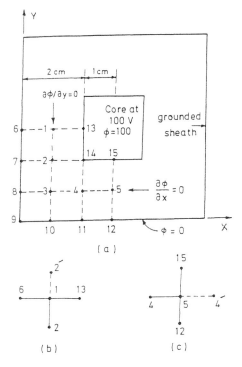

Figure 2.34 (a) Finite-difference grid for a quarter of cable; (b) and (c) fictitious nodes introduced to satisfy boundary conditions.

Thus, the set (2.33) is expressed by equations (2.99) through (2.103) which are solved simultaneously to obtain the unknown potentials at the grid nodes

$\phi_1 = 45.83$ V

$\phi_2 = 41.66$ V

$\phi_3 = 20.82$ V

$\phi_4 = 41.66$ V

$\phi_5 = 45.83$ V

It is satisfying to observe the symmetry in the calculated potentials where $\phi_1 = \phi_5$ and $\phi_2 = \phi_4$.

(10) Using the charge simulation technique, calculate the potential and field around a hemispherically capped cylindrical rod stressed with a voltage

$V = 1.0\,\text{kV}$ with respect to a ground plane (Fig. 2.35). The rod radius R is 1 cm and the gap spacing H is 1 m.

Solution: Choose a cylindrical coordinate system with the origin located at the intersection of the rod axis with the ground plane (Fig. 2.35).

The charge on the rod surface is simulated by a point charge (Q_p) located at the center of the hemispherical cap having coordinates $(0, H + R)$ in addition to a series of N semi-infinite line charges $(Q_j, j = 1, 2, \ldots, N)$ along the axis of the cylindrical portion starting at points of coordinates $(0, A_j)$, $A_j \geq R + H, j = 1, 2, \ldots, N$ (Abou-Seada and Nasser, 1968). Thus, the number of unknown simulation charges is $N + 1$.

To maintain the $Z = 0$ plane at zero potential, images of the simulation charges with respect to this plane are considered.

The potential Φ at any point (r, z) is

$$\Phi(r, z) = Q_p P(r, z) + \sum_{j=1}^{N} Q_j U(r, z, A_j) \ldots \tag{2.104}$$

where the potential coefficients

$$P(r, z) = \frac{1}{4\pi\varepsilon_0} \left[\frac{1}{\sqrt{r^2 + (H + R - z)^2}} - \frac{1}{\sqrt{r^2 + (H + R + z)^2}} \right]$$

$$= \frac{1}{4\pi\varepsilon_0} \left[\frac{1}{\sqrt{r'^2 + (H' + 1 - z')^2}} - \frac{1}{\sqrt{r'^2 + (H' + 1 + z')^2}} \right]$$

$$U(r, z, A_j) = \frac{1}{4\pi\varepsilon_0} \ln \frac{A_j + z + \sqrt{r^2 + (A_j + z)^2}}{A_j - z + \sqrt{r^2 + (A_j - z)^2}}$$

$$= \frac{1}{4\pi\varepsilon_0} \ln \frac{A_j' + z' + \sqrt{r'^2 + (A_j' + z')^2}}{A_j' - z' + \sqrt{r'^2 + (A_j' - z')^2}}$$

where r', z', H', and A_j' are, respectively, r, z, H, and A_j normalized to the radius of the rod; i.e., $r' = r/R$ and so on.

The boundary conditions at the rod surface are described as follows:
   (i) the potential at $(N - 2)$ points of the z coordinate designated $z_i \geq H + R$, selected on the cylindrical portion, is equal to the applied voltage;

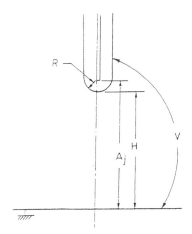

Figure 2.35 Hemispherically-capped cylindrical rod against grounded plane.

$$\Phi(R, z_i) = V; \qquad z_i \geq R + H, \qquad i = 1, 2, \ldots, N - 2 \quad (2.105a)$$

(ii) the potential at the tip of the rod on the hemispherical portion is also equal to the applied voltage;

$$\Phi(0, H) = V \qquad\qquad (2.105b)$$

To satisfy the above condition at all the hemispherical portion defined by the angle $\theta \geq \pi/2$, the even derivatives (second, fourth, ...) of $\Phi(\theta)$ with respect to θ, evaluated at the rod tip ($\theta = 0$), are set equal to zero; i.e.,

$$\frac{d^2\Phi}{d\theta^2}\Big|_{\theta=0} = Q_p P_{2\theta}(0) + \sum_{j=1}^{N} Q_j U_{2\theta}(0, A_j) = 0 \qquad (2.105c)$$

$$\frac{d^4\Phi}{d\theta^4}\Big|_{\theta=0} = Q_p P_{4\theta}(0) + \sum_{j=1}^{N} Q_j U_{4\theta}(0, A_j) = 0 \qquad (2.105d)$$

where $P_{2\theta}(0)$, $P_{40}(\theta)$, $U_{2\theta}(0, A_j)$, and $U_{4\theta}(0, A_j)$ have been expressed by Abou-Seada and Nasser (1968).

Satisfaction of equations (2.105a) at the $(N - 2)$ selected points on the rod surface and equations (2.105b) through (2.105d) at the rod tip results in $(N + 1)$ equations whose simultaneous solution determines the unknown $(N + 1)$ simulation charges. With the determination of the unknown charges, the potential at any point is obtained using equation (2.104), and the field is obtained as

$$\bar{E}(r, z) = -\left\{\left[Q_P \frac{\partial}{\partial r} P(r, z) + \sum_{j=1}^{N} Q_j \frac{\partial}{\partial r} U(r, z; A_j)\right] \bar{a}_r\right.$$

$$\left. + \left[Q_P \frac{\partial}{\partial z} P(r, z) + \sum_{j=1}^{N} Q_j \frac{\partial}{\partial z} U(r, z; A_j)\right] \bar{a}_z\right\} \qquad (2.106)$$

where $\bar{a}_r$ and $\bar{a}_z$ are unit vectors along the r and z directions.

The potential and field along the gap axis are shown in Figures 2.36 and 2.37, respectively.

(11) Consider the two-element mesh shown in Figure 2.38a. Using the finite-element method, determine the potentials within the mesh.

Solution: The element stiffness matrices are obtained using equations (2.46). For element 1, consisting of nodes 1–2–4 corresponding to local numbering 1–2–3 as in Figure 2.38b,

$$\Delta Y_1 = -1.3, \qquad \Delta Y_2 = 0.9, \qquad \Delta Y_3 = 0.4$$
$$\Delta X_1 = -0.2, \qquad \Delta X_2 = -0.4, \qquad \Delta X_3 = 0.6$$

$$A = \frac{1}{2}(0.54 + 0.16) = 0.35$$

Substituting into equation (2.46a) gives

$$[S]^{(1)} = \begin{bmatrix} 1.236 & -0.7786 & -0.4571 \\ -0.7786 & 0.6929 & 0.0857 \\ -0.4571 & 0.0857 & 0.3714 \end{bmatrix}$$

Similarly, for element 2 consisting of nodes 2–3–4 corresponding to local numbering 1–2–3 as in Figure 2.38b,

$$\Delta Y_1 = -0.6, \qquad \Delta Y_2 = 1.3, \qquad \Delta Y_3 = 0.7$$
$$\Delta X_1 = -0.9, \qquad \Delta X_2 = 0.2, \qquad \Delta X_3 = 0.7$$

$$A = \frac{1}{2}(0.91 + 0.14) = 0.525$$

Hence,

$$[S]^{(2)} = \begin{bmatrix} 0.5571 & -0.4571 & -0.1 \\ -0.4571 & 0.828 & 0.3667 \\ -0.1 & 0.667 & 0.4667 \end{bmatrix}$$

In the light of the fact that nodes 2 and 4 are of free (unknown) potentials whereas nodes 1 and 3 are of bound (known) potentials,

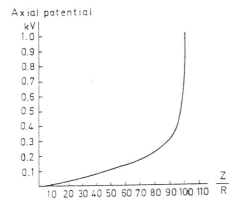

Figure 2.36   Potential along the axis of the rod–plane gap.

the submatrices $[S]_{ff}$ and $[S]_{fb}$ are expressed in terms of the global stiffness matrix

$$[S]_{ff} = \begin{bmatrix} S_{22} & S_{24} \\ S_{42} & S_{44} \end{bmatrix}$$

$$[S]_{fb} = \begin{bmatrix} S_{21} & S_{23} \\ S_{41} & S_{43} \end{bmatrix}$$

$$[\phi]_f = \begin{bmatrix} \phi_2 \\ \phi_4 \end{bmatrix}, \qquad [\phi]_b = \begin{bmatrix} \phi_1 \\ \phi_3 \end{bmatrix}$$

According to equation (2.49)

$$[S]_{ff}[\phi]_f = -[s]_{jb}[\phi]_b$$

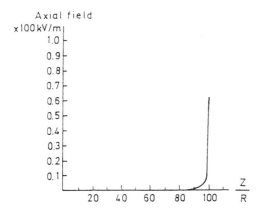

Figure 2.37   Field along the axis of the rod–plane gap.

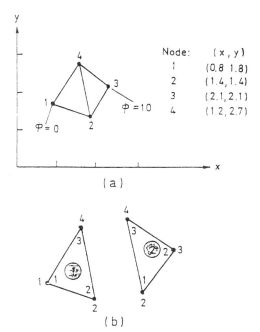

Figure 2.38 (a) Two-element mesh; (b) local and global numbering of the element.

$$\begin{bmatrix} S_{22} & S_{24} \\ S_{42} & S_{44} \end{bmatrix}\begin{bmatrix} \phi_2 \\ \phi_4 \end{bmatrix} = -\begin{bmatrix} S_{21} & S_{23} \\ S_{41} & S_{43} \end{bmatrix}\begin{bmatrix} \phi_1 \\ \phi_3 \end{bmatrix} \tag{2.49a}$$

The elements of the global stiffness matrix are obtained (Sadiku, 1994) as

$$S_{22} = S_{22}^{(1)} + S_{11}^{(2)} = 0.6929 + 0.5571 = 1.25$$
$$S_{24} = S_{23}^{(1)} + S_{13}^{(2)} = 0.0857 - 0.1 = -0.0143 = S_{42}$$
$$S_{44} = S_{33}^{(1)} + S_{33}^{(2)} = 0.3714 + 0.4667 = 0.8381$$
$$S_{21} = S_{21}^{(1)} = -0.7786$$
$$S_{23} = S_{12}^{(2)} = -0.4571$$
$$S_{41} = S_{31}^{(1)} = -0.4571$$
$$S_{43} = S_{32}^{(2)} = -0.3667$$
$$\phi_1 = 0, \qquad \phi_3 = 10$$

Substituting in equation (2.49a),

$$\begin{bmatrix} 1.25 & -0.014 \\ -0.014 & 0.8381 \end{bmatrix} \begin{bmatrix} \phi_2 \\ \phi_4 \end{bmatrix} = \begin{bmatrix} -0.7786 & -0.4571 \\ -0.4571 & -0.3667 \end{bmatrix}$$

One obtains

$$\phi_2 = 3.708 \quad \text{and} \quad \phi_4 = 4.438$$

Once the values of the potentials at the nodes are known, the potential at any point within the mesh can be determined using equation (2.43).

REFERENCES

Abdel-Salam M. J Phys D: Appl Phys 20:629,1987.
Abdel-Salam M, Abdel-Aziz E. J Phys D: Appl Phys 27:2570, 1994.
Abdel-Salam M, Abdel-Sattar S. IEEE Trans E1-26:669, 1989.
Abdel-Salam M, Al-Hamouz Z. J Phys D: Appl Phys 25:1318, 1992.
Abdel-Salam M, Al-Hamouz Z. J Phys D: Appl Phys 26:2202, 1993.
Abdel-Salam M, Al-Hamouz Z. IEE Proc-A 141:369, 1994.
Abdel-Salam M, Al-Hamouz Z. IEEE Trans 1A-31:484, 1995a.
Abdel-Salam M, Al-Hamouz Z. IEEE Trans 1A-31:447, 1995b.
Abdel-Salam M, Al-Hamouz Z. J Phys D: Appl Phys 25:1318, 1992.
Abdel-Salam M, El-Mohandes ThM. Proceedings of International Symposium on High Voltage Engineering, Braunschweig, West Germany, 1987, Paper 33.11.
Abdel-Salam M, El-Mohandes ThM. IEEE Trans IA-25, September/October 1989.
Abdel-Salam M, Ibrahim A. IEEE–PS paper A-77-131-6, 1977.
Abdel-Salam M, Khalifa M. Acta Phys 36:201, 1974.
Abdel-Salam M, Shamloul D. Proceedings of IEEE–IAS Annual Meeting, Pittsburgh, PA, 1988, p 1677.
Abdel-Salam M, Stanek EK. IEEE Trans IA-23:481, 1987a.
Abdel-Salam M, Stanek EK. IEEE Trans EI-22:47, 1987b.
Abdel-Salam M, Farghally M, Abdel-Sattar, S. IEEE Trans PAS-101:4079, 1982.
Abdel-Salam M, Farghally M, Abdel-Sattar, S. IEEE Trans EI-18:110, 1983.
Abdel-Salam M, Farghally M, Abdel-Sattar S, Shamloul D. Proceedings of 4th International Symposium on Gaseous Dielectrics, Knoxville, TN, 1984, p 492.
Abdel-Salam M, Mufti A. Elect Power Sys Res 44:145, 1998.
Abdel-Salam M, Singer H, Ahmed A. J Phys D: Appl Phys 30:1017, 1997.
Abdel-Salam et al., 1997—
Abou-Seada M, Nasser E. Proc IEEE 56:813, 1968.
Andjelic Z. Proceedings of International AMSE Conference on Modelling and Simulation, Athens, 1984, Vol 21, p 299.
Anis H, Abd-Allah M. Proceedings of Middle East Power System Conference, Cairo, 1989, pp 116–121.
Cendes ZJ, Hamann JR. IEEE Trans, PAS-100:1806, 1981.

Chen HS, Pearmain AJ. Proceedings of 4th International Symposium on High Voltage Engineering, Athens, 1983, Paper 11.07.

Comber M, Zaffanella L. IEEE Trans PAS-93:81, 1974.

Daffe J, Olsen RG. IEEE Trans PAS-98:1609, 1979.

Davis JL, Hoburg JF. J Electrost 18:1–22, 1986.

Deutsch W. Ann Phys 5:589, 1933.

El-Arabaty A, Abdel-Salam M, Mansour E. IEEE–PES paper A-77-236-3, 1977.

Foo PY, King SY. Proc IEE 123:702, 1976.

Gronwald H. Proceedings of 4th International Symposium on High Voltage Engineering, Athens, 1983, Paper 11.01.

Gutfleisch F. Berechnung elektrischer Felder durch Nachbildung der Grenzschichten mit ausgawählten Flächenelementen. Doctoral thesis, Technical University Hamburg-Harburg, Hamburg, 1989.

Gutfleisch F, Singer H, Forger K, Gomollon JA. IEEE Trans Power Deliv, 9:743, 1994.

Hoole SRH, Hoole PRP. A Modern Short Course in Engineering Electromagnetics (UK). Oxford: Oxford University Press, 1996.

Hornstein MN. Proceedings of IEEE–IAS Annual Meeting, Philadelphia, PA, 1980, p 1081.

Hornstein MN. IEEE Trans 1A-20:1607, 1984.

Ibrahim AA, Singer H. In: IEE Conference Publication, Gas Discharges and their Application. Stevenage, UK: Institution of Electrical Engineers, 1982, p 128.

Ibrahim AA, Singer H. Proceedings of 4th International Symposium on Gaseous Dialectrics, Knoxville, TN, 1984, p 106.

James ML, Smith GM, Wolford JC. Applied Numerical Methods for Digital Computation. New York: Harper & Row, 1985.

Jennings A. Matrix Computation for Engineers and Scientists. New York: John Wiley & Sons, 1977.

Khalifa M, Abdel-Salam M. Proc IEE 120:1574, 1973.

Khalifa M, Abdel-Salam M. IEEE Trans PAS-93:1699, 1974a.

Khalifa M, Abdel-Salam M. IEEE Trans PAS-93:720, 1974b.

Khalifa M, Abdel-Salam M, Aly F, Abou-Seada M. IEEE–PES paper A-75-563-7, 1975.

Khan MJ, Alexander PH. IEEE Trans EI-17:325, 1982.

McWhirter JH, Oravec JJ. Proceedings of 3rd International Symposium on High Voltage Engineering, Milan, 1979, Paper 11.14.

Misaki T, Tsuboi H, Itaka K, Hara T. IEEE Trans PAS-101:627, 1982.

Moller K, Yousef F. ETZ-Archiv 6:143, 1984.

Mukherjee PK, Roy CK. Proceedings of 4th International Symposium on High Voltage Engineering, Athens, 1983, Paper 12.09.

Oin BL, Sheng JN, Yan Z, Gela G. IEEE papers 86T and D-513-6, 1986.

Okon EE. Int J Num Math Eng 21:197, 1985.

Okubo H, Metz D. Archiv fuer Electrotechnik 60:27, 1978.

Okubo H, Ikeda M, Honda M, Tanari T. IEEE Trans PAS-101:4039, 1982.

Ryan HM, Ali SMG, Powell CW. Proceedings of 4th International Symposium on High Voltage Engineering, Athens, 1983, Paper 12.12.

Sadiku MNO. Elements of Electromagnetics. Orlando, FL: Saunders College Publishing (USA), 1994.

Scott M, Mattingley J, Ryan H. IEEE Trans EI-9:18, 1974.

Segerlind L. Applied Finite Element Analysis. New York, USA: J. Wiley & Sons, 1984.

Shen LC, Kong JA. Applied Electromagnetism. Pacific Grove, CA: Brooks/Cole, 1983.

Shephard MS, Gallagher RH. Finite Element Grid Optimization. New York: American Society of Mechanical Engineers, 1979.

Silvester PP, Ferrari RL. Finite Elements for Electrical Engineers. Cambridge, UK: Cambridge University Press, 1983.

Singer H, ETZ-Archiv 3:265, 1981.

Singer H. Archiv fuer Electrotechnik 67:309, 1984.

Singer H, Steinbigler H, Weiss P. IEEE Trans PAS-93:1660, 1974.

Steinbigler H. Proceedings of 3rd International Symposium on High Voltage Engineering, Milan, 1979, Paper 11.11.

Takuma T, Ikeda T, Kawamoto T. IEEE Trans PAS-100:2802, 1981a.

Takuma T, Kawamoto T, Fujinami H. IEEE Trans PAS-100:4665, 1981b.

Tan KX, Steinbigler H. COMPEL, 4:209, 1985.

Thannassoulis P, Comsa RP. IEEE Trans PAS-90:145, 1971.

Timascheff A. IEEE Trans PAS-94:104, 1975.

Zahn M. Electromagnetic Field Theory. New York, USA: J Wiley & Sons, 1979.

Zeitoun A, Abdel-Salam M, El-Ragheb M. IEEE–PES paper A-76-418-4, 1976.

Zienkiewicz OC. The Finite Element Method. Maidenhead, UK: McGraw-Hill, 1977.

3

Ionization and Deionization Processes in Gases

M. ABDEL-SALAM *Assiut University, Assiut, Egypt*

3.1 INTRODUCTION

In the process of partial or complete breakdown of a gas gap there are several mechanisms of electron and ion generation or annihilation in operation, either singly or in combination. In the present chapter we summarize briefly the more significant mechanisms for ionization and deionization in a gas discharge. The processes considered include:

1. Ionization by cosmic rays, x-rays, and nuclear radiation
2. Ion generation by electron impact, photoionization, interaction with metastable atoms, thermal ionization, and electron detachment
3. Deionization by recombination and by diffusion

Also, in becoming attached to a neutral particle, an electron produces a negative ion. Before discussing these processes, it would be best first to describe the behavior of gaseous dielectrics in an electric field. This calls for a review of some basic principles of the kinetic theory of gases as they pertain to gaseous ionization and deionization processes.

3.2   KINETIC THEORY OF GASES

Many of the properties of gases can be well represented by kinetic theory, which treats the gas molecules as tiny balls with no need to specify their exact size, shape, or internal construction. These properties include gas pressure, temperature, distribution of molecular speeds, and energies.

3.2.1   Kinetic Interpretation of Gas Pressure

Let us consider a gas composed of moving molecules, each with mass m_0. The cumulative effect of many molecules striking against a wall leads to a more-or-less steady force (pressure) on the wall, depending on the number of molecules hitting the wall per second. Consider the ith molecule of the gas in a cubic container; it has a velocity component v_{ix} in the X direction (Fig. 3.1). Assume that collision with the wall reverses the velocity component v_{ix}, so that the time Δt_i between two successive collisions of the ith molecule with the right-hand wall is

$$\Delta t_i = \frac{2D}{v_{ix}}$$

where D is the container-side length.

The impulse exterted on the molecule during a collision is equal to $2m_0 v_{ix}$. If an average force F_{ix} is exerted by the molecule on the wall during time Δt_i, then

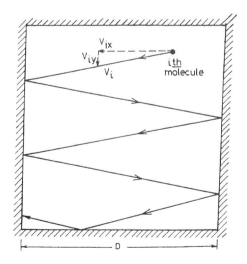

Figure 3.1   Motion of a gas molecule in a cubic container.

$$F_{ix} = \frac{2m_o v_{ix}}{\Delta t_i} = \frac{m_o v_{ix}^2}{D}$$

The total force F_x on the wall due to ND^3 molecules in the container will be the sum of the average force due to each. N is the number of molecules per unit volume.

$$F_x = \sum_{i=1}^{ND^3} F_{ix} = \frac{m_0}{D} ND^3 (\bar{v}_x^2)$$

where $\bar{v}_x^2$ is the mean square of the x-component velocities of the molecules. Since the number of molecules is very large and there is no preferred direction in the container,

$$\bar{v}_x^2 = \bar{v}_y^2 = \bar{v}_z^2 = \frac{\bar{v}^2}{3}$$

where $\bar{v}^2$ is the average of the squared velocities of the molecules. The gas pressure p then becomes

$$p = \frac{F_x}{D^2} = \tfrac{1}{3} m_0 N \bar{v}^2 = \tfrac{2}{3} N \left(\tfrac{1}{2} m_0 \bar{v}^2 \right) \tag{3.1}$$

3.2.2 Kinetic Interpretation of Gas Temperature

If a mass m of a gas of molecular weight M is confined to a volume G, the general gas law applies for the pressure p:

$$pG = \frac{m}{M} RT \tag{3.2}$$

where T is the absolute gas temperature and R is the universal gas constant [$= 8314 \, \text{J}/(\text{kg} \cdot \text{mol} \cdot \text{K})$]. Combining equations (3.1) and (3.2) and realizing that $m = NGm_0$, we have

$$T = \tfrac{1}{3} \frac{M}{R} \bar{v}^2 = \tfrac{1}{3} \frac{N_0}{R} m_0 \bar{v}^2 = \frac{2}{3k} \left(\tfrac{1}{2} m_0 \bar{v}^2 \right) \tag{3.3}$$

where the mass of molecule m_0 is the molecular weight M divided by Avogadro's number N_0, and k is Boltzmann's constant [$= 1.38 \times 10^{-23}$ J/K]. Thus, the absolute temperature of the gas is a measure of the squared velocity of the molecules within it. Thus the temperature T is directly proportional to the average translational kinetic energy of the molecules in a gas at that temperature

3.2.3   Distribution of Molecular Speeds

Not all molecules in a gas travel with equal speed. The speed variation among the molecules is usually expressed by plotting the number N_v of molecules having speeds within a unit interval ($\Delta v = 1\,\text{m/s}$) centered about the speed v (Fig. 3.2). The total number N_t of molecules is represented by the area under the curve, that is,

$$N_t = \int_0^\infty N_v \, dv$$

The speed probability distribution function $P_r(v)$ is the percentage of gas molecules having speeds within the range $v \pm dv/2$. The expression for $P_r(v)$ is

$$P_r(v) = \frac{4}{\pi} \left(\frac{m_0}{2kT} \right)^{3/2} v^2 \exp\left(\frac{-m_0 v^2}{2kT} \right) \tag{3.4}$$

Note that $P_r(v) \to 0$ only as $\text{v} \to 0$ and as $v \to \infty$. Therefore, all speeds are possible for the molecules.

3.2.4   Maxwell–Boltzmann Distribution Function

Very often interest is directed to how the gas molecules are distributed with regard to their energies. Since $P_r(v)\,dv$ is the fraction of molecules having speeds in the range dv centered on v, it is convenient to represent the

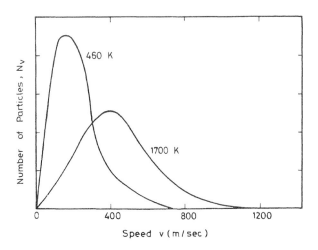

Figure 3.2   Number of molecules N_v versus speed v at 460 and 1700 K. The vertical axis is linear, in arbitrary units.

equivalent kinetic energy distribution, $P_r(U)\,dU$, as

$$U = \tfrac{1}{2}m_0 v^2 \qquad \text{and} \qquad dU = m_0 v\,dv$$

Therefore, the fraction of molecules with translational kinetic energies in the range dU centered on U is given by

$$P_r(U)\,dU = \frac{1}{\sqrt{\pi}(kT)^{3/2}}\sqrt{U}\,e^{-U/kT}\,dU \tag{3.5}$$

This distribution is shown in Figure 3.3, where the most probable energy occurs for $U = (1/2)kT$. The average value $\overline{U}$ of the kinetic energy is given by

$$\overline{U} = \int_0^\infty U P_r(U)\,dU = \tfrac{3}{2}kT \tag{3.6}$$

Since $\overline{U} = (1/2)m_0\bar{v}^2$, then $\bar{v}^2 = 3kT/m_0$, as deduced before [equation (3.3)].

The question is: What will happen if the gas is to exist in a region where its potential energy varies from place to place? The Maxwell-Boltzmann distribution law tells us how the number of molecules N changes with the total energy V. The number of moleculels N_1 that will be found in state 1, say, is related to the number N_2 in state 2 by the relation

$$\frac{N_1}{N_2} = \exp\left(\frac{V_2 - V_1}{kT}\right) \tag{3.7}$$

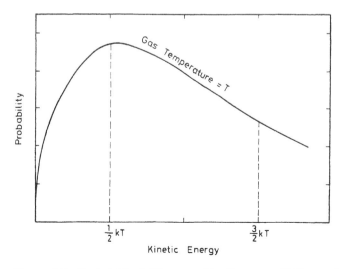

Figure 3.3 Probability $P_r(U)$ versus kinetic energy U. The vertical axis is linear.

which is known as the Maxwell-Boltzmann distribution.

3.3 BEHAVIOR OF A GASEOUS DIELECTRIC IN AN ELECTRIC FIELD

The atoms of a gas can absorb, transport, and release energy. The energy would be acquired from or given up to other atoms of the gas. The internal energy of an atom is associated with the energy levels of its electrons. Whenever an electron drops or rises from one energy level to another, a quantum of energy is released or absorbed as a photon. These photons have definite magnitudes corresponding to the specific energy levels the electrons can acquire in the atoms (McDaniel, 1964).

The electrons in the atom have a tendency to revert to their lowest possible energy levels. When all the electrons occupy their permitted energy levels, the atom has a "normal state." When some electrons are at higher energy levels but do not leave the atom, it is in an "excited state," lasting for a very short duration (10^{-8} s). The energy required is known as the "excitation energy" W_e. The atom has to release this energy to return to its normal state. The radiated photon has a wavelength λ

$$\lambda = \frac{ch}{W_e} \tag{3.8}$$

where c is the velocity of light and h is Planck's constant.

Some atoms have certain energy levels to which an internal electron may be raised but from which it cannot revert easily to its normal energy level. This state of the atom is known as a "metastable state." It ordinarily gives up its energy to other atoms in the gas, to the electrodes, or to the walls of the container. The duration of a metastable state is on the order of 0.1 s.

If an electron acquires enough energy, it gets separated from its atom by ionization. What remains of the atom after ionization is a positive ion. The electron and the ion resulting from the ionization move independently. Under the action of an applied field, they become accelerated and acquire kinetic energies. It may be noted that the ion possesses both ionization energy and kinetic energy. It will release the ionization energy only when an electron recombines with it to form a normal atom. Negative ions, formed by electrons attached to neutral atoms or molecules, have much less internal energy than do positive ions. Negative ions have a larger probability of occurrence if the gas is electronegative (i.e., its molecules have a strong affinity for electron attachment).

Electrons moving in a gas under the action of an electric field are bound to make numerous collisions with the gas molecules. The electrons are accelerated and their energies rise between collisions by amounts depending on their free paths and the applied electric field intensity. The electron velocity component along the field direction taken over several free paths per unit field intensity is termed the electron mobility.

Similarly, the positive ions receive energy from the field. When these positive ions collide with gas molecules, they lose a considerable part of their energy (about 50% on the average), as the ion and the gas molecule have closely comparable masses. Therefore, their energies are lower than those of electrons in similar situations. Thus the positive ions are less capable of ionizing the gas molecules. An ion needs to have a kinetic energy twice the ionization energy in order to ionize the atom it collides with, whereas an electron only needs a kinetic energy equal to the ionization energy. In general, ions have very large masses and small speeds. They are therefore responsible for producing space-charge clouds in the interelectrode region.

Photons have neither mass nor charge. They are thus independent of the applied field. As they give up their energy to atoms or molecules, they vanish. Photons bombarding a cathode give up their energies to help release electrons from the cathode surface. Excited atoms and metastable atoms are not influenced by the electric field because they are neutral.

3.4   ION-GENERATED PROCESSES

3.4.1   Ionization by Electron Impact

Ionization by electron impact is probably the most important ionization process for a gas discharge. Its effectiveness depends on the electron energy. When the electron collides with an atom or molecule, kinetic energy is exchanged. If no excitation or ionization results, the collision is called "elastic." If it is inelastic, the gas atom or molecule becomes excited or ionized by the energy acquired from the incident electron.This means that a portion of the kinetic energy of the electron prior to impact has been converted to potential energy of the atom or molecule. An excited atom or molecule may become ionized by a subsequent collision with another slow-moving electron. This process becomes significant only when densities of electrons are high. Very fast electrons are also poor ionizers, since the period of interaction between the electron and the atom becomes too short and therefore the amount of energy transferred from the electron to the atom decreases significantly. For every gas there exists an optimal electron energy that gives a significant ionization probability.

3.4.1.1.   Ionization Cross-section

In some models, the atoms, molecules, and electrons are assumed to be rigid solid-material balls with the aim of deriving very useful expressions such as the concept of collision cross-section and free path. Thus, when the distance between an electron and an atom is equal to the sum $(r_e + r_a)$ of their radii, a collision must have taken place (Nasser, 1971). The collision cross-section

$$\sigma = \pi(r_e + r_a)^2 \tag{3.9}$$

The values of the ionization cross-sections for various gases, resulting from collision with electrons of energies between 10 and 1000 eV, are given in Figure 3.4. If a gas has N molecules per unit volume, the number of collisions per unit length of the electron path is equal to σN and the mean free path $\bar{\lambda}$ is

$$\bar{\lambda} = \frac{1}{\sigma N} = \frac{1}{\pi N(r_e + r_a)^2} \tag{3.10}$$

The mean free path is a statistical average and indicates only the average distance between successive collisions. It varies from one gas to another (Table 3.1). Increased gas pressure p or decreased gas temperature

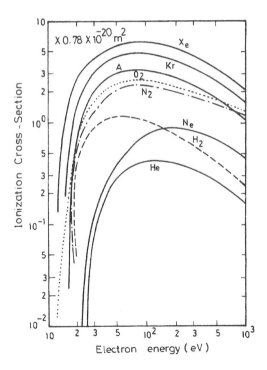

Figure 3.4 Ionization cross-sections for various gases due to electron impact. [From Rapp and Englander-Golden (1965).]

Table 3.1 Values of the Mean Free Path $\bar{\lambda}$ of some Gas Molecules[a]

Gas	H_2	He	Ne	N_2	O_2	Ar	CO_2	Kr	Xe
$\bar{\lambda}(10^{-8}\,m)$	11.77	18.62	13.22	6.28	6.79	6.66	4.19	5.12	3.76

[a]$T = 288\,K$ and $p = 101.3\,kPa$.
Source: McDaniel (1964).

proportionately increases its density and hence N. Therefore, the mean free path $\bar{\lambda}$ is inversely proportional to p at constant temperature. The probability f_x that an electron will have a collision while traveling a distance x can be given by a relationship of the form

$$f_x = \exp\left(\frac{-x}{\bar{\lambda}}\right) \tag{3.11}$$

where x is the distance traveled without a collision.

The free-path distribution function can be derived by considering a group of n_0 molecules just after a collision. When these molecules are moving in a given direction, n of them will reach a distance x without collision. Note that n is related to n_0 through the probability f_x:

$$n = n_0 f_x = n_0 \exp\left(\frac{-x}{\bar{\lambda}}\right) \tag{3.12}$$

Equation (3.12) gives the distribution of free paths where the ratio n/n_0 decays exponentially with distance x.

3.4.1.2. Ionization Coefficient

When an electron travels a distance equal to its free path λ_e in the direction of the field E, it gains an energy $= eE\lambda_e$. For the electron to ionize, its gain in energy should be at least equal to the ionization potential V_i of the gas:

$$e\lambda_e E \geq eV_i \qquad \text{or} \qquad \lambda_e \geq \frac{V_i}{E} \tag{3.13}$$

Table 3.2 lists values of V_i for the molecules of some gases and the atoms of some vapors.

The ionization coefficient α, known as "Townsend's first ionization coefficient," is the number of ionizing collisions per unit length of the path. It is equal to the number of free paths ($= 1/\bar{\lambda}_e$) times the probability of a free path being more than the ionizing length λ_{ie} (Kuffel and Abdullah, 1970). It can be expressed as

Table 3.2 Excitation and Ionization Potentials of some Gases and Metal Vapors

Material	First excitation potential (eV)	Ionization potential (eV)	
		Single ionization	Double ionization
Ar	11.56	15.8	27.5
	11.49 (metastable)		
	11.66 (metastable)		
He	20.9	24.6	54.1
	19.8 (metastable)		
Ne	16.58	21.6	40.9
	16.53 (metastable)		
	16.62 (metastable)		
Xe	8.39	12.1	21.2
	8.28 (metastable)		
	9.4 (metastable)		
H_2	11.2	15.4	
N_2	6.1	15.5	
O_2		12.2	
F_2		17.8	
SF_6		19.3	
H_2O		12.6	
CO_2		14.4	
CO		14.1	
Cl_2		13.2	
S		10.3	
K		4.3	
Cu		7.7	
Cs		3.9	

Source: Meek and Craggs (1953).

$$\alpha = \frac{1}{\bar{\lambda}_e} \exp\left(\frac{-\lambda_{ie}}{\bar{\lambda}_e}\right) = \frac{1}{\bar{\lambda}_c} \exp\left(\frac{-V_i}{\bar{\lambda}_e E}\right) \tag{3.14}$$

The mean free path $\bar{\lambda}_e$, is proportional to the reciprocal of gas density. Hence, $1/\bar{\lambda}_e = \Gamma p$, where Γ is a constant that depends on T. From equation (3.14)

$$\frac{\alpha}{p} = \Gamma \exp\left(\frac{-\Gamma V_i p}{E}\right) = \Gamma \exp\left(\frac{-Bp}{E}\right) = F\left(\frac{E}{p}\right) \tag{3.15}$$

as found experimentally. Typical values for the constant Γ of equation (3.15) are 11 253.7, 9003.0, and 3826.3 $m^{-1} \cdot kPa \cdot s^{-1}$ for air, N_2, and H_2, respectively. The corresponding values of $B = (\Gamma V_i)$ are 273 840, 256 584, and 104 134 $V \cdot m^{-1} \cdot kPa \cdot s^{-1}$. Typical values of α/p of the technically interesting gases (air, N_2, and SF_6) versus E/p at standard temperature and pressure (STP) are given in Figure 3.5.

3.4.2 Photoionization

It should be pointed out that photoionization does not occur simply due to radiation emitted from the gas itself when excited atoms return to their ground state but also takes place as a result of external radiations such as cosmic rays, x-rays, or nuclear radiations. Ionization by cosmic rays is a continuous process that produces ions and electrons everywhere since these rays are capable of penetrating most conventional walls. The fact that electrons and ions are always available in air has great significance and consequences in many applications. Without free electrons, it would not be easy to produce a spark, for example, or to ignite the combustible mixture in internal combustion engines. Measurements have shown that the intensity of cosmic rays increases sharply with altitude to reach 20 times the sea-level value at about 16 km, and then decreases gradually with further height. Its rate of ionization also increases steadily from a very small value at sea level to about 80×10^6 ion pairs per cubic meter per second at an altitude of

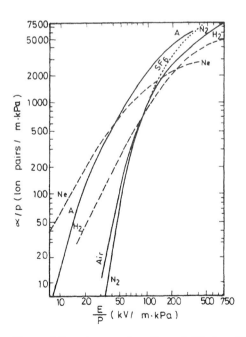

Figure 3.5 α/p as a function of E/p for different gases. [From Brown (1966); Bahalla and Craggs (1962).]

10 km (Nasser, 1971). It must be kept in mind that the insulation of high-voltage systems at high altitudes is subject to reduced air density and to increased ionization by cosmic rays.

X-rays and γ-rays ionize gases because of their high energies. Their energies range from a few electron volts to 100 MeV or more. Because of their highly energetic photons, their ionizing processes are bound to differ from those of radiations with relatively lower energies such as ultraviolet, as follows:

1. When a gas atom absorbs an x-ray photon, a loosely bound valence electron may be liberated, using part of the quantum energy. The excess energy of the photon, which is usually very high, will be converted to kinetic energy for the electrons. This electron may well ionize more atoms by collision, forming an avalanche-like ionization.

2. A high-energy x-ray photon may knock out an electron from an inner shell of the atom. The high energy absorbed may be adequate to give the electron enough momentum to leave the atom. In the ionized atom, the missing electron is substituted by another from the next-higher shell. This process is accompanied by the release of energy in the form of the emission of another x-ray of lower energy than the primary x-ray quantum. This new x-ray photon can ionize additional atoms, just as in the preceding case.

3. The atom does not completely absorb the incident x-ray quantum and the so-called Compton effect occurs, with scattering of x-rays with longer wavelengths.

If A represents a neutral atom in the gas, e^- an electron, KE the kinetic energy of the electron, and A* the excited atom of the gas, the reaction describing the excitation of the atom is

$$A + e^- + KE \longrightarrow A^* + e^- \qquad (3.16)$$

On recovering from the excited state in about 10^{-8} s, the atom radiates a photon that may in turn ionize another molecule whose ionization potential (eV_i) is equal to or less than the photon energy. The photoionization process may be described by the reaction

$$A + h\nu \longrightarrow A^+ + e^- \qquad (3.17)$$

where $h\nu$ is the photon energy, ν is the photon frequency (velocity of light c/wavelength l), and e^- is the ejected photoelectron.

The excess of the quanum $h\nu$ over eV_i may be imparted to the released electron as kinetic energy. If the photon energy is less than eV_i, it may still be absorbed by the atom, which gets excited only to a higher-energy level (i.e., a photoexcitation).

A photon emitted from an atom returning to its ground state may be absorbed by another of the same gas, exciting it. The same thing may continue to happen within a gas until the photon is lost to the boundaries. Such a state of affairs is known as "resonance" radiation and is encountered frequently in ionized gases.

Ionization by a beam of photons can be analyzed in the same way as that produced by a beam of electrons, and the concept of free path may be extended to photons. Thus the photon mean free path $\bar{\lambda}_{ph}$ is expressed as

$$\bar{\lambda}_{ph} = \frac{1}{\sigma N} = \frac{1}{\pi r_a^2 N} \tag{3.18}$$

and the photon number n_{ph} decreases with the distance x traveled, that is,

$$n_{ph} = n_{ph0} e^{-x/\bar{\lambda}_{ph}} \tag{3.19}$$

where n_{ph0} is the photon number at their origin.

Being radiation, photons get reduced in intensity by dI as they travel a distance dx at x from their origin. dI is proportional to dx and I at the distance x. Then

$$-dI = \mu I \, dx$$

where μ is the absorption coefficient of the gas. Integrating, we get

$$I = I_0 e^{-\mu x} \tag{3.20}$$

where I_0 is the intensity of photons at their origin. Comparing equations (3.19) and (3.20), we get

$$\mu = \frac{1}{\bar{\lambda}_{ph}} \tag{3.21}$$

Figure 3.6 shows μ as a function of the photon wavelength for different gases.

3.4.3   Ionization by Interaction with Metastables

For the atoms of inert gases and of some elements in group II of the periodic table, the lifetime of their metastable states extends to seconds. These states are represented by A^m and associated with energies V^m.

If V^m of an atom A^m exceeds the ionization V_i of another atom B, then, on collision, ionization may result according to the reaction

$$A^m + B \rightarrow A + B^+ + e^- \tag{3.22}$$

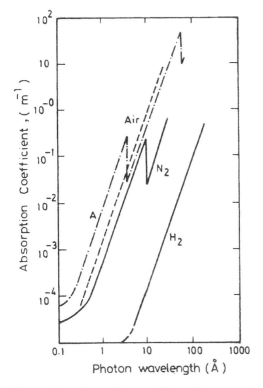

Figure 3.6 Absorption coefficient for x-rays in different gases as a function of wavelength at a temperature of 0°C and a pressure of 133 Pa.

For $V^m < V_i$, the reaction would produce only excitation of the atom B.

Another possibility for ionization by metastables is when $2V^m$ for A^m is greater than V_i for A, the reaction may proceed as follows:

$$\left.\begin{array}{l} 2A^m + A \longrightarrow A_2^* + A \\ A_2^* \longrightarrow A + A^+ + e^- \end{array}\right\} \tag{3.23}$$

where A_2^* is a diatomic molecule in the excited state. Sometimes, an atom in the metastable state may be ionized by absorbing another photon before returning to its ground state. This process is known as step ionization:

$$A^m + h\nu \longrightarrow A^+ + e^- \tag{3.24}$$

As mentioned before, ionization by metastable interactions comes into operation long after excitation. It has been shown by some research work that these reactions are responsible for the long time lags of breakdown observed in certain gases (Hartmann and Gallimberti, 1975).

3.4.4   Ionization by Nuclear Particles

Nuclear radiations include α- and β-particles, as well as photons of γ-rays. The energies of α-particles are on the order of MeV. Beta particles are fast electrons having energies on the same order of magnitude. Both types of particle make collisions with the gas atoms and cause ionization by collision. Gamma rays were discussed in Section 3.4.2.

The ionization effects of particles heavier than electrons can be considered by applying the principles of elastic and inelastic collisions. The most important particles are the proton and the α-particle. When they are injected into a gas they undergo collision with the gas atoms, thus losing their kinetic energy to the latter. Because of their heavy masses their direction of travel is not greatly affected by the collisions. They proceed in almost straight lines, leaving ionized atoms along their wakes until they finally lose their kinetic energies and come to relative rest (Fikry et al., 1978). It is interesting to note that, as in the case of β-rays, the number of ionizations produced by the particle is much greater toward the end of its path when it has slowed down appreciably, thus having a longer interaction with the gas molecules.

The distance traveled by the particle until it is stopped or "absorbed" is a well-defined range that depends on the particle's energy and mass and on the density of the absorbing medium (Table 3.3). The particle range in gases is, of course, much greater than in solids. The range of the heavy particles (α-particles and protons) increases very rapidly with energy (Table 3.3). For equal energies, the proton has a greater range because of its lower mass.

3.4.5   Thermal Ionization

If a gas is heated to a sufficiently high temperature, most of its molecules get dissociated into atoms (Fig. 3.7). Also, ionization of its atoms and molecules may result. The following are the possibilities for thermal ionization:

Table 3.3   Absorption Ranges of α-particles, Protons, and β-particles in Air at STP

Energy (MeV)	Range (m)		
	α-particle	Proton	β-particle
1	0.005	0.023	3.14
5	0.035	0.34	30.00
10	0.107	1.17	41.00

Source: Nasser (1971).

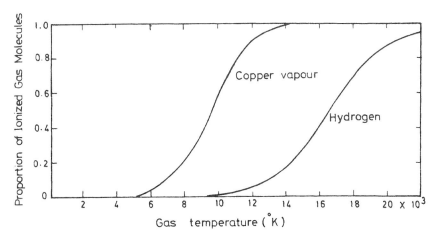

Figure 3.7 Proportion of ionized gas molecules as a function of gas temperature.

1. Ionization by collision of the gas atoms with each other. Because of the high temperature, the velocities and kinetic energies are very high and can cause ionization.
2. Photoionization resulting from thermal emission by the hot gas.
3. Ionization by collision with high-energy electrons produced by the preceding two processes. In fact, the ionization coefficient α increases with temperature (Abdel-Salam, 1976).

Thermal ionization is the chief source of ionization in flames and high-pressure arcs (see Chapter 6). The process is described by the equation

$$A + W_i \rightleftarrows A^+ + e^- \tag{3.25}$$

where A represents a neutral atom, A^+ a singly ionized atom, e^- the electron removed from the atom, and W_i the ionization energy.

3.4.6 Electron Detachment

Under certain conditions in high electric fields, electrons may become detached from negative ions. This process, however, requires a large concentration of negative ions. Loeb (1965) carried out experiments on electron detachment in oxygen and found that it occurs for $E/p = 68 \text{ V/m} \cdot \text{Pa}$. In compressed electronegative gases such as SF_6, detachment has been considered as one of the main processes involved in breakdown phenomena over a wide range of gas pressures (Maller and Naidu, 1981; Abdel-Salam et al., 1978). Gas discharges under impulse voltages may be initiated by electrons becoming detached from negative ions available in the gas (Allen et al., 1981; Sommerville et al., 1984; Abdel-Salam and Turkey, 1988). Detachment may be ascribed to the following events:

1. Photodetachment by absorption of radiation by a negative ion

$$A^- + h\nu \leftrightarrows A + e^- \qquad (3.26)$$

2. Collision of ions with a fast atom

$$A^- + A \leftrightarrows 2A + e^- \qquad (3.27)$$

3. Collision and subsequent association with neutral atoms; that is, associative detachment leading to the loss of the excessive electron

$$A^- + B \leftrightarrows AB + e^- \qquad (3.28)$$

4. Collision of molecular negative ions with fast excited atoms, leading to a vibrational excited state of the negative ion and subsequent loss of the excessive electron

$$A_2^- + B^* \leftrightarrows A_2^* + B + e^- \qquad (3.29)$$

5. Recombination of negative and positive atomic ions to form a diatomic molecule

$$A^- + B^+ \leftrightarrows AB^+ + e^- \qquad (3.30)$$

   This is a very likely process at pressures above a few hundred pascal.

3.5 DEIONIZATION PROCESSES

3.5.1 Deionization by Recombination

Whenever positively and negatively charged particles are present, there is a chance for recombination to take place. The potential energy and the relative kinetic energy of the recombining electron–ion or ion–ion pair is released as a quantum of radiation:

$$A^+ + B^- \longrightarrow AB + h\nu \qquad (3.31)$$

In this expression B^- may be an electron or a negative ion.

The rate of recombination is directly proportional to the concentrations of both positive ions and negative ions or electrons. If n_+ and n_- are the numbers of positive and negative ions per unit volume, then

$$\frac{dn_+}{dt} = \frac{dn_-}{dt} = -\zeta n_+ n_- \qquad (3.32)$$

where ζ is a constant known as the recombination coefficient. In general, $n_+ = n_- = n$, and we have

$$\frac{dn}{dt} = -\zeta n^2 \tag{3.33}$$

If recombination is allowed to take place from $t = 0$ to $t = t$ and the density of charged particles at $t = 0$ is n_0, by integrating between these limits we get

$$n = \frac{n_0}{1 + n_0 \zeta t} \tag{3.34}$$

which indicates that the density of particles decreases hyperbolically with time.

The densities n_+ and n_- no doubt increase with the gas pressure. The variation of ζ with pressure is shown in Figure 3.8, where it is evident that recombination is particularly important at high pressure. On the other hand, ζ does not show a significant increase with temperature.

From experiments in gases in which negative ions can form very easily, such as oxygen, it was found that the recombination of positive ions with negative ions takes place much faster than does the recombination of a positive ion with a free electron in a gas where negative ions cannot form, such as in the rare gases (Nasser, 1971). This is true because the two species of ions have almost equal masses and average velocities and the time they remain in close vicinity is therefore long.

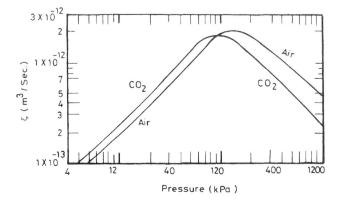

Figure 3.8 Effect of gas pressure on recombination coefficients (ζ) for air and CO_2.

3.5.2 Electron Attachment

When an electron becomes attached to a neutral molecule or atom, it produces a negative ion. Negative ions play an important role in the breakdown of technically interesting gases such as O_2, N_2, SF_6, and air. Even their presence as a small impurity can change the behavior of an ionized gas drastically. Negative-ion formation is very useful in quenching arcs in power circuit breakers in high-voltage generators and accelerators.

In electronegative gases, a general criterion for negative-ion stability can be inferred from consideration of the neutral atom, which is stable because it possesses the lowest energy level of all the possible states. Hence, for a negative ion to exist and be stable for some time, its total energy must be lower than that of the atom in its ground state. The difference in total energy between the negative ion and the unexcited neutral atom is termed the "electron affinity" of the atom. This energy is released as a photon or as kinetic energy upon attachment.

The different modes of negative-ion formation by attachment are:

Radiative attachment: takes place when the excess energy is released as a quantum upon attachment. Such a process is reversible and can be expressed as

$$e^- + A \underset{\text{detachment}}{\overset{\text{attachment}}{\rightleftharpoons}} A^- + h\nu \tag{3.35}$$

Attachment through three-body collision: takes place when the excess energy released upon electron attachment is acquired by a third particle during the collision process as kinetic energy (U). This can be expressed as

$$e^- + A + B \longrightarrow A^- + (B + U) \tag{3.36}$$

Dissociative attachment: takes place in molecular gases when the excess energy is not radiated but is used to separate the two atoms into a neutral particle and an atomic negative ion. Such a process is reversible and is expressed as

$$e^- + XY \leftrightarrows (XY)^{-*} \leftrightarrows X^- + Y \tag{3.37}$$

The suppression of electrons from an ionized gas by attachment is expressed by the attachment coefficient η. This coefficient is defined by analogy with the ionization coefficient α as the number of attachments produced in the path of a single electron traveling a unit distance in the direction of the field. Typical values of η/p

versus E/p (field-to-pressure ratio) at STP are given in Figure 3.9 for SF_6 and air (Maller and Naidu, 1981).

3.5.3 Deionization by Diffusion

During the extinction of arcs in circuit breakers, electron and ion diffusion is one of the most active deionizing agents. The faster the arc of the circuit breakers can be deionized, the higher their arc-interrupting capacity. In electrical discharges, whenever there is a nonuniform concentration of ions there will be diffusion of ions from regions of higher to regions of lower concentration. This diffusion will cause a deionizing effect in the regions of higher concentration. The flow of ions along a concentration gradient constitutes a drift velocity similar to that under electric fields. The rate of ion flow is

$$J = -D\nabla n \tag{3.38}$$

where $D = \bar{v}\,\bar{\lambda}/3\ \mathrm{m^2/s}$ is the diffusion coefficient.

As the concentration gradient varies with time, the rate of change of ion density is

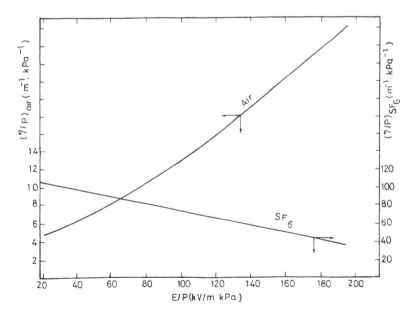

Figure 3.9 η/p as a function of E/p for air and SF_6. [From Maller and Naidu (1981); Bahalla and Craggs (1962).]

$$\frac{\partial n}{\partial t} = D\nabla^2 n = -\nabla \cdot \bar{J} \tag{3.39}$$

which is known as the continuity equation. Its solution would give the concentration of ions at any time and at any point in space. For the case of diffusion from a cylindrical concentration, the average square of displacement will be given by (Nasser, 1971)

$$(\bar{r})^2 = 4Dt \tag{3.40}$$

This equation is of greatest interest when we are dealing with discharges in cylindrical channels comprising electrons and positive ions. In the same gas, D is three orders of magnitude higher for electrons than for ions, owing to the greater average velocities $\bar{v}$.

3.6 GAS-DISCHARGE PARAMETERS AS INFLUENCED BY ATMOSPHERIC CONDITIONS

3.6.1 Ionization Coefficient α

Curve fitting of the data measured in air results in accurate expressions of α/p ($m^{-1}\,kPa^{-1}$) as a function of E/p ($V \cdot m^{-1}\,kPa^{-1}$) (Abdel-Salam and Abdellah, 1978). In SF_6, α/p is related to E/p by (Boyd and Crichton, 1971)

$$\frac{\alpha}{p} = 0.023\left(\frac{E}{p}\right) - 1230 \tag{3.41}$$

In a gas mixture, the value of the ionization coefficient α_m for a mixture of two components is equal to the sum of the two individual coefficients (α_1 and α_2) multiplied by their respective partial pressure ratios:

$$\alpha_m = \frac{p_1}{p}\alpha_1 + \frac{p_2}{p}\alpha_2 \tag{3.42}$$

Humid air can be considered as a mixture of dry air and water vapor, and thus the effect of humidity on α can be evaluated (Abdel-Salam, 1985). The dependence of α on gas temperature T is obtained from equation (3.15) simply as $\alpha/p = \alpha T/p_0 T_0$ and $E/p = ET/p_0 T_0$, where T_0 and p_0 are the normal values (i.e., $T_0 = 293\,K$ and $p_0 = 101.3\,kPa$) (Abdel-Salam, 1976).

3.6.2 Attachment Coefficient η

Measured values of η in air (Brown, 1966) made it possible to express η/p ($m^{-1}\,kPa^{-1}$) as a function of E/p ($V \cdot m^{-1}\,kPa^{-1}$) as

$$\frac{\eta}{p} = 9.735 - 0.541 \times 10^{-3}\left(\frac{E}{p}\right) + 1.16 \times 10^{-8}\left(\frac{E}{p}\right)^2 \tag{3.43}$$

In SF_6, η/p is related to E/p (Boyd and Crichton, 1971) as

$$\frac{\eta}{p} = 1170.4 - 0.004\left(\frac{E}{p}\right) \tag{3.44}$$

The effects of temperature and of gas mixtures on the attachment coefficient η_m are the same as those on α_m.

3.6.3 Photon Absorption Coefficient μ

The photon absorption coefficient $\mu(m^{-1})$ of any gas at pressure p (kPa) is expressed (Khalifa et al., 1975) as

$$\mu(p) = \frac{\mu_0 p}{101.3} \tag{3.45}$$

where μ_0 is the value of μ at STP. It equals 330 for air, 250 for nitrogen, and 600 for SF_6 (Alexandrov, 1972). The absorption coefficient $\mu_m(p)$ for a gas mixture can be evaluated using a relation similar to (3.42). Its dependence on temperature is similar to that of α.

3.6.4 Electron Drift Velocity v_e

Measured values of v_e (Naidu and Prasad, 1972) made it possible to express v_e (m/s) as a function of E/p ($V \cdot m^{-1} kPa^{-1}$) in air by Badaloni and Gallimberti (1972) and in nitrogen and SF_6 by Abdel-Salam and Abdellah (1978):

$$(v_e)_{air} = \begin{cases} 10^4\left(\dfrac{E}{p}\right)^{0.715}/113.7 & \text{for } \dfrac{E}{p} \leq 75{,}000\ V \cdot m^{-1}\ kPa^{-1} \\[2ex] 10^4\left(\dfrac{E}{p}\right)^{0.62}/39.1 & \text{for } \dfrac{E}{p} > 75{,}000\ V \cdot m^{-1}\ kPa^{-1} \end{cases} \tag{3.46}$$

$$(v_e)_{N_2} = \left[2.8 + 3.92 \times 10^{-3}\left(\frac{E}{p}\right)\right] \times 10^6 \tag{3.47}$$

$$(v_e)_{SF_6} = \begin{cases} 233.33\left(\dfrac{E}{p}\right) & \text{for } 82{,}500 \leq \dfrac{E}{p} < 127{,}500\ V \cdot m^{-1}\ kPa^{-1} \\[2ex] 245.33\left(\dfrac{E}{p}\right) & \text{for } 127{,}500 \leq \dfrac{E}{p} < 150{,}500\ V \cdot m^{-1}\ kPa^{-1} \end{cases} \tag{3.48}$$

In a gas mixture, the electron drift velocity $(v_e)_m$ is estimated, using the electron mobility in the mixture, as $(v_e)_m = (k_e)_m E$. The mobility $(k_e)_m$ is expressed in terms of the components $(k_e)_1$ and $(k_e)_2$ and the partial pressure ratios as

$$\frac{1}{(k_e)_m} = \frac{1}{(k_e)_1}\frac{p}{p_1} + \frac{1}{(k_e)_2}\frac{p}{p_2} \tag{3.49}$$

Equation (3.49) was used by Abdel-Salam (1985) to determine $(k_e)_m$ in humid air. The dependence of the electron velocity v_e on the gas temperature can be determined from equations (3.46) to (3.49) simply by replacing E/p by $ET/p_0 T_0$.

3.6.5 Electron Diffusion Coefficient D_e

A fundamental relation was developed by Loeb (1965) correlating the mobility k_e to the diffusion coefficient D_e:

$$\frac{k_e}{D_e} = \frac{e}{kT_e} = \frac{e}{z_m[(1/2)m_0\bar{v}^2]} \tag{3.50}$$

where kT_e is the energy of thermal agitation at the effective absolute temperature of the gas, k is Boltzmann's constant, and z_m is the ratio between the energy of thermal agitation and the mean thermal energy of the gas molecules. z_m was expressed (Badaloni and Gallimberti, 1972) as

$$Z_m = \begin{cases} 0.155\left(\frac{E}{p}\right)^{0.71} & \text{for } \frac{E}{p} \leq 2250\,\text{V·m}^{-1}\,\text{kPa}^{-1} \\ 0.819\left(\frac{E}{p}\right)^{0.49} & \text{for } \frac{E}{p} > 2250\,\text{V·m}^{-1}\,\text{kPa}^{-1} \end{cases} \tag{3.51}$$

The effects of pressure, temperature, and humidity on z_m are similar to their effects on α, η, and μ.

3.7 SOLVED EXAMPLES

(1) How fast is an oxygen molecule moving, on the average, in air at 27°C?

Solution:
From equation (3.3):

$$\bar{v}^2 = 3R\left(\frac{T}{M}\right) = 3(3814\,\text{J/(kg·mol·K)})\left(\frac{300\,\text{K}}{32\,\text{mol}^{-1}}\right) = 267{,}235.7\,\text{m}^2/\text{s}^2$$

M is substituted by 32 as the atomic weight of oxygen gas is 16 and oxygen is a diatomic gas (see physical constants in Appendix 3.1).

$$\therefore \bar{v} = 516.95\,\text{m/s}$$

So, the air molecules have speeds of about 500 m/s.
(2) What is the total translational kinetic energy in 2.0 g of O_2 gas at standard pressure and temperature?

Solution:
From equation (3.2):

$$G = \left(\frac{2.0 \times 10^{-3}\,\mathrm{kg}}{32\,\mathrm{mol}^{-1}}\right)(8314\,\mathrm{J/(kg \cdot mol \cdot K)})(298\,\mathrm{K})/(1.01 \times 10^5\,\mathrm{N/m^2})$$

$$= 1233.15 \times 10^{-6}\,\mathrm{m^3}$$

where $1\,\mathrm{atm} = 1.01 \times 10^5\,\mathrm{N/m^2}$ and $25°\,\mathrm{C} = 298\,\mathrm{K}$
From equation (3.1):

$$\frac{1}{2}m_0\bar{v}^2 = \frac{3}{2}p/N$$

$$= \frac{3}{2} \times 1.01 \times 10^5/N$$

Total translational kinetic energy $= NG(\frac{1}{2}m_0\bar{v}^2)$
$$= \frac{3}{2} \times 1.01 \times 10^5 \times 1533.15 \times 10^{-6}$$
$$= 232.3\,\mathrm{J}$$

(3) In a certain experiment, it is necessary for the mean free path to be at least 10 cm. Assuming the gas to be nitrogen with 1.7 Å molecular radius, find the maximum pressure in the chamber at 300 K. The mass of a nitrogen molecule is 4.8×10^{-26} kg.

Solution:
We make use of equation (3.10), considering $r_e = r_a = r_m$ (molecular radius) where the collision takes place between nitrogen molecules.

$$N = \frac{1}{4\pi(1.7 \times 10^{-10})^2(0.10)} = 46.8 \times 10^{18}$$

For $G = 1\,\mathrm{m^3}$, $m = Nm_0 = 4.8 \times 10^{-26}\,N \cong 220 \times 10^{-8}\,\mathrm{kg}$, $M = 28\,\mathrm{mol}^{-1}$, and $R = 8314\,\mathrm{J/(kg \cdot mol \cdot K)}$. Use equation (3.2) to obtain

$$p = \frac{220 \times 10^{-8}}{28} \times 8314 \times 300 = 0.196\,\mathrm{N/m^2}$$

since $T = 300$ K.
(4) At what temperature would the average kinetic energy of helium atoms (He) in gas become 1.0 eV?

Solution:

$$\frac{1}{2}m_0\bar{v}^2 = 1.0 \times 1.6 \times 10^{-19}\,\text{J} = 1.6 \times 10^{-19}\,\text{J}$$

Use equation (3.3):

$$T = \frac{2}{3 \times 1.38 \times 10^{-23}}(1.6 \times 10^{-19})$$
$$= 0.77 \times 10^4\,\text{K}$$

(5) At $0°\,\text{C}$ and 760 torr, find the volume of 1.0 kg of helium.

Solution:
Molecular weight M of $H_2 = 2.016$. Use equation (3.2):

$$G = \frac{1 \times 8314 \times 273}{2.016 \times 1.01 \times 10^5} = 11.15\,\text{m}^3$$

(6) A beam of ions is injected into a gas. The beam has an initial density n_0 ions/cm^3. Find the density of the remaining ions at a distance equal to (a) the mean free path; (b) five times the mean free path.

Solution:
Use equation (3.19):

(a) $n = n_0\,e^{-1} = 0.37\,n_0$

(b) $n = n_0\,e^{-5} = 0.0067\,n_0$

(7) At normal atmospheric pressure, the density of helium is 178.0 g/m^3. Find the mean square velocity of the helium atoms.

Solution:
Use equations (3.2) and (3.3):

$$\bar{v}^2 = \frac{3kT}{m_0} = \frac{3k}{m_0}\frac{pGM}{mR}$$

But $M/m_0 = N_A$ and $KN_A = R$

$$\therefore \bar{v}^2 = \frac{3pG}{m} = \frac{3p}{N}$$

where N is the gas density $= 178 \times 10^{-3}\,\text{kg/m}^3$

$$\therefore \bar{v}^2 = \frac{3 \times 1.01 \times 10^5}{178 \times 10^{-3}} = 0.0170 \times 10^8\,\text{m}^2/\text{s}^2$$
$$\bar{v} = 1305\,\text{m/s}$$

(8) Find the energy of a free electron in eV at $25°\,$C in atmospheric air and in a discharge tube of pressure equal to 10^{-4} torr.

Solution:
Use equation (3.3):

$$\frac{1}{2}m_0\bar{v}^2 = \frac{3}{2}kT = \frac{3}{2} \times 1.38 \times 10^{-21} \times 293 = 606.677 \times 10^{-21}\,\text{J}$$

$$= \frac{606.677 \times 10^{-21}}{1.6 \times 10^{-19}} = 3.79\,\text{eV}$$

which is independent of pressure.

(9) If the density of solid atomic hydrogen is $0.075\,\text{g/cm}^3$, find the average separation of the atoms and the average volume occupied by one atom.

Solution:
$M = 1$ for atomic hydrogen, so 1 gram of H consists of N_A atoms.

$$\text{Number of atoms/cm}^3 = N_A \times 0.075 = 6.0224 \times 10^{23} \times 0.075$$
$$= 0.454 \times 10^{23}$$

$$\text{Average volume occupied by one atom} = \frac{1}{0.454 \times 10^{23}}$$
$$= 2.24 \times 10^{-23}\,\text{cm}^3$$

$$\text{Average separation between atoms} = \sqrt[3]{2.24 \times 10^{-23}}$$
$$= 2.8 \times 10^{-8}\,\text{cm}$$

(10) A photon of wavelength $200\,\text{Å}$ is incident upon a hydrogen atom at rest. The photon gives all of its energy to the bound electron, thus releasing it from the atom (binding energy $= 13.6$ eV). Find the kinetic energy in eV and the velocity of the photoelectron.

Solution:

$$\text{Photon energy} = h\nu = \frac{hc}{l} = \frac{4.15 \times 10^{-15} \times 3 \times 10^8}{200 \times 10^{-10}} = 62.2\,\text{eV}$$

Kinetic energy of the photoelectron $= 62.2 - 13.6 = 48.6\,\text{eV}$

$$\therefore 48.6 \times 1.6 \times 10^{-19}\,\text{J} = \frac{1}{2}m_e\bar{v}_e^2$$

$$\therefore \bar{v}_e = \left[\frac{2 \times 48.06 \times 1.6 \times 10^{-19}}{9.11 \times 10^{-31}}\right]^{\frac{1}{2}} = 4.13 \times 10^6 \, \text{m/s}$$

(11) A beam of radiation has a photon energy of 0.2 MeV. When it is incident on the surface of a liquid its intensity is reduced to 1/6 at 20 cm from the surface. Find the liquid photon absorption coefficient for this wavelength.

Solution:
Use equation (3.20):

$$\mu = -\frac{1}{x}\ln\frac{I}{I_0} = -\frac{1}{20}\ln\frac{1}{6} = 0.0896 \, \text{cm}^{-1}$$

(12) The longest wavelength of radiation to cause the dissociation of a gas was found experimentally to be 1000 Å. What is the binding energy W_e of the gas?

Solution:

$$W_e = h\nu = hc/l_{max}$$

$$\therefore l_{max} = \frac{ch}{W_e}$$

or

$$W_e = \frac{3 \times 10^8 \times 4.15 \times 10^{-15}}{1000 \times 10^{-10}} = 12.45 \, \text{eV}$$

(13) A partially ionized gas has partial pressures p_e, p_i, and p_n for electrons, positive ions, and neutral particles, respectively. If α is the degree of ionization and p is the total pressure, find the values of p_e/p_n and p_i/p in terms of α.

Solution:
Assume N = total number of particles before ionization
$\quad\quad\quad N_i$ = number of ions after ionization
$\quad\quad\quad N_e$ = number of electrons after ionization
$\quad\quad\quad N_n$ = number of neutral particles after ionization

$$N = N_n + N_i = N_n + N_e$$

Use equation (3.2):

$$pG = \frac{m}{M} RT = \frac{NGm_0}{M} RT = \frac{NGm_0}{m_0 N_A} RT$$

i.e., $p = N\left(\dfrac{R}{N_A}\right) T = NkT$

Therefore,

$$p_n = N_n kT, \qquad p_e = N_e kT, \qquad p_i = N_i kT$$

Total pressure $p = p_n + p_i + p_e$

$$\frac{p_e}{p_n} = \frac{N_e}{N_n} = \frac{N_e}{N - N_i} = \frac{N_e/N}{1 - N_i/N} = \frac{\alpha}{1 - \alpha}$$

where

$$\alpha = \frac{N_e}{N} = \frac{N_i}{N}$$

$$\frac{p_i}{p} = \frac{N_i}{N_n + N_i + N_e} = \frac{N_i}{N + N_e} = \frac{N_i/N}{1 + N_e/N} = \frac{\alpha}{1 + \alpha}$$

(14) An electron was found to make an average of 85 collisions per cm in argon at $p = 1$ torr and $t = 0°$ C. Determine the corresponding value of the diameter of the argon atom.

Solution:

As derived in example (13):

$$p = NkT$$

$$N = p/kT = \frac{1.01 \times 10^5}{760} / (1.38 \times 10^{-23} \times 273)$$

$$= 3.527 \times 10^{22} \text{ atoms/m}^3$$

Use equation (3.10):

$$\bar{\lambda} = \frac{1}{\pi (r_e + r_a)^2 N} \cong \frac{1}{\pi r_a^2 N}$$

because $r_e \ll r_a$. Then

$$r_a = \left(\frac{85 \times 10^2}{\pi \times 3.527 \times 10^{22} \times 1}\right)^{1/2}$$

$$= 2.77 \times 10^{-10} \text{ m}$$

(15) The current flow through a gas dischare tube was found to be 3 A. The electron density in the tube was uniform and had a value of $1.0 \times 10^{17} \, \text{m}^{-3}$. The area of the electrodes is 8 cm² and the voltage across the electrodes is 20 V. The spacing d between electrodes is 0.8 m. Calculate the mobility of electrons, assuming that the current is mainly carried by electrons.

Solution:
Electron current density

$$J_e = I_e/A = n_e e v_e = n_e e E k_e$$

$$= n_e e \left(\frac{V}{d}\right) k_e$$

or

$$k_e = \frac{I_e}{A} \frac{d}{V n_e e} = \frac{3}{8 \times 10^{-4}} \frac{0.8}{20 \times 10^{17} \times 1.6 \times 10^{-19}}$$

$$= 9375 \, \text{m}^2/\text{s·V}$$

(16) In a cylindrical tube, there is a uniform field along the axial direction. If ions exist in the tube at a steady-state concentration gradient, determine the distribution of the ions along the axis of the tube. Disregard the field of the ions.

Solution:
The ion-flow upwards due to diffusion should be equal to the ion-flow downwards due to the electric field E in order to achieve a steady-state concentration gradient.

i.e., $J_{\text{diff}} = J_E$

Using equation (3.38):

$$J_{\text{diff}} = -D_+ \frac{dn_+}{dz}$$

$$= k_+ n_+ E$$

$$\int_{n_{+0}}^{n_+} \frac{dn_+}{n_+} = -\int_0^z \frac{k_+ E}{D_+} dz$$

$$\int_{n_{+0}}^{n_+} \frac{dn_+}{n_+} = -\int_0^z \frac{k_+ E}{D_+} dz$$

$$\therefore \ln\frac{n_+}{n_{+0}} = \frac{-k_+}{D_+}Ez$$

$$n_+ = n_{+0}e^{-k_+Ez/D_+} \tag{3.52}$$

(17) If the electric field E in the above problem is 5 V/m and the ion density at a certain location is 10^{11} ions/m^3, determine the ion density 0.02 m away in both directions at 25° C.

Solution:
Similar to equation (3.50), one can write for positive ions

$$\frac{D_+}{k_+} = \frac{kT}{e} \tag{3.53}$$

Then, equation (3.52) can be written as

$$n_+(z) = n_{+0}e^{-eEz/kT}$$

$$n_+(0.02) = 10^{11}e^{-1.6\times10^{-19}\times5\times0.02/(1.38\times10^{-23}\times293)} = 1.9\times10^9 \text{ ions/m}^3$$

$$n_+(-0.02) = 10^{11}e^{-1.6\times10^{-19}\times5\times0.02/(1.38\times10^{-23}\times293)} = 52.3\times10^{11} \text{ ions/m}^3$$

(18) A spherical cloud of ions is allowed to drift under the effect of a uniform field of 250 V/m. If the initial diameter of the cloud is 0.3×10^{-3} m, determine its diameter after drifting a distance of 0.05 m at 25° C.

Solution:
Similar to equation (3.40), one can write for 3-dimensional drift

$$\bar{r}^2 = 6D_+t \text{ and } t = \frac{z}{v_+} = \frac{z}{k_+E}$$

From equation (3.53):

$$D_+ = \frac{kT}{e}k_+$$

Then,

$$\bar{r}_1^2 = 6\frac{kT}{e}k_+\frac{z}{k_+E} = 6\frac{kT}{e}\frac{z}{E}$$

Before drift:

$$\bar{r}^2 = 6\frac{kT}{e}\frac{z_1}{E}$$

After drift:

$$\bar{r}_2^2 = 6\frac{kT}{e}\frac{z_2}{E}$$

$$\bar{r}_1^2 - \bar{r}_2^2 = \frac{6kT}{eE}(z_2 - z_1)$$

$$= \frac{6 \times 1.38 \times 10^{-23} \times 293 \times 0.05}{1.6 \times 10^{-19} \times 250} = 29.3 \times 10^{-6}$$

$$\bar{r}_2^2 = \bar{r}_1^2 + 0.1 \times 10^{-6} = 0.09 \times 10^{-6} + 29.3 \times 10^{-6} = 29.39 \times 10^{-6}$$

$$\bar{r}_2 = 5.42 \times 10^{-3}\,\text{m}$$

(19) (a) Calculate the mean free path of electrons in nitrogen at $p = 500$ pascals (1 atm = 760 torr = 101.3 kPa).

(b) Calculate the ionization potential of nitrogen.

Solution:
(a) Reference is made to equations (3.14) and (3.15):

$$\frac{\alpha}{p} = \Gamma e^{-v_i/E} \quad \bar{\lambda} = \Gamma e^{-Bp/E}$$

$$\bar{\lambda} = \frac{V_i}{Bp} = \frac{1}{\Gamma p}$$

$$\Gamma = 9003\,\text{m}^{-1}\cdot\text{kPa}^{-1} \text{ and } B = 256{,}584\,\text{V/m·kPa}$$

$$\bar{\lambda} = \frac{1}{9003 \times 0.5} = 0.2222 \times 10^{-3}\,\text{m}$$

(b) $V_i = \dfrac{B}{\Gamma} = 256{,}584/9003 = 28.5\,\text{V}$

REFERENCES

Abdel-Salam M. J Phys D: Appl Phys 9:L-148, 1976.
Abdel-Salam M. IEEE Trans IA-21:35, 1985.
Abdel-Salam M, Abdellah M. IEEE Trans IA-14:516, 1978.
Abdel-Salam M, Turkey A. IEEE Trans IA-24:1031, 1988.
Abdel-Salam M, Radwan R. Ali Kh. IEEE–PES paper A-78-601-7, 1978.
Alexandrov GN. Proceedings of IEE Gas Discharges Conference, London, 1972, pp 398–401.
Allen NL. Berger G, Dring D, Hahn N. Proc IEE 128A:565, 1981.

Badaloni S, Gallimberti I. Basic Data of Air Discharges. Padova, Italy: Università di Padova, 1972.

Bahalla MS, Craggs JD. Proc Phys Soc (Lond) 80:151, 1962.

Boyd HA, Crichton GC. Proc IEE 118:1872, 1971.

Brown SC. Basic Data of Plasma Physics. Cambridge, MA: MIT Press, 1966.

Fikry L, Abdel-Salam M, Zeitoun A, Goher M. IEEE Trans IA-14:510, 1978.

Hartmann G, Gallimberti I. J Phys D: Appl Phys 8:670, 1975.

Khalifa M, El-Debeiky S, Abdel-Salam M. Proceedings of International High Voltage Symposium, Zurich, Switzerland, 1975, pp 343–347.

Kuffel E, Abdullah M. High Voltage Engineering. Oxford: Pergamon Press, 1970.

Loeb LB. Electrical Coronas: Their Basic Physical Mechanism. Berkeley, CA: University of California Press, 1965.

McDaniel EW. Collision Phenomena in Ionized Gases. New York: John Wiley & Sons, 1964.

Maller VA, Naidu MS. Advances in HV Insulation and Arc Interruption in SF_6 and Vacuum. Oxford: Pergamon Press, 1981.

Meek JM, Craggs JD. Electrical Breakdown of Gases. Oxford: Clarendon Press, 1953.

Naidu MS, Prasad P. J Phys D: Appl Phys 5:1090, 1972.

Nasser E. Fundamentals of Gaseous Ionization and Plasma Electronics, New York: John Wiley & Sons, 1971.

Rapp D, Englander-Golden P. J Chem Phys 43:1464–1470, 1965.

Sommerville IC, Farish O, Tedford DJ. Proceedings of 46th International Symposium on Gasous Dielectrics, Knoxville, TN, 1984, pp 137–144.

APPENDIX 3.1
Some Physical Constants

Name	Symbol	Value	Dimension
Electron charge	e	-1.602×10^{-19}	coulomb (c)
Electron rest mass	m_e	9.108×10^{-31}	kg
Proton rest mass	m_p	1.672×10^{-27}	kg
Electron charge to mass ratio	e/m	1.7588×10^{11}	c/kg
Electron volt	eV	1.602×10^{-19}	joule (J)
Planck's constant	h	6.6257×10^{-34}	J·s
		4.15×10^{-15}	eV·s
Boltzmann's constant	k	1.3806×10^{-23}	J/K
Velocity of light in vacuum	c	2.99793×10^8	m/s
Avogadro's number	N_A	6.0224×10^{26}	molecules per kg molecular weight
Universal gas constant	R	8.314×10^3	J/kg·mol·K
Gas density at NTP ($t = 0°$ C and $p = 760$ torr)	N_N	2.685×10^{25}	molecules/m^3
Gas density at $t = 20°$ C and $p = 1$ torr	N	3.56×10^{22}	molecules/m^3
Normal atmospheric pressure	p_{at}	760	torr
		1.013×10^5	N/m^2
		1.0336×10^4	kg/m^2
Permittivity of free space	ε_0	8.85×10^{-12}	F/m
Permeability of free space	μ_0	$4\pi \times 10^{-7}$	H/m
Product of photon wavelength, in Å, and energy, in eV	λw	12,398	Å·eV

4

Electrical Breakdown of Gases

M. ABDEL-SALAM *Assiut University, Assiut, Egypt*

4.1 INTRODUCTION

Air has been the insulating ambient most commonly used in electrical installations. Among its greatest assets, in addition to its abundance, is its self-
restoring capability after breakdown. Liquid and solid insulants in use often
contain gas voids that are liable to break down. Therefore, the subject of
electrical breakdown of gases is indispensable for designers and operators of
electrical equipment.

An electrical discharge in a gas gap can be either a partial breakdown
(corona) over the limited part of the gap where the electrical stress is highest
or a complete breakdown. The complete breakdown of an entire gap initially takes the form of a spark requiring a high voltage, and through it a
relatively small current flows. Depending on the source and the gas-gap
conditions, the spark may be either extinguished or replaced by a highly
conductive conducting arc. Depending on the circumstances, the arc may be
maintained, as in arc furnaces, or extinguished, as in circuit breakers. In this
chapter we present a brief discussion of the breakdown mechanisms in uniform and nonuniform fields, under DC, AC, and impulse voltages, of air
and SF_6 at various pressures and temperatures.

4.2 PRE-BREAKDOWN PHENOMENA IN GASES

4.2.1 Generation of Electron Avalanches

Ionization by electron impact is probably the most important process in the breakdown of gases. However, as will be shown later, this process alone is not sufficient to produce breakdown. Consider a uniform field between electrodes immersed in a gas. Electrons may, for example, originate from the cathode as a result of ultraviolet radiation, or from the gas volume by ionization of neutral molecules, as noted in Chapter 3. They could also be produced at a later stage by photons from the discharge itself. If an electric field E is applied across a gap, the electrons present will be accelerated toward the anode, gaining energy with distance of free travel.

As mentioned in Chapter 3, if the electron acquires enough energy, it can ionize a gas molecule by collision. Leaving a positive ion behind, the new electron, together with the primary electron, proceed along the field and repeat the process. At a distance x from the cathode, the number of electrons will thus have increased to n_x. A further increase dn_x is

$$dn_x = \alpha n_x dx \tag{4.1}$$

where α is Townsend's first ionization coefficient. In a uniform field (i.e., constant α) with an initial number of electrons n_0 emitted from the cathode, their number at x will be

$$n_x = n_0 \exp(\alpha x) \tag{4.2}$$

It is well to emphasize at this stage the statistical fluctuating nature of the impact ionization process and the fact that α is only an average value for the number of ionizations per unit length of electron drift along the field. It can be shown (Raether, 1964) that the size of an electron avalanche from a single starting electron follows the exponential distribution

$$p_r(n) = \frac{1}{\bar{n}} \exp\left(\frac{-n}{\bar{n}}\right) \tag{4.3}$$

where $P_r(n)$ is the probability of occurrence of an avalanche with size n, n is the average size, $n = e^{\alpha d}$, and d is the gap length.

The head of the avalanche is built up of electrons while its long tail is populated by positive ions (Fig. 4.1). The track is wedge-shaped, due to the diffusion of the drifting electron swarm. Because of the difference in drift velocities of electrons (v_e) and ions (v_+), the latter are virtually stationary during the time required for the electrons to reach the anode. The transit time τ_e for the electrons of the avalanche to cross the gap is of the order of nanoseconds.

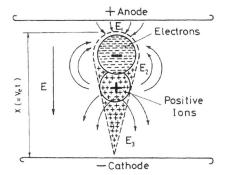

Figure 4.1 Distribution of charge carriers in an avalanche and their contribution to the applied uniform electric field. $E_1 > E; E_2 < E; E_3 > E$.

Figure 4.2 shows the components of the current pulse produced by an avalanche started by one electron leaving the cathode. Its initial part is carried by electrons moving rapidly toward the anode, whereas the latter part is carried by positive ions drifting slowly toward the cathode.

4.2.2 Avalanches with Successors

The "primary avalanche" process as described above is completed when the ions have entered the cathode. If, however, the amplification of the avalanche ($e^{\alpha d}$) is increased, the probability of additional electrons being liber-

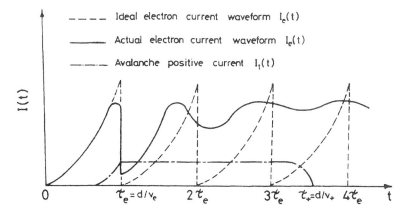

Figure 4.2 Current due to electrons and positive ions of an avalanche. Electron current waveform when secondary electrons are produced by photons received by the cathode.

ated in the gap by other (secondary) mechanisms is increased, and these electrons can initiate new avalanches. The secondary ionizing agents—positive ions, excited atoms, photons, metastables (Section 3.4)—are presented quantitatively by a coefficient defined as the average number of secondary electrons produced at the cathode corresponding to one ionizing collision in the gap. γ is called Townsend's second ionization coefficient. It is a function of E/p in the same manner as α but has a much smaller magnitude (compare Figures 4.3 and 3.5) (Naidu and Kamaraju, 1996).

The number of secondary electrons produced at the cathode during the life of the first avalanche is given by

$$\beta = \gamma(e^{\alpha d} - 1)n_0 \tag{4.4}$$

These new electrons start the second generation of avalanches. The generation interval after which the succeeding avalanche starts depends on the secondary process. It is common in air discharges that the secondary process is photoelectric and the new avalanches start after most $\tau_e = d/v_e$ (Fig. 4.2). The dashed line shows the idealized case, which assumes that all of the radiation producing the photo affect at the cathode is produced at the moment the electrons enter the anode. In fact, however, the radiation intensity is proportional to $e^{\alpha v_e t}$, and secondary electrons are liberated from the

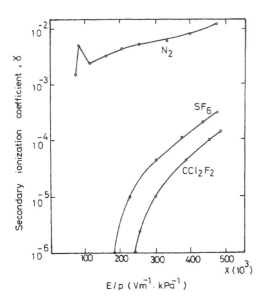

Figure 4.3 Secondary ionization coefficient γ as a function of E/p for different gases.

cathode during the transit time of the avalanche electrons, which results in the solid line of Figure 4.2.

Such generations of rapidly succeeding avalanches can produce a space charge of slow positive ions in the gap. These space charges enhance the electric field somewhere in the interelectrode space, with a subsequent rapid current growth, leading to breakdown.

4.3 BREAKDOWN IN STEADY UNIFORM FIELDS

Two typical gas breakdown mechanisms have been known: the Townsend mechanism and the streamer (or channel) mechanism. For several decades there has been controversy as to which of these mechanisms governed spark breakdown. It is now widely accepted that both mechanisms operate, each under its own most favorable conditions. The avalanche process described above is basic for both mechanisms of breakdown.

4.3.1 Townsend Mechanism

Townsend observed that the current through a uniform-field air gap at first increased proportionately with the applied voltage in the region $(0–V_1)$ and remained nearly constant at a plateau value I_{01} (Fig. 4.4). I_{01} corresponds to the photoelectric current produced at the cathode by external irradiation. At voltages higher than V_2, the current rises above I_{01} at a rate that increases rapidly with the voltage until a spark results. If the illumination level at the

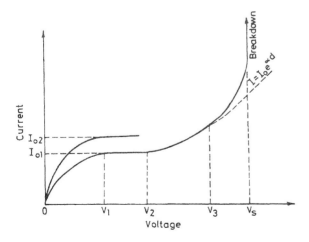

Figure 4.4 Average current growth to breakdown as a function of the applied voltage.

cathode is increased, the plateau I_0 rises proportionately but the voltage V_s at which sparking occurs remains unaltered, provided that there is no space-charge distortion for the electric field between the electrodes. The increase of current in the region V_2–V_3 is ascribed to ionization by electron impact, whereas the secondary (γ) process accounts for the sharper increase of current in the region V_3–V_s and eventual spark breakdown of the gap.

4.3.1.1 Current Growth Equations

It has been proved in Section 4.2.1 that each electron leaving the cathode produces on the average ($e^{\alpha d} - 1$) new electrons (and positive ions) while traversing the distance d. Having n_0 electrons produced at the cathode by the external source of radiation, let

> n_0' = number of secondary electrons produced at the cathode
>
> n_0'' = total number of electrons leaving the cathode

and so

$$n_0'' = n_0 + n_0' \tag{4.5}$$

Each electron leaving the cathode produces (on the average) ($e^{\alpha d} - 1$) collisions in the gap. Therefore, the number of ionizing collisions in the gap equals $n_0''(e^{\alpha d} - 1)$. By definition,

$$n_0' = \gamma n_0''(e^{\alpha d} - 1)$$

thus

$$n_0'' = \frac{n_0}{1 - \gamma(e^{\alpha d} - 1)} \tag{4.6}$$

The number of electrons arriving at the anode is

$$n_d = n_0'' e^{\alpha d}$$

so that

$$n_d = \frac{n_0 e^{\alpha d}}{1 - \gamma(e^{\alpha d} - 1)}$$

In the steady state, the circuit current I will be given by

$$I = \frac{I_0 e^{\alpha d}}{1 - \gamma(e^{\alpha d} - 1)} \tag{4.7}$$

In the absence of Townsend's secondary mechanism, i.e., $\gamma = 0$,

$$I = I_0 e^{\alpha d} \tag{4.7a}$$

4.3.2 Townsend's Criterion for Spark Breakdown

Equation (4.7) describes the growth of the average current in the gap before spark breakdown occurs. At low field strengths V/d, $e^{\alpha d} \to 1$, so that $I = I_0 e^{\alpha d}$ in the region $V_2–V_3$. As V increases, $e^{\alpha d}$ and $\gamma e^{\alpha d}$ increase until $\gamma e^{\alpha d}$ approaches unity and I approaches infinity [Equation (4.7)]. In this case the current will be limited only by the resistance of the power supply and the conducting gas. This condition is defined as "breakdown" and its "Townsend criterion" is thus

$$\gamma(e^{\alpha d} - 1) = 1 \tag{4.8}$$

Normally, $e^{\alpha d} \gg 1$, and the expression becomes simply $\gamma e^{\alpha d} = 1$.

Thus, for a given gap d, breakdown occurs when γ and α acquire the corresponding critical values. Both coefficients are functions of the electric field strength E. For a given photoelectric current I_0 and applied field E_1, say, the circuit current I varies with the gap d according to equation (4.7) and as shown in Figure 4.5. A maximum length d_{1s} is reached when the criterion (4.8) is fulfilled (i.e., spark breakdown occurs and the current I rises sharply).

At lower values of d, $\gamma e^{\alpha d} \ll 1$ in equation (4.7), so that $I \simeq I_0 e^{\alpha d}$, and the plot of $\ln(I/I_0)$ versus d is linear, with slope α. As d increases, $\gamma e^{\alpha d}$ increases and upcurving occurs until, when $\gamma e^{\alpha d} \simeq 1$, I approaches infinity and sparking occurs at $d = d_s$. The spark breakdown voltage corresponding to each value of E can easily be calculated (Abdel-Salam and Stanek, 1988).

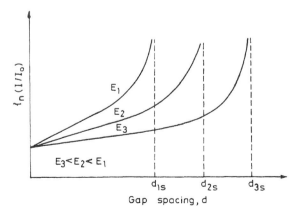

Figure 4.5 Townsend-type plot of $\ln(I/I_0)$ versus d.

4.3.3 Paschen's Law

Remembering from Chapter 3 that coefficients α and γ are both functions of the gas pressure p and the electric field $E = V/d$, thus $\alpha = pF_1(E/p)$, $\gamma = F_2(E/p)$, equation (4.8) can be rewritten as

$$\left(F_2 \frac{V}{pd}\right)\left[\exp\left(pdF_1 \frac{V}{pd}\right) - 1\right] = 1 \tag{4.8a}$$

Equation (4.8a) gives the breakdown voltage V implicitly in terms of the product of gas pressure p and electrode separation d. The breakdown voltage V_s is the same for a given value of the product pd

$$V_s = F(pd) \tag{4.9}$$

which is the well-known Paschen's law.

The relation between V_s and pd is plotted in Figure 4.6. To explain the shape of the curve, consider a gap of fixed spacing. As the pressure decreases from a point to the right of the minimum, the gas density decreases and the electron free path increases. Consequently, an electron makes fewer collisions with gas molecules as it travels toward the anode. Since each collision entails some loss of energy, it follows that a lower electric field would still furnish the electrons with kinetic energies sufficient for ionizing collisions.

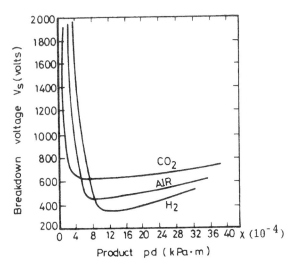

Figure 4.6 Breakdown voltage V_s versus the product pd for different gases. [Courtesy of F. Llewellyn-Jones (1957) and Oxford University Press.]

When the minimum of the curve is reached, the density is low and there are relatively few collisions. It is necessary now to take into account the fact that an electron does not necessarily ionize a molecule on colliding with it, even if the energy of the electron exceeds the ionization energy. The electron has a finite chance of ionizing, which depends on its energy (Chapter 3). If the density and hence the number of collisions decreases, breakdown can occur only if the chance of ionizing is increased by an increase in voltage, which is shown to the left of the minimum. The minimum breakdown voltages and the corresponding pd values are given in Table 4.1 (Naidu and Kamaraju, 1996).

The discussion so far has ignored temperature variations. A more general statement of Paschen's law is therefore $V = F(\rho d)$, where ρ is the gas density, which takes into account the effect of temperature (Abdel-Salam, 1976). The validity of Paschen's law has been confirmed experimentally up to 1100°C (Alston, 1968). Further increase in temperature ultimately results in failure of Paschen's law because of significant thermal ionization above 2000 K, as shown in Chapter 3.

Empirical relations have been suggested by many workers to express the breakdown voltage of uniform-field air gaps at atmospheric pressure (Alston, 1968).

$$V_s = 2440d + 61\sqrt{d} \text{ kV} \tag{4.10}$$

where d is the gap length in meters.

Table 4.1 Minimum Breakdown Voltages for Various Gases

Gas	V_s (min) (V)	pd at V_s (min) (Pa·m)
Air	327	0.754
Ar	137	1.197
H_2	273	1.530
He	156	5.320
CO_2	420	0.678
N_2	251	0.891
N_2O	418	0.665
O_2	450	0.931
SO_2	457	0.439
H_2S	414	0.798

Source: Naidu and Kamaraju (1996).

4.3.4   Streamer Mechanism

The very short time lags of spark breakdown, measured when high over-voltages are applied to uniform field gaps, are not consistent with the Townsend mechanism, which is based on the generation of a series of successive avalanches. Also, it was difficult to envisage how the Townsend mechanism would apply for long gaps where the sparks are observed to branch and to have an irregular character of growth. As a consequence, and anticipating the experimental results of Raether (1964), the streamer theory of the spark was proposed by Loeb (1955) and Meek for the positive streamer, and independently by Raether for the negative streamer. According to both versions, spark discharge develops directly from a single avalanche which the space charge transforms into a plasma streamer. Thus the conductivity grows rapidly, and breakdown occurs through its channel.

The principal features of both versions are the photoionization of gas molecules in the space ahead of the streamer and the local enhancement of the electric field by the space charge at its tip. The space charge produces a distortion of the field in the gap, as evident from Figure 4.1.

4.3.4.1   Version by Loeb and Meek

The positive streamer from the anode toward the cathode in uniform-field gaps is explained as follows. When the avalanche has crossed the gap, the electrons are swept into the anode, the positive ions remaining in a cone-shaped volume extending across the gap (Fig. 4.7a). A highly localized

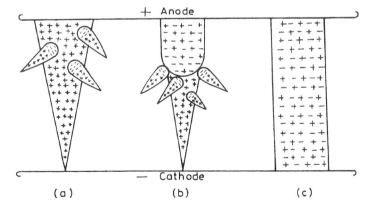

Figure 4.7   Cathode-directed streamer as envisaged by Meek and Loeb: (a) first avalanche crossing the gap; (b) streamer extending from the anode; (c) streamer crossing the gap.

space-charge field is produced near the anode, but elsewhere in the gap the ion density is low. However, in the gas surrounding the avalanche, photoelectrons are produced by photons emitted from the densely ionized gas constituting the avalanche stem. These photoelectrons initiate auxiliary avalanches, which are directed by both the space-charge field and the externally applied field. Evidently, the greatest multiplication in these auxiliary avalanches will occur along the axis of the main avalanche where the space-charge field supplements the applied field. Positive ions left behind by these avalanches effectively branch, lengthen, and intensify the space charge of the main avalanche toward the cathode. The process thus develops a self-propagating streamer, which effectively extends the anode toward the cathode (Fig. 4.7b). Ultimately, a conducting filament of highly ionized gas bridges the whole gap between the electrodes (Fig. 4.7c).

The transition from an electron avalanche into a streamer is considered to occur when the radial field E_r produced by the positive ions at the head of the avalance is of the order of the externally applied field E, that is,

$$E_r = \kappa E \tag{4.11}$$

where $\kappa \simeq 1$ and

$$E_r = \frac{5.27 \times 10^{-7} \alpha \exp(\alpha x)}{(13{,}500 x/p)^{1/2}} \text{ V/m}$$

with α in m^{-1}, x in meters, and p the gas pressure in pascals. Equation (4.11) expresses the criterion of breakdown by Loeb and Meek. The breakdown voltage is that at which the avalanche grows to the limit that its field E_r becomes equal to E.

4.3.4.2 Raether's Version

A slightly different criterion was proposed independently by Raether for negative anode-direced streamers. He postulated that streamers would develop when the initial avalanche produces a sufficient number of electrons ($e^{\alpha x}$) such that its space-charge field E_r is comparable to the applied field E. The total field, thus enhanced, would promote secondary anode-directed avalanches ahead of the initial one, forming a negative streamer. These secondary avalanches are initiated by photoelectrons in the space ahead of the streamer (Fig. 4.8).

Experiments for breakdown in air proved the validity of the Townsend mechanism in uniform fields with values of pd at least up to about 15 kPa·m (Alston, 1968). At higher pd values, the streamer mechanism plays the dominant role in explaining breakdown phenomena.

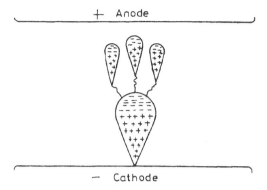

Figure 4.8   Anode-directed streamer as envisaged by Raether.

The streamer mechanism has been widely used to explain breakdown in nonuniform fields, such as in point–plane or point–point gaps. Similar but more practical situations are encountered in cases of imperfections on high-voltage conductors such as a nick of a strand or from airborne substances such as insects, dust or leaf particles, bird droppings, and other nonmetallic materials. Details of breakdown in nonuniform field gaps are discussed in Sections 4.6 and 4.7.

In long gaps, if the voltage gradient at the stressed electrode exceeds the corona onset level, the ionization activity in the gap increases. As a result, a highly ionized and luminous filamentary channel, called the leader channel, develops at the electrode and propagates toward the other electode. At the tip of the leader channel, filamentary branches called leader streamers exist where most of the ionization activity takes place (Fig. 4.9). The electrons produced due to ionization in the leader streamers feed through the leader channel into the stressed electrode. Depending on the value of the instantaneous voltage gradient at the stressed electrode and the leader channel length, the leader streamer either stops after having crossed a part of the gap, or reaches the plane electrode, causing a return ionizing wave to develop there. In the latter case, the ionizing wave advances toward the leader channel tip and leads to the final jump. At this stage, the leader channel tip advances very rapidly, bridging the entire gap and causing a complete breakdown.

4.4   BREAKDOWN IN ELECTRONEGATIVE GASES AND THEIR MIXTURES

The last three decades have witnessed intense research into electronegative gases such as SF_6 (sulfur hexafluoride) and CCl_2F_2 (Arcton-12). For such

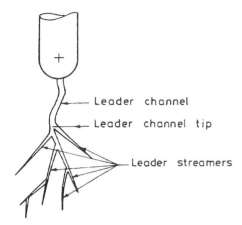

Figure 4.9 Leader channel and leader streamers.

gases the electron attachment coefficient η is significant and their dielectric strength is considerably higher than that of air. Those studies have led to the development of modern gas-insulated systems (GIS). Thus a high insulation strength can be achieved without severely raising the gas static pressure (see Chapter 10).

Townsend's equation can be modified to include the effects of both ionization and attachment processes. With the attachment coefficient being quite significant in such gases, Townsend's criterion for breakdown, equation (4.8), is modified to

$$\gamma \frac{\alpha}{\alpha - \eta} [e^{(\alpha - \eta)d} - 1] = 1 \tag{4.12}$$

For $\eta > \alpha$, and for large gaps, this criterion is approximated by

$$\alpha = \frac{\eta}{1 + \gamma} \tag{4.13}$$

This is a condition that depends only on E/p, and sets a limit for E/p below which no discharges should be possible, whatever the value of d. As γ is usually much smaller than unity ($\gamma \simeq 10^{-4}$), equation (4.13) can safely reduce to $\alpha = \eta$. The limiting values of E/p below which no discharges are possible are 86.3 V/m·Pa for SF_6 and 89.3 V/m·Pa for CCl_2F_2, compared to 23.6 V/m·Pa for air. With an increase in gas electronegativity (i.e., with an increase in η), the limiting value of E/p for breakdown increases. This is why the breakdown voltage measured in electronegative gases such as SF_6 is substantially higher than that measured in air (Fig. 4.10). It is widely used

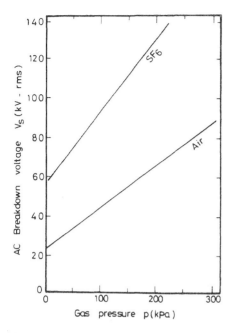

Figure 4.10 Breakdown voltages of air and SF$_6$ for 0.75 m diameter quasi uniform-field electrodes having 2.1 cm gap spacing, as a function of pressure.

as an insulating as well as arc-quenching medium in high-voltage equipment such as circuit breakers and metal-clad switchgear (GIS).

Breakdown voltages in gases normally show anomalies in their increase with gas pressure. In SF$_6$, the breakdown voltage of a positive point–plane gap increases with gas pressure, reaching a maximum value at a pressure P_m (Fig. 4.11). Above P_m, the breakdown voltage drops with further increase of pressure until a critical value P_c is reached. At pressures above P_c, breakdown occurs without any preceding corona and the breakdown voltage resumes a slow increase with pressure (Abdel-Salam et al., 1978). The high breakdown voltage at P_m is attributed to the so-called "corona stabilization" (i.e., a field modification arising from space-charge effects). The pressure P_m decreases with an increase in gap spacing. The decrease in breakdown voltage at pressures above P_m is attributed to the enhanced corona-streamer propagation across the gap as a result of the effectiveness of photoionization at higher gas densities. Above P_c, field emission from the cathode becomes the significant factor for breakdown to occur unpreceded by corona (Khalifa et al., 1977). The dielectric strength of highly electronegative gases is drastically

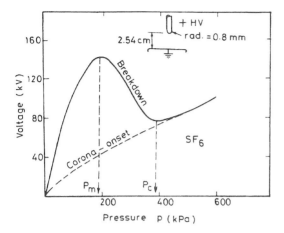

Figure 4.11   Positive DC breakdown and corona-onset voltages in SF_6, as functions of pressure.

reduced by the presence of foreign conducting particles, as discussed in Chapter 10.

4.5   EFFECTS OF GAS PARAMETERS

The breakdown voltage of a given gap depends on the gas parameters $(\alpha, \eta, \mu, \cdots)$, which in turn are functions of the electric field and of such factors as the gas pressure, temperature, and humidity (Section 3.6). Therefore, the breakdown voltage of a given gas gap decreases at higher temperatures and increases with the static pressure. The combined factor is the relative air density (Abdel-Salam and Stanek, 1988). The breakdown voltage decreases only slightly at higher humidities.

In many gases, and at pressures around the atmospheric level, little difference has been observed between the breakdown voltages obtained with electrodes of different materials, provided that the surface remains clean and smooth. At high pressures, however, surface cleanliness and smoothness considerably improve the breakdown characteristics (Chapter 10). Point discharges that exist at the irregularities of a rough surface cause a considerable pre-breakdown current and result in a decrease in positive breakdown voltage. Field emission from irregularities on metallic cathodes contributes significantly to the pre-breakdown current (Khalifa et al., 1977). Therefore, the breakdown versus pressure characteristics show saturation (i.e., Paschen's law ceases to apply beyond a certain pressure). With successive sparking, microscopic protrusions on the electrode surfaces tend to get burned off and breakdown voltage rises to a higher plateau.

4.6   BREAKDOWN IN NONUNIFORM DC FIELDS

Most practical gas gaps have nonuniform fields. Examples include wire-to-plane gaps and coaxial cylinders, where the applied field and the first Townsend coefficient α vary across the gap. Electron multiplication is thus governed by the integral of α over the path. Townsend's criterion for the spark takes the form

$$\int_0^d \alpha \, dx = \ln\left(1 + \frac{1}{\gamma}\right) \tag{4.14}$$

where d is the gap length.

Meek's criterion for the spark is obtained from equation (4.11) after modification as

$$\alpha(x) \exp\left(\int_0^x \alpha \, dx\right) = G(x, p) \tag{4.15}$$

where $G(x, p)$ is evaluated from uniform-field data at the same gas pressure p.

In a nonuniform field gap, if the maximum field occurring at the highly stressed electrode is less than five times the average field, the discharge phenomena will be similar to those in a uniform field. In more divergent fields, however, a different phenomenon sets in (Waters, 1978). At certain voltages below breakdown level, ionization may be maintained locally at the highly stressed electrode (Chapter 5). No criterion has yet been well established for the advance of streamers in nonuniform fields, although computations and explanations for streamer growth have been reported (Abdel-Salam et al., 1976).

Once corona starts, the electric field becomes distorted by space charge, and here the dependence of the breakdown voltage on the electrode configuration is much more complex than the dependence of the corona onset voltage. In sphere–plane and point-to-point gaps, if the stressed electrode is positive, the space charge acts as an extension of the electrode, but if the sphere is negative, the space charge acts as a screen that decreases the field in its vicinity and thus tends to raise the breakdown voltage (Fig. 4.12). Because of these polarity effects, characteristics obtained with the pointed electrode positive are usually more important for practical applications. Figure 4.13 illustrates schematically the dependence of the voltage on the gap length and sphere diameter. There are three main regions:

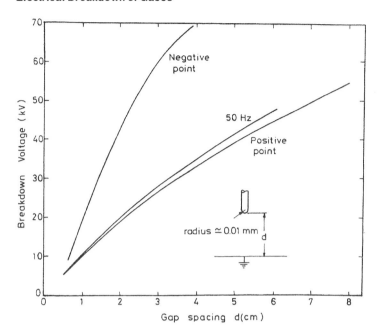

Figure 4.12 Breakdown voltage of a point-to-plane gap in atmospheric air with DC voltage of both polarities. Breakdown voltage with 50 Hz voltage is included for comparision.

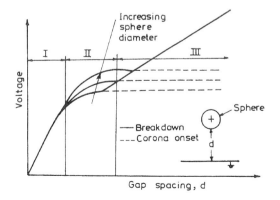

Figure 4.13 Breakdown and corona-onset voltages versus gap spacing for spheres of different diameters.

Region I. At small gaps, depending on the sphere diameter, the field is almost uniform, and the breakdown voltage depends mainly on the gap length.

Region II. For moderate gap lengths, the field shows significant non-uniformity. Therefore, the breakdown voltage increases with the sphere diameter as well as with the gap length. The effect of the sphere diameter on the field magnitude becomes more significant as the gap increases.

Region III. For gaps exceeding twice the sphere diameter, breakdown is preceded by corona. The maximum field, and therefore the corona onset voltage, are influenced by the sphere diameter, whereas the breakdown voltage depends mainly on the gap length.

4.7 BREAKDOWN IN NONUNIFORM AC FIELDS

The breakdown process gets completed in an interval on the order of 10^{-6} to 10^{-8} s. This represents an extremely small fraction of half a cycle of the power frequency. Therefore, the mechanism of breakdown is essentially the same as under DC. The only difference is that the ions in the gas will be subjected to a slowly alternating field. If the applied AC voltage magnitude is such that at the voltage peak the discharge onset conditions are reached, electron avalanches will be produced in the same way as under DC. The space charges produced will have ample time to leave the gap before the field reverses polarity. The maximum gap spacing (L_{max}) in which this is possible is the distance the ions move under such conditions. In uniform fields, this distance is about 1 m in atmospheric air under power frequency. Spacings encountered in high-voltage power transmission systems exceed L_{max} and the ions are constrained to the vicinity of the conductors; recombination occurs among the outgoing and returning ions (Abdel-Salam et al., 1984).

4.8 BREAKDOWN UNDER IMPULSE VOLTAGES

Impulse overvoltages arise in power systems due to lightning or switching surges. They represent a principal factor in the design of equipment insulation (Chapter 14).

It is therefore important to appreciate the fact that the breakdown mechanim under impulse voltages is different from cases of steady DC or low-frequency AC. Under impulse, a time lag is observed between the instant the applied voltage is sufficient to cause breakdown and the actual event of breakdown, if it occurs. The two basic relevant phenomena are the appearance of electrons for initiating the avalanches and their ensuing temporal growth.

In the case of steady or slowly varying fields, there is usually no difficulty in finding an initiatory electron from natural sources (e.g., cosmic rays or detachment from negative ions) (Allen et al., 1981). However, under an impulse voltage of short duration ($\simeq 10^{-6}$ s), depending on the gap volume, natural resources may not be sufficient to provide the initiating electron at the appropriate site in time for the breakdown to occur. The probability of breakdown increases from zero to 100% over a suitable voltage range. The time t_s that elapses between the application of a voltage greater than V_s, the gap's static breakdown voltage, and the appearance of a suitably placed initiatory (seed) electron is called the statistical time lag t_s of the gap because of its statistical nature (Berger, 1973). After such a seed electron appears, the subsequent time t_f required for the breakdown of the gap to materialize is known as the formative time lag. The sum ($t_f + t_s$) is the total time lag, or the time to breakdown (TBD).

For example, in a positive point-to-plane gap, if the seed electron is too close to the point it develops an avalanche of insufficient size for streamers to develop. Also, streamers cannot form if the seed electron is too far from the anode, where the attachment coefficient exceeds the ionization coefficient. Thus a critical volume is defined (Fig. 4.14a), which at the static breakdown voltage V_s should theoretically be reduced to a point on the axis of the electrode system. The critical volume grows axially and laterally with the increase in applied voltage (Fig. 4.14a) (Abdel-Salam and Turkey, 1988). The probability that a negative ion appears in the critical volume to give birth to an electron expresses the distribution of the statistical time lag t_s shown in Figure 4.14b.

For breakdown to occur, the applied impulse voltage V must be greater than V_s. Breakdown on the front of an applied voltage wave is shown in Figure 4.15. The overvoltage $V - V_s = \Delta V$ clearly depends on ($t_s + t_f$) and on the rate of rise of the applied voltage.

The differences among the volt–time characteristics for different air-gap geometries, for internal and external insulation of equipment, and for overvoltage protective devices provide the basis of insulation coordination. This important subject is discussed in Chapter 15.

4.8.1 Statistical Time Lags

It is interesting to examine some of the factors controlling t_s. If θ is the rate at which electrons are produced in the gap by external irradiation, P_{r1} the probability of an electron appearing in the critical volume where it can lead to a spark, and P_{r2} the probability that such an electron will lead to a spark, then the average statistical time lag t_s equals $1/(\theta P_{r1} P_{r2})$. Also, if a gap has survived breakdown for a period t, then the probability that it will break

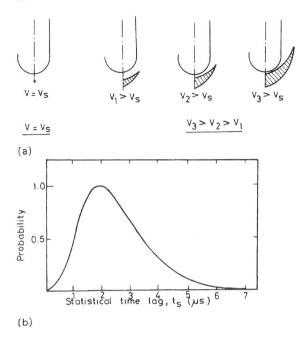

Figure 4.14 Initiation of positive corona under surge voltages: (a) variation of critical volume with applied voltage; (b) distribution of time lag (t_s) for the initiation of positive corona. [From Goldman and Goldman (1978).]

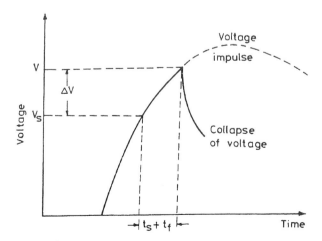

Figure 4.15 Breakdown on the front of the applied impulse voltage wave.

down in the next time interval dt is $\theta P_{r1} P_{r2}\, dt$. This will be independent of t if $\theta P_{r1} P_{r2}$ is independent of t. The probability of breakdown with a time lag t is $\exp(-t/t_s)$.

4.8.2 Formative Time Lags

Once an initiatory electron is made available in the gas gap, it will, under the applied voltage, start an electron avalanche with the subsequent processes, such as secondary avalanches, streamers, and leaders, eventually culminating in the sparkover of the gap. The formative time lag t_f consumed in these processes was measured experimentally by Kohrman and others (Nasser, 1971). For overvoltages of less than 1% for air gaps, relatively long time lags $t_f (\geq 20\,\mu s)$ were observed, which led to the belief that at such small overvoltages a Townsend mechanism may be at work. At higher overvoltages t_f was of the order of $1\,\mu s$, which supported the streamer mechanism.

4.8.3 Breakdown of Long Gaps under Switching Impulses

In gaps with a highly nonuniform field, the space charge resulting from prebreakdown corona causes a marked distortion of the applied electrostatic field. The establishment of such space charge takes a finite time; consequently, the breakdown characteristics will be affected by the rate at which the voltage rises. It has been observed that, with wavefronts of 50 to $300\,\mu s$, positive breakdown voltages of non-uniform-field gaps were minimum (Fig. 4.16). With shorter wavefronts, corona will develop and produce a large space charge in the gap, which delays the advancement of the leader stroke and leads to an increase in the breakdown voltage. With much longer fronts, the space charge of the corona gradually fills the zone near the stressed electrode and reduces the voltage gradient there, again raising the breakdown voltage. Thus, the breakdown voltage depends upon the rate of rise of voltage in an impulse and on the electrode spacing. These dependencies are manifest in the existence of the so-called "U-curves". From Figure 4.16 it is noticed that the critical front duration corresponding to the minimum breakdown voltage increases with gap length. Harada et al. (1973) have given an empirical formula for the critical front duration:

$$T_c = 40 + 35d\,\mu s \tag{4.16}$$

where the gap length d is in meters.

Because of the time and expense involved in high-voltage tests on large structures, any method of extrapolating known results is economically attractive and important for the design of EHV systems. Given the breakdown voltage of a rod–plane gap, the breakdown voltage for any other gap length can be estimated by

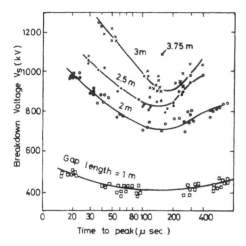

Figure 4.16   Dependence of the breakdown voltage of a rod-gap on the voltage front duration for various gap lengths. [From Bazelyan et al. (1961).]

$$V_{s+} = 500K_g d^{0.5} \, \text{kV} \tag{4.17}$$

where K_g is the gap factor for the configuration. This is a rule-of-thumb used for estimating the breakdown voltage of gaps of various geometries, based on the breakdown voltage of a rod–plane gap of the same length. Paris et al. (1973) deduced the gap factors for various electrode configurations, ranging from 1.15 for conductor–plane gaps, to 1.3 for rod–rod gaps, and up to 1.9 for large conductor-to-rod gaps. The gap factor method has also been applied to lightning impulses (Paris et al., 1973).

Semiempirical models have been proposed by researchers to calculate the breakdown voltage of long air gaps of up to 50 m (e.g., Alexandrov and Podporkyn, 1979; Waters, 1978). Very recently, Rizk (1989) developed a mathematical model for calculating the continuous leader inception and breakdown voltages of long air gaps under positive switching impulses with critical time-to-crest. The model is based on assuming the presence of the following factors:

1. Axial propogation of the leader.
2. Constant charge injection during propagation ($q_1 = 45 \, \mu\text{C/m}$).
3. Constant velocity of leader propagation ($v = 1.5 \, \text{cm/}\mu\text{s}$).
4. Resemblance between the leader and the electrical arc with a conductance that varies exponentially with the lifetime of the leader (time constant $\tau = 50 \, \mu\text{s}$). Subsequently, the voltage gradient

within the leader varies from an initial value E_i ($=$ 400 kV/m) to an ultimate value E_∞ ($=$ 50 kV/m).

5. Constant voltage gradient through the leader corona streamer ($E_s = 400$ kV/m).

The agreement between the values predicted by the model, including the height of the final jump, leader voltage drop, and 50% breakdown voltage, and those measured experimentally is excellent in light of the aforementioned assumptions and the unique values of q_1, v, E_i, E_∞, and E_s. The model dealt with rod–, sphere–, and conductor–plane gaps.

For rod–plane gaps, Rizk (1989) developed an expression for the breakdown voltage V:

$$V = \frac{1556 + 50d}{1 + 3.89/d} + 78\,\text{kV} \qquad \text{for } d > 4\,\text{m} \tag{4.18}$$

which predicts breakdown voltages that agree reasonably well with findings of the Renardier group (Paris et al., 1973).

4.9 HIGH-FREQUENCY BREAKDOWN

In uniform-field gaps, breakdown starts with an avalanche process, as described above for the cases of DC and low-frequency AC. As the frequency f of the applied field increases to very high levels, the discharge behavior starts to differ. To explain this point, let the field in the gap be expressed as $(V/d)\sin 2\pi ft$, V being the applied peak voltage and d the gap length. Then the maximum distance L_{max} that any positive ion can travel during one half-cycle is

$$L_{max} = \int_0^{1/2f} k_+ \frac{V}{d} \sin 2\pi ft\, \mathrm{d}t = \frac{k_+ V}{\pi fd} \tag{4.19}$$

k_+ being the positive ion mobility. For $d > L_{max}$, most positive ions will not be able to reach the cathode before the applied voltage reverses sign. In other words, for a given d the critical frequency f_c at which all positive ions can just be cleared from the gap during one half-cycle is

$$f_c = \frac{k_+}{\pi d^2} V \tag{4.20}$$

At frequencies $f > f_c$ the cloud of positive-ion space charge will oscillate between the electrodes while new avalanches grow and add to its density and size until instability and breakdown occur. This accumulating space charge will no doubt distort the field in the gap. Therefore, breakdown occurs at lower field strengths for high-frequency AC than under DC (Fig. 4.17). At frequencies $f < f_c$ the breakdown conditions are quite similar to those of static fields.

The foregoing analysis of ion motion can be extended to the electrons oscillating between the electrodes. In analogy with f_c for ions, there will be a critical frequency above which electrons would have no time to reach the opposite electrode. They will oscillate in the gap and collide with the gas molecules. When the field is adequately high, they will produce more and more electrons until breakdown is completed with no participation from the electrodes (MacDonald, 1966). This critical frequency, f_{ce}, depends on the electron mobility k_e, the electrode spacing d, and the magnitude of the applied voltage V.

$$f_{ce} = \frac{k_e V}{\pi d^2} \tag{4.21}$$

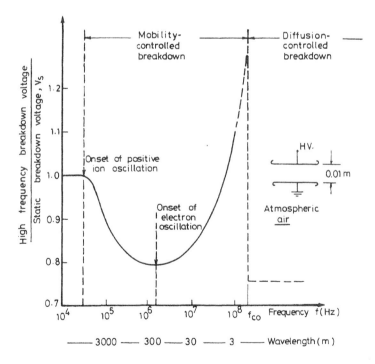

Figure 4.17   Ratio of high-frequency breakdown voltage to static breakdown voltage as a function of frequency for a uniform air gap. [From Gänger (1953).]

Since electron drift velocities are two orders of magnitude higher than those of positive ions under the same conditions, the magnitude of f_{ce} is about two orders of magnitude higher than f_c (Fig. 4.17) (Nasser, 1971).

The shape of the curve in Figure 4.17 can be explained as follows. The decrease in breakdown voltage in the lower part of the frequency range above f_c is caused by the distortion of the electric field in the gap by the positive-ion space charges accumulating in the gap. At the higher frequencies, however, the electrons in the gap oscillate at increased frequencies and some of them would fail to reach the anode during the half-cycle in which they were created. They would thus remain in the gap and partly neutralize the positive-ion space charge and the distortion in the field—hence the parabolic part of the curve.

The breakdown voltage even exceeds its static level at the higher frequencies where the electrodes do not contribute to the ionization process as they do under DC or power-frequency AC. At a certain frequency f_{co} the breakdown mechanism is controlled by diffusion; with the electrodes playing no role, the breakdown voltage drops sharply.

When $f < f_{co}$, electrons are lost by virtue of their mobility to the electrodes and the breakdown mechanism is called the mobility-controlled mechanism. When $f > f_{co}$, the electrons are lost by diffusion and the breakdown mechanism is known as the diffusion-controlled breakdown mechanism. At extremely high frequencies, electrons and ions oscillate in the gap, giving rise to higher currents whose phase relationship to the applied voltage is controlled by the rate of electron–ion recombination relative to the frequency. The breakdown field strength can be determined by relating the field-dependent ionization rate to that of charge loss by diffusion (MacDonald, 1966).

All this discussion has been confined to uniform-field gaps. Nonuniform-field gaps also show differences between their high- and low-frequency performances. Such nonuniform fields are experienced in cases of transmitter antennas and their insulation. For point-to-plane gaps, the high-frequency breakdown voltages increase with gap length, as at low frequency, but are different in magnitude.

4.10 SOLVED EXAMPLES

(1) In an experiment to measure α for a given gas, it was found that the steady-state current is 2.7×10^{-8} A at a voltage of 10 kV and a spacing of 0.005 m between the plane electrodes. With the spacing increased to 0.01 m, the current increases to 2.7×10^{-7} A for the same electric field between the electrodes. (a) Calculate α. (b) Calculate the number of electrons emitted from the cathode per second.

Solution:
(a) Using equation (4.7a):

$$I_1 = I_0 e^{\alpha d_1}, \qquad I_2 = I_0 e^{\alpha d_2}$$

$$\alpha = \frac{1}{d_2 - d_1} \ln \frac{I_2}{I_1} = \frac{1}{0.01 - 0.005} \ln \frac{2.7 \times 10^{-7}}{2.7 \times 10^{-8}} = 460.5 \, \text{m}^{-1}$$

(b) $I_0 = I_1 e^{-\alpha d_1} = 2.7 \times 10^{-8} e^{-460.5 \times 0.005} = 2.7 \times 10^{-9}$

$$n_0 = \frac{I_0}{1.6 \times 10^{-19}} = 1.687 \times 10^{10} \, \text{electrons/s}$$

(2) At the conditions of example (1), determine the electrode spacing that would result in an electron multiplication of 10^9.

Solution:

$$10^9 = e^{\alpha d}$$

$$\therefore \alpha d = 20.72$$

and $d = 20.72/460.5 = 0.045 \, \text{m}$

(3) In a nonuniform field near a cathode, α is expressed as

$$\alpha = a - b\sqrt{x} \, \text{m}^{-1}$$

where $a = 4 \times 10^4$, $b = 15 \times 10^5$, and x is measured from the cathode surface in meters. If an electron starts its motion at the cathode, calculate the size of the developed avalanche at a distance of 0.0005 m.

Solution:
Rewrite equation (4.2) for nonuniform fields:

$$n = n_0 e^{\int \alpha \, dx}$$

$$\int_0^{0.0005} \alpha \, dx = \int_0^{0.0005} (a - b\sqrt{x}) dx = \left[ax - \frac{2}{3} bx^{3/2} \right]_0^{0.0005}$$

$$= \left[a(0.0005) - \frac{2}{3} b(0.0005)^{3/2} \right] = 8.82$$

Avalanche size $= e^{8.82} = 6768$

(4) If an electron starts at a distance of 0.001 m from a cathode in a field where $\alpha = a - bx \text{ m}^{-1}$, find the distance it must travel to produce an avalanche of 10^9 electrons. Use the a and b values given in example (3).

Solution:
Rewrite equation (4.2) for nonuniform fields:

$$n = n_0 e^{\int \alpha \, dx} = 10^9$$

$$\int_{0.001}^{x} (a - bx) \, dx = \left[ax - \frac{bx^2}{2} \right]_{0.001}^{x} = 9 \ln 10$$

$$a(x - 0.001) - \frac{b}{2}(x^2 - 10^{-6}) = 9 \ln 10$$

$$7.5 \times 10^5 x^2 - 4 \times 10^4 x + 59.97 = 0.0$$

A quadratic equation in x, so two values of x are expected
$x = 0.0518$ m away from the cathode
(to be disregarded where α becomes negative for $x > 0.027$ m)
or $x = 0.001\,54$ m away from the cathode.

(5) In example (4), determine the minimum distance measured from the cathode at which an electron may start an avalanche having a size of 10^{19}.

Solution:
Rewrite equation (4.2) for nonuniform fields:

$$n = n_0 e^{\int \alpha \, dx} = 10^{19}$$

$$\int_{0.0}^{x} (a - bx) dx = 19 \ln 10 = \left[ax - \frac{bx^2}{2} \right]_{0.0}^{x}$$

$$= ax - \frac{bx^2}{2}$$

or $7.5 \times 10^5 x^2 - 4 \times 10^4 x + 43.75 = 0.0$

$x = 0.0011$ m or $x = 0.0522$ m (to be disregarded)

Minimum distance $= 0.0011$ m

(6) Derive an expression for the total number of electrons n reaching the anode when photoionization is produced uniformly by x-rays in the gas between two parallel plates spaced d meters apart. Photoionization

occurs at a rate of n_{ph} ionization/m^3 per second. The plates are stressed by a voltage enough to cause ionization by collision with α electrons produced per electron per unit length along the field direction between the plates. Assume the area of the plates is equal to $1\,m^2$.

Solution:
Let n = rate of electron production at distance x.
    dn = number of electrons produced in a slab dx at a distance x measured from the cathode (Fig. 4.1).

$$dn = n\alpha\,dx + n_{ph}\,dx = (n\alpha + n_{ph})\,dx$$

$$\int_0^n \frac{dn}{n\alpha + n_{ph}} = \int_0^d dx$$

$$\frac{1}{\alpha}[\ln(n\alpha + n_{ph})]_0^n = d$$

$$\frac{1}{\alpha}\ln\left(\frac{n\alpha + n_{ph}}{n_{ph}}\right) = d$$

$$\therefore \frac{n\alpha + n_{ph}}{n_{ph}} = e^{\alpha d}$$

$$n = \frac{n_{ph}}{\alpha}(e^{\alpha d} - 1)$$

If $n_{ph} = 0$, $n = 0$, which means that there are no initiatory electrons for electron-multiplication by collision.

(7) The breakdown of air between two parallel plates spaced by a distance of 0.002 m is 9 kV. Calculate the total secondary coefficient of ionization γ at NTP. Use Γ and B values listed following equation (3.15) for evaluating α.

Solution:
Using equation (3.15),

$$\frac{\alpha}{p} = \Gamma e^{-Bp/E}$$

$$E = \frac{9 \times 10^3}{0.002} = 4.5 \times 10^6\,V/m$$

$$\Gamma = 11{,}253.7\,m^{-7}\cdot kPa^{-1}, \qquad B = 273{,}840\,V/m\cdot kPa$$

$$p = 1\,\text{atm} = 101.3\,\text{kPa}$$

$$\alpha = 101.3 \times 11{,}253.7\,e^{-27{,}380 \times 101.3/4.5 \times 10^6}$$
$$= 2396.25\,\text{m}^{-1}$$

Breakdown criterion equation (4.8), is

$$\gamma(e^{\alpha d} - 1) = 1$$

$$\gamma(e^{2396.25 \times 0.002} - 1) = 1$$

$$\therefore \gamma = 0.008\,36$$

(8) Three measurements of the current between two parallel plates were 1.22, 1.82, and 2.22 of the initiating photocurrent I_0 at distances 0.005, 0.015 04, and 0.019 m, respectively. E/p and p were maintained constant during the measurements. Calculate: (a) the first ionization coefficient, (b) the secondary ionization coefficient.

Solution:
(a) Using equation (4.7a),

$$I_1 = I_0 e^{\alpha_1 d_1} = 1.22 I_0$$

$$\therefore \alpha_1 d_1 = \ln 1.22 = 0.199$$

$$\alpha_1 = 0.199/0.005 = 39.8\,\text{m}^{-1}$$

Similarly,

$$\alpha_2 d_2 = \ln 1.82 = 0.5988$$

$$\alpha_2 = 0.5988/0.015\,04 = 39.8\,\text{m}^{-1}$$

and

$$\alpha_3 d_3 = \ln 2.22 = 0.7975$$

$$\alpha_3 = 0.7975/0.019 = 41.97\,\text{m}^{-1}$$

At d_3, the γ-mechanism must be acting since α should not change as long as E/p is constant.
(b) At d_3:

$$I_3/I_0 = \frac{e^{\alpha d}}{1 - \gamma(e^{\alpha d} - 1)} \qquad \text{[equation (4.7)]}$$

$$2.22 = \frac{e^{39.8 \times 0.019}}{1 - \gamma(e^{39.8 \times 0.019} - 1)}$$

$$\therefore \gamma = 0.0354$$

(9) Keeping E/p and p constant at 12,000 V/m·kPa and 0.133 kPa respectively in example (8), determine the distance and voltage at which transition to a self-sustained discharge takes place.

Solution:
At the transition to a self-sustained discharge (equation 4.8)

$$\gamma(e^{\alpha d} - 1) = 1$$

or

$$d = \frac{1}{\alpha}\left[\ln\left(\frac{1}{\gamma} + 1\right)\right]$$

Distance at the transition to a self-sustained discharge

$$= \frac{1}{39.8}\left[\ln\left(\frac{1}{0.0354} + 1\right)\right] = 0.085 \, \text{m}$$

As $E/p = 12,000$ V/m·kPa and $p = 0.133$ kPa

$$\therefore E = 1596 \, \text{V/m}$$

Voltage at the transition to a self-sustained discharge

$$= 1596 \times 0.085 = 135.66 \, \text{V}$$

(10) Calculate the breakdown voltage using (a) Raether's criterion and (b) Meek and Loeb's criterion for a uniform atmospheric air gap having a spacing $d = 0.001$ m. Use Γ and B values listed following equation (3.15) for evaluating α.

Solution:
(a) *Using Raether's criterion*:
The radial component of the space-charge field at distance r from the avalanche head where the size is $e^{\alpha x}$

$$E_r = \frac{e e^{\alpha x}}{4\pi\varepsilon_0 r^2} \tag{4.22}$$

$$r^2 = 3D_e t \quad \text{(as assumed by Raether, 1964)}$$

$$= 3D_e \frac{x}{k_e E}$$

According to equation (3.50)

$$\frac{D_e}{K_e} = \frac{kT}{e}$$

$$r^2 = 3\frac{kT}{e}\frac{x}{E}$$

Following equation (3.3), the electron energy eV is equal to $\frac{3}{2}kT$ (Raether, 1964)

$$\therefore r^2 = \frac{2Vx}{E}$$

Substitute in equation (4.22)

$$E_r = \frac{ee^{\alpha x}E}{4\pi\varepsilon_0(2Vx)}$$

At breakdown, $E_r = E$ and $x = x_c$

$$e^{\alpha x_c} = \frac{8\pi\varepsilon_0 V x_c}{e}$$

$$\alpha x_c = \ln\frac{8\pi\varepsilon_0 V}{e} + \ln x_c$$

The numerical value of V was considered equal to 1.5 V (Raether, 1964).

$$\alpha x_c = 17.7 + \ln x_c \tag{4.23}$$

For uniform fields, equation (4.23) takes the form

$$\alpha d = 17.7 + \ln d$$

As $d = 0.001\,\text{m}$, $\alpha = 10{,}792.2\,\text{m}^{-1}$
As $p = 1\,\text{atm} = 101.3\,\text{kPa}$,

$$\alpha/p = 106.54\,\text{m}^{-1}\text{kPa}^{-1}$$

Use $\Gamma = 11{,}253.7\,\text{m}^{-1}\,\text{kPa}^{-1}$ and $B = 273{,}840\,\text{V/m·kPa}$
As $\alpha/p = \Gamma e^{-Bp/E}$, $E/p = 58764.81\,\text{V/m.kPa}$

$$E = 58764.81 \times 101.3 = 595.29 \times 10^4\ \text{V/m}$$

The breakdown voltage

$$V_s = Ed = 595.29 \times 10^4 \times 0.001 = 5.95\,\text{kV}$$

(b) *Using Meek and Loeb's criterion*:
   Using equation (4.11),

$$E_r = 5.27 \times 10^{-7}(p/d)^{1/2}\alpha e^{\alpha d} \qquad (4.24)$$
$$= E \text{ at breakdown} \qquad (4.25)$$
$$p = 1\,\text{atm} = 101.3\,\text{kPa}$$

Solve equations (4.24) and (4.25) iteratively. For each assumed value of E, obtain α at atmospheric pressure ($\alpha = p\Gamma e^{-Bp/E}$) and substitute at the right-hand-side of equation (4.24) to check its equality to E. If not, change the assumed value of E and repeat the procedure until equations (4.24) and (4.25) are satisfied simultaneously at $E = 468 \times 10^4\,\text{V/}$ m.

The breakdown voltage $V_s = 468 \times 10^4 \times 0.001 = 4.68\,\text{kV}$.

(11) An oscillogram was recorded for the electron current of an avalanche in a uniform-field gap of 0.05 m spacing. If the current reaches a peak and declines abruptly in 0.2 μs after the release of the initiating electrons at the cathode, determine the electron drift velocity. If the exponential rise of the current has a time constant of 35 ns, calculate the first Townsend's ionization coefficient.

Solution:
Transit time of electrons

$$\tau = 0.2\,\text{μs} = 0.2 \times 10^{-6}\,\text{s}$$

Electron drift velocity

$$v_e = d/\tau = 0.05/0.2 \times 10^{-6}$$
$$= 0.25 \times 10^{-6}\,\text{m/s}$$

$$\text{Time constant} = \frac{1}{\alpha v_e} = 35 \times 10^{-9}\,\text{s}$$

$$\alpha = \frac{1}{35 \times 10^{-9} \times 0.25 \times 10^{-6}}$$
$$= 114.3\,\text{m}^{-1}$$

(12) An alternating voltage of 200 kV (rms) is applied to a 10 cm uniform gap in atmospheric air. (a) If the frequency of the voltage is 50 Hz and the mobility of positive ions is $1.4 \times 10^{-4}\,\text{m}^2/\text{s·V}$, determine the travel time of positive ions from one electrode to the other. (b) What is the maximum frequency that can be applied and still just permits the clearing of all positive ions?

Solution:

(a) Let the alternating field E be described by $E_a \cos \omega t$

$$x = \int_0^t k_+ E \, \mathrm{d}t = \int_0^{\omega t} k_+ \frac{E_a}{\omega} \cos \omega t \, \mathrm{d}(\omega t) = \frac{k_+ E_a}{\omega} \sin \omega t \qquad (4.26)$$

$$E_a = \frac{200 \times \sqrt{2} \times 10^3}{0.1} = 2828.4 \times 10^3 \, \mathrm{V/m}$$

$$x = 0.1 = \frac{1.4 \times 10^{-4} \times 2828.4 \times 10^3}{2\pi \times 50} \sin(2\pi \times 50t) = 1.26 \sin(3.14t)$$

$$314t = 0.079 \, \mathrm{rad}$$
$$t = 0.079/314 = 0.253 \, \mathrm{ms}$$

(b) From equation (4.26):

$$f = \frac{k_+ E_a}{2\pi x} \sin \omega t$$

The maximum time available before the voltage reverses is at $\omega t = \pi/2$ and $x = d$

$$f_{\max} = \frac{k_+ E_a}{2\pi d}$$
$$= \frac{1.4 \times 10^{-4} \times 2828.4 \times 10^3}{2\pi \times 0.1} = 630.2 \, \mathrm{Hz} \qquad (4.27)$$

REFERENCES

Abdel-Salam M. J Phys D: Appl Phys 9:L-149, 1976.

Abdel-Salam M, Stanek EK. IEEE Trans IA-24:1025, 1988.

Abdel-Salam M, Turkey A. IEEE Trans IA-24:1031, 1988.

Abdel-Salam M, Khalifa M, Hashem A. Proceedings of IEEE–IAS Annual Meeting, Chicago, IL, 1976, pp 507–512.

Abdel-Salam M, Radwan R, Ali Kh. IEEE-PES paper A-78-601-7, 1978.

Abdel-Salam M, Farghally M, Abdel-Sattar S, Shamloul D. Proceedings of 4th International Symposium on Gaseous Dielectrics, Knoxville, TN, 1984.

Alexandrov GN, Podporkyn GV. IEEE Trans PAS-98:597, 1979.

Allen NL, Berger G, Dring D, Hahn R. Proc IEEE 128:565, 1981.

Alston LL. High Voltage Technology. Oxford: Oxford University Press, 1968.

Bazelyan EM, Brago EN, Stekolnikov IS. Sov Phys-Dokl 52:101, 1961.

Berger G. PhD thesis, Université de Paris-Sud, Paris, 1973.

Gänger B. Der elektrische Durschlag von Gasen. Berlin: Springer-Verlag, 1953.

Goldman M, Goldman A. Corona discharges. In: Hirsh MN, Oskam HJ, eds. Gaseous Electronics. New York: Academic Press, 1978, pp 219–290.

Harada T, Aihara Y, Aoshima Y. IEEE Trans PAS-92:1085, 1973.

Khalifa M, Abdel-Salam M, Radwan R, Ali Kh. IEEE Trans PAS-96:886, 1977.

Llewellyn-Jones F. Ionization and Breakdown of Gases. London: Methuen, 1957.

Loeb LB. Basic Processes of Gaseous Electronics. Berkeley, CA: University of California Press, 1955.

MacDonald AD. Microwave Breakdown in Gases. New York: John Wiley & Sons, 1966.

Naidu MS, Kamaraju V. High Voltage Engineering. New Delhi: Tata McGraw-Hill, 1996.

Nasser E. Fundamentals of Gaseous Ionization and Plasma Electronics. New York: John Wiley & Sons, 1971.

Paris L, Tashchini A, Schneider KH, Weck KH. Electra 29:29, 1973.

Raether H. Electron Avalanches and Breakdown in Gases. London: Butterworth, 1964.

Rizk F. IEEE Trans PD-4:596–606, 1989.

Waters RT. Spark breakdown in nonuniform fields. In: Meek JM, Craggs JD, eds. Electrical Breakdown of Gases. New York: John Wiley & Sons, 1978, pp 385–532.

5

The Corona Discharge

M. KHALIFA *Cairo University, Giza, Egypt*

M. ABDEL-SALAM *Assiut University, Assiut, Egypt*

5.1 INTRODUCTION

"Corona" literally means the disk of light that appears around the sun. The term was borrowed by physicists and electrical engineers to describe generally the partial discharges that develop in zones of highly concentrated electric fields, such as at the surface of a pointed or cylindrical electrode opposite to and at some distance from another. This partial breakdown of air is quite distinct in nature and appearance from the complete breakdown of air gaps between electrodes. The same applies for other gases.

The corona is also distinct from the discharges that take place inside gas bubbles within solid and liquid insulation, although the underlying phenomena of gas discharges are the same. The corona discharge is accompanied by a number of observable effects, such as visible light, audible noise, electric current, energy loss, radio interference, mechanical vibrations, and chemical reactions. The chemical reactions that accompany corona in air produce the smell of ozone and nitrogen oxides.

Corona has long been a main concern for power transmission engineers because of the power loss it causes on the lines and the noise it causes in radio and TV reception. On the other hand, corona does have several beneficial applications, as in Van de Graaff generators, electrostatic precipitators, electrostatic printing, electrostatic deposition, ozone production, and ionization counting (Berg and Hauffe, 1972; Landham et al., 1987; Seelentag, 1979), as described in detail in Chapter 19.

5.2 MECHANISM OF CORONA DISCHARGE

The discharge process depends on the polarity of the applied voltage. Therefore, it will be discussed first for each polarity under DC.

5.2.1 Positive Corona

At the onset level, and slightly above, there exists a small volume of space at the anode where the field strength is high enough to cause ionization by collision. When a free electron is driven by the field toward the anode, it produces an electron avalanche (Chapter 3). The cloud of positive ions produced at the avalanche head near the anode forms an eventual extension to the anode. Secondary generations of avalanches get directed to the anode and to these dense clouds of positive ions (Fig. 5.1). This mode of corona consists of what are called onset streamers. If conditions are favorable, the high field space at the anode may suit the formation of streamers extending tangentially onto the anode; these are called "burst-pulse streamers" (Loeb, 1965).

At slightly higher voltages a cloud of negative ions may form (Fig. 5.1) near the anode surface such that the onset-type streamers become very numerous. They are short in length, overlap in space and time, and the discharge takes the form of a "glow" covering a significant part of the HV conductor surface (Fig. 5.2b). The corresponding current through the HV circuit becomes a quasi-steady current (Fig. 5.2). This is in contrast with current pulses corresponding to onset streamers (Fig. 5.2) (Giao and Jordan, 1967; Khalifa, 1979). The positive current pulse corresponds to a succession of generations of electron avalanches taking place in the ionization zone at the anode (Khalifa and Abdel-Salam, 1974a).

At still higher voltages the clouds of negative ions at the anode can no longer maintain their stability and become ruptured by violent pre-breakdown streamers, corresponding to irregular, high-amplitude current pulses (Figs. 5.1 and 5.2). If we continue to raise the voltage, breakdown eventually occurs across the air gap. Figure 5.3 presents the range for each type of corona discharge with positive DC voltage applied across a point-to-plane gap (Nasser, 1971).

5.2.2 Negative Corona

At the onset level and slightly higher, the corona at the cathode has a rapidly and steadily pulsating mode; this is known as Trichel pulse corona. Each

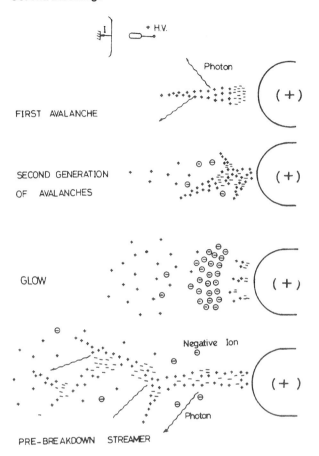

Figure 5.1 Development of the first and subsequent generations of avalanches in positive corona discharges.

current pulse corresponds to one main electron avalanche occurring in the ionization zone (Fig. 5.4) (Khalifa and Abdel-Salam, 1974b; Zeitoun et al., 1976). In this case the ionization zone extends from the cathode surface outward and as far as the point where the field becomes too weak for ionization by collision to compensate for electron attachment. Beyond such a point, more and more of the avalanche electrons get attached to gas molecules and form negative ions which continue to drift very slowly away from the cathode. During the process of avalanche growth, some photons radiate from the avalanche core in all directions (Fig. 5.4). The photoelectrons thus produced can start subsidiary avalanches that are directed from the cathode. The motion of the electrons and negative ions away

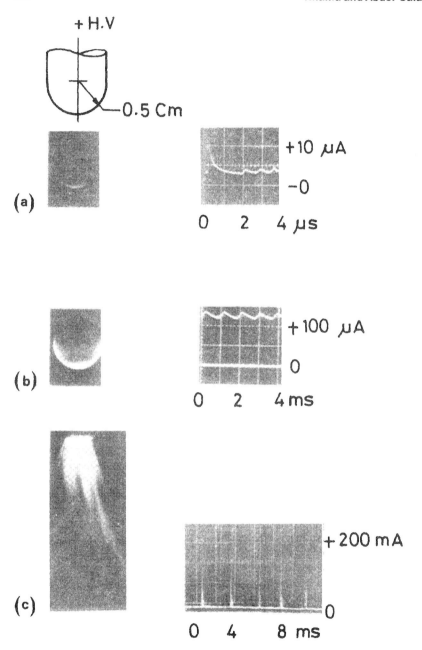

Figure 5.2   Photographs and corresponding current oscillograms of different modes of positive corona. [From Giao and Jordan (1967).]

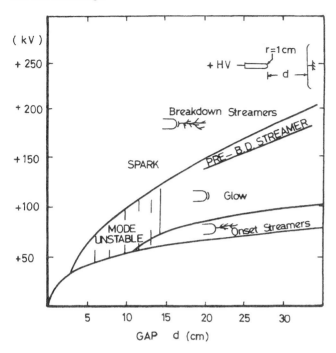

Figure 5.3 Onset voltages of various positive corona modes and sparkover voltage as functions of point-to-plane gap spacing. [Courtesy of E. Nasser (1971) and Wiley-Interscience.]

from the cathode and that of positive ions toward it correspond to the corona current pulses flowing through the high-voltage circuit, as shown in Figure 5.5a, and can easily be computed.

With an increase in applied voltage, the Trichel pulses increase in a repetitive rate up to a critical level at which the negative corona gets into the steady "negative glow" mode (Figs. 5.4 and 5.5b). At still higher voltages pre-breakdown streamers appear, eventually causing a complete breakdown of the gap (Figs. 5.4 and 5.5c).

5.2.3 AC Corona

The basic difference between AC and DC coronas is the periodic change in direction of the applied field under AC, and its influence on the residual space charge left over from the discharge during preceding half-cycles (Fig. 5.6). Thus positive onset streamers and burst-pulse streamers may appear only over an extremely small range of voltage at onset, followed by a posi-

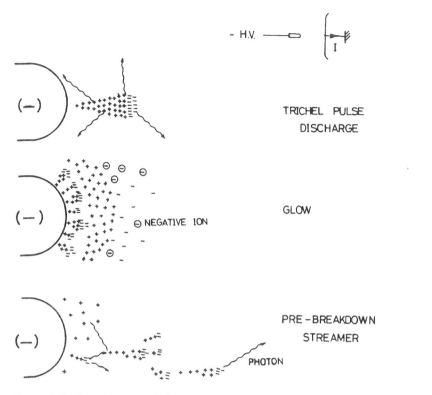

Figure 5.4 Development of electron avalanches in negative corona discharges.

tive glow. Both negative Trichel pulses and negative glow can be observed in an AC corona. If the applied voltage has a suitable magnitude, then, depending on the electrode geometry, both positive and negative glows and streamer coronas can be observed in each cycle.

5.2.4 Impulse Corona

Under impulse voltages, the corona starts in an air gap almost clear of any space charge. Therefore, electron avalanches and streamers extend over significant distances. The onset streamers produced and their branches can easily form traces on photographic films in contact with the anode or cathode, in what is known as Lichtenberg figures (Fig. 5.7). Such a figure cannot be produced by DC because of the choking effect of accumulating clouds of space charges. Under AC a very large number of traces get superimposed on each other.

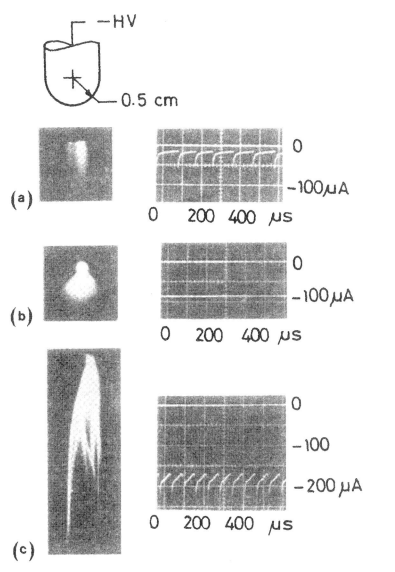

Figure 5.5 Photographs and corresponding current oscillograms of various modes of negative corona. [From Giao and Jordan (1967).]

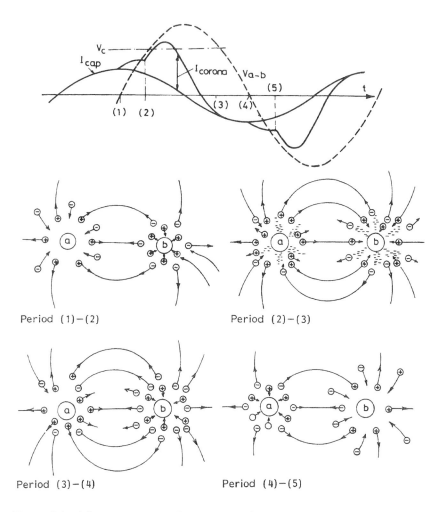

Figure 5.6 AC corona current for a symmetrical gap between two parallel conductors a and b. Note that when the voltage V_{a-b} is below the corona-onset level V_c, the corona current corresponds to the motion and recombination of residual ions in the gap between the high-voltage conductors during the periods shown in the oscillogram.

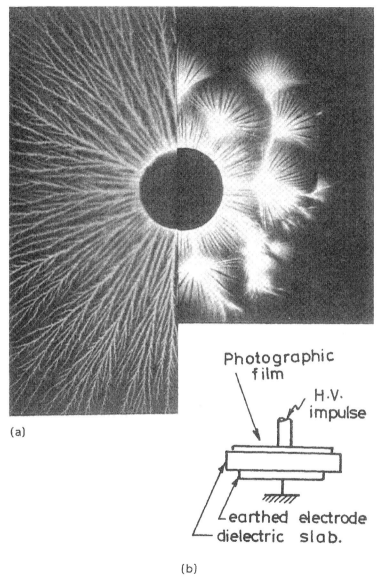

(a)

(b)

Figure 5.7(a) Photographic traces of corona discharge under impulse voltage (Lichtenberg figures): (left) positive polarity; (right) negative polarity; (b) test setup.

On long HV transmission lines, the current corresponding to corona discharges under traveling extra-high-voltage surges has the beneficial effect of reducing the surge peak and front steepness, as shown by experiment and computation (Chapter 14). Thus their stresses on the power system insulation are relieved.

5.3 THE CORONA ONSET LEVEL

It has been well established by experiment and by computation that corona discharge starts at the surfaces of HV electrodes and conductors when their surface voltage gradients reach a critical value E_0 (Abdel-Salam and Khalifa, 1977). The magnitude of E_0 depends on the voltage polarity and on the pressure and temperature of the ambient gas. Humidity has a minor effect. The air pressure p (kPa) and temperature θ (°C) are usually combined into one factor δ, the relative air density, referred to STP. Thus

$$\delta = \frac{2.94p}{273 + \theta} \tag{5.1}$$

with δ ranging between 0.9 and 1.1. Experimentally measured E_0 values fit the following relations:

For AC:

$$E_0 = 30\delta \text{ kV}_{\text{peak}}/\text{cm} \tag{5.2}$$

For DC:

$$E_{\pm} = A_{\pm}\delta + B_{\pm}\sqrt{\frac{\delta}{r}} \text{ kV/cm} \tag{5.3}$$

A_+ and A_- are in the respective ranges 31 to 39.8 and 29.4 to 40.3; B_+ and B_- are, correspondingly, 11.8 to 8.4 and 9.9 to 7.3.

Under AC, a slightly higher field strength E_v corresponds to corona being clearly visible on the conductor surface and can be expressed as $E_v = 30\delta(1 + 0.3/\sqrt{\delta r})$ kV$_{\text{peak}}$/cm, r being the conductor radius in centimeters.

5.3.1 The Corona Onset Voltage

This can easily be calculated using E_0 or $E_{\pm}$ once the conductor arrangement and dimensions are known. Methods of field calculations were discussed in Chapter 2. It should be realized that field calculations would normally be based on the assumption of perfectly clean, smooth conductors; this is different from practical conditions. At any point of microroughness on a practical conductor surface the field will be highly concentrated and the critical field strength for corona onset will be reached there, while the average field strength over the entire surface remains considerably lower. A corresponding surface factor should be taken into account while estimating the corona onset voltage V_0 for the conductor arrangement. This factor is usually taken as about 0.6 for new rough-stranded conductors and about 0.85 for weathered conductors.

5.3.2 Computation of DC Corona Onset Voltages

Onset voltages of DC corona can be estimated using the relation (5.3). They can also be computed according to an algorithm based on the ionization and deionization process acting in a corona discharge of either polarity (Khalifa and Abdel-Salam, 1974a).

5.3.2.1 Case of Positive Polarity

The first avalanche proceeds toward the anode and ends at its surface. It develops a cloud of positive ions, and photons are emitted from its core. Due to the photo-ionization process, these photons generate photoelectrons to start successor avalanches (Fig. 5.8). At corona onset voltage, the number of electrons N_1 in the first avalanche is equal to the number N_2 in the

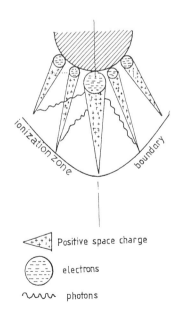

Positive space charge

electrons

photons

Figure 5.8 Development of the primary avalanche and its successors in the vicinity of a positive stressed electrode.

successor avalanches, and the first avalanche becomes unstable. This instability results in the formation of luminous filamentary streamers (Loeb, 1965; Nasser, 1971). Therefore, it is implied that the level of ionization in the successor avalanches is such that sufficient photons and hence photoelectrons will be generated to make the ionization self-sustained when

$$N_2 \geq N_1 \tag{5.4}$$

The onset voltage does not appear explicitly in the relation (5.4), and the onset voltage is the critical value which fulfills the equality (5.4) (Khalifa and Abdel-Salam, 1974a). The computed onset voltages are in good agreement with experiment (Abdel-Salam, 1985).

5.3.2.2 Case of Negative Polarity

When the applied electric field strength near the cathode surface reaches the threshold value for ionization of gas molecules by electron collision, an electron avalanche starts to develop along the direction away from the cathode (Fig. 5.9). With the growth of the avalanche, more electrons are developed at its head, more photons are emitted in all directions, and more positive ions are left in the avalanche wake.

For a successor avalanche to be started, the preceding avalanche should somehow provide an initiating electron at the cathode surface, pos-

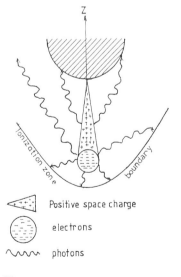

Positive space charge

electrons

photons

Figure 5.9 Development of an avalanche in the vicinity of a negative stressed electrode.

sibly by photoemission, positive ion impact, metastable action, or field emission. Field emission is possible only at field strengths exceeding 5×10^7 V/m (Rein et al., 1977). Electron emission by positive ion impact is more than two orders of magnitude less probable than photoemission (Loeb, 1965). Metastables have been reported to have an effect approximately equal to that of positive ion impact (Nasser, 1971). Therefore, only the first mechanism (electron emission from the cathode by photons) was considered in the mathematical formulation of the onset criterion, where at least one photoelectron is emitted by the photons of the first avalanche to keep the discharge self-sustaining, i.e.,

$$N_{eph} \geq 1 \tag{5.5}$$

where N_{eph} is the number of electrons photoemitted from the cathode.

The onset voltage does not appear explicitly in the relation (5.5), and the onset voltage is the critical value which fulfills the equality (5.5) (Khalifa and Abdel-Salam, 1974a). The computed onset voltages are in good agreement with experiment (Khalifa and Abdel-Salam, 1974b; Abdel-Salam and Stanek, 1988).

It is known that the onset voltage gradient for a positive corona is slightly lower than that for a negative corona. Therefore, the voltage level corresponding to positive corona onset can be calculated according to the method described above for a monopolar corona. At the negative conductor, which has not yet reached its corona onset level, there are positive ions drifting toward it and coming from the positive corona discharge at the positive conductor. These ions enhance the field intensity at the negative conductor and cause a corona to start at a voltage slightly lower than the value calculated for the monopolar case. This slight difference can also be computed (Abdel-Salam and Khalifa, 1977; Al-Hamouz et al., 1998).

5.3.3 Computation of DC Corona Pulse Characteristics

5.3.3.1 Case of Positive Polarity

Positive pulsating corona discharges occur near the onset level as onset streamers and at much higher voltages as pre-breakdown streamers (Nasser, 1971). The streamer is built up by a number of successive generations of electron avalanches which take place in the ionization zone around the positive HV electrode (Khalifa and Abdel-Salam, 1974b).

The ionization zone around the HV electrode is defined as the space where the resultant field strength is so high that the first Townsend's coefficient of ionization is greater than the coefficient of electron attachment. With successions of avalanches and accumulation of ion space charges, the ionization zone expands and its contour changes.

The electrons are driven to the electrode by the resultant of the field of the applied voltage and the field of the space charge. They become involved with air molecules in exciting and ionization collisions to produce an avalanche. Photons are emitted by the avalanche in various directions. Photoelectrons are produced in various locations from which they start a new generation of avalanches. Each photoelectron produced within the ionization zone will be accelerated by the prevailing field from its point of origin and start one avalanche of the second generation. Most of the new avalanches of the succeeding generation get started almost at the instant of termination of the previous avalanche when its population of electrons is highest.

The corona current through the HV circuit corresponds to the motion of electrons and ions in the space between the HV conductors. The motion of ions can be safely neglected as compared to that of electrons during the corona pulse. Thus, the instantaneous current of a given generation of avalanches is calculated as the sum of the components corresponding to the individual avalanches.

In experiment, the capacitance of the HV and measuring circuits smooths the variation of the current from instant to instant during the life of each generation, which is of the order 1 ns, and the current is averaged over the life of each generation. The calculations are continued over successive generations until the magnitude of the current falls below 1% of the maximum value reached for any generation. The current pulse is then considered to have ended.

5.3.3.2 Case of Negative Polarity

Different from the positive case, the negative corona pulse corresponds to one primary avalanche accompanied by a series of auxiliary avalanches which develop in the discharge zone around the HV electrode during the life of the primary avalanche (Khalifa and Abdel-Salam, 1974b).

The primary avalanche is initiated by an electron driven by the electric field outwards from a point of microroughness on the conductor surface. By ionizing and exciting collisions with the air molecules, more and more electrons, photons, and a cloud of positive ions are produced in the primary avalanche.

The photons reaching the conductor emit new electrons of number N_p which start the successor avalanches. When N_p reaches unity (i.e., when the first electron leaves the conductor) the first successor avalanche gets launched. It grows in the same way as the primary avalanche and contributes to the irradiation of the conductor surface, leading to the emission of more and more photoelectrons which start new successor avalanches, and so on.

As the electrons drift away from the conductor in the decreasing field, more and more of them get attached to air molecules and form negative ions. The positive and negative ion clouds grow with time as more avalanches are formed. The space where the field strength is high shrinks and eventually no more avalanches can develop.

The magnitude of the corona pulse current at any instant is calculated as the sum of the current components corresponding to the individual avalanches.

The shapes and durations of the calculated positive and negative pulses agree quite closely with experiment. Different from positive pulses, the calculated negative pulse amplitudes are almost equal to those measured (Khalifa and Abdel-Salam, 1974b).

5.3.3.3 Pulse Amplitudes and Repetition Rates

The amplitudes of the positive and negative corona pulse are evaluated during the process of calculations as the maximum values obtained in the computations of the current pulse. Also, charge contents of the pulses are easily evaluated by integrating the current waves. To calculate the pulse repetition rates, one has to define the conditions to be fulfilled after the termination of a pulse in order that the subsequent pulse can be initiated (Khalifa and Abdel-Salam, 1974b).

In the case of a positive pulse, the favorable condition is taken as the positive ion clouds being swept so far from the high-voltage electrode that their residual field at the boundary of the ionization zone becomes negligible, i.e., 0.1% of the field strength produced there by the applied voltage.

For negative pulses, the new pulse will not start until both positive and negative ion space charges have left the discharge zone. The positive ions are swept to the HV electrode, where they are neutralized. The negative ions move so far from the electrode that their residual field strength at the boundary of the discharge zone becomes negligible, i.e., 0.1% of the applied field strength. The discharge-zone boundary is defined as the place where 99.9% of the electrons in the avalanches have become attached to air molecules and formed negative ions. The calculated pulse repetition rates are closer to those measured. The bigger the charge content of the calculated pulses, the longer the time they take to disappear from the ionization zone and allow for the initiation of the next pulse, and the smaller the repetition rate (Khalifa and Abdel-Salam, 1974b).

5.3.4 Possible Corona in Compressed Air and SF$_6$

Because sulfur hexafluoride is an electronegative gas, it has a high affinity for electron attachment. This makes it more difficult for electron avalanches to grow. Thus corona and sparkover occur at voltages considerably higher than those in air. Above a certain critical gas pressure sparkover occurs across the gas gap without any preceding corona (Section 4.4). At such high pressures the coefficient of ionization by collision becomes lower than the coefficient of electron attachment: $\alpha < \eta$ (Chapter 3). Thus electrons produced at the cathode by photoemission and field emission would have to contribute more substantially to the discharge in order to maintain its stability. Taking both these processes into account has made possible the computation of breakdown voltages of gaps in compressed air and SF_6 (Khalifa et al., 1977; Abdel-Salam, 1978).

5.4   CORONA POWER LOSS

5.4.1   Corona Loss Formulas

Empirical formulas have been suggested for evaluating corona losses on AC lines and on both monopolar and bipolar DC lines.

5.4.1.1   AC Lines

Empirical formulas were suggested early in this century by Peek and Peterson for estimating P_c, the fair-weather corona losses of overhead transmission lines (Begamudre, 1986). Because of several flaws, Peek's formula was superseded by that of Peterson, which takes the form

$$P_c = \frac{3.73K}{(D/r)^2} fV^2 \times 10^{-5} \, \text{kW/conductor} \cdot \text{km} \tag{5.6}$$

where f is the frequency, V the line voltage, and D and r the phase conductor separation and radius. K is a factor depending on the ratio of the operating voltage V to the corona onset line voltage V_0 (Fig. 5.10).

A much more recent and more scientific approach was the computer program developed by Abdel-Salam et al. (1984) on the basis of the physical phenomena of corona discharge (Shamloul, 1989).

In properly designed transmission lines, the corona loss in fair weather is usually insignificant. Typical values measured range from 0.3 to 1.7 kW/conductor·km for 500 kV lines and from 0.7 to 17 kW/conductor·km for 700 kV lines (Electrical Power Research Institute, 1979).

Effect of Conductor Bundling.   For transmission lines with bundled conductors, Peterson's formula can be modified by including the capacitance geometric mean radius of the bundle instead of the single-conductor radius. Naturally, the separation between subconductors in the bundle has an effect on the amount of corona loss. There is an optimal separation between subconductors in the bundle that corresponds to minimum corona loss.

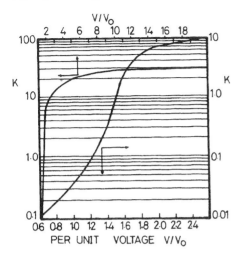

Figure 5.10 Factor K to be used in Peterson's corona loss formula as a function of the per-unit operating voltage referred to the corona-onset voltage.

Effects of Weather. The principal weather parameters are air temperature, pressure, wind, humidity, rain, snow, and dust. The air temperature and pressure are included in the factor δ. No perceptible effect of wind on the AC corona can be noticed. The same goes for humidity unless it approaches 100%. With condensation, and much more seriously with rain, the corona losses increase 10-fold or more, depending on the rate of rainfall (Electrical Power Research Institute, 1979). The increase in corona loss due to rain and dust on conductors of large diameter is much greater than with smaller conductors, as the coronating points are more numerous in the former case.

5.4.1.2 DC Lines

In unipolar DC lines, we have ions of only one polarity in the space between conductors, and between the conductors and ground. With bipolar lines, however, we have both positive and negative ions in the interconductor space and there is a high probability of ion recombination. This particular phenomenon accounts for the corona loss on bipolar DC lines being much higher than on monopolar lines, and even higher than on AC lines, as computed by Abdel-Salam et al. (1982).

In comparison with AC lines having equal per-unit effective voltages with respect to the corona onset levels, the fair-weather corona loss on a bipolar line was found to be about twice that on a three-phase AC line. On the other hand, increases in bipolar line corona loss with voltage and in foul weather are not as rapid as in the case of AC lines.

The effects of atmospheric humidity and wind can be measured and computed (Khalifa and Abdel-Salam, 1974b). The corona loss of monopolar lines can increase by about 20% if humidity increases from 60% to approaching 100%. A much more significant increase occurs if wind blows across the conductors of bipolar lines (Fig. 5.11).

5.4.2 Computation of Corona Power Loss

5.4.2.1   AC Lines

As the AC voltage applied to a single conductor reaches a critical value $V_{0\pm}$, the electric field at the conductor surface reaches the corona onset value $E_{0\pm}$. Corona starts when the electric field exceeds the onset value, and charges of the same sign as that of the conductor potential are emitted into space and move away from the conductor in the form of shells. The charge in a given shell raises the field in the space outside the shell, thus absorbing a part of the applied voltage, while reducing the field at the coronating conductor surface by provoking an increase of the induced opposite charge there (Clade et al., 1969; Abdel-Salam et al., 1984). Therefore, the emission is regulated in such a way that any increase in the space charge due to a new emission causes a reduction of the surface field and therefore a slowing down of emission. It has been realized by experiment (Waters et al., 1972) and theory (Khalifa and Abdel-Salam, 1973) that the electric field at the

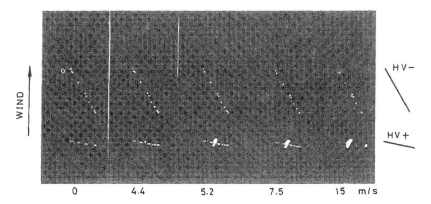

Figure 5.11   Appearance of bipolar DC corona on two conductors, 6.5 mm in diameter and 45 mm apart, at ±63 kV and at different wind speeds. The conductor arrangement is indicated.

surface of a coronating conductor assumes a value less than the onset value.

The voltage $V_{0\pm}$ is the onset voltage at which corona starts in the first positive or negative half-cycle. In the succeeding half-cycle, corona starts at voltages $V_{i\pm} < V_{0\pm}$ due to the effect of the residual space charge. $V_{i\pm}$ are termed the ionization onset voltages. It has been assumed (Clade et al., 1969) that the ionization during each cycle terminates at the voltage peak (V_p) and starts at voltage $V_{i\pm}$, determined by a simple relation

$$V_p - V_{0\pm} = V_{0\pm} - V_{i\pm} \tag{5.7}$$

Of course, $V_{i\pm} = V_{0\pm}$ for the first positive or negative half-cycles. Experiment showed that the corona terminates after the voltage peak of the positive and negative half-cycles at $V_{e\pm}$.

Many experiments have been done to study the performance of AC corona. The $V - Q$ relation is the most important outcome of these experiments to evaluate the AC corona loss (Abdel-Salam and Shamloul, 1992). The corona starting and ending voltages are prerequisites for drawing the $V - Q$ curve, whose area expresses the corona loss per cycle.

Single-phase Transmission Lines. Consider a single-phase conductor-to-plane arrangement as shown in Figure 5.12a. The conductor is simulated by a set of unknown line charges uniformly distributed inside the conductor and coaxial. The magnitudes of these line charges are determined to satisfy the boundary conditions (the potential calculated at the conductor surface

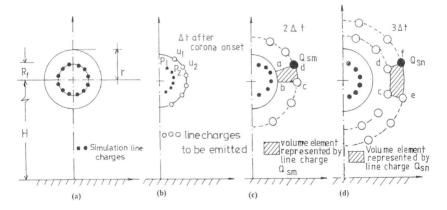

Figure 5.12 Charge development during the first half-cycle: r = conductor radius, R_f = radius of a fictitious cylinder where the simulation charges are located, and H = conductor height above ground plane.

must equal the applied voltage) at a number of boundary points equal to the number of unknowns.

When the applied voltage reaches the positive onset value V_{0+} for the first time, the magnitudes of the unknown charges, denoted as the positive onset values, are $Q_{0j+}, j = 1, 2, \ldots$.

When the applied voltage exceeds V_{0+}, ion emission starts from the conductor. The magnitudes of the unknown charges are $Q_{j+}, j = 1, 2, \ldots$.

On the assumption of constant conductor surface charge, the magnitudes of the emitted space charge Q_{sj+} are assumed (Abdel-Salam and Shamloul, 1992) to be equal to the difference between the line charges Q_{j+} and the corresponding onset values Q_{0j+}.

It is quite reasonable to model the ion emission from the wire in a discrete form; each time interval the conductor generates an infinitely thin cylindrical shell of charges. Each shell was treated (Clade et al., 1969) as a block moving away from the wire or returning back to it. Due to the presence of the ground plane, the space line charges simulating the shell are not concentric with the conductor. Even the charges $Q_{j+}, j = 1, 2, \ldots$, are determined at the points $P_1, P_2, \ldots$ (Fig. 5.12b). They start their journey at the points $u_1, u_2, \ldots$ on the conductor surface under the prevailing field (Fig. 5.12c).

The ion emission, being assumed discrete, dictates that each space line charge represents the charge over a volume element per unit length of the wire. Figures 5.12c and d show the volume elements abcd and dcef represented by the mth line charge and the nth line charge of the first and second emission respectively. The size of the volume element changes from time to time as the line charge departs away from or returns back to the conductor.

For the first negative half-cycle, ion emission starts at V_{i-} and terminates at V_{e-}, the same as explained before for the first positive half-cycle. The liberated negative space charges move away from the stressed conductor. At the same time, the positive space charge returns to the conductor and meets the outgoing negative ones. Some of the positive ions are neutralized through recombination. The space line charges continue to move in opposite directions and will undergo further recombination processes once they meet the opposite charges. Sometimes, some line charges disappear totally due to many successive recombination processes.

All the procedures described above for emission during the first cycle are repeated for the succeeding cycles until a steady-state ion emission is achieved. Corona loss depends on both the amount of charges liberated in space and displacement of these charges through the prevailing field E_f in the wire neighborhood.

The energy dissipation through a complete cycle is calculated as

$$W = \sum_A \int_C Q_{si} E_f \mathrm{d}r \qquad (5.8)$$

where the summation range A is over all positive and negative space line charges Q_{si}, and C the line-charge trajectory (since its emission from the conductor surface) over a cycle. The corona power loss is simply the energy dissipation W per cycle times the frequency of the applied voltage.

The calculated corona power-loss values agreed with those measured experimentally (Abdel-Salam and Shamloul, 1992).

Three-phase Transmission Line. The proposed method of calculating the corona power loss on three-phase transmission lines is similar to that for single-phase lines. Based on the conductor radius, the positive and negative onset field values are defined for each phase conductor. Each phase conductor of the three-phase transmission line is simulated by a set of infinite line charges. The onset charge for each phase conductor will be the summation of those simulation charges located inside it (Abdel-Salam and Shamloul, 1994; Abdel-Salam and Abdel-Aziz, 1994b).

The charges simulating each phase conductor surface charge are obtained by setting the potential at each boundary point equal to the respective phase voltage. This results in a system of simultaneous linear algebraic equations whose solution evaluates the simulation line charges of each phase. The emission law is similar to that in the single-phase case in which, whenever the magnitude of the conductor charge of any phase exceeds the onset value, the surplus charge is emitted as space line charges. Space line charges are displaced under the electric field due to the conductor as well as the space line charges for the three phases. Also, recombination between positive and negative space line charges is taken into account. Corona energy loss is calculated by

$$W = \sum_{3\text{phases}} \sum_A \int_C Q_{si} E_f \mathrm{d}r \qquad (5.9)$$

Corona power loss is calculated by multiplying the corona energy loss by the system frequency. The calculated values of corona power loss agreed satisfactorily with those measured (Abdel-Salam and Shamloul, 1994; Abdel-Salam and Abdel-Aziz, 1994b). More detailed calculations of three- and six-phase corona power loss are reported elsewhere (Abdel-Salam and Abdel-Aziz, 2000).

5.4.2.2 DC Lines

The corona power loss on monopolar and bipolar DC lines was computed using finite-element and charge-simulation techniques. However, the computation of corona power loss on AC lines was limited to the charge-simulation technique.

(a) Using the Charge-simulation Technique. At voltages below the onset value, the field in the inter-electrode spacing is Laplacian. At voltages above the onset value, ions are produced, forming a drifting space charge of the same polarity as that of the coronating electrode. The ions flow towards the other electrode, filling the inter-electrode spacing, where the field is Poissonian. In symmetrical arrangements, such as parallel plates or coaxial cylinders, the ions move along the Laplacian flux lines, which means that the Laplacian and Poissonian fields have the same direction. In asymmetrical arrangements, such as transmission-line arrangements, the space charge causes a deviation between the two patterns, depending on the spatial distribution of ions in the inter-electrode spacing. Thus, Deutsche's assumption, stated in Section 2.6 is no longer valid.

For the Poissonian field around DC lines, the space charge is simulated by a set of fictitious discrete charges, the same as the surface charges on the conductors as described in Section 2.4.3 (Horenstein, 1984; Elmoursi and Castle, 1986; Abdel-Salam and Abdel-Sattar, 1989; Elmoursi and Speck, 1990; Abdel-Salam and Abdel-Aziz, 1994). However, the locations of the simulation charges are located in the space outside the coronating conductor(s) where the field is to be calculated. Therefore, the locations of the simulation charges are singularity points in evaluating the potential and the electric field there. An attempt was made to evaluate the potential and field at the singularity points in a parallel-plate arrangement (Abdel-Salam and Abdel-Aziz, 1992).

As described in Section 2.6.3, the two sets of discrete charges have to be chosen to satisfy the boundary conditions of Poisson's equation (2.58). Images of these charges with respect to the ground plane are considered.

To accommodate the boundary conditions, boundary points are chosen on the coronating conductors.

Satisfaction of the boundary conditions at the boundary points results in a system of nonlinear equations whose solution determines the unknown simulation charges. The technique to solve these nonlinear equations is iterative in nature, where the solution is initialized and the electric field is updated until a self-consistent solution of the electric field and the space line charges is obtained.

The volume charge density is calculated by dividing each discrete line charge by the volume it occupies, and the current density is calculated using equation (2.58). The corona current is obtained by integrating the current density over the periphery of the coronating conductor. Multiplying the corona current by the applied voltage determines the corona power loss.

The calculated corona currents agreed with those measured on lines with single conductors (Abdel-Salam and Abdel-Aziz, 1994a) and on lines with bundled conductors (Abdel-Salam and Mufti, 1998).

(b) Using the Finite-element Technique. The first step in the finite-element formulation of Poisson's equation is to divide the area of interest (around transmission-line conductors) into triangular elements forming a grid (Aboelsaad et al., 1989; Abdel-Salam and Al-Hamouz, 1992). Poisson's equation is then transformed into a set of linear equations by minimizing the energy functional expressed by equation (2.35) over an element. Summing up for all the elements of the grid results in a set of simultaneous equations whose solution determines the unknown potentials at the nodes. The electric field can be determined; see equation (2.58).

However, determination of the electric field in the presence of space charge must satisfy the continuity of the current density, equation (2.58). Therefore, the ionized field is divided into flux (stream) tubes. Two field lines define a flux tube as field lines never cross. The continuity equation can therefore be extended to all the flux tubes.

The finite-element formulation of Poisson's equation and the continuity equation of current density are solved iteratively for the electric potential and the space charge-density. As the space-charge density is known at the coronating-conductor surface, the current density is calculated using equation (2.58). The corona current is obtained by integrating the current density over the periphery of the coronating conductor(s). Multiplying the corona current by the applied voltage determines the corona power loss.

The calculated corona current agreed with those measured for monopolar (Abdel-Salam and Al-Hamouz, 1993, 1994, 1995b; Abdel-Salam et al., 1997; Al-Hamouz and Abdel-Salam, 1999) and bipolar lines (Abdel-Salam and Al-Hamouz, 1995a; Al-Hamouz et al., 1999).

5.5 CORONA NOISE

Corona noise includes interference with radio, television, and other wireless reception, caused by corona. Also, audible noise is experienced near EHV lines and substations. Corona undoubtedly interferes also with carrier signals transmitted along EHV lines.

The main source of corona radio noise is from positive streamers (Fig. 5.2), as their amplitudes are much higher than those of the negative Trichel pulses. As the pulses are random in amplitude, duration, and repetition rate, their noise is felt over a continuous spectrum. The noise level decreases at higher frequencies; this has been shown by both measurement and computation.

In measurements of the radio noise level and its lateral profile near bipolar HV DC lines, the highest level was recorded under the positive conductor whereas the noise contributed by the negative conductor was rather insignificant. For AC transmission lines, the lateral decay of radio noise is less steep than for DC lines.

In the case of AC corona, positive streamer pulses occur during the positive half-cycles and are also the main source of noise. At voltages slightly above the onset level, positive corona tends to take the form of a steady glow, assisted by the negative ions produced during the preceding negative half-cycles. Therefore, HV DC lines are usually more noisy than AC lines at voltage gradients slightly above the onset levels, particularly in fair weather.

5.5.1   Effect of Line Conductor Size

Radio interference (RI) measurements under both AC and DC EHV lines have shown that the RI level rises with the voltage gradient (i.e., field strength E_{max} at the HV conductor surface) according to the relation

$$RI = C(E_{max})^n \text{ dB} \tag{5.10}$$

C being a constant. For DC lines, the exponent n has a value of 5 to 7 in fair weather and 1.5 to 3.5 in rain. For AC lines, on the other hand, the exponent n is about 7 to 8 in both fair and foul weather.

For the same conductor voltage gradient, in both AC and DC lines, the RI level was found to increase with the conductor radius according to the relation

$$RI = C_1 r^2 \text{ dB} \tag{5.11}$$

This relation was found to be independent of conductor bundling.

5.6   SOLVED EXAMPLES

(1) Two parallel plates are spaced a distance d and are placed in SF_6 gas. One of the plates is stressed with a voltage V with respect to the other plate which is grounded. What would be the number of electrons in the avalanche growing between the plates? Neglect the effect of self-space charge of the avalanche on its growth.

Solution:
According to equations (3.41) and (3.44),

$$(\alpha - \eta)/p = A(E/p) + C, \qquad E = V/d$$

where

$A = 0.027$ and $C = -2400.4$

$$(\alpha - \eta) = AE + Cp \tag{5.12}$$

$$\text{Avalanche size} = \exp[(\alpha - \eta)d]$$
$$= \exp[AV + Cpd] \tag{5.13}$$

(2) In problem (1), consider $d = 1\,\text{mm}$ and the gas pressure $p = 1\,\text{atm}$. Calculate the breakdown voltage V_s
(a) according to equation (4.13),
(b) corresponding to an avalanche size of 10^8,
(c) according to the criteria expressed by equations (5.4) and (5.5).

Solution:
(a) As γ is much smaller than unity, equation (4.13) is reduced to $\alpha = \eta$. According to equations (3.41) and (3.44), $\alpha = \eta$ occurs when

$$0.027(E/p) = 2400.4 \tag{5.14}$$

where

$E = V_s/d$

As $p = 101.3$ kPa,

$\therefore V_s = 9005.9\,\text{V} = 9\,\text{kV}$

(b) $p = 101.3$ kPa. According to equation (5.13)

$10^8 = \exp[AV + Cpd]$

where

$Cp = -2400.4 \times 101.3 = -24.32 \times 10^4$

$18.42 = 0.027V_s - 24.32 \times 10^4 \times 10^{-3}$

$V_s = 9689.6\,\text{V} = 9.70\,\text{kV}$

(c) As the avalanche self-space charge is neglected, the breakdown voltage will be the same irrespective of the polarity of the stressed plate.

According to the criteria expressed by equations (5.4) and (5.5)

$V_{s+} = 9.4\,\text{kV}, \qquad V_{s-} = 9.2\,\text{kV}$

(3) Repeat problem (2) for gas pressures of 3 and 5 atmospheres and comment on the results obtained.

Solution:
(a) According to equation (5.14),

$$V_s = 2400.4\,pd/0.027$$
$$= 2400.4 \times 3 \times 101.3 \times 10^{-3}/0.027\,\text{V}$$
$$= 27\,\text{kV at }p = 3\,\text{atm}$$

$$V_s = 2400.4 \times 5 \times 101.3 \times 10^{-3}/0.027\,\text{V}$$
$$= 45.03\,\text{kV at }p = 5\,\text{atm}$$

(b) According to equation (5.13),

$$10^8 = \exp[AV + Cpd]$$

$$18.42 = 0.027V_s - 24{,}004 \times 3 \times 101.3 \times 10^{-3}$$
$$V_s = 27.7\,\text{kV at }p = 3\,\text{atm}$$

$$18.42 = 0.027V_s - 24{,}004 \times 5 \times 101.3 \times 10^{-3}$$
$$V_s = 45.71\,\text{kV at }p = 5\,\text{atm}$$

(c) According to the criteria expressed by equations (5.4) and (5.5),

$$V_{s+} = 27.5\,\text{kV}, \qquad V_{s-} = 27.73\,\text{kV at }p = 3\,\text{atm}$$
$$V_{s+} = 45.2\,\text{kV}, \qquad V_{s-} = 45.5\,\text{kV at }p = 5\,\text{atm}$$

It is quite clear that the increase of gas pressure improves the dielectric strength of the gas since the breakdown voltage increases with gas pressure.

(4) Two concentric spheres have radii a and b, $b > a$. The interelectrode spacing is filled with SF_6 gas. The inner sphere is stressed with a voltage V with respect to the outer sphere, which is grounded. Determine the thickness of the ionization zone surrounding the inner sphere and the maximum size of the avalanche growing in this zone. Neglect the effect of self-space charge of the avalanche on its growth.

Solution:
At the ionization zone boundary $\alpha - \eta = 0$ where $E = E_i$. According to equation (5.12), $E_i = -Cp/A$ when $\alpha - \eta = 0$. With reference to Table 2.1,

$$E = \frac{V_{ab}}{r^2(b-a)} \tag{5.15}$$

$$-Cp/A = \frac{V_{ab}}{r_i^2(b-a)}$$

where r_i is the radius defining the ionization zone boundary.

$$r_i = \sqrt{\frac{V_{ab}A}{(b-a)Cp}} \tag{5.16}$$

Substituting E from equation (5.15) into equation (5.12),

$$\alpha - \eta = \frac{AV_{ab}}{r^2(b-a)} + Cp$$

The maximum avalanche size occurs when the avalanche grows over the whole thickness of the ionization zone.

$$\begin{aligned}
\text{Maximum avalanche size} &= \exp\left[\int_a^{r_i} (\alpha - \eta)dr\right] \\
&= \exp\left[\int_a^{r_i} \left(\frac{AV_{ab}}{r^2(b-a)} + Cp\right)dr\right] \\
&= \exp\left\{\frac{AV_{ab}}{(b-a)}\left[\frac{1}{a} - \frac{1}{r_i}\right] + Cp(r_i - a)\right\}
\end{aligned} \tag{5.17}$$

(5) In problem (4), consider $a = 0.1\,\text{cm}$ and $b = 2.1\,\text{cm}$ and the gas pressure $p = 1\,\text{atm}$. Calculate the corona onset voltage

(a) corresponding to avalanche size of 10^8,
(b) according to the criteria expressed in equations (5.4) and (5.5).

Solution:
(a) Avalanche size is expressed by equation (5.17) which depends on the applied voltage V and the corresponding thickness r_i of the ionization zone; the latter depends also on V, as expressed by equation (5.16).

In equation (5.12) relating to $(\alpha - \eta)$,

$$A = 0.027, \qquad C = -2400.4$$

Given $a = 0.1 \times 10^{-2}\,\text{m}$, $b = 2.1 \times 10^{-2}\,\text{m}$, $p = 101.3\,\text{kPa}$, solve equations (5.16) and (5.17) iteratively to get V_0 (corresponding to an avalanche size of 10^8) $= 13.92\,\text{kV}$.

(b) According to the criteria expressed in equations (5.4) and (5.5),

$$V_{0+} = 13.1\,\text{kV and } V_{0-} = 13.7\,\text{kV}$$

(6) Repeat problem (5) for gas pressures of 3 and 5 atmospheres and comment on the results obtained.

Solution:
(a) For $p = 3 \times 101.3\,\text{kPa}$, solve equations (5.16) and (5.17) iteratively to get V_0 (corresponding to an avalanche size of 10^8) = 34.6 kV.
   For $p = 5 \times 101.3\,\text{kPa}$, solve equations (5.16) and (5.17) iteratively to get $V_0 = 54\,\text{kV}$.
(b) According to the criteria in equations (5.4) and (5.5),

$$V_{0+} = 34.9\,\text{kV}, \qquad V_{0-} = 35.1\,\text{kV at } p = 3\,\text{atm}$$
$$V_{0+} = 53.8\,\text{kV}, \qquad V_{0-} = 54.5\,\text{kV at } p = 5\,\text{atm}$$

It is quite clear that the increase of gas pressure results in increasing the corona onset voltage.

(7) Different flat arrangements of a three-phase transmission line are shown in Fig. 5.13. Each phase of the line has

(a) single conductor, Fig. 5.13a;
(b) bundle-2 conductor comprising two subconductors arranged in horizontal or vertical configuration, Fig. 5.13b;
(c) bundle-3 conductor with the three subconductors placed at the vertices of an upright or inverted triangle, Fig. 5.13c;
(d) bundle-4 conductor with the four subconductors placed at the vertices of a square, Fig. 5.13d;
(e) bundle-4 conductor with the four subconductors placed at the vertices of a diamond-form square, Fig. 5.13e.

Conductor radius = r, subconductor-to-subconductor spacing = s, phase-to-phase spacing = D, phase voltage = V. Calculate the corona onset voltage.

Solution:
It is evident that the surface field strength is higher at the middle phase than at the outer phases. The difference is normally about 7% for practical line dimensions.

   The maximum field E_m values at the subconductor surface of the middle-phase of the line arrangements of Fig. 5.13 have already been expressed (Jha, 1977), where the effect of ground is disregarded (conductor height above ground $\gg$ phase-to-phase spacing D). The corona onset voltage is determined by the onset field E_0.

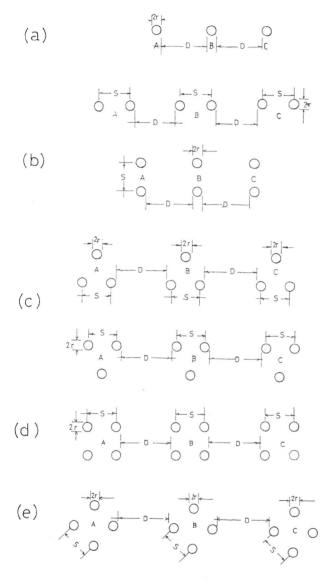

Figure 5.13 Three-phase transmission-line arrangements: (a) using single conductor; (b) using bundle-2 conductor arranged horizontally or vertically; (c) using bundle-3 conductor arranged at vertices of an upright or inverted triangle; (d) using bundle-4 conductor arranged at vertices of a square; (e) using bundle-4 conductor arranged at vertices of a diamond-form square.

(a) $E_m = \dfrac{V}{r \ln(D/r)}$

$\therefore$ Corona onset voltage $V_0 = E_0 r \ln(D/r)$ (5.18a)

(b) $E_m = \dfrac{V\left(1 + \dfrac{2r}{s}\right)}{2r \ln \dfrac{D}{\sqrt{sr}}}$

$\therefore$ Corona onset voltage $V_0 = 2E_0 r \ln \dfrac{D}{\sqrt{sr}} \Big/ \left(1 + \dfrac{2r}{s}\right)$ (5.18b)

(c) $E_m = \dfrac{V\left(1 + \dfrac{3\sqrt{3}r}{s}\right)}{3r \ln \dfrac{D}{\sqrt[3]{rs^2}}}$

$\therefore$ Corona onset voltage $V_0 = 3E_0 r \ln \dfrac{D}{\sqrt[3]{\sqrt{2}rs^2}} \Big/ \left(1 + \dfrac{3\sqrt{3}r}{s}\right)$

(5.18c)

(d) $E_m = \dfrac{V\left(1 + \dfrac{4\sqrt{3}r}{s}\right)}{3r \ln \dfrac{D}{\sqrt[4]{\sqrt{2}rs^3}}}$

$\therefore$ Corona onset voltage $V_0 = 4E_0 r \ln \dfrac{D}{\sqrt[4]{\sqrt{2}rs^3}} \Big/ \left(1 + \dfrac{4\sqrt{2}r}{s}\right)$

(5.18d)

(e) $E_m = \dfrac{V\left(1 + \dfrac{3\sqrt{2}r}{s}\right)}{4r \ln \dfrac{D}{\sqrt[4]{\sqrt{2}rs^3}}}$

$\therefore$ Corona onset voltage $V_0 = 4E_0 r \ln \dfrac{D}{\sqrt[4]{\sqrt{2}rs^3}} \Big/ \left(1 + \dfrac{3\sqrt{2}r}{s}\right)$

(5.18e)

(8) For the arrangements of problem (7), calculate the positive and negative corona onset voltages at natural temperature and pressure if the
subconductor radius $r = 1\,\text{cm}$, subconductor-to-subconductor spacing
$s = 40\,\text{cm}$, and phase-to-phase spacing $D = 5\,\text{m}$. Use the empirical formula reported in Section 5.3.

Solution:
According to the empirical formula

$$E_0 = 30\delta(1 + 0.3/\sqrt{\delta r})\,\text{kV}_{\text{peak}}/\text{cm}$$

where r is the conductor radius in cm.
At standard temperature and pressure, $\delta = 1$.

$$E_{0+} = 39\,\text{kV}_{\text{peak}}/\text{cm}$$

Using equations (5.18), the positive and negative corona onset voltages
are calculated as:

(a) $V_0 = 6.2 \times 39 = 241.8\,\text{kV}_{\text{peak}}$
              $= 171\,\text{kV}$

(b) $V_0 = 8.32 \times 39 = 342.48\,\text{kV}_{\text{peak}}$
              $= 242.2\,\text{kV}$

(c) $V_0 = 9.97 \times 39 = 388.83\,\text{kV}_{\text{peak}}$
              $= 274.9\,\text{kV}$

(d) $V_0 = 11.79 \times 39 = 459.81\,\text{kV}_{\text{peak}}$
              $= 325.1\,\text{kV}$

(e) $V_0 = 12.16 \times 39 = 474.24\,\text{kV}_{\text{peak}}$
              $= 335.3\,\text{kV}$

It is quite clear that the onset voltage increases with the increase of the
number of subconductors per bundle.

(9) The conductors of a three-phase transmission line are placed at the
vertices of an upright triangle of sides 5 m, 5 m, and 8.66 m. The radius
of conductors $r = 1\,\text{cm}$. Calculate the corona onset voltage at standard
temperature and pressure.

Solution:
According to equation (5.18a)

$$V_0 = E_0 r \ln(D_{eq}/r)$$

D_{eq} = mean geometric distance between the conductors
$$= \sqrt[3]{5 \times 5 \times 8.66} = 6\,\text{m}$$

The corona onset field $E_0 = 30\delta(1 + 0.3/\sqrt{\delta r})$
At $\delta = 1$ at standard temperature and pressure,

$$E_0 = 39\,\text{kV}_{peak}/\text{cm}$$

$$V_0 = 39 \ln 600 = 249.5\,\text{kV}_{peak}$$
$$= 176.4\,\text{kV} (= 305.5\,\text{kV line-to-line})$$

(10) For the three-phase transmission line of problem (9), calculate corona power loss and the corona current when the line is operating at 50 Hz, 275 kV. Assume smoothness m_1 and weather m_2 coefficients are equal to 0.92 and 0.95, respectively. Atmospheric pressure and temperature are respectively equal to 75 cm Hg and 35° C.

Solution:

Relative air density $\delta = \dfrac{3.92p}{273 + t}$

where t is the temperature in °C and p is the pressure in cm Hg.

$$\delta = \frac{3.92 \times 75}{273 + 35} = 0.95$$

The corona onset field $E_0 = 30\delta(1 + 0.3/\sqrt{\delta r})m_1 m_2$
$$= 30 \times 0.95(1 + 0.3/\sqrt{0.95})0.92 \times 0.95$$
$$= 32.58\,\text{kV}_{peak}/\text{cm}$$

$$V_0 = 32.58 \ln 600 = 208.4\,\text{kV}_{peak}$$
$$= 147.37\,\text{kV}(255.25\,\text{kV line-to-line})$$

$$V_0 = 275/255.25 = 1.078$$

The corresponding value of the K factor (Fig. 5.10) = 0.05. According to Peterson's formula (5.6),

$$\text{Corona power loss } P_c = \frac{3.73 \times 0.05}{(D_{eq}/r)^2} fV_{ph}^2 \times 10^{-5} \text{ kW/(cond.km)}$$

$$= \frac{3.73 \times 0.05 \times 50 \times (275 \times 10^3/\sqrt{3})^2}{(600)^2} \times 10^{-5}$$

$$= 6.53 \text{ kW/(cond.km)} = 19.6 \text{ kW/km}$$

$$\text{Corona current } = P_c/V_{ph} = 6.53 \times 10^3/(275 \times 10^3/\sqrt{3}) \text{ A}$$
$$= 41.1 \text{ mA/km}$$

(11) A single conductor, 3.175 cm radius, of a 525 kV line is strung 13 m above ground. Calculate (a) the corona onset voltage and (b) the effective radius of the corona envelope, at a voltage of 2.5 p.u. (i.e., 2.5 times rated voltage). Consider smoothness m_1 and weather m_2 factors are both equal to 0.9. Relative air-density factor $= 1$.

Solution:

(a) $E_0 = 30\delta(1 + 0.3/\sqrt{\delta r})m_1 m_2 = 30(1 + 0.3/\sqrt{3.178})0.9 \times 0.9$
$$= 28.4 \text{ kV}_{peak}/\text{cm}$$
$$= 20 \text{ kV/cm}$$

$$V_0 = E_0 r \ln(2h/r) = 426 \text{ kV}$$
$$= 426\sqrt{3} = 737.8 \text{ kV (line-to-line)}$$

At 525 kV line-to-line operating voltage, there is no corona present.
(b) 2.5 p.u. voltage $= 2.5 \times 525 = 1312.5$ kV (line-to-line), which is higher than the onset voltage, so corona is present on the conductor. The corona envelope around the conductor is considered an extension to its radius.

When considering the effective radius of the corona envelope r_e, one may assume a smooth surface of the envelope.

$$1312.5\sqrt{2}/\sqrt{3} \text{ kV}_{peak \text{ per phase}} = E_0 r_e \ln(2H/r_e)$$
$$= 30(1 + 0.3/\sqrt{r_e})r_e \ln(2600/r_e)$$

where r_e is the envelope radius in cm.
A trial and error solution yields $r_e = 5$ cm. This represents an increase in conductor radius r by about 57%.

REFERENCES

Abdel-Salam M. ETZ-Archiv 99(5):271–275, 1978.
Abdel-Salam M. IEEE Trans IA-21:35–40, 1985.
Abdel-Salam M, Abdel-Aziz EZ. Proceedings of IEEE–IAS Annual Conference, Houston, TX, 1992, pp 1627–1631.
Abdel-Salam M, Abdel-Aziz EZ. J Phys D: Appl Phys 27:807–817, 1994a.
Abdel-Salam M, Abdel-Aziz EZ. J Phys D: Appl Phys 27:2570–2579, 1994b.
Abdel-Salam M, Abdel-Aziz EZ. Electric Power Sys Res 50, 2000 (to appear).
Abdel-Salam M, Abdel-Sattar S. IEEE Trans EI-24:669–679, 1989.
Abdel-Salam M, Al-Hamouz Z. J Phys D: Appl Phys 25:1551–1555, 1992.
Abdel-Salam M, Al-Hamouz Z. J Phys D: Appl Phys 26:2202–2211, 1993.
Abdel-Salam M, Al-Hamouz Z. Proc IEE–A 141:369–378, 1994.
Abdel-Salam M, Al-Hamouz Z. IEEE Trans IA-31:447–483, 1995a.
Abdel-Salam M, Al-Hamouz Z. IEEE Trans IA-31:484–493, 1995b.
Abdel-Salam M, Khalifa M. Proceedings of 13th International Conference on Phenomena in Ionized Gases, Berlin, 1977, Part I, pp 435–436.
Abdel-Salam M, Mufti AH. Electric Power Sys Res 44:145–154, 1998.
Abdel-Salam M, Shamloul D. IEEE Trans EI-27:352–361, 1992.
Abdel-Salam M, Shamloul D. COMPEL, 12:143–156, 1994.
Abdel-Salam M, Stanek EK. IEEE Trans IA-24:1025–1030, 1988.
Abdel-Salam M, Farghally M, Abdel-Sattar S. IEEE paper 82WM-212-9, 1982.
Abdel-Salam M, Farghally M, Abdel-Sattar S, Shamloul D. Proceedings of 4th International Symposium on Gaseous Dielectrics, Knoxville, TN, 1984, pp 492–497.
Abdel-Salam M, Al-Hamouz Z, Mufti AH. J. Electrostatics 39:129–144, 1997.
Aboelsaad MM, Shafai L, Rashwan MM. Proc IEE-A 136:33–40, 1989.
Al-Hamouz Z, Abdel-Salam M. IEEE Trans IA-35:380–386, 1999.
Al-Hamouz Z, Abdel-Salam M, Mufti AH. IEEE Trans IA-34:301–309, 1998.
Begamudre RD. Extra High Voltage AC Transmission Engineering. New York: John Wiley & Sons, 1986.
Berg W, Hauffe K. Current Problems in Electrophotography. Berlin: Walter de Gruyter & Company, Mouton Publishers, 1972.
Clade JJ, Gary CH, Lefevre CA. IEEE Trans PAS-88:695–703, 1969.
Electrical Power Research Institute. Transmission Line Reference Book, 345 kV and Above. Palo Alto, CA: Electrical Power Research Institute, Project UHV, 1979.
Elmoursi AA, Castle GSP. IEEE Trans IA-22:80–85, 1986.
Elmoursi AA, Speck CE. IEEE Trans IA-26:384–392, 1990.
Giao T, Jordan J. IEEE publication 31-C-44, 1967, pp 5–15.
Horenstein MH. IEEE Trans IA-20:1607–1612, 1984.
Jha Rs. A Course in High Voltage Engineering. Delhi-Jullundur: Dhanpat Rai & Sons, 1977.
Khalifa M. Proceedings of 3rd International Symposium on High Voltage Engineering, Milan, Paper 53-05.
Khalifa M, Abdel-Salam M. Proc IEE 120:1574–1575, 1973.
Khalifa M, Abdel-Salam M. IEEE Trans PAS-93:720–726, 1974a.

Khalifa M, Abdel-Salam M. IEEE Trans PAS-93:1693–1699, 1974b.

Khalifa M, Abdel-Salam M, Radwan R, Ali Kh. IEEE Trans PAS-96:886–895, 1977.

Landham E, Dubard J, O'Brien M, Lindsey C, Plulle W. Proceedings of IEEE Annual Meeting on Electrostatic Processes, Atlanta, GA, 1987, Paper 8-2.

Loeb L. Electrical Coronas—Their Basic Physical Mechanisms. Berkeley, CA: University of California Press, 1965.

Nasser E. Fundamentals of Gaseous Ionization and Plasma Electronics. New York: Wiley-Interscience, 1971.

Rein A, Arnesen A, Johansen I. IEEE Trans PAS-96:945–954, 1977.

Seelentag W. Electrostatic imaging. In: Watson W, ed. IEEE Medical Electronics Monographs 28–33. Stevenage, UK: Peter Peregrinus, 1979.

Shamloul D. PhD thesis, Assiut University, Assiut, Egypt, 1989.

Waters, RT, Richard TE, Stark WB. Proc IEE 119:717–723, 1972.

Zeitoun A, Abdel-Salam M, El-Ragheb M. IEEE paper A-76-418-4, 1976.

6

The Arc Discharge

M. KHALIFA *Cairo University, Giza, Egypt*

M. ABDEL-SALAM *Assiut University, Assiut, Egypt*

6.1 INTRODUCTION

Arc discharge is encountered in the everyday use of power equipment. Whenever a circuit breaker or a load-break switch is opened while carrying a current, an arc strikes between its seperating contacts. A persistent fault in a transformer, machine, or cable would eventually involve an arc. Therefore, information about the characteristics of arcs, the contact erosion they cause, and the factors conducive to their extinction is essential for the proper design, operation, and protection of such high-voltage equipment. Applications of the electric arc and its plasma include arc-discharge lamps, some furnaces, and processes in the manufacture of pure metals and some electronic devices.

6.2 ARCS IN CIRCUIT BREAKERS

When opening a switch or breaker, the contacts move apart and the contact area decreases rapidly until finally the contacts are physically separated. When the contact area decreases to a very small spot, the contact resistance increases considerably while the flowing current becomes highly concentrated. For a circular spot of contact with radius r, the contact resistance equals $\rho/2r$, ρ being the resistivity of the contact material. Thus, for a circular spot of radius $10\,\mu m$ and a current of $10\,A$, the current density reaches about $3 \times 10^6 A \cdot cm^{-2}$. If the contacts are made of pure copper, the contact resistance can exceed $0.1\,\Omega$. The corresponding power loss at such a microscopic spot would suffice for melting and even evaporating the hemisphere of metal at the contact spot within a period on the order of $1\,\mu s$.

The metal vapor filling the space between the parting contact spots would thus furnish the conducting medium for the circuit current to flow in the form of an arc. If the electric circuit is such that no stable arc can exist, the molten metal bridging the microscopic gap between the contact spots will soon be broken as the spots part further (Fig. 6.1). Also, when the breaker or switch is closed under voltage, a spark will occur because the contacts get very close. The very short arc that ensues soon gets extinguished.

6.3 REGIONS OF THE ARC

In air circuit breakers the arc burns mainly in an atmosphere of air; it burns mainly in hydrogen in oil circuit breakers; and it burns in composites of sulfur and fluorine in SF_6 circuit breakers. It burns mainly in an atmosphere of metal vapor in the case of vacuum circuit breakers. Under all these conditions the arc is composed of three principal regions: the cathode and anode regions and the arc column (Fig. 6.2), no matter what the total arc length is. Through all three regions the current is carried by electrons and ions. In a steady arc, a balance is struck between power input and losses.

6.3.1 Cathode Region

Electrons emitted from the cathode spot can be produced mainly by thermionic emission if the cathode is made of a high-melting-point metal (e.g., carbon, tungsten, or molybdenum). With cathodes of low melting point, electrons can be supplied by field emission from points of microroughness where the electric field is highly concentrated. An additional important source of electrons at the cathode is ionized metal vapor. This occurs in vacuum and other types of circuit breaker, and in mercury-vapor lamps.

The current is also carried partially by positive ions drifting slowly to the cathode from the plasma of the arc column. In the space between the surface of the cathode spot and the cloud of positive ions there is a high electric field (Fig. 6.2). Therefore, a significant cathode voltage drop builds up over the cathode region. Its magnitude and the width of the region depend on the arc current, the medium in which it is burning, and the cathode material.

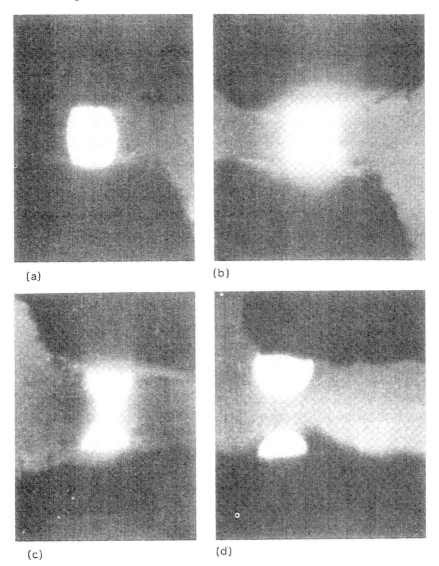

(a)

(b)

(c)

(d)

Figure 6.1   Development of a molten metal bridge between separating iron contacts while carrying 40 A. Steps: (a), (b), (c), (d). Contact separation in step (a) is 0.5 mm. [Courtesy of F. Llewellyn-Jones (1957) and Oxford University Press.]

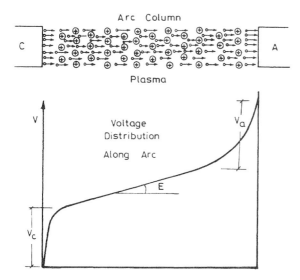

Figure 6.2 Electrons and positive ions in the arc and its longitudinal voltage distribution.

6.3.2 Anode Region

The electrons from the arc plasma bombarding the anode spot and delivering all their energies keep the anode spot at a very high temperature. Positive ions are produced at the anode by thermal ionization of the gas and of any metal vapor near the anode spot. These ions drift slowly away from the anode into the arc plasma. Space-charge distribution near the anode produces a nonlinear field in the area and an anode voltage drop (Fig. 6.2).

Because of the different distributions of space charges at the cathode and anode, the cathode drop takes place across a considerably shorter distance than the anode drop (Fig. 6.2). The cathode region, however, covers a sufficient number of free paths for the electrons leaving the cathode to reach the level of ionization by collision. They thus liberate more electrons from the gas and metal vapor in the arc column.

6.3.3 Arc Column

At an arc's very high temperature, on the order of 10^4 K, the gas molecules are mostly dissociated. Many of their atoms and those of the metal vapor present will be ionized (Chapter 3). The degree of thermal ionization ζ of a gas depends on its temperature T, pressure p, and ionization potential V_i according to the well-known Saha relation

$$\frac{\zeta^2}{1-\zeta^2}p = A T^{2.5} \exp\frac{-eV_i}{kT} \tag{6.1}$$

where e is the electronic charge and k is Boltzmann's constant. The magnitude of the constant A depends on the units used.

Inside the column the densities of positive and negative charge carriers are equal on the average and are comparable to that of neutral gas molecules. Therefore, the arc plasma exerts no electrostatic field. It has a significant electrical conductivity. Measurements have indicated that the conductivities of several gases (e.g., N_2, H_2, SF_6) are insignificant up to about 5000 K, rise steadily to about 30 S·cm^{-1} at 10^4 K, and are about 80 S·cm^{-1} at 2×10^4 K (Flurscheim, 1975). No doubt the extreme temperatures and high conductivities are confined to the core of the arc column. Both decrease sharply at some radius beyond which there is no current conduction to speak of. This effective radius of the arc column is a function of the arc current and the ambient gas and its pressure. The temperature distributions over the cross-sections of 80 A arcs burning in O_2, N_2, and SF_6 are shown schematically in Figure 6.3. The interaction between the arc column and its surrounding ambient takes the form of diffusion of charge barriers and heat transfer.

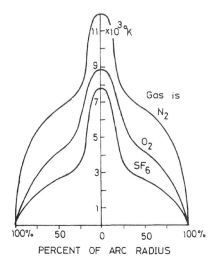

Figure 6.3 Radial temperature distribution over the cross-section of an arc as dependent on the ambient gas. [From Frind (1960).]

6.4 ENERGY BALANCE IN A STEADY ARC

If the arc carries a DC current of constant or slowly varying magnitude and is not subjected to turbulence in its environment, it can be described as a steady arc. Let us now look at the power balance for each region of this arc.

6.4.1 Cathode Region

The cathode spot receives both electrical and thermal power inputs, P_1 and P_2.

$$P_1 = aI_i(V_c + V_i) \tag{6.2}$$

where V_c is the cathode drop (Fig. 6.2), V_i the gas ionization potential, I_i the current component delivered by ions, and the factor a represents the fraction of the ions' energy delivered to the cathode spot and not to the arc plasma.

$$P_2 = G\beta(T_g - T_c) \tag{6.3}$$

where G is a geometrical factor, β the coefficient of heat transfer, and T_g and T_c are the temperature of the arc column near the cathode and that of the cathode spot, respectively. These two inputs are balanced by four components of power loss from the cathode spot: electron emission, heat conduction into the cathode volume, metal evaporation, and radiation from the cathode spot.

6.4.2 Anode Region

The input to the anode spot is the sum of the potential energy of the electrons falling through the anode drop, their kinetic energy delivered on impact, and the heat input to the anode spot. This input is consumed as heat conduction, radiation, and evaporation of metal from the anode spot. The sizes of the anode and cathode spots are functions of the arc current, the electrode material, and the ambient gas pressure.

6.4.3 Arc Column

Here the input is purely electrical. The power input per unit length of the arc column is $E(I_e + I_i)$, E being the voltage gradient along the arc, and I_e and I_i the components of the arc current carried by electrons and ions. The current is mostly electronic since ion mobilities are about three orders of magnitude lower than that of electrons. The densities and mobilities of these charge carriers are functions of the arc column temperature.

The arc column loses some of its charge carriers by diffusion from its surface. It also loses heat by radiation and convection. The components of heat dissipation depend on the ambient. For example, in oil circuit breakers, energy is consumed in boiling some oil, dissociating some of its molecules to produce hydrogen and hydrocarbon gases, expanding the gas bubble, and dissociating and ionizing some of the gas molecules.

Along the arc column, the power input maintains the ionization in the arc plasma, and compensates for the losses at the periphery and for any changes in the heat stored in the column. Because of the different ionization potentials and thermal conductivities and other thermal parameters, it was noted that, for the same current in the range 10 to 1000 A, the arc column gradient E in hydrogen was about 10 times its value in oxygen when the arc was contained in a tube 2 cm in diameter. The current density varies along the axis of the arc column. This induces gas flow and static pressure gradients along the arc. This in turn can lead to ejection of molten metal and vapor from the electrode spots.

It can be shown that there is an interdependence between the arc current, column temperature, and radius. The arc always adjusts itself so that the system in equilibrium has minimum entropy. The arc column radius r and temperature T are such that the heat loss is minimal. Thus for a given arc current the column voltage gradient E is also a minimum consistent with the power balance mentioned above. Thus

$$\frac{\mathrm{d}E}{\mathrm{d}r} = 0 \qquad \text{and} \qquad \frac{\mathrm{d}E}{\mathrm{d}T} = 0$$

In other words, for a given arc current I, if the arc column cross-section were to decrease, its conductance would decrease and the voltage gradient E would increase. If, on the other hand, the cross-section were to increase, its heat loss to the surroundings would increase, its temperature would decrease and so would its conductivity, and thus E would increase.

Experimental tests have shown that the arc column radius r is proportional to I^n, with n ranging between 0.25 and 0.6, depending on the type of cooling for the arc. In air-blast breakers it is volumetric cooling, whereas for a stationary arc cooling is effected at its column surface.

6.5 STEADY-STATE ARC CHARACTERISTICS

As is evident from the previous discussion, the arc is by no means a simple element. The relation between the arc voltage and current depends on the arc length, electrode material, and ambient. The vacuum arc voltage was shown to depend principally on the cathode metal (Reece, 1975).

For arcs of fixed length in atmospheric air, the V–I characteristics follow inverse relations of the form shown in Figure 6.4. An approximate relation between the arc column voltage gradient E and its current I can be obtained as follows:

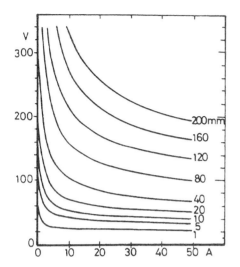

Figure 6.4 DC voltage–current characteristics for arcs of different lengths burning in air between copper electrodes. [From Rieder (1967).]

$$I = E[\pi r^2 (N_e k_e + N_i k_i)] \tag{6.4}$$

The electron and ion densities N_e and N_i and their respective mobilities k_e and k_i are functions of temperature. Thus equation (6.4) can be rewritten as

$$I = E(\pi r^2) F_1(T) \tag{6.4a}$$

Under steady conditions, the input power is balanced by heat dissipation from the arc column surface, another function of temperature $F_2(T)$. Thus

$$E^2(\pi r^2) F_1(T) = 2\pi r F_2(T) \tag{6.5}$$

Eliminating r and differentiating, we get

$$E = CI^{-1/3} \tag{6.6}$$

where C is a constant.

Many experimental results on DC arcs in air fit relations between the total arc voltage V and current I of the form

$$V = a + bl + (c + dl)I^{-1} \tag{6.7}$$

For arc currents up to 20 A and lengths l of about 5 cm, the constants of this relation have the following values: $a = 17\,\text{V}$, $b = 22\,\text{V/cm}$ $c = 20\,\text{W}$, and $d = 180\,\text{W/cm}$.

For the same current and length, the arc voltage increases with ambient pressure above atmospheric. In a rarefied atmosphere of about 10 Pa, however, and with a high enough source voltage, the arc becomes unstable and turns into a glow discharge. We thus get the well-known $V–I$ characteristic shown in Figure 6.5 (Nasser, 1971). This is experienced in some gas-discharge lamps and tubes.

As the arc current is reduced to a very small value, a limit is reached when the arc input power is no longer sufficient to maintain the column temperature. The electron and ion densities in the arc plasma decrease so drastically that there is not enough conductivity to carry the current through the arc, and it is extinguished.

6.6 MAGNETIC PHENOMENA IN ARCS

The arc is influenced by two magnetic fields, its own field and that of its feeding circuit. The circumferential field produced by the arc's current does exert a pressure to squeeze the arc column. The pressure can easily be calculated as being equal to $I^2/\pi r^2$. At the anode and cathode spots the arc radius is much smaller than anywhere along the column. Therefore,

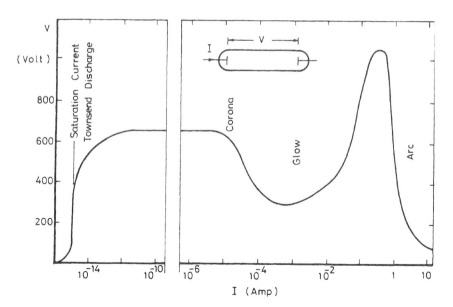

Figure 6.5 Typical voltage–current characteristic for the various discharges in a gaseous gap. Neon is at a pressure of 130 Pa. The gap is 50 cm between disk electrodes 2 cm in diameter. (Courtesy of E. Nasser and Wiley-Interscience.)

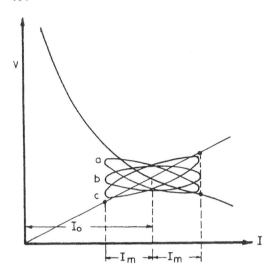

Figure 6.6 Voltage–current characteristic of a DC arc with a superposed current modulation of amplitude I_m: (a) low frequency; (b) medium frequency; (c) high frequency.

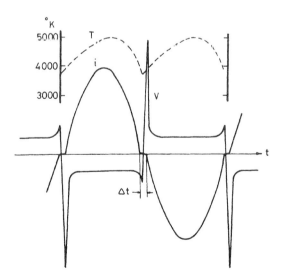

Figure 6.7 Time variations of current, voltage, and temperature for a 50 Hz, 10 A arc, 3 mm long, in air between copper electrodes.

there is a high axial gradient of that pressure which can set up jets of plasma and metal vapor from these spots with velocities reaching 1 km/s (Barrault et al., 1972). Also, as current-carrying circuits under their own magnetic fields tend to increase their self-inductance, the arc will bow outward into a larger loop. Such phenomena are usually exploited in the design of DC circuit breakers for accelerating arc extinction (Chapter 11).

6.7 DYNAMIC ARC CHARACTERISTICS

Because of the thermal capacity of the arc column, a sudden rise in the magnitude of an otherwise steady arc current will have to be accompanied by an initial rise in the arc voltage so as to furnish the extra energy needed for building up the column ionization to the level corresponding to the new current magnitude. After some time, termed the thermal time constant θ of the arc, the arc voltage settles down to a steady value according to the V–I characteristic. Of course, the opposite will occur if the arc current suddenly drops. This can be visualized by thinking of the power balance for a unit length of the arc column. There, the input power

$$P = EI = H + \frac{dQ}{dt} \tag{6.8}$$

where H is the heat loss per second and Q is the heat stored in the thermal capacity.

If the current variation is an AC modulation superposed on the DC current magnitude I_0, the corresponding voltage variation will be such that the operating point will follow a loop (Fig. 6.6). The loop shape depends on the frequency f of the current modulation. At extremely low frequencies the variation will almost follow the static characteristic. On the other hand, at extremely high frequencies the arc behaves like a linear resistance. Its thermal capacity prevents its temperature from varying at any significant fraction of such frequencies.

6.7.1 AC Arc Characteristics

Near the peak value of the arc current the voltage necessary to maintain it is relatively low (Fig. 6.7). As the current approaches zero, a higher and higher voltage is needed to maintain the arc. When the voltage across the arc is not high enough, it will be unstable and will become extinguished even before the zero point of the sinusoid (Fig. 6.7). For the current to flow in the opposite direction in the following half-cycle, the arc gap has to break down again under a sufficiently high voltage. Therefore, the arc voltage at the beginning of the half-cycle is considerably higher than that at its end.

There is also a period Δt of effectively zero current around the virtual zero point of the sinusoid (Fig. 6.7). The arc column temperature also varies during the AC cycle; the peak temperature lags behind the peak current because of the arc's thermal capacity.

6.8   THE ARC AS A CIRCUIT ELEMENT

Electric circuits subjected to analysis sometimes contain arcs, as in faulty power systems or when arc furnaces are included. Accurate analysis requires a truly representative circuit to take the place of the arc. The arc is by no means a simple circuit element. This applies to both AC and DC circuits. According to DC arc characteristics such as the ones in Figure 6.4, the ratio V/I has a positive magnitude (the same as for a metal resistance), and both consume power. However, the arc equivalent resistance varies with the current and dV/dI is negative. Therefore, a stabilizing impedance must be included in series with the arc in both AC and DC circuits (Fig. 6.8).

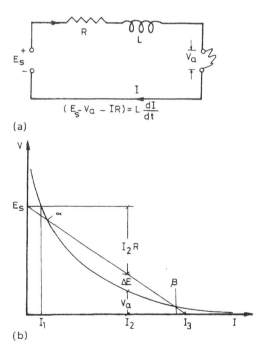

(a)

$$(E_s - V_a - IR) = L \frac{dI}{dt}$$

(b)

Figure 6.8   Arc as an element in a DC circuit: (a) arc in the circuit; (b) voltages across the arc and the circuit (the arc is stable at the point β but not at α).

Looking at the arc's *V–I* characteristic, it can easily be proved that stable operating conditions are represented by point β (Fig. 6.8b). If the current happens to swing below point α it will become extinguished.

For arcs of very small length, the voltage does not vary significantly over a wide current range, which means that the arc can be represented by a fixed back voltage opposing the source (Fig. 6.9). The equivalent circuit also comprises a resistance and an inductance that are functions of the arc length *l*, its current *I*, and the current's rate of change with time *I′*. The magnitudes of these parameters can be obtained from experimental tests. Such a circuit (Fig. 6.9a) would apply for both DC and AC circuits during periods when the arc is burning. During periods of effective current zero, Δt (Fig. 6.7), however, the arc can be represented by a high resistance that increases with time. The nonlinear and irregularly time-varying characteristics of arcs produce current harmonics and voltage ripples in the feeding network, which can be represented approximately.

6.9 ARC INTERRUPTION

The situation with DC is basically different from that under AC.

6.9.1 DC Case

In the DC case the arc current is forcibly brought down to zero. To do that, as will be described in Chapter 11, the arc is driven rapidly by its thermal buoyancy and magnetic field to extend its length and to subdivide into a number of partial arcs. Thus the voltage required to maintain it would increase rapidly. The time for arc interruption can be estimated given the

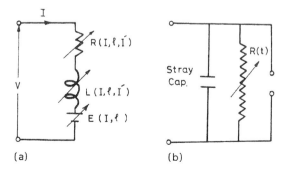

(a) (b)

Figure 6.9 Equivalent circuit of the arc: (a) cases of DC and AC during the burning period (*R*, *L*, and *E* are variables); (b) case of AC during the extinction period (Δt in Figure 6.7).

electric circuit and breaker characteristics. The voltage and current oscillograms would look like those represented in Figure 6.10. It should be noted
that the circuit inductance acts to delay the arc interruption. On the other
hand, a resistance or capacitance across the arc would accelerate its
interruption, as it would draw an additional current through the circuit
resistance.

6.9.2 AC CASE

Due to its thermal capacity, the conductance of the postarc gap cannot
vanish instantly at current zero. Some time will be needed for the electrode
spots and the gases in the postarc gap to cool down and for the electrons
and ions of the arc plasma to recombine and/or diffuse. The conductance of
the gap drops during the first few microseconds after arc extinction, and the
gap's dielectric strength builds up in the subsequent tens of microseconds
(Buttler and Whittaker, 1972).

The behavior of the postarc gap was originally represented by Cassie
and Mayr and later improved by Browne and others (Thomas et al., 1995).

6.9.2.1 Cassie and Mayr Models

Cassie (1932) and Mayr (1934), working independently, formulated two
differential equations based on rather different concepts of the physical
nature of the arc column. These are abridged as follows.

Cassie assumed an arc column within which the temperature is fixed
and uniform in space and time, having constant resistivity, power loss, and
energy content, and a cross-sectional area variable with current and time.
This requires a constant voltage in steady state given by $V = E_0$, where V is
the arc voltage.

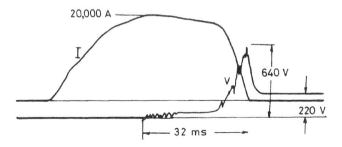

Figure 6.10 Voltage and current variations during arc interruption. Note the
overvoltage caused by the circuit inductance and the capacitance across the arc.

Mayr assumed a cylindrical arc column of constant cross-sectional area within which the temperature varies with the arc's radial dimension and with time. He also assumed the arc behavior to be governed by a characteristic energy quantity leading to a constant power loss given by $IV = N_0$, where I is the arc current.

Cassie's equation can be written in the form (Browne, 1978)

$$\frac{1}{R_a}\frac{dR_a}{dt} = \frac{1}{\theta}\left(1 - \frac{V^2}{E_0^2}\right) \tag{6.9}$$

Mayr's equation can be written in the form (Browne, 1978):

$$\frac{1}{R_a}\frac{dR_a}{dt} = \frac{1}{\theta}\left(1 - \frac{VI}{N_0}\right) \tag{6.10}$$

where θ represents the ratio of a characteristic amount of stored energy to the power loss for both models and is called the arc time constant. Thus, for a steady arc, its equivalent $d(1/R)/dt = 0$ in equation (6.10) (i.e., its $VI =$ constant). This tends to corroborate the shape of the $V–I$ charactristic of DC arcs (Fig. 6.4).

Mayr's model represents the arc around the current zero and indicates how the postarc column resistance continues to increase. Some further refinements for this model were introduced by Rizk (1985) and others (Buttler and Whittaker, 1972). The constant θ was shown to depend on the magnitude of the arc current. It cannot actually be defined as a time constant in the rigid mathematical sense, once we realize how complex the arc phenomena are.

A comparative study between Cassie and Mayr models has been reported (Nakanishi, 1992).

6.9.2.2 Cassie–Mayr Composite Model

It has been observed (Browne, 1948; Leeds et al., 1975; Urbanek, 1971) that the Cassie model, with its steady-state arc voltage E_0, fits best the voltage and current waveforms for the arc during the period prior to current zero. The Mayr model, with its constant loss N_0, describes best the characteristics of a low-current arc, such as exists during the critical few microseconds just after current zero, even during interruption of a large alternating current. This is because during the time before current zero the energy loss is mainly caused by turbulence-enhanced thermal conduction (Browne, 1978). Mayr's equation becomes a better approximation only when the arc temperature falls to a level just above that at which electrical conductance due to thermal ionization practically disappears and the energy is transferred mainly by radial heat conduction (Browne, 1978). Therefore, to use Mayr's equation alone to represent the whole interaction period of the arc may be an avoid-able approximation.

The composite model of Cassie–Mayr, as well as most of the models for arc interruption, is valid only for a brief period of time corresponding to the region of a few microseconds around current zero. The exact time when the transition from Cassie to Mayr models occurs is relatively unimportant so long as it occurs near enough to current zero that the arc volt–ampere product is small compared to the continuing power loss from the arc (Browne, 1978).

The transition time is normally taken as the time when the arc resistance reaches the value given by (Browne, 1978)

$$R_0 = \frac{E_0}{I\omega\theta} \tag{6.11}$$

wher R_0 = the value of the arc resistance at about current zero, I = the rms value of the interrupted current in amperes, and ω = the angular frequency of the source waveform in radians/second.

The value of the characteristic constant of Mayr's model N_0 can be shown to be given by (Browne, 1948)

$$N_0 = E_0 I\omega\theta \tag{6.12}$$

Only two parameters, E_0 and θ, need to be estimated in order to implement the composite model. The ratio E_0/θ determines the failure or the success of interruption.

Browne (1978) and Frost (1976, 1977) have validated this composite model by further comparisons between experimental data and computed results. The model has shown especially good agreement with SF_6 interrupter data.

6.9.2.3   Cassie–Mayr–Cassie Composite Model

Additionally, Browne (1978) and Frost (1976, 1977) proposed the use of Cassie's equation again, after Mayr's equation, to model arc reignition. Hence, the composite model in the case of reignition is a Cassie–Mayr–Cassie model. The transition time from Mayr's to Cassie's equation occurs when the arc resistance computed by Mayr's equation reaches a maximum value R_m. Thereafter, the arc is modeled by Cassie's differential equation again but with a new steady-state voltage constant E_0, given by (Browne, 1978)

$$E_0 = \sqrt{N_0 R_m} \tag{6.13}$$

To summarize, the composite model of Cassie–Mayr–Cassie is as follows:

1. Use Cassie's differential equation (6.9) to model the arc behavior from a few arc time constants θ before current zero until the arc resistance reaches the value given by equation (6.11).
2. Use Mayr's differential equation (6.10) to model the arc behavior with Mayr's constant N_0, given by equation (6.12).
3. If the arc resistance computed by Mayr's equation starts to decrease, indicating resignation, use Cassie's different equation (6.9) from the point the arc resistance has reached a maximum value, with a new value for Cassie's constant E_0 given by equation (6.13).

The temporal variation of the arc parameters of a circuit breaker, including arc resistance, power, and dielectric strength, was studied (Ahmed, 1997) following the Cassie–Mayr and Cassie–Mayr–Cassie models. Before current zero, a satisfactory agreement was found between calculated and measured temporal variations of the arc voltage and current for the same short circuit current. The arc parameters were found to be inversely proportional to the arc time constant and differ according to the arc-interrupting medium, confirming that arc extinction in SF_6 gas is faster than in air for the same short-circuit current.

6.9.2.4 Arc Extinction

To study the arc extinction in a circuit breaker, one needs detailed information about the thermal energy of the arc, the kind of arc chamber, etc.

As is well known, the air blast or gas flow in the arc interruption chamber of the circuit breaker is to push the ionized medium away from the breaker contacts, thus reducing the current to be extinguished. Therefore, the effect of the air blast or gas flow may be simulated by a shunt resistance across the breaker contacts (Ahmed, 1997) which reduces the current to be extinguished. The smaller the value of the shunt resistance, the larger is the current to be bypassed away from the contacts. The effect of the shunt resistance on the arc behavior at successive current zeros was investigated. The shunt resistance changes its value from one half-cycle to the next half-cycle to simulate the continuously decreasing current that flows between the breaker contacts.

The shunt resistance across the breaker contacts, denoted as R_{sh}, was taken as equal to β times R_0, i.e., the arc resistance at about current zero:

$$R_{sh} = \beta R_0 \qquad (6.14)$$

The multiplier β changes from one half-cycle to the next and the rate of its change is governed by the speed of the breaker in interrupting the arc.

However, the initial conditions of the arc were considered to be nearly unchanged from half-cycle to next half-cycle according to the measurements of Rao (1992).

The temporal variation of the arc characteristics during its extinction was studied by Ahmed (1997) following the Cassie–Mayr and Cassie-Mayr–Cassie models. The arc extinction following the Cassie–Mayr–Cassie model was found faster than that of the Cassie–Mayr model.

6.9.2.5 Current Chopping

The circuit to be interrupted by the circuit breaker inevitably has distributed capacitance and inductance. As the power-frequency current through the breaker arc falls toward zero, the arc column diameter shrinks and exhibits a negative V–I characteristic. This effectively negative resistance of the arc would accentuate negatively damped oscillations of the local LC circuit at a frequency on the order of 10 kHz.

Such oscillations could be initiated by any sudden drop in the arc voltage due to any change in its conditions. With the high-frequency component superposed on that at the power frequency, the total arc current drops rapidly and prematurely to zero (Fig. 6.11). This phenomenon is called current chopping. It induces excessive overvoltages in the circuit, as discussed in Chapters 11 and 14.

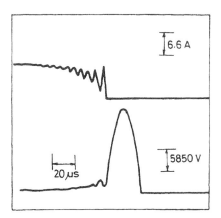

Figure 6.11 Current and voltage variations during AC circuit interruption by an air-blast circuit breaker. Note the superposed oscillations in the current and its chopping before its natural zero. [From Rizk (1963).]

6.9.2.6   Arc Reignition

The voltage that appears across the postarc gap after its extinction may or may not reignite it. If it does, the current will continue to flow for another half-cycle. If it does not, the circuit breaker or switch is considered to have opened the circuit successfully.

Arc reignition is decided by a race against time between the voltage appearing across the postarc gap and the deionization of its gases. Ionization will continue to be partly thermal because of the residual heat in the new cathode spot and in the gases in the gap. The applied electric field will also accelerate electrons, giving them energies that would probably be high enough for ionization by collision, as explained in Chapter 3.

During the first few microseconds after the virtual current zero, reignition is governed by the energy balance. If the applied voltage per unit length of the postarc gap is E', and this causes a current i to flow as a result of the gap's residual conductance, the input power per unit length of the gap is $E'i$. If this exceeds the heat loss, the arc will reignite.

During a subsequent period arc reignition will be dielectric rather than thermal. It will be governed by the applied voltage exceeding the dielectric strength of the gap. As the electrodes have reversed polarity, a dense cloud of positive ions remains near the new cathode. This cloud will severely distort the field near the cathode, and the dielectric strength of the postarc gap will be relatively low. However, after some microseconds the cloud of positive ions will be so diffuse that its influence on the electric field in the postarc gap will become much less pronounced. Thus the breakdown voltage of the gap rises fast initially, but afterward slowly approaches its ultimate value.

6.10   ARC EROSION

Erosion of the contacts by arcing is caused partly by evaporation of metal from electrode spots during the arc. Some droplets of molten metal may also be removed from the electrode spots by high differential gas pressures in the gap and by excessive electric fields, as explained above.

Tests and analyses of erosion phenomena have revealed that the energy of the arc, rather than its charge, is the deciding factor. The current wave shape also has an effect on the rate of arc erosion (Fig. 6.12). For otherwise similar conditions, the rate of contact erosion is lower for contact metals with high melting points, greater latent heats of evaporation, and higher thermal conductivities.

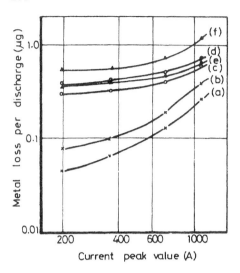

Figure 6.12 Rates of arc erosion from electrodes of different materials. DC arc charge, 73 mC; arc length, 1 mm: (a) anode and (b) cathode of molybdenum, (c) anode and (d) cathode of steel, (e) anode and (f) cathode of W–Cu sintered mixture.

6.11 APPLICATIONS

The applications of electric arcs include arc-discharge lamps, arc furnaces, arc welders and cutters, and plasma torches. Arcs are also used in the production of high-quality metal strips and for ion implantation on silicon wafers in the production of very large scale integrated circuits.

Arc furnaces are widely used in metal-producing plants. For smelting aluminum, for instance, DC arc furnaces are preferred, particularly when electric power is available economically. Also, furnaces fed from three-phase AC are used for melting and purifying metals.

Glow discharge has been used in a plasma spray for applying a metallic protective coating to important machine parts, such as turbine blades, under vacuum (Shankar, 1981). It has also been used more recently for excessive heating of strips of metal powder in a rarefied atmosphere to produce high-quality stainless steel ribbons. This method was proved superior to the conventional method involving successive rolling and heat treatments (Millar, 1987).

Modern techniques for the production of very large scale integrated circuits involve isolating islands of p-type silicon with submicrometer layers of SiO_2. An effective method employs a plasma beam of high-energy positive oxygen ions (O^+) (Dettmer, 1987).

6.12 PROBLEMS

(1) Explain how the arc is initiated in lightning flashover of transmission-line insulators, electric welders, circuit breakers, and fuses.

(2) Sketch the voltage distribution along a stable arc in a DC circuit and comment how the arc is sustained in the cathode and anode regions as well as in the arc column.

(3) Explain how to reduce the current level at which chopping takes place in circuit breakers. Why is current chopping not a serious problem with vacuum circuit breakers?

(4) In arc welding, explain why a drooping characteristic of the supply voltage is essential for maintaining a steady arc and describe how this characteristic is obtained in case of (i) a DC source, (ii) an AC source.

(5) Make a comparative study between the different models of an AC arc: Cassie, Mayr, composite Cassie–Mayr, and composite Cassie–Mayr–Cassie models.

(6) Compare the problem of interrupting 1000 A in a 120 V AC circuit with that of interrupting 100 A in a 1200 V AC circuit.

(7) Sketch and comment on the waveform of voltage and current in a short arc (a) between electrodes of high-melting point, (b) between electrodes of low melting point.

(8) Describe the principle of magnetic blow-out in arcs.

REFERENCES

Ahmed MAM. Restriking voltage as influenced by arc modeling in gas blast circuit breakers. MSc thesis, Assiut University, Egypt, 1997.

Barrault M, Mackburn T, Edels H, Satyanarayana P. IEE conference publication 90, 1972, pp 221–223.

Browne Jr TE. AIEE Trans Pt III 67-I:141–153, 1948.

Browne Jr TE. IEEE Trans PAS-97:478–484, 1978.

Buttler T, Whittaker D. Proc IEE 119:1295–1300, 1972.

Cassie AM. Arc rupture and circuit severity: A new theory. Conference International des Grands Reseaux Electriques à Haute Tension, Paris, France, 1932.

Dettmer R. IEE Electron Power 33(4):273–277, 1987.

Flurscheim CH (ed). Power Circuit Breakers—Theory and Design. Stevenage, UK: Peter Peregrinus, 1975.

Frind G. Z Angew Phys 42:515, 1960.

Frost LS. Dynamic circuit breaker test analysis and its use in the generalized specification of short line fault performance. IEEE Conference Record-Abstract, IEEE International Conference on Plasma Science, Austin, TX, 1976.

Frost LS. Dynamic arc analysis of short line fault tests for circuit breaker specification. IEEE Power Engineering Society Summer Meeting, Mexico City, Mexico, 1977.

Leeds WM, Browne Jr TE, Strom AP. AIEE Trans 76-III:906–909, 1975.

Llewellyn-Jones F. The Physics of Electrical Contacts. Oxford: Oxford University Press, 1957.

Mayr O. Beitrag zur Theorie der statischen und der dynamischen Lichtbogens. Archiv Elektrotech (Berlin) 37:588–608, 1934.

Millar R. IEE Power Eng J 1(5):257–265, 1987.

Nakanishi E. Switching Phenomena in High Voltage Circuit Breakers. New York: Marcel Dekker, 1992.

Nasser E. Fundamentals of Gaseous Ionization and Plasma Electronics. New York: Wiley-Interscience, 1971.

Rao S. Switchgear and Protection. Delhi-6, India: Khanna, 1992.

Reece M. In: Flurscheim CH, ed. Power Circuit Breakers—Theory and Design. Stevenage, UK: Peter Peregrinus, 1975, Ch 2.

Rieder W. Plasma und Lichtbogen. Braunschweig, Germany: Friedr Vieweg & Sohn, 1967.

Rizk F. PhD thesis, Chalmers Technical University, Göteborg, Sweden, 1963.

Rizk F. IEEE Trans PAS-104:948–955, 1985.

Shankar S, Koenig D, Dardi L. J Metals 33(10):13–20, 1981.

Thomas DWP, Pereira FT, Christopoulos C, Howe AF. IEEE Trans. Power Deliv 10:1829–1835, 1995.

Urbanek J. IEEE Proc 59:502–508, 1971.

7

Insulating Liquids

M. KHALIFA and H. ANIS *Cairo University, Giza, Egypt*

7.1 INTRODUCTION

Since the turn of the century, oils have been in use for insulating cables, transformers, and circuit breakers. Insulating liquids are broadly classified as organic, mineral or synthetic. Organic oils started in use late in the nineteenth century, whereas mineral oils were introduced in about 1910 with the development of petroleum refineries. Synthetic liquids with a wide spectrum of properties started to be developed by the petrochemical industry in about 1960. They include synthetic hydrocarbons, halogenated hydrocarbons, silicones, and synthetic esters.

7.2 TYPES OF OILS

7.2.1 Organic Oils

This group of liquids includes vegetable oils, rosin oils, and esters. Natural esters are produced by the chemical reaction between a vegetable acid and an alcohol. Molecules of esters contain atoms of carbon, hydrogen, and a considerable number of oxygen atoms, about 20% (Breuer and Hegemann, 1987). The reaction is helped by a catalyst such as sulfuric acid. An inert water entrainer is also included to help in removing the water formed in the reaction.

Phosphate esters are manufactured from such raw material as coal tar "phenol" in a chemical reaction with phosphorus oxychloride, thus:

phenol + phosphorus oxychloride $\longrightarrow$ triphenol phosphate + hydrochloric acid

Phosphate esters have very good fire resistance compared to mineral oils. However, their poor oxidation stability limits their use to special applications. With the inevitable depletion of mineral reserves in the world becoming more of a concern, several attempts have been made recently to use organic liquids in transformers and other applications where mineral oil has been unparalleled and where synthetic oils were being introduced during the last 20 years (Sankaralingam and Krishnaswamy, 1987; Marinho et al., 1987; Chan, 1987).

7.2.2 Mineral Oils

Petroleum is known to comprise a variety of molecular species. It is therefore broadly classified as of a paraffinic, naphthenic, aromatic, or intermediate group of molecules. The paraffin base is characterized by the chemical formula C_nH_{2n+2}, the naphthenes by the formula C_nH_{2n}, and the aromatics by the formula C_nH_n. The number n of monomers per molecule varies over the range of tens to a few hundreds depending on the type of crude, the distillation and purification processes, and its subsequent treatment and life in service.

Oil of the naphthenic group has traditionally been favored for impregnating paper to be used in insulating HV and EHV cables, because of its good gas-absorbing properties. As naphthenic crudes represent only 5 to 10% of the world's total production, which is inevitably declining, alternative liquids have been explored by research laboratories. Synthetic liquids have proved to be the answer. Their properties can, in a sense, be tailored to match each specific application.

7.2.3 Synthetic Oils

Thanks to intensive research and developments in the petrochemical industries, we now have several synthetic liquids covering a wide range of properties, including nonflammability. Available are askarels, olefins, silicon oils, phosphate esters, and other liquids.

7.2.3.1 Askarels

The synthetic oils commercially known as askarels are manufactured by chlorinating polychlorobiphenyls (PCBs), aromatic hydrocarbons available as byproducts in some petrochemical industries. For example, hexachloro-

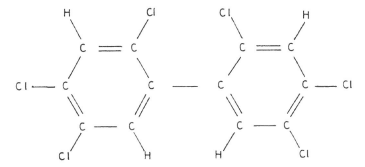

Figure 7.1 Example of the atomic arrangement in a molecule of hexachlorobi-phenyl (askarel).

biphenyl is produced from benzene and has an atomic arrangement of molecules, such as the one shown in Figure 7.1.

7.2.3.2 Silicon Oils

The molecules of silicon oils contain silicon and oxygen atoms in addition to their many carbon and hydrogen atoms. As an example, a molecule of poly(dimethyl siloxane) is represented in Figure 7.2, with the number n varying from 10 to over 1000. Typically, n is about 50 for silicon oils used in transformers.

7.2.3.3 Other Synthetic Liquids

In addition to the phosphate esters mentioned above, in recent years we have witnessed the development of organofluoric liquids of compositions such as $(C_4F_9)_3N$ and $(C_4F_9)_2O$. These have very high chemical stability.

7.3 BASIC PROPERTIES OF INSULATING OIL

The following properties relate to routine and/or acceptance testing of dielectric (insulating) oils (Claiborne and Pearce, 1989; Sierato and Rungis, 1995).

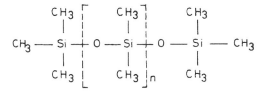

Figure 7.2 Example of the atomic arrangement in a molecule of poly(dimethyl siloxane) (silicon oil). n could be 10 to 1000.

Color and appearance.   For new oils the maximum acceptable color value (color number) is 0.5, and a bright and clear appearance is required.

Density is not very significant in determining the quality of insulating oils but may be useful for type identification and evaluation of an oil's suitability for use. Specific gravity is important when there is a concern about water in oil which may freeze and rise to float on top of the oil.

Viscosity is important for the dissipation of heat. Changes in viscosity with temperature depend on the oil constitution, being most pronounced for aromatic and least for paraffinic hydrocarbons.

Pour point is a measure of the ability of the oil to flow at low temperature. It is an important parameter that always has to be considered when low ambient temperatures of working equipment are expected.

Flash point is an indication of the flammability of oil.

Interfacial tension.   A high value of interfacial tension for new oil, not less than 0.040 N/m, indicates the absence of undesirable polar contaminants. This makes the test a useful screening method for new oils exposed in transport to soaps, acids, varnishes, and solvents.

Neutralization number (acidity) is a measure of the trace amount of acidic or alkaline contaminants in the oil. The test is normally accepted for both vendor qualification and delivery acceptance purposes.

Water content increases electric conductivity and dissipation factor and worsens electric strength. However, it may impede charge generation in the oil. Oils with a moisture content below 10 ppm are considered to be dry and otherwise require drying before use.

Oxidation stability.   In many cases a good oil oxidation stability may be achieved by a suitable content of the naturally occurring aromatics. However, a nitrogen and sulphur compound-free oil is required.

Inhibitor content is a parameter used for oil vendor qualification.

Gassing tendency is evaluated in new oils to satisfy specification requirements, but it is not usually used as a prime indicator of oil characteristics. Some standards recommend the measurement of the total gas content of new oil. A dissolved gas analysis can identify low energy discharges as well as high thermal stresses and disruptive discharge conditions.

Dielectric breakdown voltage at power frequency is the most often controlled parameter describing the oil's function as an insulant. Electric strength is strongly dependent on the amount of contamination in the oil and on the oil constituents. It is sensitive to the amount of specific hydrocarbons and increases considerably for oils with high aromatic content. The preparation of the oil sample may greatly influence the breakdown characteristics of the oil. Dry and clean oils exhibit high breakdown voltages, which can be reduced dramatically when solid particles and dissolved water are present. Most of the breakdown voltage specifications require a minimum of 30 kV with a 2.5 mm gap when measured wtih disc electrodes.

Impulse strength is rarely considered in the various oil test programs.This property is sensitive to both polarity and electrode geometry, but it also reflects the oil composition and can be lowered when the aromatic hydrocarbon content is increased.

Dissipation factor is an important parameter describing the oil's function as a dielectric. The dissipation factor of an oil at power frequency provides the same information as the oil resistivity and is affected by the same factors and test conditions.

Contaminant content affects many oil properties.

Chemical composition analysis. Slight changes in the proportions of any of the oil compounds may affect oil properties. Infrared (IR) spectroscopy has proved useful over the years for the analysis of insulating oils. It can be used to determine the percentage of the aromatic content and the hydrocarbon composition proportions, as well as the presence of inhibitors. High resolution mass spectrometry provides a detailed analysis of hydrocarbons and detects low levels of oxygen and nitrogen compounds.

Electrostatic charging tendency (ECT) describes the property of an oil to become charged when in relative motion against a solid surface. The degree of frictional electrification determines whether or not this property can lead to electrical discharges and dielectric breakdown in a transformer.

Resistance to partial discharges. The term "stability to electrical stress" has been used to relate to the gassing properties of oils under discharge conditions. In general, the significance of partial discharge parameters, such as inception voltage, the magnitude of discharges, and the dynamics of discharge characteristics in insulating oils, still remains a matter of debate.

Electric strength of oil in motion. The electric strength of oil flowing in electrically stressed regions is a function of flow rate, electric field, temperature, moisture content, and particulate contaminants. In the oil velocity range of 5 cm/s to 200 cm/s, the mean breakdown strength may increase, but it decreases at higher velocities.

7.3.1 Effect of Temperature on Viscosity

The variations in the viscosities of some typical mineral oils—a silicon oil, an ester, an olefin, and chlorinated hydrocarbon—are depicted in Figure 7.3. In the figure are shown three grades (I, II, and III) of mineral oil prescribed by the International Electrotechnical Commission for use in transformers and similar equipment (IEC, 1982a). In Figure 7.3, it is noted that the viscosity of hexachlorobiphenyl is considerably higher than those of other synthetic and mineral oils, except at excessive temperatures. The viscosities of silicon oil and olefin are distinctly less sensitive to temperature variation.

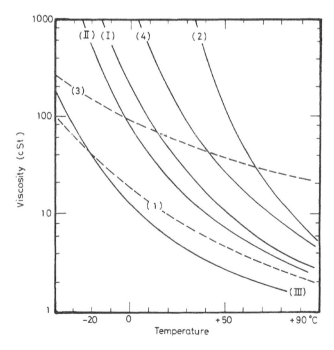

Figure 7.3 Variation of viscosity with temperature for (1) low-viscosity polybutene (olefin), (2) hexachlorobiphenyl (askarel), (3) silicon oil for transformers, (4) trixylenyl phosphate ester. I, II, III: mineral oils graded by the International Electrotechnical Commission.

In applications where the fire hazard is a concern, mineral oils are usually replaced by askarels or phosphate esters because of their remarkable fire resistance. Esters are preferred because of their low toxicity and the fact that they are biodegradable. Also, organofluoric liquids are stable up to temperatures as high as 500°C and are ideal for cooling electronic equipment of the sealed-off types that have large power ratings. Because of their low boiling points, they are used in evaporative cooling of power equipment.

7.3.2   Water Solubility in Oils

Water is molecularly soluble in small quantities in mineral oils, about 50 ppm. Its solubility depends on the molecular composition and on temperature. Water is more soluble in silicon oils and much more in esters, particularly phosphate esters, where the solubility exceeds 600 ppm. Water can be absorbed by the oil from the ambient, or can be produced by oxidation processes of the oil itself, as explained in Section 7.4.

Above the saturation level extra water becomes emulsified in the form of finely dispersed droplets of diameter 1 to 10 μm suspended in the oil. If the water content of the mineral oil exceeds about 400 ppm the droplets tend to flocculate, becoming larger than 10 μm and settling to the bottom of the container. In liquids such as askarels, which have specific gravities exceeding unity, water would collect at the surface.

The water-absorbing property of phosphate esters presents a serious problem for their users. Unless extreme care is taken in their handling, their water content can reach levels as high as 0.1%, with deleterious effects on their properties, particularly the dielectric strength.

7.4   CHEMICAL REACTIONS

Under thermal and electrical stresses oils become oxidized in the presence of oxygen. The gas can be either in contact with the oil or actually dissolved in it. Oxidation is accelerated in the presence of metals such as copper, which act as a catalyst.

The chemical reactions involved may produce lighter molecules (scission) or heavier molecules (polymerization), depending on the type of oil and the reaction conditions. Among the products of oxidation are hydrogen, water, acids, and wax. The wax deposits in the form of sludge. Examples of such reactions are given in Figure 7.4. Some of the oxidation products inhibit the reaction, which thus stabilizes with time. Some synthetic chemicals have recently been developed (Cookson, 1987) which, when added in small quantities to the oil, inhibit the oxidation reaction and thus tend to extend the oil's life in service.

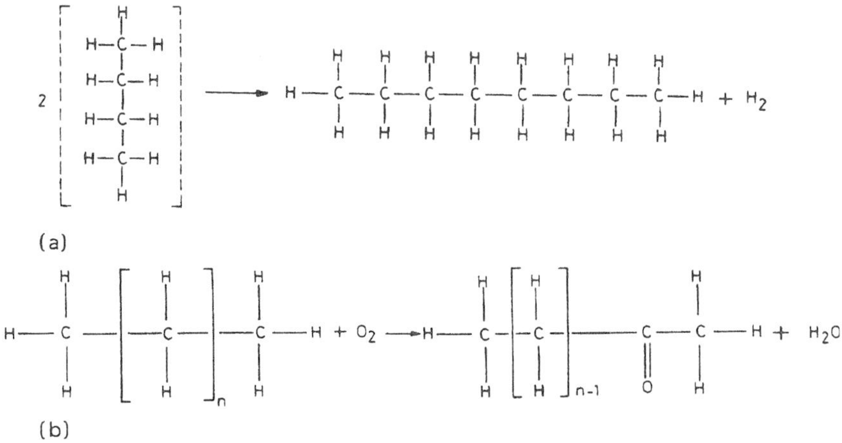

(a)

(b)

Figure 7.4   Examples of chemical reactions involved in the oxidation of mineral oils: (a) polymerization; (b) formation of water.

Compared to mineral oils, natural esters, askarels, and some olefins oxidize more rapidly. With oxidation, olefins produce acids and polymerize, forming resins. On the other hand, silicon oils at temperatures below 150° C are very stable. Above 200° C, however, they oxidize, forming water, carbon oxides, and acids.

Traces of acid in the oil could be produced by oxidation reactions. Acids could result from the refining process or be present originally in the crude and not be completely eliminated during distillation. In esters, acids form as a result of their hydrolysis and thermal decomposition. These acids may chemically attack the solid insulation and/or enamel or paint of the immersed equipment.

7.4.1   Chemical Reactions Enhanced by Electrical Discharge

Electrical discharge in oil-immersed equipment can be either a complete or a partial breakdown (i.e., a silent discharge). In the case of silent discharges there is usually a gas bubble, with enhancement of applied electric fields at its boundaries. Under high electric fields some oil molecules at the surface of the bubble get dissociated, releasing hydrogen.

The ionized hydrogen gas produced by the discharge would be chemically active. Some chemical reactions enhanced by gas discharges may include gas absorption, with hydrogen, oxygen, or nitrogen accepted in some unsaturated molecules of aromatic oils. Such degassing action would no doubt help to maintain the stability of the oil and extend its life under silent electrical discharges. The following is an example:

$$(CH_2—CH_2) + H_2 \xrightarrow{\text{electrical discharge}} (CH_3—CH_3)$$

The activity of the chemical reaction and its products depend heavily on the type of oil, the gases present, the temperature and pressure, and the presence of catalysts.

In the case of complete breakdown of an oil gap and of immersed solid insulation, an arc is involved. In such cases gases such as acetylene, ethylene, and ethane get evolved, resulting from the chemical decomposition of some oil molecules. Also, carbon monoxide results from decomposition of the solid insulation.

Under severe electrical or thermal stresses, askarels evolve acidic gases, mainly hydrogen chloride. The hydrochloric acid thus formed causes severe electrolytic corrosion of insulation and metals immersed in the askarel. Also, as is known, askarels are nonbiodegradable and are thus ecologically unacceptable. Esters do not suffer from these disadvantages.

7.5 ELECTRICAL PROPERTIES

Electrical properties include the electrical conductivity of the oil, its permittivity, dissipation factor, and dielectric strength.

7.5.1 Electrical Conductivity

It has been noted that when liquids known to be insulants are subjected to direct voltage, they actually conduct a current, however small. Even a simple insulating liquid such as hexane conducts a current that varies with the electric stress applied. It has been suggested that under low electric fields the conduction is due mainly to positive and negative ions belonging to dissociated molecules. Dissociation and recombination are in dynamic equilibrium, thus:

$$A—B \underset{\text{recombination}}{\overset{\text{dissociation}}{\rightleftarrows}} A^+ + B^-$$

However, when an electric field exceeding $1\,kV/cm$ is applied, the rate of dissociation exceeds that of recombination and increases at higher fields (Nelson and Lee, 1987). This state of affairs is basically different from the case of conduction through gases, where below the corona onset level the number of charge carriers is limited and independent of the applied field strength (Chapter 4). This explains why the saturation current is constant in gas gaps, whereas it increases in oil gaps with applied voltage (Cross and Jaksts, 1987).

At still higher fields the current grows at a rapidly increasing rate. This growth is accounted for both by field emission of electrons from the cathode and field-aided dissociation of liquid molecules. As the ions drift through the liquid, they gain energy from the field and lose some of it in collisions with the liquid molecules. The energy thus gained by the liquid molecules accounts for extra vibrations. The vibrations of C—H and C—C bonds among the carbon and hydrogen atoms result in some of them being broken, thus producing extra ions.

Under impulse voltages conduction currents reach much higher values because the choking effect of space charges accumulating at the electrodes is insignificant compared to the DC case, as in gas discharges.

In liquids of commercial purity, additional conduction currents will be caused by the impurities. These may include a wide variety of particles: droplets of water and acids, resins, and cellulose fibers. Their different conductivities, permittivities, and affinities for ionization will have widely varying effects on conduction through the insulating liquid.

Conductivity rises sharply with temperature as a result of both the increased dissociation of the liquid molecules and its decreasing viscosity. At a temperature T the number of n of dissociated molecules per unit volume is related to the total number per unit volume N in terms of W, the dissociation energy, as

$$n = N \exp(\ W/kT) \tag{7.1}$$

k being Boltzmann's constant. Thus the conductivity of the liquid due to dissociated molecules is

$$\sigma = ne(\mu_+ + \mu_-) \tag{7.2}$$

Here e is the electronic charge and μ is the ion mobility. It is evident that the conductivity depends exponentially on temperature, as observed experimentally (Fig. 7.5).

Under AC the liquid conductivity represents a considerable part of its dielectric losses. The other part of the loss under AC is due to hysteresis in the polarization of the liquid molecules. The losses are expressed as a dissipation factor (tan δ) where δ is the loss angle. The dissipation factor increases only slightly with applied electric stress at moderate levels. Under very high stresses, however, the dissipation factor increases considerably with stress, at an ever-increasing rate, until the liquid breaks down (Denat et al., 1983).

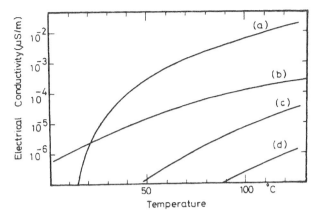

Figure 7.5 Electrical conductivities of some insulating liquids as functions of temperature: (a) PCB (askarel), (b) commercial transformer oil, (c) purified transformer oil, (d) highly purified transformer oil.

7.5.2 Dielectric Constant

The dielectric constants (i.e., relative permittivities) of insulating liquids range from about 2.2 for mineral oils to about 5.3 for askarels. Askarels have therefore been favored for applications in capacitors. The dielectric constants of mineral and most synthetic oils vary only slightly with temperature.

7.6 THEORIES OF DIELECTRIC BREAKDOWN

Unlike the case of gases, there is no single theory that has been unanimously accepted for the breakdown of insulating liquids. There are two main reasons for this. One is that the liquid phase is much less amenable to theoretical analysis than the gas phase. Second, insulating liquids are almost never absolutely pure, if not because of the ingress of contaminants, then due to traces left in the liquid by the processes of manufacture and purification. Each of these ingredients plays a role in the breakdown of the bulk of the liquid. Thus it is an accepted fact that the breakdown strength of a sample of insulating oil depends on its impurity content rather than its molecular composition. The three principal theories that have been proposed for the breakdown of insulating liquids are (a) the electronic breakdown theory, (b) the bubble theory, and (c) the suspended particle theory.

7.6.1   Electronic Breakdown Theory

According to this theory, electrons are emitted from the cathode (Kuffel and Zaengl, 1984). On their way to the anode, the electrons collide with atoms of liquid molecules. If enough energy is transferred during such collisions, some electrons are knocked off their atoms and drift with the original electrons toward the anode. Thus electron avalanches such as those occurring in gas discharges develop in the liquid and ultimately lead to breakdown. Evidence supporting this theory, even under fast impulse voltages, is rather scarce.

7.6.2   Bubble Theory

Leakage currents are not distributed evenly over electrode surfaces. As directed by the applied electric field, they tend to concentrate at points of microroughness where the field lines converge (Fig. 7.6a). Near such points, the highly concentrated leakage currents, with corresponding Joule heating of a microscopic volume of liquid, will cause a rapid rise in temperature to high above the boiling point. With very limited heat convection and conduction, such a temperature rise could occur over short periods on the order of 1 µs. Thus a bubble of vapor forms at a point of microroughness on the electrode surface (Fig. 7.6a).

Three other alternatives were proposed to account for formation of the gas bubble (Korobejnikov et al., 1983): release of occluded gases from micropores in electrode surface layers; cavitation caused by mechanical strain of the liquid under the highly concentrated electric field, with corresponding electrostrictive pressure differential; and electrochemical dissociation of some liquid molecules, with the release of gases. Furthermore, in the case of oil-impregnated paper, as in transformer windings, cables, and capacitors, repeated expansion and contraction under cyclic loads would cause cavitation and form gas bubbles. Under each alternative, the concentrated electric field at points of microroughness on electrode surfaces or somewhere along the gap would play a basic role in the production of gas bubbles.

Tests on oil gaps under high-voltage pulses of nanosecond duration could provide the data on development of gas bubbles at the electrodes and their initiating breakdown (Korobejnikov et al., 1983; Lesaint and Tobazéon, 1987). Because of the difference between the permittivities of the gas and the liquid, the electric field inside the bubble would be considerably higher than that in the liquid. When the bubble reaches a critical volume, and under high enough fields, the gas in the bubble would become ionized and electron avalanches would develop into streamers. The space charges formed at the ends of the bubble would act to deform it under the electric field by Coulomb forces. Also, the electrons and ions accelerated by

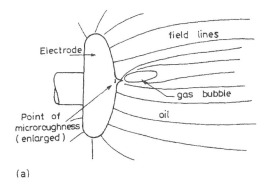

(a)

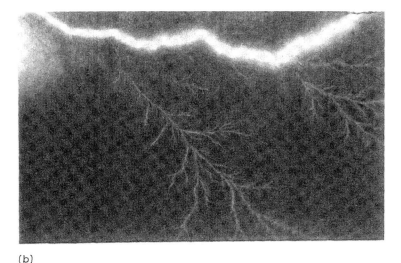

(b)

Figure 7.6 (a) Formation of a gas bubble at a point of microroughness on the electrode surface in an oil gap; (b) gas discharge in the bubble, developing it into branched streamers with ultimate breakdown.

the field inside the bubble would impinge against areas of its walls, helping to decompose some of the liquid molecules and to evaporate others. The gas bubble grows longitudinally (Fig.7.6a), which would lead eventually to complete breakdown of the entire liquid gap (Fig. 7.6b).

The energy supplied to the gas bubble is consumed in vaporizing some additional liquid (on the order of 10^6 J/m^3), ionizing and expanding the gas bubble (10^5–10^8 J/m^3), and decomposing some oil molecules into relatively simpler compounds (10^7 J/m^3). The speed of expansion of gas bubbles was measured in the range 100 m/s to 10 km/s (Felici, 1987).

Negative streamers start from sharp points on the cathode. Their growth is aided by electrons bombarding the gas–oil interface at the streamer's tip. The positive streamer propagates from the anode and electrons fall into its tip from the oil molecules located ahead of it, which get dissociated by the local intense electric field (Chadband and Sadeghzadeh-Araghi, 1987). With the growth of the streamer, its space-charge field increases exponentially. It may reach the order of 10 mV/cm, exceeding greatly that of the applied Laplacian field at sites away from the electrodes, which explains lateral branching of these streamers (Fig. 7.6b).

The bubble theory is supported by the observed dependence of breakdown strength on the applied static pressure. If the liquid is degassed, its dielectric strength becomes less dependent on the static pressure. Also, gas-absorbing additives cause a rise in the breakdown strength.

7.6.3 Suspended Particle Theory

The particle could well be a fiber, probably soaked with moisture, or it may be even a droplet of water. Under an applied electric field, and because their dielectric constant ε_{rf} is much higher than that of oil ε_{ro}, fiber particles get polarized and move along converging fields (Fig. 7.7).

Assuming hemispherical tips for the fiber particles, with radius r, the charge $\pm q$ at either of its ends would be

$$\pm q = \pm \pi r^2 (\varepsilon_{rf} - \varepsilon_{ro}) \varepsilon_0 E \tag{7.3}$$

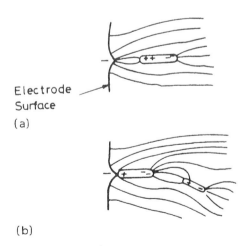

Electrode
Surface
(a)

(b)

Figure 7.7 Steps of collection of fiber or other particles in suspension to eventually bridge the oil gap. They become polarized and directed by the field.

where ε_0 is the permittivity of free space and E is the field strength at either end of the fiber, assumed equal. If the field in the oil gap is nonuniform, as is usually the case, the resultant force will move the fiber (Fig. 7.7). For a short fiber of length equal to radius r, and with $\varepsilon_{rf} \gg \varepsilon_{ro}$, the driving force will be

$$F_E = r^3 \varepsilon_0 E \text{ grad } E \tag{7.4}$$

As this particle moves with velocity v in the oil of viscosity η, there will be an impeding force with a magnitude of $6\pi r \eta v$. When a particle reaches either electrode, its outward tip will act as an extension to the electrode and attract more fibers (Fig. 7.7).

It is noted from equation (7.4) that the driving force on the fiber increases in relation to its size and with the degree of nonuniformity of the field. Thus in a sphere gap they will migrate toward the higher field at its axis and gradually line up and bridge the gap (Fig. 7.8) (Kind, 1978). Such a relatively conducting path would in effect short-circuit the gap, resulting in breakdown. Figure 7.9 shows how a water drop gets distorted under the electric field in an oil gap and ultimately leads to its breakdown.

Figure 7.8 Fiber particles bridging a gap under high voltage. [Courtesy of D. Kind (1978) and F. Vieweg & Sohn.]

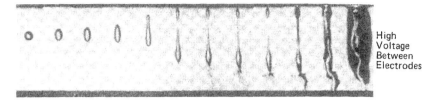

High
Voltage
Between
Electrodes

Figure 7.9 Water drop being gradually elongated by an electric field and ultimately causing breakdown. The water is evaporated by the discharge.

The evidence in support of this theory includes the increased time required to reach breakdown of the liquid with increased viscosity. Evidently, this phenomenon could occur under DC and power-frequency AC, but not under high-frequency or fast impulse voltages.

7.6.4 Streamers in Liquid Insulation

The study of pre-breakdown and breakdown phenomena in liquid dielectrics has been improved by the use of fast digitizing oscilloscopes for voltage and current measurements, fast optical detecting systems for measurements of density gradients using shadowgraph and Schlieren photography coupled with an image converter camera, electric field distributions using electric field-induced birefringence of the Kerr effect, and optical spectroscopy of emitted light (Beroual et al., 1998). It is therefore possible to follow the different stages leading to breakdown and to establish that the breakdown of liquids is generally preceded by some events called "streamers." The optical refractive index of streamers is different from that of the surrounding liquid and the structure is similar to that observed in the breakdown of gas and solid insulation (Saker and Atten, 1996).

Streamers in dielectric oils are characterized by different structures according to the experimental conditions. They produce typical shapes of current transients or emitted light signals. They are accompanied by shock waves and their conductivity depends on the mechanisms involved in their propagation. The streamer stops when the electric field becomes too small, producing a string of microbubbles that dissolve in the liquid. Kerr effect measurements of the electric field found that streamers were highly conducting, with a voltage drop across the streamers that was less than 10% of the total voltage across the electrodes (Jayaram, 1996).

Streamers are generally classified as slow and "bushy" for streamers emanating from the negative electrode, or fast and "filamentary" for streamers emanating from the positive electrode. Positive streamers are often about ten times faster than negative streamers, although transformer oil is an exception, with positive and negative streamer velocities being in the same range.

Streamers can be classified in different modes depending on their velocity and the polarity of the voltage. Each propagation mode corresponds to a given inception electric field. Therefore, the electrode geometry and the amplitude of the applied voltage have a significant influence on the structure, the velocity, and the mode of propagation of the streamers. However, such a classification versus the polarity is inconclusive when considering liquids of specific molecular structure or containing selective additives.

The characteristics of streamers depend on the chemical composition and physical properties of the liquid (pure, or containing a small concentration of selected additives); pressure and temperature; the electrode geometry; the voltage magnitude, polarity, and shape; and contaminants of air, moisture, particles, and other trace impurities.

Figure 7.10 depicts the propagation of positive streamers in a nonuniform oil gap (Torshin, 1995). First, positive primary streamers are formed, with intensified branching at about 2 MV/cm. Streamers then propagate at 2–3 km/s. Under higher voltages, a positive secondary streamer is formed as an extension of the primary structure beyond 12 MV/cm field level. Their velocities may reach 32 km/s. At the streamer front, shock waves were recorded.

7.7   FACTORS INFLUENCING THE DIELECTRIC STRENGTH OF INSULATING LIQUIDS

The dielectric strength of insulating liquids has been observed to depend on the liquid temperature, applied static pressure, size of the gap, electrode material and surface conditions, and impurities contained in the oil.

7.7.1   Temperature and Pressure

The effect of temperature on the dielectric strength of an insulating liquid depends on its type and degree of purity (Calderwood and Corcoran, 1987). For example, the dielectric strength of dry transformer oil is insensitive to temperature except slightly below the boiling point. There, the dielectric strength decreases drastically, probably because of the formation of vapor bubbles and their growth, which are aided by the decrease at such temperatures of the oil's viscosity and surface tension. The dielectric strengths of oils that have a trace of moisture are sensitive to temperature variations over the full range from about −20° C up to their boiling point of about 250° C.

The dielectric strength of an insulating liquid under DC and power-frequency AC increases significantly with applied static pressure. Raising the

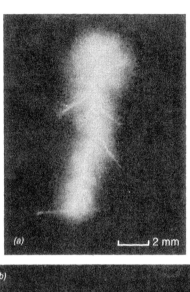

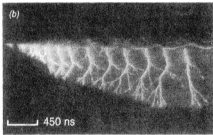

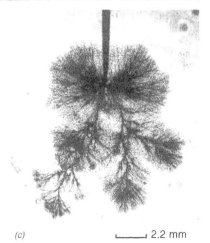

Figure 7.10 Propagation of fast positive streamers in a point–plane geometry under 0.5/85 μs impulse voltage: (a) still photograph; (b) streak photograph; (c) Schlieren photograph (Torshin, 1995).

pressure from atmospheric to 10 times higher increases the dielectric strength by about 50%, depending on the type of liquid. Another effect of pressure is the suppression of pre-breakdown discharges. These observations support the bubble theory of liquid breakdown. Under very fast impulse voltages of duration less than 0.05 µs, breakdown voltage is insensitive to both pressure and temperature, as observed by Korobejnikov et al. (1983).

7.7.2 Electrode and Gap Conditions

The breakdown voltage of an oil gap depends on its width as well as the electrode shape and material. For gaps with highly nonuniform fields, for example, such as that of a point-to-sphere gap, there is a polarity effect. The negative DC breakdown voltage is lower than the positive voltage up to a critical gap length above which the relation reverses (Yehia, 1984). This critical gap length depends on the liquid and the electrode material. There seems to be no simple explanation for these phenomena. However, the material of the cathode surface layer determines the electric stress necessary for electron emission. These electrons play a decisive role in the conduction and breakdown processes.

The size and shape of electrodes determine the volume of liquid subjected to high electric stress and the degree of field nonuniformity. The bigger this volume, the higher the probability of its containing impurity particles. The more of these particles that are present, the lower would be the breakdown voltage of the liquid gap. The effect of moisture content is shown in Figure 7.11.

The sensitivity of liquid breakdown to these factors is logically higher under DC and power-frequency AC than under fast impulse voltages (Yehia, 1984; Krueger, 1987). Thus, the impulse ratios of highly nonuniform gaps of contaminated or technically pure liquids can reach about 7, much higher than the gas gaps of similiar geometries.

It has also been shown that stressing the oil gap under high voltage for a long time, and repeated sparks of limited energy, tend to raise the breakdown voltage of the gap (Dawoud, 1986). This is called conditioning the oil gap. Particles in suspension collect at zones of field concentration. Points of microroughness on the electrodes get eroded by concentrated discharge currents. A film of discharge byproducts gradually covers the discharge areas of both electrodes. In the case of silicon oil, repeated breakdowns tend to cover the electrodes with a film of gel and solid decomposition products (Krueger, 1987). If a high-energy arc is allowed to take place in the liquid gap, the arc products cause the liquid properties to deteriorate.

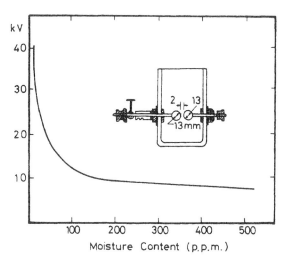

Figure 7.11 Dielectric strength of mineral oil as a function of its moisture content.

7.7.3 Impurities

Impurities include solid particles of carbon and wax, by-poducts of aging and discharges, cellulose fibers, residues of filtration processes, water, acids, and gases. Impurities usually cause a reduction in the dielectric strength of an insulating liquid, the largest effect being that of the simultaneous presence of moisture and fibers. Cellulose fibers are known to be hygroscopic. Thus, floating moist fibers tend to bridge the oil gap, as explained earlier.

Under both AC and DC the effect of a trace of moisture is drastic on meticulously dried liquids, much greater than that of commercial liquids (Fig. 7.11). The effect of moisture is less pronounced in the case of oil gaps with strongly nonuniform fields and with liquids containing no fibers. Because water solubility is considerably higher in silicon oil and phosphate esters than in mineral oil, they need to be much more carefully dried and kept.

Metal particles may be present in dielectric liquids, particularly those used in transformers and circuit breakers. Their presence reduces the dielectric strength of oil by as much as 70%. Figure 7.12 illustrates this phenomenon, in which longer and thinner particles contribute more to the reduction of the oil's breakdown strength.

7.7.4 Insulating Oil in Motion

The behavior of transformer oil and other dielectric fluids used for the cooling and insulation of power system equipment is significantly influenced

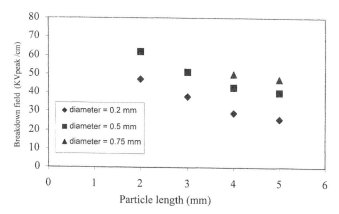

Figure 7.12 Influence of particle dimensions on the mean breakdown fields for free steel particles under AC voltage.

by motion enforced by the action of circulating pumps. Two important factors affect the situation. First, charges generated by streaming electrification in critical parts of the hydraulic circuit having high velocity and/or turbulence can accumulate to distort the electric field in positions where dielectric integrity is prejudiced. Also, the dielectric strength of the fluid is altered by the actions of the flow (Nelson, 1994). Charge separation at interfaces between a moving fluid and a solid boundary can give rise to the generation of substantial electric fields. Either alone or in combination with the existing electric fields imposed by the energization of the equipment, these can give rise to insulation failure. The initial response of apparatus manufacturers has been to reduce design velocities and curtail the operation of pumps.

In apparent contrast, during standard oil testing, the continuous flow of oil was found to increase the mean dielectric strength, as seen in Figure 7.13 (Danikas, 1990). The increase depends on the electrode material and is larger with steel electrodes than with brass. The increase of dielectric strength can be explained by assuming either that the oil flow impedes the entry of impurities into the gap, or that the oil motion delays the establishment of particle bridges between the electrodes. The change in dielectric strength was significant with an oil velocity of 3 cm/s, although a much higher velocity is normally needed to have such an effect. To further complicate the picture, excessive increase in oil velocity causes the flow to become turbulent, where gas bubbles may then be created which lead to a reduction in dielectric strength.

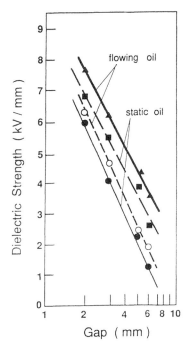

Figure 7.13 Gap spacing effect for large steel electrodes. The plotted points are for electrode areas of 61.92 cm² (points plotted as triangles and squares for flowing and static oil, respectively), and 82.56 cm² (open circles for flowing oil, solid circles for static oil) (Danikas, 1990).

7.8 AGING

When an insulating liquid has been kept at elevated temperatures and/or under electric stress and exposed to oxygen, its properties degrade. The amount of moisture content and the acidity level rise, while resistivity and dielectric strength drop. Oxidation of the insulating liquid is accelerated at higher temperatures, with more oxygen available and in the presence of a catalyst such as copper. As low-viscosity, highly refined oils have higher tendencies to dissolve air and oxygen, they become oxidized faster than do oils of lower grades and higher viscosities. Among petroleum oils, those of the naphthenic groups are considerably more stable than the paraffinics and aromatics, and therefore have dominated applications with transformers, switchgear, and cables.

While the insulating liquid is in service or being stored in contact with oxygen, some impurity particles which inevitably are present gradually flocculate. Thus particle complexes grow in size. Silent discharges of the corona type and concentrated leakage currents help the formation of water, resins, acids, and the evolution of hydrogen. Disruptive discharges such as arcs and intensive localized heating of the liquid and solid insulation produce particles of carbon, and gases such as carbon monoxide, carbon dioxide, and acetylene. Wax often forms by polymerization of oil at the walls of gas bubbles when the bubbles become ionized. The aggressive acidic products of oxidation and discharges attack the solid insulation, iron, and copper immersed in the liquid. Also, incompletely cured varnishes on oil-immersed windings dissolve in the oil and polymerize. Solid particles of carbon, wax, corroded iron, and polymerized liquid settle as sludges. Thus the physical, chemical, thermal, and electrical properties of the insulating liquid deteriorate.

To maintain the qualities of an insulating liquid in service, its important physical, chemical, and electrical properties have to be regularly checked. Before the qualities of the insulating liquid change beyond permissible levels, special measures should be taken to reclaim the liquid.

7.8.1 Additives

To prolong the life of an insulating liquid in service, action can be taken along two fronts. First, the actual process of aging—which is mainly oxidation—should be inhibited. Second, the oxidation products should be treated so as to minimize their deleterious effects on the liquid properties.

To start with, to minimize the rate of oxidation, the amount of oxygen dissolved in, or in contact with, the liquid should be minimized. An example is the case of sealed equipment with a nitrogen cushion above the oil. Otherwise, oxidation inhibitors are dissolved in the liquid. These are scavengers that react with oxidation products and thus break the oxidation chain reaction. There are also passivators that react with metal salts which otherwise would act as catalysts for the oxidation reaction. The amount of salt added to the liquid is on the order of 0.1%, although the exact amount for optimal results depends on the salt and base liquid composition (Kamath and Murthy, 1987).

Oxidation inhibitors react with the free radicals and peroxides produced by the oxidation process and thus break their chain reaction mechanism, which would otherwise give momentum to the oxidation process (Dawoud, 1986). By reacting with peroxides of radicals, passivators prevent the formation of naphthanates of copper and iron which are the usual catalysts for the oxidation reaction. For example, for petroleum oils, established additives include di-*tert*-butyl-*para*-cresol (DBPC), dimethylaniline (DMA), quinones, anthracenes, and phenyls, whereas for askarels, anthraquinone acts as a scavenger for the hydrogen chloride, which is the most chemically aggressive decomposition product.

7.9   TESTS ON INSULATING LIQUIDS

To ensure the qualities of insulating liquids and their compatibility with the equipment in which they are to be used, numerous tests for these liquids have been prescribed. The International Electrotechnical Commission has issued publications describing testing methods and the criteria to be used in accepting insulating liquids (e.g., IEC, 1978, 1982a,b). The electrical tests include measurement, under controlled conditions, of the dielectric strength, dielectric dissipation factor, resistivity, and permittivity. Tests for the liquid's physical characteristics include measurement of its viscosity, pour point, flash point, and moisture content. Chemical tests include measuring its degree of acidity, oxidation, stability, and gassing characteristics (IEC 1963, 1974a,b).

The moisture content of an insulating liquid is measured by heating a sample to a set temperature in a rarefied atmosphere. The water vapor developed in the test vessel is a function of the water content in the liquid. Also, the degree of acidity of the oil can be measured by noting the amount of potassium hydroxide that just neutralizes the acids in the sample. It has been observed that the acidity of oil affects its dissipation factor much more sensitively than it does its dielectric strength (Krueger, 1987).

There are other tests for measuring and identifying the gases dissolved in the liquid. Equipment has been developed, based on the principle of fuel cells, for on-line monitoring of the amount of hydrogen dissolved in transformer oil while in service (Webb, 1987). Identifying the gases dissolved in the insulating liquid of equipment and measuring their relative quantities can assist greatly in monitoring the performance of equipment and in diagnosing any incipient fault at an early stage. Thus a major equipment fault could be prevented and the corresponding capital loss avoided.

7.10   RECONDITIONING OF INSULATING LIQUIDS

Insulating liquids normally remain in service as long as their qualities have not deteriorated beyond levels set by standard specifications and general experience. For instance, their breakdown voltage, across the specified test gap of 2 mm, should not decrease below 30–55 kV; the magnitude depends on whether the liquid is used in low-voltage or high-voltage equipment. For the latter, insulation is usually designed to undergo higher stresses. The corresponding limits for the resistivity of the liquid are 3 and 10 GΩ·m, and for the dissipation factor they are 0.2 and 0.05 at 90°C. In more highly stressed equipment, such as cables, the limit for the dissipation factor is set as low as 0.001.

The level of acidity neutralization factor of insulating liquids can be measured as the amount of potassium hydroxide sufficient to neutralize the acids in 1 g of the liquid. The acceptable level is 0.5 mg KOH/g. The acceptable level of water content is 15–30 mg/kg; the lower figure is for power equipment. The level is set even lower (0.1 mg/kg) for EHV cables, as their working electrical stresses are very high (IEC, 1982a).

As the level set for any of the foregoing characteristics is approached, measures for reconditioning the liquid should be taken. There is portable equipment now available for reconditioning insulating oils while in service (Fig. 7.14). A method well known and in use is that of filtering and vacuum drying. While under a vacuum of about 1 kPa, the liquid is heated to about 30–60°C, well above the boiling point of water at such a reduced pressure. With large oil surfaces exposed to the vacuum, it becomes freed from its moisture content and dissolved gases. In the process it also gets freed from acids and particulate matter. The process could include replenishing inhibitors and other additives in the liquid. Reconditioned oils usually have about half the life of new ones.

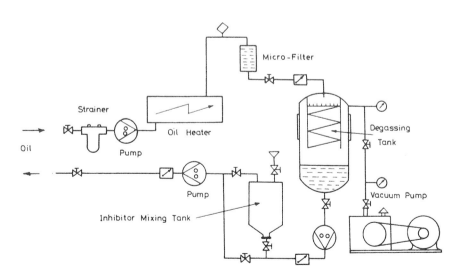

Figure 7.14 Setup into which insulating oil is drawn, filtered, dried under vacuum, degassed, and its inhibitors replenished before it is returned to the equipment.

7.11 PROBLEMS

(1) In an experiment for determining the breakdown strength of transformer oil the following observations were made. Derive a mathematical formula to relate the gap spacing and the applied voltage for the oil. Use the result to determine the breakdown strength of oil according to standard specifications.

Gas spacing (mm)	3	6	9	12
Voltage at breakdown (kV)	68	140	190	255

(2) What are the main types of insulating oils? Outline the merits and disadvantages of each type.

(3) List the basic properties of insulating oils and comment on the practical relevance of each property.

(4) Define—analytically—the electrical conductivity of insulating oil. How is this conductivity affected by temperature?

(5) Enumerate the various modes of breakdown in insulating liquids. Describe the development of streamers in oils, showing their similarities to streamers in gases.

(6) Explain how the dielectric strength of insulating oils is affected by the presence of metallic particles, describing the effect of particle dimensions.

(7) Explain how the motion of oil inside electric apparatus affects the dielectric properties of the oil.

REFERENCES

Beroual A, Zahn M, Badent A, Kist K, Schwabe AJ, Yamashita H, Yamazawa K, Danikas M, Chadband WG, Torshin Y. IEEE Electr Insul Mag 14:6–17, 1998.

Breuer W, Hegemann G. Proceedings of CIGRE Symposium on New and Improved Materials for Electrotechnology, Vienna, 1987, Report 500.09.

Calderwood J, Corcoran P. Proceedings of 9th International Conference on Conduction and Breakdown in Dielectric Liquids, Salford, UK, 1987, pp 124–128.

Carraz F, Rain P, Tobazéon R. IEEE Trans DEI-2:1052–1063, 1995.

Chadband W, Sadeghzadeh-Araghi M. Proceedings of 9th International Conference on Conduction and Breakdown in Dielectric Liquids, Salford, UK, 1987, pp 325–330.

Chan J. IEEE Trans EI-3(3):10–11, 1987.

Claiborne CC, Pearce HA. IEEE Electr Insul Mag 5:16–19, 1989.

Cookson AH. Proceedings of CIGRE Symposium on New and Improved Materials for Electrotechnology, Vienna, 1987, Report 500.00.

Cross J, Jaksts A. Proceedings of 9th International Conference on Conduction and Breakdown in Dielectric Liquids, Salford, UK, 1987, pp 271–279.

Danikas MG. IEEE Electr Insul Mag 6:27–34, 1990.

Dawoud R. MS thesis, University of Alexandria, Alexandria, Egypt, 1986.

Denat A, Cosse B, Cosse J. Proceedings of 4th International Symposium on High Voltage Engineering, Athens, 1983, Paper 24.02.

Felici N. Proceedings of 9th International Conference on Conduction and Breakdown in Dielectric Liquids, Salford, UK, 1987, pp 30–35.

IEC. Method for the Determination of the Electrical Strength of Insulating Oil. Publication 156. Geneva: International Electrotechnical Commission, 1963.

IEC. Methods for Assessing the Oxidation Stability of Insulating Liquids. Publication 474. Geneva: International Electrotechnical Commission, 1974a.

IEC. Methods for Sampling Liquid Dielectrics. Publication 475. Geneva: International Electrotechnical Commission, 1974b.

IEC. Measurement of Relative Permittivity, Dissipation Factor and D.C. Resistivity of Insulating Liquids. Publication 247. Geneva: International Electrotechnical Commission, 1978.

IEC. Specification for Unused Mineral Insulating Oils for Transformers and Switchgear. Publication 296. Geneva: International Electrotechnical Commission, 1982a.

IEC. Determination of Water in Insulating Oil, Oil-Impregnated Paper and Pressboard. Publication 733. Geneva: International Electrotechnical Commission, 1982b.

Jayaram S. IEEE Trans DEI-3:410–416, 1996.

Kamath K, Murthy T. Proceedings of CIGRE Symposium on New and Improved Materials for Electrotechnology, Vienna, 1987, Report 500.07.

Kind D. An Introduction to High Voltage Experimental Technique. Braunschweig: Friedr Vieweg & Sohn, 1978.

Korobejnikov S, Yanshin K, Yanshin E. Proceedings of 4th International Symposium on High Voltage Engineering, Athens, 1983, Paper 24.09.

Krueger M. Proceedings of 9th International Conference on Conduction and Breakdown in Dielectric Liquids, Salford, UK, 1987, pp 487–501.

Kuffel E, Zaengl W. High Voltage Engineering Fundamentals. Oxford: Pergamon, 1984.

Lesaint O, Tobazéon R. Proceedings of 9th International Conference on Conduction and Breakdown in Dielectric Liquids, Salford, UK, 1987, pp 343–347.

Marinho A, Sampaio E, Monteiro M. Proceedings of CIGRE Symposium on New and Improved Materials for Electrotechnology, Vienna, 1987, Report 500.06.

Nelson JK. IEEE Electr Insul Mag 10:16–27, 1994.

Nelson J, Lee M. Proceedings of 8th International Conference on Conduction and Breakdown in Dielectric Liquids, Salford, UK, 1987, pp 298–302.

Saker A, Atten P. IEEE Trans DEI-6:784–791, 1996.

Sankaralingam S, Krishnaswamy K. Proceedings of CIGRE Symposium on New
and Improved Materials for Electrotechnology, Vienna, 1987, Report 500.01.
Sierato A, Rungis J. IEEE Electr Insul Mag 11:8–19 1995.
Torshin YuV. IEEE Trans DEI-2:167–179, 1995.
Webb N. IEE Power Eng J 1(5):295–298, 1987.
Yehia S. PhD thesis, University of Alexandria, Alexandria, Egypt, 1984.

8

Solid Insulating Materials

M. KHALIFA and H. ANIS *Cairo University, Giza, Egypt*

8.1 INTRODUCTION

Solid insulating materials are encountered in every electric and electronic device and piece of equipment, both large and small. They are vital for isolating conductors and for their proper performance. Solid insulants cover a very wide range, including organic and inorganic, natural and synthetic, and simple, bonded, and impregnated materials. Their common feature is that they are compounds—not pure chemical elements. They have extremely high electrical resistivities and high dielectric strengths below certain temperatures. They differ in their electrical, physical, and chemical properties, which are used as a guide in selecting the material appropriate for each application. After a discussion of the important electrical, physical, and chemical properties common for solid insulating materials, and the theories of their electrical breakdown, some of the most widely used will be briefly described.

8.2 ELECTRIAL PROPERTIES

These electrical properties are essentially the dielectric constant, electrical resistivity, and dielectric strength of the material. The dielectric constant (i.e., the relative permittivity) is determined by the phenomenon of polarization that takes place inside the material when under an electric field. The polarizability of a dielectric is an intrinsic property that depends on its molecular composition.

The electrical resistivity of the material, together with any hysteresis that might take place under AC fields, determines its dielectric losses, expressed by power engineers as the loss factor or dissipation factor, and by elecronics engineers in terms of the quality factor of the dielectric. The dielectric strength of an insulator depends on its material, its shape and size, its ambient, the type and duration of the applied voltage, the presence of field concentration, and other factors as discussed in Section 8.4.

Solid dielectrics are compounds, as a rule; their properties may acquire magnitudes depending on the direction in which they are measured. In the special cases where this happens the material is nonisotropic. This phenomenon of nonisotropicity appears in some crystalline materials, including quartz.

8.2.1 Relative Permittivity

Most dielectrics are chemical compounds of positive and negative ions. Thus, under electric fields, each ion tends to migrate toward the electrode of opposite polarity, as is well known (e.g., Seely and Poularikas, 1979). This polarization induces bound charges on the electrodes (Fig. 8.1). The electric flux density D is related to the field strength E by the well-known relation $D = \varepsilon E = \varepsilon_0 \varepsilon_r E$, where ε_0 and ε are, respectively, the permittivities of free space and the dielectric. The relative permittivity (i.e., the dielectric constant) of the material $\varepsilon_r = \varepsilon / \varepsilon_0$. Thus

$$D = \varepsilon_0 E + P \tag{8.1}$$

where P is the partial charge density on the dielectric surface resulting from its polarization. P is also defined as the dipole moment per unit volume of the dielectric. From equation (8.1) it is evident that

$$\varepsilon_r = 1 + \frac{P}{\varepsilon_0 E} = 1 + \chi \tag{8.2}$$

χ being the susceptibility of the dielectric.

It can be easily shown that when a dielectric is charged, as in a capacitor, its energy stored per unit volume is $(1/2)\varepsilon_0 \varepsilon_r E^2$. Therefore, dielectrics with higher permittivities are favored by capacitor manufacturers. Typical values of the dielectric constants of insulating materials in common use are listed in Table 8.1.

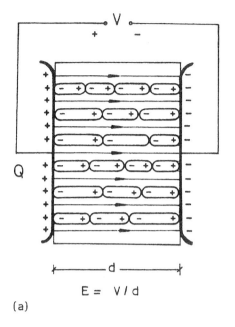

$$E = V/d$$

(a)

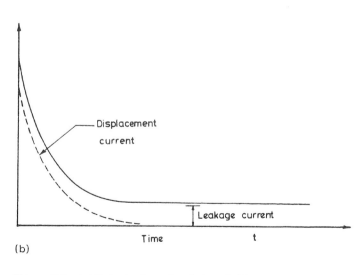

(b)

Figure 8.1 (a) Polarization of a dielectric block under DC electric field; (b) current components fed to a chapter as they vary with time from the instant it is connected to a DC source.

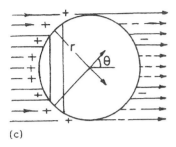

(c)

Figure 8.1 (continued) (c) field caused by surface charge on a spherical cavity
within the dielectric block of (a).

Table 8.1 Electrical Properties of Some Insulating Materials

Material	ε_r	$\tan\delta$ at 25°C, 1 MHz (average)	ρ_v at 25°C ($\Omega \cdot$ cm)	Breakdown stress (kV/mm)
Quartz glass	3.8	5×10^{-4}	10^{17}	
Lead glass	9			
Soda-lime glass	7	0.01		10–40
Glass ceramic	5–7	0.001–0.1	10^{12}–10^{14}	20–80
Porcelain	5	0.04	10^{12}–10^{15}	20–30
Zirconium ceramic	7–12	0.003		20–30
Alumina ceramic	10	0.0005		
Barium titanate ceramic	2000–8000	0.1		
Micanite	5–11	0.003	10^{14}–10^{16}	50 (for 1 mm specimen) 100–200 (for 0.1 mm specimen)
PVC	6	0.1		
PE	2.3	0.0001	10^{15}–10^{18}	40
Polystyrene	2.6	0.0001	10^{16}–10^{18}	30–40
PTFE	2	0.0002	10^{16}–10^{18}	25
Bakelite	4.5	0.1	10^{13}	
Vulcanized natural rubber	3.5	0.05		
Methyl methacrylate	3.6	0.01		

8.2.1.1 The Field inside a Cavity

The insulator body may contain particles of impurities or gas voids (in, for example, plastics, glass, or paper). In such a case the field strength inside the void is bound to differ from that outside, depending on the relative permittivities of the dielectric and the void. Take, for example, the enlarged spherical gas void inside the dielectric shown in Figure 8.1. The field inside it exceeds that outside it (E) by the component E_c produced by the charges on its surface. The density of this surface charge would vary with θ according to a cosine curve, with a minimum equal to P. Thus

$$E_c = \oint d(E_c \cos \theta) = \int_0^\pi \frac{P(\cos^2 \theta)2\pi r^2 \sin \theta}{4\pi\varepsilon_0 r^2} d\theta$$
$$= \frac{P}{3\varepsilon_0} \tag{8.3}$$

Substituting for P from equation (8.2), the total field in the void is

$$E_t = \frac{2 + \varepsilon_r}{3} E \tag{8.4}$$

In the case that the void has a different shape, this expression would differ. For example, if the void approaches a capillary tube along the field extending to the electrodes, $E_t = E$. On the other hand, if the void approaches a thin disk normal to the field, $E_t = \varepsilon_r E$, as can easily be shown.

8.2.2 Dielectric Polarization

The majority of dielectrics are nonpolar; that is, the centers of charges of the positive and negative ions in the molecule coincide. In a polar dielectric such as poly(vinyl chloride) (PVC), C_2H_3Cl, the size and charge of the chlorine atom are quite different from those of hydrogen atoms (Fig. 8.2). Thus a net dipole forms, which is different from the case of polyethylene (C_2H_4) where the four hydrogen atoms of the monomer are balanced. Under an electric field, the dipoles of a polar dielectric tend to become oriented along the field. In addition to this orientational polarization, the applied electric field exerts a force on each positive charge (be it a positive ion or even a nucleus of an atom) and on each negative charge (negative ions and electrons) and thus slightly displaces them along the field. Displacement polarization results in all dielectrics, both polar and nonpolar (Fig. 8.3). Also, impurity particles with permittivities and conductivities that differ from those of the parent material become charged and tend to migrate along the applied field (migrational polarization).

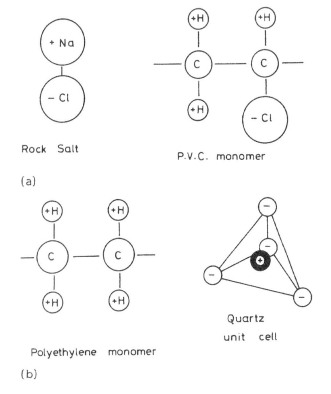

Rock Salt

P.V.C. monomer

(a)

Polyethylene monomer

Quartz

unit cell

(b)

Figure 8.2 Molecular representation of some polar (a) and nonpolar (b) dielectrics.

Thus, in a unit volume of the dielectric, there are N_e atoms in N_m molecules, each with a dipole moment p_e and p_m, respectively. They add up to

$$P = N_e p_e + N_m p_m = (N_e \alpha_e + N_m \alpha_m) E_t \tag{8.5}$$

where α is the polarizability of each under the local field E_t. From equations (8.4) and (8.5) it can easily be shown that

$$\frac{3\varepsilon_0(\varepsilon_r - 1)}{\varepsilon_r + 2} = N_e \alpha_e + N_m \alpha_m \tag{8.6}$$

which is the Clausius–Mosotti equation (Hummel, 1985).

It is obvious that each component of polarization needs time to materialize fully (Figs. 8.1 and 8.3). The fastest is electronic polarization, with a relaxation time of the order of 10^{-16} s whereas the slowest is migrational

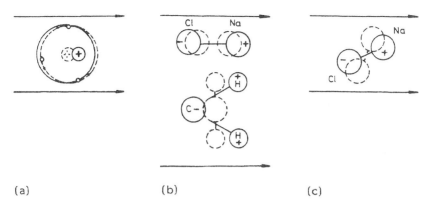

(a)   (b)   (c)

Figure 8.3 Displacement polarization of an atom (a) and a molecule (b), and orientational polarization of a molecule (c). Arrows indicate field direction.

polarization for which the relaxation time extends to seconds, minutes, hours, or in some cases even longer. The relaxation time for orientational polarization reaches the order of 10^3 s for glass.

The relaxation time for polarization manifests itself in the fact that, if a step-function voltage is applied across a capacitor with a solid dielectric, its polarization will take some time, however small, to build up, as shown in Figure 8.4. In studies concerned with periods much longer than the electronic and ionic relaxation times, and if the other polarizations are assumed to have one equivalent relaxation time τ, the polarization of the dielectric under study varies with time according to a relation of the form

$$P(t) = P_\infty - (P_\infty - P_0)\exp\left(-\frac{t}{\tau}\right) \tag{8.7}$$

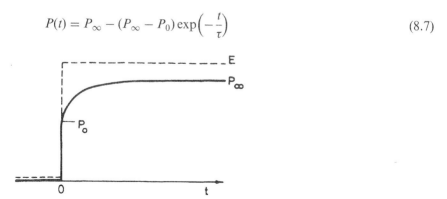

Figure 8.4 Buildup of polarization P in a dielectric under the effect of a step-function-imposed electric field E. P_0, initial polarization; P_∞, polarization value after infinite time.

If the initial and ultimate magnitudes of polarization P_0 and P_∞ are significantly different, this means that the relative permittivity of the material has two correspondingly different values, according to equation (8.2): ε_{r0} and $\varepsilon_{r\infty}$.

Now, if a capacitor with such a dielectric is subjected to a sinusoidally alternating voltage with angular frequency ω, the polarizaton will vary with the voltage but will lag slightly behind it, depending on the difference $(P_\infty - P_0)$ and on ω as compared to $1/\tau$. The polarization and permittivity would thus be complex quantities, that is,

$$P = P' - jP'' \tag{8.8}$$

and

$$\varepsilon_r = \varepsilon_r' - j\varepsilon_r'' \tag{8.9}$$

It can easily be shown that

$$\varepsilon_r' = \varepsilon_{r\infty} - \frac{\varepsilon_{r\infty} - \varepsilon_{r0}}{1 + \omega^2\tau^2} \tag{8.10}$$

and

$$\varepsilon_r'' = \frac{\omega\tau}{1 + \omega^2\tau^2}(\varepsilon_{r\infty} - \varepsilon_{r0}) \tag{8.11}$$

These two components depend on the dielectric material, its relaxation time, and the frequency of the applied electric field. Both are shown schematically in Figure 8.5.

Under DC, and also at very low frequencies, $\varepsilon_r' \longrightarrow \varepsilon_{r0}$ and ε_r'' almost vanishes, as noted from Figure 8.5 and equations (8.10) and (8.11). With increase in frequency, the inertia of the dipoles will cause the lag of the orientational polarization behind the field alteration to become more and more significant, until this type of polarization ceases to contribute to the whole. Thus ε_r' drops. It can easily be shown that at $\omega\tau = 1$, $d\varepsilon_r'/d\omega$ is maximum and ε_r'' reaches a peak. This explains the variations of the two curves in Figure 8.5 at the first and subsequent resonant frequencies. The variation of ε_r' with frequency is usually termed the "dispersion" curve, whereas the variation of ε_r'' is known as the "resonance absorption characteristic," as ε_r'' represents dielectric loss (i.e., energy absorption).

The dispersion of dielectric materials can be measured by tests described by national and international standard specifications (IEC, 1973). A capacitor containing the dielectric is charged from a DC source with a voltage V_0 and then momentarily short-circuited and its residual open-circuit voltage V_1 is measured; the dispersion is expressed as $V_1/(V_0 - V_1)$.

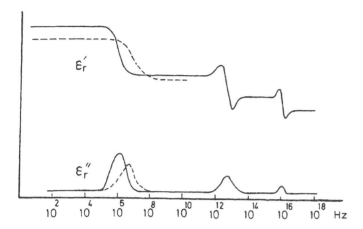

Figure 8.5 Schematic showing the components ε_r' and ε_r'' of the complex permittivity of a dielectric as they vary with the frequency of the applied electric field: ———, at a temperature T_1; — — — — — —, at a higher temperature T_2. [From Tareev (1980).]

The effect of temperature on the permittivity and polarization of nonpolar dielectrics is negligible. In polar materials, however, the random thermal motion of the dipoles is heavily influenced by temperature. The higher the temperature, the lower will be the relaxation time τ. Hence the magnitudes of ε_r' will be lower and the resonant frequencies of ε_r'' will be higher (Fig. 8.5).

8.2.3 Electrical Conduction

Practical insulating materials have admittedly high resistivities but these are, nonetheless, finite. For wood, marble, and asbestos the volume resistivity ρ_v is in the range 10^6–$10^8\,\Omega\cdot m$, whereas for polystyrene and polyethylene it is in the range 10^{14}–$10^{16}\,\Omega\cdot m$. Quartz crystals are nonisotropic; ρ_v is about $10^{12}\,\Omega\cdot m$ along its optical axis and about $10^{14}\,\Omega\cdot m$ in directions normal to it. The volume and surface resistances of an insulator are two paths in parallel for the leakage currents I_v and I_s it draws when connected to a power source (Fig. 8.6). Their relative values can vary drastically, depending on the insulator material and its surface conditions.

Electrical conduction through dielectrics is undertaken by ions, and is basically different from the process in conductors where electrons are responsible for the conductivity. The reason is that in dielectrics the energy necessary for dislodging ions from their positions in the atomic lattice is on the order of that of their thermal vibrations at normal room temperature.

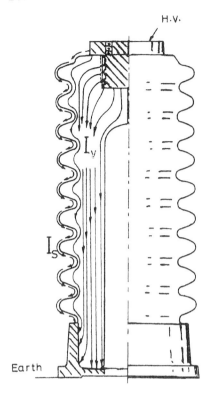

Figure 8.6 Insulating bus support in half-section showing the two components of its leakage current: I_v, through its volume, and I_s, creeping on its surface.

On the other hand, the energy needed to liberate electrons in a dielectric is at least one order of magnitude higher. For example, in a simple dielectric such as rock salt (NaCl), the energy required to remove an ion from its lattice position is 0.8 eV, whereas electrons need at least 6 eV for their liberation.

It is evident that, with increased temperature, more ions could be dislodged from their positions in the atomic lattice and could contribute to the electrical conduction. Also, their mobilities increase with temperature exponentially. Therefore, for insulating materials the volume resistivities decrease sensitively with increased temperature.

8.2.4 Dielectric Losses

Practical insulating materials draw leakage currents, however small. There is additional energy loss in the process of polarizing the material. The latter loss is significant under AC with frequencies approaching the resonant frequencies shown in Figure 8.5. Thus, under AC the current drawn by a practical capacitor leads the applied voltage by an angle slightly less than 90°; the difference δ is termed the loss angle. Power engineers refer to the factor (tan δ) as the loss factor, whereas electronics engineers prefer to use cot δ, the quality factor of the dielectric.

8.2.4.1 Effect of Temperature

There are two factors to be considered here. First, in a homogeneous dielectric the conductivity contributed by the jumping of ions from one site to another under the applied electric field becomes easier with higher vibrational energies (i.e., at higher temperatures). The volume resistivities of insulating materials therefore drop considerably with temperature. Distinguished examples are plastics and glass. With PVC, for example, the resistivity drops and the loss factor rises about 10-fold for a temperature rise from 80°C to 120°C. Second, in a heterogeneous material, such as paper, a trace of moisture accentuates the effect of temperature. For oil-impregnated paper, for instance, 3% by weight of moisture causes the loss factor to rise from 0.5% at 20°C to about 30% at 99°C.

8.2.4.2 Effect of Electric Stress

The loss factor for an insulating material is not much affected by the magnitude of the applied electric stress except at levels high enough for secondary phenomena to set in. A well-known example is the ionization that occurs in gas voids within the material. The corresponding power loss is analogous to the corona loss (Chapter 5). It causes a considerable rise in the loss factor above the discharge-onset voltage (Fig. 8.7). At higher voltages the gases in additional voids get ionized. A well-designed and well-made cable should have the minimum number of gas voids (Chapter 12). Measuring the loss angle δ is one of the important tests to be carried out on test samples of high-voltage cables (Chapter 18).

8.3 THERMAL, MECHANICAL, AND CHEMICAL PROPERTIES

Heat is conducted through insulating materials by phonons. The thermal vibrations in a crystal or a solid continuum can be represented as elastic waves propagating through the solid with the velocity of sound. Thermal conduction of insulating materials is the one important factor determining heat transfer in cables, capacitors, and some electronic devices. In transformers and machines, on the other hand, heat is removed primarily by convection. Evaporative cooling is sometimes used to cope with hot spots.

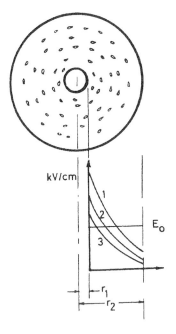

Figure 8.7 Cross-section of a coaxial paper-insulated cable with gas voids scattered in its insulation. E_0 is the gas discharge onset field strength. The curves represent the radial field distributions at different applied voltages $V_1 > V_2 > V_3$.

Thermal conductivities of plastics range between 0.15 and 0.3 W/m·K, whereas for porcelain and glass it is about 1.2 to 1.7 W/m·K. For alumina and magnesia it is as high as 35 W/m·K, which recommends both materials for making special ceramics.

8.3.1 Thermal Classes

Insulating materials commonly used in machines, transformers, capacitors, and cables have been classified according to the maximum safe temperature each material or combination of materials can endure for a very long time under normal service conditions (IEC, 1984). The code letter and temperature limit assigned to each class are given in Table 8.2. Examples of materials belonging to each class are given below.

Class Y: paper, cotton, and natural silk, when not impregnated with oil, varnish, or other insulating liquid. This class also includes vulcanized natural rubber and thermoplastics limited by their softening temperature to

Table 8.2 Thermal Classes of Solid Insulating Materials

Class code[a]	Y	A	E	B	F	H	200	220	250
Maximum temperature for continuous operation (°C)	90	105	120	130	155	180	200	220	250

[a]Before 1984, the IEC had grouped the three top classes in one group termed class C, withstanding temperatures > 180°C.
Source: IEC (1984).

be within this class, such as polyethylene (PE), cross-linked polyethylene (XLPE), and poly(vinyl chloride) (PVC).

Class A: paper, cotton, and natural silk when impregnated with oil, varnish, or resin. This class also includes some organic enamels, varnishes, and thermoplastics with suitable softening temperatures.

Class E: laminated and molded plastics with thermosetting binders, epoxy enamels, and varnishes of suitable thermal endurance (e.g., phenol-formaldehyde, melamine-formaldehyde, and polyethylene terephthalate fibers).

Class B: inorganic fibrous and flexible materials (e.g., mica, glass, asbestos) when bonded or impregnated with an organic binder such as shellac, alkyd, or rosin with mineral fillers.

Class F: materials of class B but bonded or impregnated with resins of suitable thermal endurance (e.g., epoxy and silicon alkyd).

Class H: materials of class B but bonded or impregnated with suitable inorganic resin, such as silicon resin. This class includes silicon rubber.

Class C: inorganic materials (e.g., mica, ceramics, fused quartz) when used without binders or with binders suitable for the required thermal endurance. This class also includes polytetrafluoroethylene (PTFE). In 1984 this class was split (IEC, 1984) into three subclasses, coded 200, 220, and 250, corresponding to their safe operating temperatures.

It is realized that insulating materials in general can withstand temperatures significantly higher than the limit assigned to their class but only for a very short duration—a fraction of a second for a short circuit cleared rapidly by a circuit breaker. At too high a temperature, insulating materials inadmissibly degrade in their electrical and mechanical properties. Their electrical conductivities and losses increase. Their mechanical strength and elasticity limit decrease considerably. Some insulating materials, such as plastics and cellulosic materials, are apt to melt, decompose, shrink, craze, char, or even burn at elevated temperatures.

8.3.2 Effects of Moisture

Paper and other organic fibrous materials are bound to contain at least traces of moisture. The cellulose they contain, like hydroxyl, carboxyl, and amine groups, has an affinity for water. Also, being polymers, most plastics have a measurable permeability for water. Moisture has been found to permeate through micropores in plastic sheaths of underground cables. Once inside the cable and under electric stress, these water droplets aim at sites of higher field concentration by electrophoresis. At their new sites they will highly distort the field, become ionized, and start water treeing, which may eventually cause breakdown of the cable, as discussed in Section IV. The deleterious effects of moisture are worsened by salts, acids, and other such easily ionizable ingredients if dissolved in water. A trace of 3% moisture absorbed in dry paper reduces its resistivity by about six orders of magnitude. Therefore, meticulous care is taken during the manufacture of paper and similar cellulosic materials when designed for use as electrical insulation.

8.3.3 Tests

Tests have been prescribed for ensuring various qualities of insulating materials. Thermal endurance tests are by no means straightforward, because the degradation of an insulating material due to exposure to a high temperature cannot be expressed by a single variable. However, a thermal endurance test that has been widely adopted is to measure the loss in weight of a specimen after exposure to a certain high temperature for several hundred hours (Steffens, 1986; IEC, 1981, 1983, 1987).

8.4 DIELECTRIC BREAKDOWN

The process of breakdown of solid dielectrics is much more complex than that of gases, for example. Although solid insulating materials have been the subject of numerous investigations, no single theory fully explains the process of breakdown and predicts the breakdown stress of a given insulator. One of the principal reasons for this state of affairs is the dependence of the breakdown on numerous factors, including the material's temperature, voltage duration, and ambient conditions. However, three primary processes of dielectric breakdown have been discerned: intrinsic, thermal, and electrochemical. Each occurs under a certain set of conditions conducive to it. They are all discussed briefly in this section.

8.4.1 Intrinsic Breakdown

Intrinsic breakdown occurs purely due to the electronic behavior of the dielectric, with no effect of ambient or temperature rise. It is therefore sometimes called electronic breakdown. It is an ideal, extremely difficult to identify in practice by eliminating all secondary causes of breakdown. Care should be taken not to allow breakdown due to thermal instability of the dielectric with temperature rise caused by leakage currents, nor should the dielectric fail under electromechanical stresses, nor should there be impurities or gas discharges to initiate failure due to electrochemical decomposition.

It is thus evident that intrinsic breakdown would be possible for idealized crystalline materials. In 1947, Fröhlich proposed a criterion for electronic breakdown as "the electric stress at which the dielectric acquires a field-enhanced conductivity." He developed a theory for crystalline dielectrics, such as potassium chloride, relating the breakdown stress to six constants of the crystal lattice. A detailed discussion of his theory is beyond the scope of this section. As his crystal constants are not measurable with any accuracy for simple crystalline dielectrics, to say nothing of practical amorphous materials, a brief qualitative review will suffice.

Cosmic and other radiations would set a few electrons free in the crystalline dielectric. These would be accelerated by the applied electric field and would collide with some lattice sites on their way. The energies of these electrons differ according to Boltzmann's distribution (Chapter 3). Depending on the energy of each individual electron, its energy gain from the electric field and loss during the interaction with the lattice site it collides with vary as shown schematically in Figure 8.8. Curves 1, 2, and 3 represent the rates of energy gain by an electron from fields with different strengths $E_1 > E_2 > E_3$, whereas curve S represents the energy loss. The limit W_i is

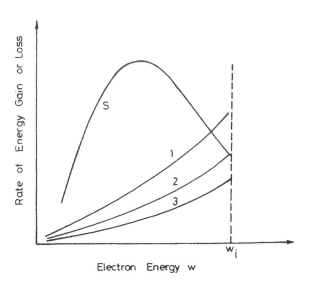

Figure 8.8 Rates of energy gain by electrons from the applied electric field E and the rate of their energy loss S to the crystal lattice as functions of the electron energy. W_i is the ionization energy for the atoms in the lattice: $E_1 > E_2 > E_3$.

the energy sufficient for the electron to ionize the atom it interacts with at the lattice site.

For intrinsic breakdown to occur, the energy gain should just exceed the loss. This would result in the field-enhanced conductivity suggested by Fröhlich. Thus the critical field strength is E_2, as shown in Figure 8.8. Attemps have been made to measure intrinsic breakdown voltages of dielectric specimens in the laboratory under carefully controlled conditions. Care was taken to eliminate or greatly suppress secondary causes of breakdown. The highest breakdown stress reached for mica, for example, was 10 MV/cm and for polyethene it was 8 MV/cm, obtained with extreme care to prevent secondary effects from causing breakdown at much lower stresses.

8.4.2 Thermal Breakdown

Thermal breakdown results from disruption of thermal equilibrium in the dielectric: the rate of heat generation within the material exceeds that of cooling. In such a case the temperature starts to rise monotonically instead of leveling off at a steady value (Fig. 8.9a). With temperature rise, the volume resistivity of the insulating material decreases and the dielectric losses increase. Thus, with a monotonic temperature rise, a runaway process sets in, leading eventually to breakdown.

Attempts at mathematical analysis of this situation were made by Whitehead in 1951 (Kuffel and Zaengl, 1984). For simplicity, consider the linear heat flow to the ambient together with the dielectric losses in a capacitor of a large area A under DC (Fig. 8.10). The heat balance of the elemental volume ($\Delta A \Delta x$) at a distance x from the center plane is

$$\frac{E^2(x)}{\rho_v(x)} \Delta A \Delta x = -k \Delta A \Delta x \frac{d^2 T}{dx^2} \tag{8.12}$$

where $E(x)$ is the electric field strength at x; k the thermal conductivity of the dielectric; and ρ_v its volume resistivity, known to decrease almost exponentially with temperature. Realizing that the density of leakage current through the dielectric $i = E(x)/\rho_v(x)$ is independent of x, equation (9.12) can be rearranged to

$$-k \frac{d^2 T}{dx^2} = iE(x) \tag{8.13}$$

Integrating both sides over the distance O–x, we get

$$-k \frac{dT}{dx}\bigg|_O^x = i[V(x) - V(O)] \tag{8.14}$$

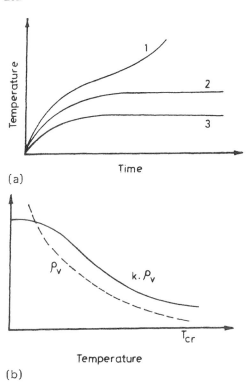

(a)

(b)

Figure 8.9 (a) Temporal increase of temperature of dielectric in a capacitor when subjected to different step voltages: $V_1 > V_2 > V_3$. (b) Schematic showing the variations of the volume resistivity ρ_v of a dielectric and of the product of its volume resistivity and its thermal conductivity k. T_{cr} is the critical temperature at which the dielectric fails.

as $dT/dx = V(O) = 0$ at $x = 0$, the center plane of the dielectric block. Thus, rearranging and integrating equation (8.14), we get

$$\frac{V^2}{8} = -\int_{T_i}^{T_a} k\rho_v \, dT \tag{8.15}$$

where V is the total voltage across the dielectric block.

The critical conditions are reached when the temperature at the center of the dielectric $T_i = T_{cr}$, the critical temperature at which the thermal runaway process sets in. Thus, according to Whitehead's theory, the thermal breakdown voltage of a dielectric block is independent of its thickness and can be obtained by integrating the product $(k\rho_v)$ over the range from the

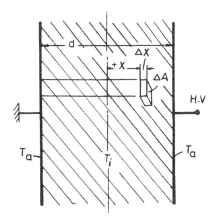

Figure 8.10 Heat flow to the ambient from a dielecric block. The heat is caused by the dielectric loss under voltage. The maximum temperature T_i occurs at the center plane, whereas T_a is the ambient temperature.

ambient to the material's critical temperature (Fig. 8.9b). Computations of this area were made for numerious insulating materials and their voltages for thermal breakdown ranged between 1 and 5 MV (Alston, 1968). Experimental measurements, however, yielded voltages up to only 400 kV. The discrepancy is explained at least partially by the nonhomogeneity of most dielectric materials and inaccurate values of their physical characteristics at higher materials.

One other important point to note here is that there is a limit for the thermal breakdown voltage, no matter how thick the insulation is made. This is to be kept in mind when designing high-voltage machines and EHV cables. The relation (8.15) eventually does not hold for very small thicknesses. A theoretical account can be developed if the surface temperature of the specimen is allowed to exceed that of the ambient, which is more realistic. In this case the heat balance equation per unit area over the entire thickness of the specimen (Fig. 8.10) takes the form

$$\frac{V^2}{\rho_v d} = HF(T_i)$$

or

$$\frac{V^2}{d} = H\rho_V F(T_i) \tag{8.16}$$

The function $F(T_i)$ relates the surface temperature to the highest temperature inside the specimen, and the product $HF(T_i)$ gives the rate of heat dissipation to the electrodes and ambient. As the dielectric resistivity ρ_v is another function of temperature, the right-hand side of equation (8.16) can be considered as a constant for a given critical temperature at which the dielectric fails thermally. Thus, this simple approximate approach yields the relation between the thermal breakdown voltage and the small thickness d of a dielectric specimen as

$$V = cd^{1/2} \tag{8.17}$$

c being a constant.

Although the discussion above was concerned with a DC voltage applied across the dielectric specimen, the same result could be reached had we considered instead an AC voltage. In that case the dielectric loss would be expressed as ($\omega CV^2 \tan \delta$).

8.4.3   Treeing in Plastics

Treeing began to receive wide attention in the mid-1970s as polymers started replacing oil-impregnated paper to insulate high-voltage cables. Electric treeing initiated by gas discharges at microscopic sites, if not inhibited, was seen to lead to slow, yet complete, breakdown.

Although specimens of polyethylene and cross-linked polyethylene tested under AC of short duration break down at not less than 700–800 kV/mm, and the macroscopic working stresses in cables with such an insulation range from about only 3 kV/mm (in low-voltage cables) to 20 kV/mm (for 250 kV cables), some cables have actually failed in service as a result of treeing (Deschamps et al., 1984; Hosokawa et al., 1982). Treeing begins at microscopic sites of imperfections in the molecular structure of polymers. There may be metallic particles, cavities, or thermo-oxidated or deteriorated particles of material, all of which are extremely difficult to eliminate. Also, water can diffuse through microscopic pores and cracks in the insulation and through sheaths of polymer-insulated cables. Such cracks might be 10 nm in width in the amorphous regions between crystalline lamellae of semicrystalline polyers (Sletbak and Ildstad, 1984; Marton, 1987; Filippini et al., 1987). Water, with its soluble ions, would migrate by electrophoresis under the electric field to sites of higher intensity. When water is involved, the phenomenon is termed water treeing.

A concentrated electric field at the tip of a tubule, pulsating under AC, combined with hydrostatic pressure caused by water inside it being heated by concentrated leakage currents and chemical decomposition, would result in pulsating mechanical forces at the tip of the tubule. This would cause extension and branching of the tubule in a treelike shape in steps of micro-fractures of the material (Fig. 8.11).

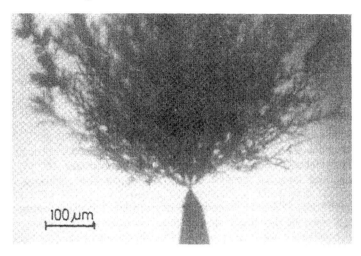

Figure 8.11   Tree developed at the tip of a needle electrode inserted into an epoxy dielectric. (Courtesy of A. Rasmy and O. Gouda.)

The oxidation theories of water treeing describe the growth of water trees as driven by electro-oxidation of the polymer, which takes place in the direction of the local electric field and in a polar amorphous region of the polymer. As a consequence of the electro-oxidation, polymer chains are broken and a "tree path" is formed. The local electro-oxidation of the polymer along this track converts the track region from hydrophobic to hydrophilic, which results in the condensation of water molecules dispersed in the polymer matrix to form liquid water in the track. The liquid water in a track promotes the transport of ions that mediate further the electro-oxidation of the polymer at the tip of the track. Thus a track becomes self-propagating in a manner similar to the self-propagation of an electrical tree or a gas breakdown channel, although on a very different time scale (Meyer and Filippini, 1979; Sletbak, 1979; Fischer et al., 1987; Xu, 1994; Fan and Yoshimura, 1996). The water tree region contains a wide range of chemical species, with various forms of carbonyl and metal ions and water being dominant. Among carbonyl species, carboxylate ions are dominant, followed by trace amounts of esters and ketones. The water tree region contains a wide range of chemical species, with various forms of carbonyl and metal ions and water being dominant.

8.4.3.1 Effect of Aging Conditions on the Type of Water Treeing

The composition of water trees and, in turn, their breakdown strength depends on the aging conditions. The type of water tree growth mechanism varies considerably with the applied aging conditions. The water treeing processes determine the properties of the resulting water trees. The characteristics of water trees differ from those of unaged polyethylene. It has been shown that water trees grown under different conditions can be quite different entities, having different properties (Ross, 1993).

8.4.3.2 Influence of Crystalline Morphology on the Growth of Water Trees in Polyethylene (PE)

The growth of water trees is faster in low-density polyethylene samples in which spherulites are larger and the number of the spherulites is smaller. Spherulites are one of the main constituents of PE; they take the form of small rounded bodies with radiate fibrous structure. On the contrary, water trees grow more slowly in low-density polyethylene samples in which spherulites are smaller and the number of the spherulites is higher. The spherulite boundary regions, whose character is hydrophilic, can greatly inhibit the growth of water trees, while other two kinds of boundary region (weak electrolyte; hydrophobic) have little influence on their growth of water trees (Fan and Yoshimura, 1996).

8.4.3.3 Role of Semiconducting Compounds in Water Treeing of XLPE Cable Insulation

Semiconductor shields are used in power cables. The initiation of water trees at a semiconductor–polyethylene interface requires ions which are generally present in and on the semiconducting material. The concentration of water-soluble ions on the surface of semiconducting material can be reduced through simple water extraction, and such reduction has a substantial effect on the number and length of water trees.

The growth of water trees does not depend solely on ions originating from the semiconducting shields. Ground water will generally carry substantial ionic contamination. The tree length and size distribution differs significantly for the semiconducting compounds tested, with the newer technology products which incorporate the cleanest carbon black showing a trend toward reduced tendency for the growth of vented water trees (Boggs and Mashikian, 1994).

8.4.3.4 Mechanical Aspects of Electrical Treeing in Solid Insulation

The mechanical properties of insulation affect the initiation and growth of electrical trees. The relationships between treeing phenomena and specific mechanical properties can be masked by the simultaneous dependence on more than one mechanical property, and the outcome will depend on the dominance of one parameter over others under a given physical condition. For example, in a specimen containing initial strain, plasticization effects on fracture toughness are masked by the reduction in initial strain with decreased tree growth, contrary to expectations. The true dependence of tree growth on fracture toughness can be observed only under conditions of low internal strain (Auckland et al., 1993). In reality, strain-free insulation may be impracticable, due to the manufacturing processes and/or the inability to carry out post-manufacture annealing, in which case plasticized insulation may limit initial strain and enhance tree resistance in a material. Strain-free material, however, is preferred for having less initiation and slower growth, in which case plasticization is to be avoided. The ensuing reduction in tensile strength, modulus of elasticity, and fracture toughness, produced by plasticization, is detrimental from the point of view of tree inhibition.

In homogeneous resin, the tensile stength, modulus of elasticity, and fracture toughness all show a pronounced decrease in value as the temperature is increased. The material's resistance to crack propagation therefore becomes lower with the application of heat. The main reason for the degradation in mechanical properties is that the material passes through its glass transition point within the range of test temperatures. The treeing resistance displays a similar fall as the temperature exceeds this critical value.

8.4.3.5 Water Tree-based Electrical Trees

The water tree structure consists of a large number of voids, which are filled with water when the tree is water saturated. Channels may be present within the inner regions of the water tree, but the structure of the outer regions is a collection of voids. It is in, or even beyond, this outer region that an electrical tree is initiated. The electrical field associated with a water tree depends on the shape of the tree, the dielectric constant, and conductivity within the tree. The electrical tree forms at the end of the water tree.

8.4.3.6 Length Distribution of Water Trees

The length distribution of water trees—rather than the breakdown voltage—is in many cases used to assess the degradation in insulation. An inverse relationship exists between the size of the water trees causing the breakdown and the breakdown stress level for cables aged at moderate stresses (Ross, 1993).

When the duration of aging is also taken into account, the growth rate of water trees can be determined. Variables reflecting the length distribution can be: the largest water trees per unit volume, the standard deviation of tree lengths, the density of trees, etc.

8.4.3.7 Detection of Water Tree Growth using Transient Current Measurement

In a study by Li et al. (1995), a time window for transient current measurements was chosen so that reliable low-frequency dielectric data could be obtained. With water-treed low-density polyethylene (LDPE) samples aged in the laboratory, measurements of transient currents, with subsequent transformation into the frequency domain, showed a loss peak at 10^{-4} to 10^{-3} Hz when moisture was retained in the samples.

Even with free water removed, a higher dielectric loss is observed in water-treed samples than in virgin samples. The relaxation behavior may serve as an indicator of water treeing in polymeric cable insulation. The results of both transient measurements were said to be affected significantly by the treatment conditions of the samples.

8.4.4 Tracking in Electrical Insulation

Insulation surfaces may be capable of meeting the overvoltage criteria when the equipment is installed. However, as time goes on and contaminants such as salt, dust, and atmospheric chemical agents collect on these surfaces, the integrity and reliability may be jeopardized under humid conditions by a form of surface failure called "tracking." Solid/air interfaces—which are subject to tracking—exist in bushings or terminations, on support structures, or in the end-windings of a rotating machine. Moreover, polymeric materials contain carbon atoms in their molecular structure and are therefore prone to tracking. This is particularly alarming with the increasing use of dielectric organic materials such as polyethylene, rubbers, epoxies, and polyesters for reasons of economy, strength, and ease of fabrication to good tolerances.

8.4.4.1 Failure Mechanism

A moisture film on a polluted surface, whether formed by condensation or precipitation, will conduct leakage current with a magnitude determined by the type and extent of the pollution. As the leakage current flows between live parts, or from a live part to ground, the conductive film will heat nonuniformly and begin to evaporate nonuniformly, distorting the stress distribution over the surface. In particular, "dry-bands" will form, resulting in regions of very high resistivity between the edges of the remaining wet film. Nearly the total surface voltage will appear across this dry-band, causing flashover of the gap. The arcing will continue until the gap widens sufficiently by further evaporation to the point where the arc can no longer be maintained.

When this occurs on the surface of an organic material, the temperature of the arc may be sufficient to decompose the material locally, releasing free carbon. Dry-banding, arcing, and the formation of free conducting carbon will proceed in a relatively random manner until a continuous conducting path forms between the live parts, or at least bridges a sufficient portion of the surface, resulting in flashover.

The "track" on the surface usually has a characteristic branched appearance. Similar surface failures have been observed in other circumstances, such as on structures in oil-filled transformers, or on the surface of polymeric (filled epoxy) spacers in gas-insulated gear. Whereas tracking on headboards or barriers in oil-filled transformers is also usually associated with the presence of excessive moisture, the tracks on GIS spacers are more frequently associated with damage done by internal flashovers caused by the presence of particles, and by severe transient overvoltages. Figure 8.12 shows a model of the tracking breakdown process (Yoshimura et al., 1982).

8.4.4.2 Mitigation of Tracking

Tracking may be prevented by keeping the surfaces in question clean and dry, where this is possible. The installation of heaters in outdoor metalclad switchgear compartments has solved many problems. Preventive maintenance in the form of frequent and thorough cleaning of insulator surfaces, support panels, machine end-windings, and the like, is highly recommended. Such maintenance is obviously very costly. It thus makes sense to spend more money, if necessary, to specify track-resistant plastic insulating materials in the first place, and even to modify such materials, where possible, during repairs after a tracking failure. In some cases a temporary solution may be achieved by cleaning or filing away the carbonized matter and applying a good grade of electrical varnish such as an alkyd. It is far better to replace with a better material as soon as possible (Kurtz, 1987).

Some polymers are inherently more resistant to tracking than others. In most cases the performance of a material such as glass-reinforced polyester resin can be greatly improved by the addition of materials such as hydrated alumina. Some epoxies are better than others. Most can be improved by the addition of inert fillers as well as hydrated alumina, which acts to oxidize free carbon as it is formed.

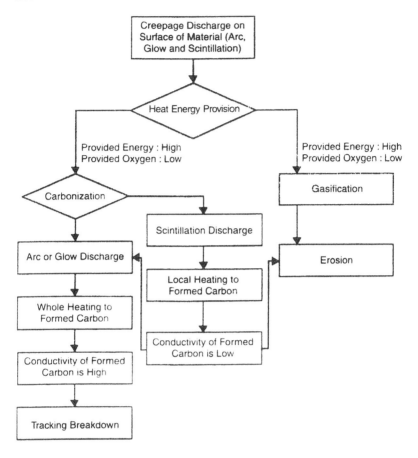

Figure 8.12 A model of tracking breakdown.

8.4.4.3 Tracking Tests

It is important to evaluate the tracking resistance of such materials as high temperature vulcanized (HTV) and room temperature vulcanized (RTV) silicone rubbers, and ethylene vinyl acetate (EVA). A number of very effective test methods have been developed to quantify and rank insulating materials in terms of their resistance to failure by surface tracking. In general, the test procedures expose samples of the (organic) insulating material to an environment representing an extreme exposure to voltage stress, contamination, and moisture. In most cases, the tests rank materials in roughly similar order, so that virtually any of them may be used to select a "superior" material; however, some tests may be better suited to distinguish between materials at one end of the "severity of exposure range" than at the other.

A laboratory test for evaluation of tracking and erosion resistance of high voltage insulating materials has been described (Gorur et al., 1997). This test is based on combining some features of the dust and fog test and the inclined plane test. The materials evaluated included various formulations of HTV silicone rubber and polyolefin polymers. The test was used for screening materials and obtaining a relative ranking of the tracking and erosion resistance of various materials. Measurements of the leakage current were made via a computerized data acquisition system, and the discharge activity was monitored with a high-speed camera. It is claimed that materials that pass the 8-hour test without initiation of tracking or significant erosion under conditions that are more severe than field conditions will be suitable for outdoor HV insulation applications.

8.4.4.4 Environmental Effects on Tracking

The environmental effects on tracking are considered, because of the increasing use of organic insulating materials in areas that are heavily contaminated, at high altitudes, and/or in highly humid regions. Of those effects the following are considered to be relevant:

- *Ultraviolet rays*: Outdoor insulators are subjected to ultraviolet (UV) radiation from the sun. Ultraviolet rays are one of the most severe aging factors for polymer insulators (Du et al., 1996). The lifetime of insulators is affected by exposure to UV radiation, as seen in Figure 8.13 (Kamosawa et al., 1988).
- *Low atmospheric pressure*: The tracking resistance of polycarbonates was reported to increase with the decrease in atmospheric pressures such as that encountered at higher altitudes. In contrast, another material—polyethylene terephthalate—showed a decrease in tracking resistance with decreased atmospheric pressure (Du et al., 1995).
- Other environmental factors that may have effects on tracking resistance of polymers include exposure to *gamma radiation* and to *acid rain*.

8.5 INSULATING MATERIALS

Traditional materials include cellulosics such as paper and organic fabrics. They also include mica, glass, and ceramics. Cellulosic materials suffer from hygroscopicity and are bound to contain gas voids. Therefore, to be usable as good insulating materials, they have to be specially treated and impreg-

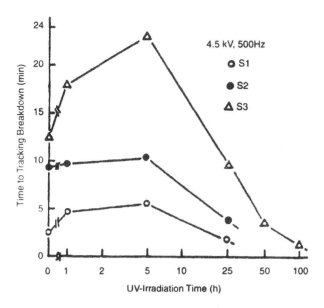

Figure 8.13   Effect of ultraviolet radiation on the time to tracking breakdown.

nated with oil or varnish. The insulating characteristics can be improved drastically by combining paper with polypropylene (PPP) (Samm, 1987; Schaible, 1987; Mark, 1984).

Mica has traditionally been used where high temperatures and/or surface discharges are experienced (e.g., in vacuum tubes, heaters, and DC machine commutators). Mica powder mixed with finely ground glass in a hot-pressing process yields a dense machinable material with excellent electrical qualities, particularly suitable for insulators in high-voltage high-frequency equipment. Recently, synthetic mica has been developed in which the OH groups are replaced by fluorine atoms, enabling the mica to withstand much higher temperatures.

Porcelain and glass have traditionally been used in insulating overhead power lines and busbars (Chapter 9). However, since the 1960s much lighter insulators with excellent electrical and mechanical characteristics have been made of epoxy resins reinforced with glass fibers. Depending on the composition of ceramics, they can be manufactured for use as insulators with almost any application, as magnets, or as semiconductors with resistivities sensitively varying with temperature, humidity, or electric stress.

Several newly developed insulating materials fall within the classification as polymers. Their molecular arrangements are either linear or cross-

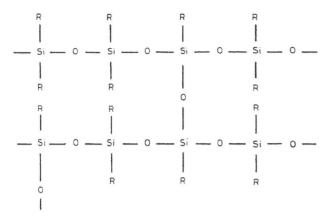

Silicone Resin

Figure 8.14 Representation of molecular arrangements of a silicon resin.

linked, the former including elastomers and thermoplastics, and are generally very flexible and yield with time under a given mechanical load. The higher the temperature, the greater the yield. They include polyethylene (PE), poly(vinyl chloride) (PVC), and polyamides (nylons). Crosslinking has considerably improved the electrical qualities of PE. Thus XLPE has been favored as an insulant in both low-voltage and medium-voltage cables. As an exception to the rule, polytetrafluoroethylene (PTFE), which is a linear polymer, is not flexible nor does it yield and melt at high temperatures. Because of the high binding energies of fluorine atoms, the polymer is about 95% crystalline. At temperatures as high as 400°C, it decomposes. It enjoys thermal and mechanical endurance and chemical stability, much higher than that of PE and PVC (Margolis, 1985).

Depending on their production process, polysiloxanes (silicon resins) can be either elastomers or thermosets. The latter have a molecular arrangement such as that represented in Figure 8.14. The organic radical R could be a methyl (CH_3), ethyl (C_2H_5), or phenyl (C_6H_5). The link of silicon with oxygen, Si—O, is stronger than that of carbon, C—C, which accounts for the considerably higher thermal endurance of silicon polymers.

8.6 ACTIVE DIELECTRICS

In addition to the passive dielectrics employed solely as electrical insulators, there are dielectrics that have other applications. In such "active" dielectrics there are mutual effects between imposed electric fields and mechanical stresses, heat influx or penetrating light. They are used in piezoelectric and pyroelectric transducers, nonlinear capacitors, ferroelectric memory devices, electro-optic voltage sensors, electric light shutters, vidicons, and electrets. Their far-reaching applications are extending rapidly in many areas of engineering (CIGRE, 1987).

8.7 PROBLEMS

(1) A solid specimen of dielectric has a dielectric constant of 4.2, and tan $\delta = 0.001$ at a frequency of 50 Hz. If it is subjected to an alternating field of 50 kV/cm, calculate the heat generated in the specimen due to the dielectric losses.

(2) A solid dielectric specimen of dielectric constant 4.0, shown in the figure, has an internal void of thickness 1 mm. The specimen is 1 cm thick and is subjected to a voltage of 80 kV (rms). If the void is filled with air, and if the breakdown strength of air can be taken as $30\,kV_{peak}/$ cm, find the voltage at which an internal discharge can occur.

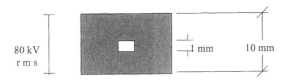

(3) A coaxial cylindrical capacitor is to be designed with an effective length of 20 cm. The capacitor is expected to have a capacitance of 1000 pF and to operate at 15 kV. If high density polyethylene is selected as the capacitor's insulating material, whose dielectric constant is 2.3 and which breaks down at 50 kV/cm, estimate suitable dimensions for the capacitor's electrodes.

REFERENCES

Alston L. High Voltage Technology. Oxford, UK: Oxford University Press, 1968.
Auckland DW, Taha A, Varlow BR. Proceedings of Conference on Electrical Insulation & Dielectric Phenomena (CEIDP), Pocono Manor, PA, USA, 1993, pp 636–641.
Boggs SA, Mashikian MS. IEEE Electr Insul Mag 10:23–27, 1994.
CIGRE. Proceedings of CIGRE Symposium on New and Improved Materials for Electrotechnology, Vienna, 1987.
Deschamps J, Michel R, Lepers J. CIGRE report 21-08, 1984.

Du B, Suzuki A, Kishi K, Kobayashi S. Trans IEE Japan 115A:1284–1293, 1995.

Du B, Suzuki A, Kobayashi S. Trans IEE Japan 116A:170–176, 1996.

Fan ZH, Yoshimura N. IEEE Electr Insul Mag 3:849–858, 1996.

Filippini J, Poggi Y, Rahamarimalala V, de Bellet J, Matey G. Proceedings of CIGRE Symposium on New and Improved Materials for Electrotechnology, Vienna, 1987, Report 620.04.

Fischer P, Peschke E, Schroth R, Farkas A. Proceedings of CIGRE Symposium on New and Improved Materials for Electrotechnology, Vienna, 1987, Report 620.06.

Fröhlich H. Proceedings of the Royal Physics Society A188:521–532, 1947.

Gorur RS, Montesions J, Varadadesikan L, Simmons S, Shah M. IEEE Trans DEI-4:767–774, 1997.

Hosokawa K, Kojma K, Toshida N, Kasahara T, Kaneko R. CIGRE report 21-09, 1982.

Hummel R. Electronic Properties of Materials. Berlin: Springer-Verlag, 1985.

IEC. Specifications of Insulating Materials Based on Mica—Definitions and General Requirements. Publication 371-1.

IEC. Methods of Tests. Publication 371-2. Geneva: International Electrotechnical Commission, 1981.

IEC. Specifications for Individual Materials. Publication 171-3. Geneva: International Electrotechnical Commission, 1983.

IEC. Recommendations for the Classification of Materials for the Insulation of Electrical Machines and Apparatus in Relation to their Thermal Stability in Service. Publication 85. Geneva: International Electrotechnical Commission, 1984.

IEC. Guide for the Determination of Thermal Endurance Properties of Electrical Insulating Materials—General Guidelines for Ageing Procedures and Evaluation of Test Results. Publication 261-1. Geneva: International Electrotechnical Commission, 1987.

Kamosawa T, Yoshimura N, Nishida M, Noto F, Masui M. Trans IEE Japan 108A:397–404, 1988.

Kuffel E, Zaengl W. High Voltage Engineering Fundamentals. Oxford, UK: Pergamon, 1984.

Kurtz M. IEEE Electr Insul Mag 3:12–14, 1987.

Li HM, Fouracre RA, Crichton BH. IEEE Trans DEI-2:866–874, 1995.

Margolis J. Engineering Thermoplastics: Properties and Applications. New York: Marcel Dekker, 1985.

Mark R, ed. Handbook of Physical and Mechanical Testing Paper and Paperboard. New York: Marcel Dekker, 1984.

Marton K. Proceedings of CIGRE Symposium on New and Improved Materials for Electrotechnology, Vienna, 1987, Report 620.08.

Meyer C, Filippini J. Polymers 20:1186–1187, 1979.

Rasmy A, Gouda O. Proceedings of IEE International Conference on Dielectric Materials, Measurements and Applications, Manchester, 1992, pp 484–487.

Ross R. IEEE Electr Insul Mag 9:7–16, 1993.

Samm R. IEEE Electr Insul Mag 3(4):41–42, 1987.

Schaible M. IEEE Electr Insul Mag 3(1):8–12, 1987.

Seely S, Poularikas A. Electromagnetics—Classical and Modern Theory and Applications. New York: Marcel Dekker, 1979.

Sletbak J. IEEE Trans PAS-98:1358–1365, 1979.

Sletbak J, Ildstad E. Record of IEEE International Symposium on Electrical Insulation, Montreal, 1984, pp 29–32.

Steffens H. IEEE Electr Insul Mag 2(6):39–40, 1986.

Tareev B, ed. Electrical and Radio Engineering Materials. Moscow: Izdatelstvo Mir, 1980.

Xu J, Boggs S. IEEE Electr Insul Mag 5(10):29–36, 1994.

Yoshimura N, Nishida M, Noto F. Proc Inst Electrostat Japan 6:72–79, 1982.

9

High-Voltage Busbars

A. EL-MORSHEDY *Cairo University, Giza, Egypt*

9.1 INTRODUCTION

The substation or switching station functions as a connection and switching point for transmission lines, subtransmission feeders, generating units, and transformers. The substation design aims to achieve a high degree of continuity, maximum reliability, and flexibility, and to meet these objects with the highest possible economy.

The substations used in distribution systems operate at voltage levels up to 69 kV. Transmission substations serving bulk power sources operate at voltages up to 765 kV. Voltage classes used for major substations include 69, 115, 138, 161, 230, and 287 kV, considered as high voltage, and 345, 500, and 765 kV, considered as extrahigh voltages (EHV). Higher voltage classes, including 1100 and 1500 kV, are in various stages of planning or construction. These are referred to as ultrahigh voltages (UHV).

In this chapter we deal with the conventional types of substation where atmospheric air is the insulating medium. Gas-insulated switchgear (GIS) is discussed in Chapter 10.

Substation busbars (or buses) are a most important part of the station structure since they carry high amounts of energy in a confined space and their failure would have very drastic repercussions on the continuity of power supply. Therefore, the bus system must be built to be electrically flexible and reliable enough to give continuous service. It must have adequate capacity to carry all loads, and robust construction to withstand foreseeable abnormal electromechanical forces.

9.2 BUSBAR ARRANGEMENTS

Most substations conform to one or the other of the following basic arrangements as a result of different factors: (a) single bus; (b) double bus with double breaker; (c) double bus with single breaker; (d) main and transfer bus; (e) ring bus; and (f) breaker-and-a-half with two main buses. The choice of the arrangement to be used depends on the relative importance assigned to such items as safety, reliability, simplicity of relaying, flexibility of operation, first cost, ease of maintenance, available ground area, location of connecting lines, ease of rearrangement, and provision for expansion.

9.2.1 Single Bus

The single bus is the one in common use and has the simplest design. It is generally used in small outdoor stations having relatively few outgoing or incoming feeders and lines. The single-bus design (Fig. 9.1) is not normally used for major substations. Dependence on one bus can cause a serious outage in the event of bus failure.

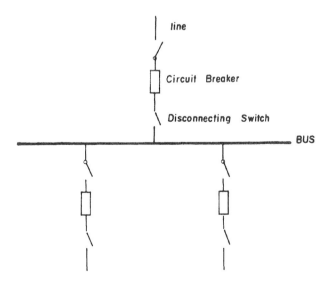

Figure 9.1   Single busbar arrangement.

9.2.2 Double Bus with Double Breaker

The double bus with double breaker design (Fig. 9.2) is used with large switching stations. It requires two circuit breakers for each feeder circuit. Normally, each circuit is connected to both buses, which presents a high order of reliability.

9.2.3 Double Bus with Single Breaker

This scheme (Fig. 9.3) uses two main buses, and each circuit includes two disconnecting switches. A bus-tie circuit breaker is connected to the main buses. When closed, it allows transfer from one bus to the other without de-energizing the feeder circuit; only the bus disconnecting switches need then be operated.

9.2.4 Main-and-Transfer Bus

The main-and-transfer bus (Fig. 9.4) is used for most distribution work. One circuit breaker serves each circuit. The transfer bus is a standby for emer-

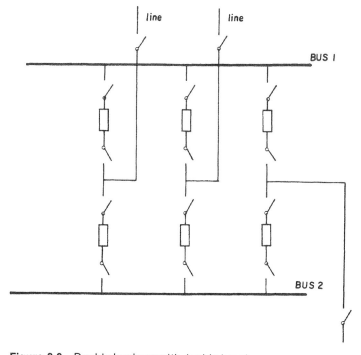

Figure 9.2 Double busbars with double breakers.

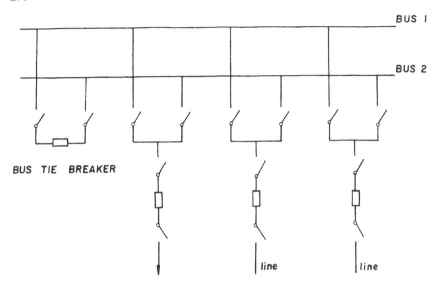

Figure 9.3 Double busbars with single breaker.

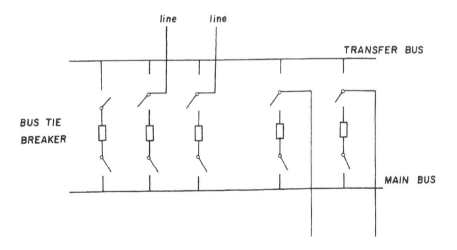

Figure 9.4 Main and transfer busbars.

gency use. Also, a bus-tie circuit breaker is provided to tie the main and transfer buses together when the need arises.

9.2.5   Ring Bus

The advantage of the ring bus scheme (Fig. 9.5) is that it requires the use of only one circuit breaker per circuit. In addition, each outgoing circuit has two sources of supply. This system is economical in cost, reliable, and very flexible.

9.2.6   Breaker-and-a-Half with Two Main Buses

This scheme (Fig. 9.6) has three breakers in series between the main buses. Two circuits are connected between the three breakers, hence the term breaker-and-a-half. Under normal operating conditions all breakers are closed and the two main buses are energized. To trip a circuit, the two associated circuit breakers must be opened.

9.2.7   Summary

It has been found that the two commonest busbar arrangements at all voltage levels are the single busbar and the double busbar with single breaker. The single busbar arrangement predominates up to 70 kV, whereas the double busbar with single breaker is more commonly used above 70 kV.

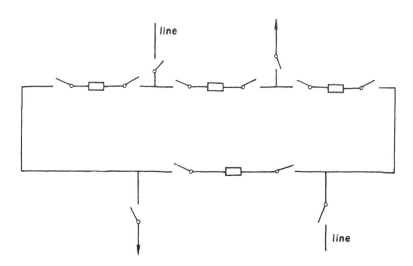

Figure 9.5   Ring bus.

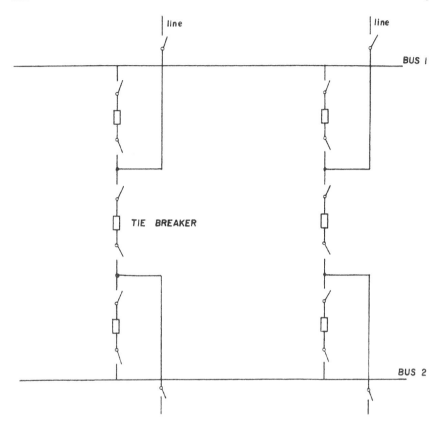

Figure 9.6   Breaker-and-a-half with two main buses.

The ring bus and the breaker-and-a-half arrangements have been gaining popularity among designers for voltages above 160 kV.

9.3   RIGID BUS VERSUS STRAIN BUS

Usually, low-voltage and medium-voltage busbars are of the rigid bus type, using copper or aluminum bars or tubings as the phase conductors, with pedestal-type insulator supports. Those of higher voltages use strain insulators and stranded aluminum (ACSR) or copper conductors.

Substations of the EHV class are usually of the strain bus design. Medium-voltage substations normally use the rigid-bus approach and enjoy the advantage of low station profile and ease of maintenance and operation.

Combinations of rigid and strain bus constructions are sometimes employed in conventional arrangements, up to 765 kV, so as to benefit from both types.

The advantages of the rigid bus compared with the strain bus design are as follows:

1. The rigid bus design employs less steel and simple, low-level structures.
2. The rigid conductors are not under constant strain.
3. The pedestal-type bus supports are usually more accessible for cleaning.
4. The rigid bus has a low profile, which provides good visibility of the conductors and apparatus.

The disadvantages of the rigid bus design are:

1. It is comparatively expensive, due to the higher cost of tubing and connections.
2. It usually needs more supports and insulators.
3. It is more sensitive to structural deflections, which may lead to possible damage.
4. It usually requires more ground space than the strain bus.

9.4 BUS CONDUCTOR MATERIALS

Several factors determine the proper choice of busbar conductor material. These factors are determined largely by the type of installation and the size of electrical load to be carried. Factors to be considered when making this choice are voltage drop, power loss, current-carrying capacity, ice and wind loads, short-circuit loads, and conductor corrosion.

The busbar materials in general use are aluminum and copper. Heat-treatable aluminum alloys, especially in tubular shapes, are most widely used in HV and EHV outdoor substations. They combine high strength and good conductivity.

Aluminum is one-third the weight of copper for a specified length. Aluminum and its alloys require little maintenance. For a given current rating and for equal temperature rise, the aluminum bus would have a 33% larger cross-section than the equivalent copper bus. The resulting deflection for the copper bus is about 31% greater than for the aluminum bus. As aluminum has several advantages over copper, most rigid bus installations use tubings of aluminum or its alloys.

9.5 BUSBAR CLEARANCES

For safety and reliability it is essential that adequate clearances to live parts be provided. The minimum phase-to-ground and phase-to-phase air clearances in substations are prescribed to guarantee the withstand levels of switching and lightning impulses considered necessary by the system designer (see Chapter 14).

In fact, in substations up to 200 kV, air insulation is predominantly dictated by lightning overvoltages. There the minimum phase-to-phase clearances are usually taken about 15% longer than those to grounded metal frames. In certain cases larger phase-to-phase clearances have been preferred, as dictated merely by the requirements of equipment maintenance.

The dielectric stresses in the air gaps between phases and to ground due to switching overvoltages have, on the other hand, very different characteristics (see Chapters 4 and 14). In particular, the stress between the phases is greater than the stress to ground. Moreover, the behavior of air gaps under switching impulse is affected by electrode shapes, which in the case of clearances between phases are usually different from the case of clearances to ground.

9.5.1 Phase-to-Ground Clearances in Substations

The insulation levels for systems and equipment with voltages rated at 245 kV and above are combinations of two components: the rated switching impulse withstand voltage and the rated lightning impulse withstand voltage (Section 18.3).

The air clearances between phase conductors and between them and grounded metal frames can be estimated by knowing the dielectric strengths of such air gaps under AC and impulse voltages. Numerous tests have been carried out in high-voltage testing stations in many countries in dry air and under simulated rain conditions for air gaps between rods and planes, between parallel conductor bundles, between conductors and planes, and between conductors and portal frames. The results have been expressed by empirical relations and can also be predicted by a generalized formula (see Section 4.8). For rod-to-plane and similar gaps, positive impulse flashover voltages are lower than those with negative polarity. Rain has no significant influence on positive impulse flashover voltages (Rizk, 1976).

These relations can be used to estimate the 50% breakdown voltages of such air gaps under switching and lightning impulses. The standard deviations σ of these breakdown voltages were observed, by experiment, to be about 6% for switching surges and 3% for lightning surges. Based on the evaluated 50% impulse breakdown voltages of air gaps, their impulse withstand voltage V_w can be estimated by the formula

$$V_w = V_{50\%}(1 - 1.3\sigma) \tag{9.1}$$

This withstand voltage represents the level at which the probability of flashover is less than 10%. In substations, three main categories of phase-to-ground clearances are present:

1. Distances between conductors and portals
2. Distances between live parts of apparatus and portals
3. Distances between conductors and ground

Table 9.1 gives the air clearances according to the recommended rated impulse withstand voltages.

9.5.2 Clearances between Conductors

The distances between different live conductors are determined not only by their voltages but also by the necessity, in some cases, to carry out maintenance work while neighboring conductors are alive. Lightning surge stresses between phases will normally be within the magnitude of stresses to ground. On the other hand, the phase-to-phase overvoltage peaks are sub-

Table 9.1 Phase-to-Ground Air Clearance, Dependent on the Rated Switching and Lightning Impulse Withstand Voltages

Rated switching impulse withstand voltage (kV)	Air clearance (m)	Rated lightning impulse withstand voltage (kV)	Air clearance (m)
650	1.15	750	1.35
750	1.45	850	1.55
850	1.79	950	1.73
950	2.16	1050	1.92
1050	2.55	1175	2.14
1175	3.07	1300	2.37
1300	3.64	1425	2.60
1425	4.24	1550	2.83
1550	4.87	1800	3.28
		1950	3.56
		2100	3.83
		2400	4.38

jected to considerable variations, depending on the wave shapes of the two phase-to-ground components, possibly having opposite polarities, and on the ratio and time delay between these two components (El-Morshedy, 1982). The 50% flashover voltage of phase-to-phase air clearances is dependent on the wave shapes used in laboratories for the two phase-to-ground components, their times to crest, and the time delay between their crest values representing expected field conditions. It has been recommended that phase-to-phase insulation tests be carried out with two equal switching impulses of opposite polarities, equal crest values, and synchronous peaks. The crest value of each impulse should be half the rated switching impulse withstand voltage between phases. Besides being representative of the real stresses, only this method allows testing of phase-to-phase insulation under actual geometrical substation conditions without causing undue flashover on phase-to-ground insulation.

The air clearances in substations may be generally identified as:

1. Clearances between conductors
2. Clearances between conductors and apparatus
3. Clearances between poles of the same phase

Most electrode configurations are characterized by highly inhomogeneous and sometimes symmetric field distributions. Rod gaps, rod-to-plane gaps, and gaps between ring-shaped electrodes have generally been accepted as representing practical geometrical conditions typical for the high-voltage and extra-high-voltage (EHV) ranges.

Figure 9.7 shows the test results for the 50% flashover voltages using two synchronous ($\Delta t = 0$) switching impulses with equal crest values U_1 and U_2. The time to crest for the positive component has been found equal to that value which gives the lowest flashover voltage (Sections 4.8 and 15.2). Since the electrical field distribution is symmetric for ring and rod gaps, no polarity effect is observed. Only for nonsymmetric electrode configurations is the influence of the polarity remarkable. For a rod–plane gap, Figure 9.7 shows that the lower values are obtained with the positive polarity at the rod, where the field concentration is highest.

Table 9.2 gives the recommended minimum clearances between phases as represented by the rod–rod and ring–ring configurations. These clearances were calculated from the 50%-interphase flashover voltages.

As mentioned above, the degree of inhomogeneity of the electric field varies with the voltage level under consideration. Therefore, the determination of the necessary interphase clearance is more related to the results for the rod–rod configuration in the lower voltage range. In the EHV range, however, the ring–ring configuration becomes more and more representative because of the sizes of the hardware used at such voltage levels.

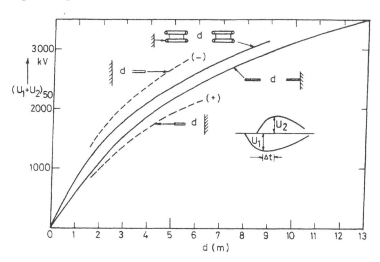

Figure 9.7 Fifty percent flashover voltages of a gap d between conductors of the shape shown, under two simultaneous equal and opposite switching surges U_1 and $U_2(\Delta t = 0)$.

Some trends in the design practices of North American 500 kV and 750 kV AC substations are indicated by the clearances in Table 9.3 (IEEE Working Group, 1969, 1970, 1983). It is noted that the minimum phase-to-ground and phase-to-phase clearances are greater than those recommended in Tables 9.1 and 9.2, and that in EHV substations rigid, strain, or a combination of both bus constructions are in use.

9.6 THERMAL RATING

In the design of buses the temperature rise of conductors above ambient while carrying current is very important. Buses are generally rated on the basis of the temperature rise that can be permitted without danger of overheating equipment terminals, bus connections, and joints.

The permissible average temperture rise for plain copper and aluminum buses while carrying their normal loads is usually limited to 30°C above an ambient temperature of 40°C. This value is the accepted standard for IEEE, NEMA, and ANSI. This is the overall temperature rise and a maximum or hot-spot temperature rise of 35°C is permissible. Above this temperature oxidation increases rapidly and would give rise to cumulative and excessive heating at joints and contacts. Many factors influence the heating of the bus, such as the type of material used and the bus's size and shape

Table 9.2 Correlations between the Switching Impulse Withstand Voltage-to-Ground and between Phases and Minimum Phase-to-Phase Air Clearances

Rated switching impulse withstand voltage to ground (kV)	Ratio between phase-to-phase and phase-to-ground switching impulse withstand voltages	Switching impulse withstand voltage between phases (kV)	50% switching impulse flashover voltage between phases (kV)	Minimum phase-to-phase air clearance (m)	
				Ring–ring	Rod–rod
650	1.62	1050	1139	1.50	2.05
750	1.57	1175	1274	1.75	2.35
850	1.53	1300	1410	2.05	2.65
950	1.63	1550	1681	2.65	3.35
1050	1.71	1800	1952	3.40	4.20
1175	1.66	1950	2115	3.90	4.75
1300	1.73	2250	2440	5.20	6.20
1425	1.79	2550	2766	6.80	7.95
1550	1.74	2700	2928	7.65	8.95

Table 9.3 Examples of Electrical Station Design

	American Electric Power (AEP) (500 kV)	American Electric Power (AEP) (765 kV)	Ontario Hydro (500 kV)		Hydro Quebec (735 kV)
Minimum clearance (metal-to-metal)					
Phase-to-phase (m)	6.1	10.7	6.5		12.2
Phase-to-ground (m)	3.7	6.1 (horizontal)	4		5.6
Above roadway inside substation (m)	7.9	23.5 to roadway, 11 min. clearance to grade	10		16.7
Main bus parameters					
Current rating	3360 A	3730 A	2770 A		2000 A
Type (rigid/strain)	Strain	Rigid	Rigid		Strain/rigid
Phase spacing (center-to-center, m)	7	13.7	7.6		15.3
Span (maximum distance between supports, m)	30.5 and 110	13.7	10.2		10
Basic insulation level (BIL) (kV)	1550	2050	1800		2200
Conductors					
Number per phase	2	1	1		2
Diameter (cm)	4.19	14.7	11.43		5.87
Type	ACSR	Al tube	Al tube		Al
Subconductor spacing (cm)	45.7	–	–		38.1
Height above ground at support (m)	8.9	23.5	14.3		25.9 and 10.7
Bay bus			High level	Low level	
Current rating	3360 A	5350 A	2860 A	1930 A	2000 A
Type (rigid/strain)	Strain	Strain	Strain	Rigid	Strain/rigid
Bay span (m)	30.5	32.9	30.5	–	79.3
Bay conductors					
Number per phase	2	2	2	1	2
Diameter (cm)	4.19	6.1	4.45	7.62	5.87
Type	ACSR	Al conductor	Al conductor	Al tube	Al conductor
Subconductor spacing (cm)	45.7	45.7	33	–	38.1
Height above ground at support (m)	8.9	38.1 and 12.2	28.4	7	25.9 and 10.7

(Conway, 1979). The maximum continuous current-carrying capacity is important in selecting the proper conductor material, size, and shape of cross-section.

9.6.1 Conductor-rated Current

The ampacity (i.e., current-carrying capacity) of a busbar is determined by its ratio of surface area (for heat dissipation) to its cross-sectional area. This means that for a single shape there are limits to the current-carrying capacity with respect to its dimensions. For large currents it is better to provide several parallel on-edge flat bars than to use a single bar of equivalent total area and the same height. With the arrangement of multiple bars, the proximity and skin effects should be considered in estimating their AC losses (Weiss and Csendes, 1982). Figure 9.8 shows comparative ampacity ratings at power frequency for various conductor arrangements, each with a cross-sectional area of 25.8 cm^2 (Simpson and Greenfield, 1971). Evidently, the thick single bar is impracticable for high currents, especially for AC use. The ampacity of a given busbar can be estimated using the following simple relation:

$$I^2 R = (W_c + W_r)A_s \qquad \text{watts} \qquad (9.2)$$

where I is the conductor current (A), R the conductor AC resistance (Ω/m), W_c heat dissipated by convection (W/cm^2), W_r heat dissipated by radiation (W/cm^2), and A_s surface area (cm^2 per meter of busbar length).

For a circular conductor, the skin effect causes its AC resistance to exceed its DC resistance, their ratio being

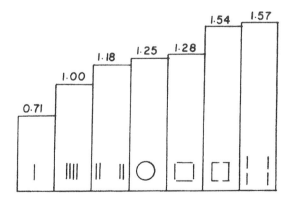

Figure 9.8 Comparative ampacity ratings for various arrangements and shapes of busbar conductors.

$$\frac{R_{AC}}{R_{DC}} = k\sqrt{\frac{\mu_r f}{\rho}} \tag{9.3}$$

where k is a constant listed in tables (Weiss and Csendes, 1982), f is the frqeuency (Hz), μ_r the relative permeability, and ρ resistivity (Ω/m^2). A more elaborate method is followed for computing the proximity effect, using the same basic physical phenomena (Salama and Hackam, 1984).

Evidently, tubes have lower AC resistance than solid or flat conductors of the same net cross-section. Tubes with thin walls are affected the least by skin effect. Because of the higher resistivity of aluminum, aluminum conductors are affected less by skin effect than are copper conductors of similar cross-sections, as seen from equation (9.3).

The heat dissipated by convection W_c and radiation W_r can be estimated from the following relationships

$$W_c = \frac{5.73 \times 10^{-3}\sqrt{pv}}{T_a^{0.123}\sqrt{d}}\Delta\theta \qquad W/cm^2 \tag{9.4}$$

$$W_r = 5.7E\left[\left(\frac{T}{1000}\right)^4 - \left(\frac{T_0}{1000}\right)^4\right] \qquad W/cm^2 \tag{9.5}$$

where

p = pressure in atmospheres ($p = 1.0$ for atmospheric pressure)
v = surrounding air velocity (m/s)
T_a = average absolute temperature of conductor and ambient air (K)
d = outside diameter of cylindrical conductor (cm)
T = absolute temperature of the conductor (K)
T_0 = absolute temperature of the ambient (K)
$\Delta\theta$ = temperature rise of conductor above ambient ($T - T_0$) (K)
E = relative emissivity of conductor surface
= 1.0 for black body and about 0.5 for average oxidized copper and aluminum

For other conductor shapes, d is the outside diameter of an equivalent circular conductor of the same surface area. By calculating ($W_c + W_r$), A_s, and R, it is then possible to determine I from equation (9.2). This method is generally applicable to both copper and aluminum conductors. Tests have shown that aluminum conductors dissipate heat at about the same rate as copper conductors of the same outside diameter when the temperature rise is the same.

9.6.2   Transient Ampacity of a Busbar

The current-carrying ampacity of a busbar element is limited by the maximum operating temperature. A differential equation relating the current to temperature was derived by Coneybeer et al. (1994), applying the law of conservation of energy. The energy stored in the busbar equals the net heat transfer into the bus plus the heat generated due to Joulean heating. The net energy into the bus is the absorbed solar energy minus the convective loss and the energy radiated to the surroundings. Neglecting the internal thermal resistance of the busbar, the differential equation is used to determine the relationship between the busbar temperature and the current it carries. Thus the transient ampacity of the busbar is obtained (Coneybeer et al., 1994). If the differential term is set to zero, the resulting algebraic equation governs the steady state ampacity of the busbar.

In fact, transient ratings better reflect the thermal performance of a busbar than steady-state values. This is because the bus temperature will not change instantaneously when the current changes. The temperature changes gradually as the metal stores energy due to its thermal capacitance.

9.6.3   Considerations for Short-Circuit Currents

An adequate conductor size is necessary for carrying the fault current caused by an external short circuit without its temperature exceeding the safe limit during the period from the instant of initiation of the fault until its interruption. For both copper and aluminum bus connectors carrying DC, the following equation was suggested

$$I_f = \frac{\kappa A}{6.45\sqrt{t}} \qquad \text{A} \tag{9.6}$$

where I_f is the DC fault current, t the time (s) from initiation to clearing of the fault, and A the sectional area of the busbar (mm^2). κ is a coefficient ranging between 600 and 750, depending on the conductor material, its maximum allowable temperature, and the ambient temperature (Simpson and Greenfield, 1971). For the case of copper busbars, the value of coefficient κ is about 70% higher. For AC the value of I_f obtained should be multiplied by the ratio $(R_{AC}/R_{DC})^{-1/2}$.

9.7   MECHANICAL STRESSES ON BUSBAR CONDUCTORS

A bus conductor carrying a short-circuit current is subjected to electromagnetic forces that tend to pull adjacent bars to each other if they carry currents in the same direction and push them apart if the currents are in opposite directions (Fig. 9.9). This force increases quadratically with the

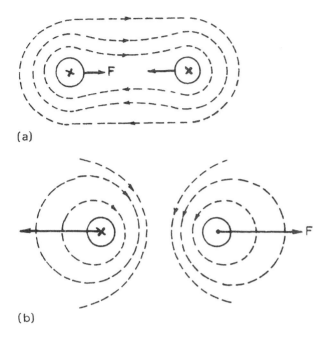

(a)

(b)

Figure 9.9 Magnetic fields and electrodynamic forces acting on two parallel conductors carrying currents in the same direction (a) and when the currents are in opposite directions (b).

current magnitude and depends on the shape and arrangement of conductors and the natural frequency of the complete assembly, including mounting structure and insulators (Craig and Ford, 1980; Awad and Huestis, 1980).

The electromagnetic force on either of the two conductors carrying currents of instantaneous values I_1 and I_2 and separated by a distance d can easily be evaluated as

$$F = \frac{2I_1I_2}{10^5d} \qquad \text{N/m} \tag{9.7}$$

Whether both forces are those of attraction or repulsion depends on the relative directions of the currents I_1 and I_2 (Fig. 9.9).

As a result of the electromagnetic forces between the bus conductors there will be a mechanical reaction on the bus supports. These forces are far greater than their ordinary loading caused by the conductors' weights and wind. The insulators should have enough mechanical strength and resilience.

The short-circuit current may reach excessive peak values, depending on the instant the short circuit occurs with respect to the AC voltage wave and the X/R ratio of the faulted circuit. This causes an offset in the current wave (Fig. 9.10). For example, if the X/R ratio is 20, the first current peak will be about 2.183 times the sustained rms fault current.

The electromagnetic forces set up between parallel conductors in which current varies as in Figure 9.10 produce force pulsations; the component at power frequency dominates for the first few cycles, but afterward the dominant component is at double the power frequency. A bus installation should be designed so that its natural frequency of oscillation, or that of any of its parts or supporting structures, is not close to the power frequency or its multiples, to evade resonant pulsations that might otherwise endanger the mechanical structure.

9.7.1   Electrodynamic Force Magnitudes

There are considerable differences between short-circuit forces, depending on the type of fault and the bus conductor arrangement. The probability of a short circuit occurring on all three phases is very remote, yet it should be considered in the design of busbars. Phase-to-ground and double-phase-to-ground faults are more likely. Also, during the operation of a circuit breaker clearing a short circuit, one phase current gets interrupted first and thus a three-phase fault turns into a double-line fault. The maximum instantaneous short-circuit lateral forces acting on bus conductors in various arrangements are listed in Table 9.4.

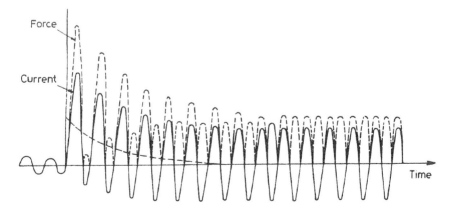

Figure 9.10   Asymmetrical short-circuit current in a circuit with a high X/R ratio, and the corresponding electrodynamic force acting on the busbars.

Table 9.4 Maximum Instantaneous Short-Circuit Lateral Force Acting upon Bus Conductors

Busbar arrangement	Type of circuit and kind of fault	Force (N/m)
	DC or single-phase AC. Symmetrical	$F = 4I_{rms}^2/(10^5 d)$
	Single phase AC. Fully offset asymmetrical wave	$F = 16I_{rms}^2/(10^5 d)$
	3-phase AC conditions of asymmetry to give maximum force on any phase	$F = 13.9I_{rms}^2/(10^5 d)$
	3-phase AC: 1. Fully offset asymmetrical wave in phase A 2. Conditions of asymmetry to give maximum force on A or C 3. Conditions of asymmetry to give maximum force on B	$F = 12I_{rms}^2/(10^5 d)$ $F = 13I_{rms}^2/(10^5 d)$ $F = 13.9I_{rms}^2/(10^5 d)$

I in A, d in cm.

9.7.2   The Effect of Magnetic Field on Optimal Design of a Rigid Bus

Recently, regulatory guidelines have been set in several countries limiting the maximum level of magnetic field near transmission substations to which the public can be exposed. Several regulatory authorities have imposed a limit of 10 T on the allowable magnitude of magnetic fields at the edge of rights of way or substation property.

The effect of magnetic field constraint on the optimal design of a 500 kV rigid bus substation was examined by Anders et al. (1994), using a mathematical model.

The inclusion of the magnetic field constraint in the station design procedure will produce a tendency to decrease interphase spacing in comparison with the standard design. With a permitted magnetic field strength of 15 T, the magnetic field constraint is negligible. The larger the magnitude of the current flowing in the buses, the more important the magnetic field constraint becomes. The optimal interphase spacing decreases by a factor of 2 when the current magnitude increases by the same factor. The influence of the magnetic field constraint on the optimal interphase spacing decreases as the cost of land increases (Anders et al., 1994).

9.8   INSULATORS

Air-insulated busbars are supported if rigid, or, if a strain type, suspended by insulators. These insulators have traditionally been made of special porcelain or toughened glass. With the developments in polymeric materials, traditional insulators have for several decades been experiencing serious competition from insulators made of epoxy reinforced with glass fibers. Polymeric insulators usually weigh less than half or even one-fourth of their corresponding porcelain insulators for the same mechanical and electrical strengths (Karady and Limmer, 1979).

Busbar insulators should have adequate electrical and mechanical strength under the expected ambient and operating conditions of temperature variations, rain, pollution, and normal and abnormal loadings. The abnormal loadings include electrodynamic forces occurring during short circuits on the busbars, as described above. Acceptable electrical and other qualities of insulators are usually set by national and international standard specifications (see, e.g., IEC, 1979).

Insulators for outdoor use should pass high-voltage tests under simulated rain conditions. Also, insulators to be employed in sites of significant atmospheric pollution should be designed and tested accordingly, as discussed below.

For the higher operating voltages, more insulator units are stacked or connected in a chain, whereas for greater mechanical strength more insulator chains are bolstered together "in parallel." "Long rod" insulators can replace chains of short "cap-and-pin" units.

9.8.1 Insulator Pollution

When pollution accumulates on insulator surfaces and its soluble ingredients dissolve in the moisture accumulated from fog or dew, a conducting path, however resistive, gets established on the insulator surface all the way from the high-voltage conductor to the grounded metal frame. Thus leakage currents flow. With nonuniformities in pollution and leakage currents naturally occurring over insulator surfaces, hot spots form. A localized spot dried up by concentrated leakage currents pushes the current paths sideways and thus extends laterally to form a dry band. Such a dry band, with its drastically increased resistance, takes a major fraction of the total insulator voltage. A partial arc can strike across the band. With partial arcs burning on one unit of an insulator chain, additional voltage stresses appear over other units. The voltage distribution over the chain comes to be far from uniform (Aoshima et al., 1981). This may eventually develop into a complete flashover of the whole insulator chain under normal operational voltage. Alternatively, the current through the partical arcs may be too small to maintain them. In such a case, the wet contaminated areas of the insulator surfaces dry up gradually and the partial arcs become extinguished. Intensive research has been going on in many countries for several decades (Khalifa et al., 1988; Houlgate et al., 1982; Pei-Zhong and Cheng-Dong, 1982).

Rain does not always provide the cleaning needed to combat pollution, particularly under conditions of heavy marine or industrial pollution or when rain is not regular enough; desert pollution is a case in point (El-Koshairy et al., 1982). Measures usually taken to improve the insulation strength in polluted atmospheres include using "anti-fog" insulators with extended leakage paths and insulators with semiconducting glaze. Also, greasing their surfaces prevents the formation of continuous conducting films. Combating pollution of high-voltage insulators has also been achieved successfully by regular washing of the insulators while under high voltage, thus maintaining power supply continuity. To help improve system performance, pollution monitors have been developed so that the insulator cleaning operation would be carried out only when necessary (Khalifa et al., 1988).

9.10   PROBLEMS

(1) Give a single line diagram of connections showing the layout of the busbar sections in a power station. Include the necessary isolators and circuit breakers for at least two alternators and two feeders.

(2) A 15 cm × 15 cm × 0.6 cm square tube of a certain alloy, area 35.4 cm^2, is to withstand 150 kA fault current. R_{AC}/R_{DC} is 1.08. How long can this fault be sustained while the temperature rises to 300°C above 50°C ambient? Assume $k = 700$ for this alloy.

(3) It is required to select a bar of a certain alloy that will withstand 100 kA fault current if the top temperature limit is 250°C and ambient is 70°C and maximum time is 40 cycles, assuming R_{AC}/R_{DC} is 1.08 and $k = 650$ for this alloy.

(4) What are the advantages and disadvantages of the rigid bus design?

(5) Calculate the heat dissipated by convection and radiation for a cylindrical busbar of 5 cm outside diameter. The pressure is 1 atm, the surrounding air velocity is 10 m/s, the conductor temperature is 130°C and the temperature of the surroundings is 40°C. The relative emissivity is 0.8.

(6) Describe the process leading to pollution flashover. Discuss the methods which may be used to reduce the probability of pollution flashover in polluted locations.

REFERENCES

Anders GJ, Ford GL, Horrocks DJ. IEEE Trans PWRD-9:1384–1390, 1994.
Aoshima Y, Tatsuya H, Keiji K. IEEE Trans PAS-100:948–955, 1981.
Awad, MB, Huestis HW. IEEE Trans PAS-99:480–487, 1980.
Coneybeer RT, Black WZ, Bush RA. IEEE Trans PWRD-9:1822–1829, 1994.
Conway BJ. IEEE Trans PAS-98:1384–1402, 1979.
Craig B, Ford GL. IEEE Trans PAS-99:434–442, 1980.
El-Koshairy MAB, El-Sharkawi E, Awad M, Zarzoura H, Khalifa M, Nosseir A. CIGRE paper 33–09, 1982.
El-Morshedy AMK. Proc IEE 129:199–205, 1982.
Houlgate RG, Lambeth PJ, Roberts WJ. CIGRE paper 33-01, 1982.
IEC. Tests on Indoor and Outdoor Post Insulators for Systems with Nominal Voltages Greater than 1000 V. Publication 168. Geneva: International Electrotechnical Commission, 1979.
IEEE Committee Report. IEEE Trans PAS-88:854–861, 1969.
IEEE Committee Report. IEEE Trans PAS-89:1521–1524, 1970.
IEEE Working Group. IEEE Trans PAS-102:513–520, 1983.
Karady G, Limmer HD. Proceedings of the American Power Conference, pp 1120–1125, 1979.
Khalifa M, El-Morshedy A, Gouda OE, Habib SED. Proc IEE 135C:24–30, 1988.

Pei-Zhong H, Cheng-Dong X. CIGRE paper 33-07, 1982.
Rizk F. IEEE Trans PAS-95:1892–1900, 1976.
Salama M, Hackam R. IEEE Trans PAS-103:1493–1501, 1984.
Simpson TW, Greenfield EW. Electrical Conductor Handbook. New York: The Aluminum Association, 1971.
Weiss J, Csendes Z. IEEE Trans PAS-101:3796–3803, 1982.

.

10

Gas-Insulated Switchgear

H. ANIS *Cairo University, Giza, Egypt*

10.1 INTRODUCTION

Gas-insulated switchgear (GIS) that uses compressed sulfur hexafluoride (SF_6) gas overcomes many of the limitations of the conventional open-type HV switchgear, as it offers the following advantages:

1. The space occupied by the switchgear is greatly reduced.
2. It is totally unaffected by atmospheric conditions such as polluted or saline air in industrial and coastal areas, or desert climates.
3. It possesses a high degree of operational reliability and safety to personnel.
4. It is easier to install in difficult site conditions (e.g., on unstable ground or in seismically active areas).
5. In addition to having a dielectric strength much greater than that of air, SF_6 has the advantages of being nontoxic and nonflammable.

10.2 CHEMICAL AND PHYSICAL PROPERTIES OF SF_6

At atmospheric pressure, SF_6 has a dielectric strength two to three times that of air, and this ratio increases wtih increasing pressure. At 3 atm, the dielectric value is about the same as that of transformer oil. In addition to this superior dielectric property, SF_6 has the following basic physical and chemical properties:

1. It is nontoxic, having almost the chemical properties of a noble gas. It can, however, cause suffocation if it displaces oxygen from an area because of its much higher density than that of air.
2. It is nonflammable.
3. It is noncorrosive. It does not react with other materials, as it is inert. Furthermore, it inhibits surface erosion and oxidation. Only when SF_6 gas is heated above 500°C will it decompose, and then its decomposition products react with hydrogen and other materials. With special filters, however, it is possible to alleviate this effect (Maller and Naidu, 1981).
4. It has a high partial vapor pressure at both normal and low temperatures.
5. It has excellent heat transfer characteristics. Its high molecular weight and low viscosity enable it to transfer heat by convection more effectively than do common gases.

Furthermore, SF_6 gas exhibits excellent properties for arc quenching and is therefore also used as an interrupting medium in circuit breakers instead of oil or air. The use of SF_6 in circuit breakers is discussed in Chapter 11. Table 10.1 summarizes some of the basic physical properties

Table 10.1   Physical Properties of SF_6

Molecular weight	146.05
Melting point	−50.8°C
Sublimation temperature	−63.9°C
Density (liquid)	
at 50°C	1.98 g/ml
at 25°C	1.329 g/ml
Density (gas at 1 bar and 20°C)	6.164 g/l
Critical temperature	45.6°C
Critical pressure	36.557 atm
Critical density	0.755 g/ml
Surface tension (at −50°C)	11.63 dyn/cm
Thermal conductivity ($\times 10^4$)	3.36 (cal/s)/(cm·K)
Viscosity (gas at 25°C $\times 10^4$)	1.61 poise
Boiling point	−63.0°C
Specific heat (at 30°C)	0.143 cal/g
Relative density (air = 1)	5.10
Vapor pressure (at 20°C)	10.62 bar

Source: Maller and Naidu (1981).

of SF_6. Mixing a small amount of SF_6 gas with a relatively inexpensive gas such as nitrogen, hydrogen, carbon dioxide, or air increases the dielectric strength of the latter substantially.

10.3   LAYOUT OF GAS-INSULATED SWITCHGEAR

In GIS all live parts are enclosed in a compressed-gas system which is divided into a number of compartments. This division enables the isolation of one compartment for maintenance or repair purposes while the other compartments remain pressurized. In Figure 10.1a the single-line diagram of a double-busbar arrangement is shown. In Figure 10.1b the diagram is redrawn in the form of a typical gas GIS circuit breaker bay. Basic components that make up any one GIS bay are the circuit breaker, disconnectors, earthing switches, busbars, and current and voltage transformers. The implementation of the foregoing arrangement into a real GIS depends on the voltage level. Figures 10.2 and 10.3 show sectional views of the general arrangements of two different gas-insulated switchgear bays belonging to the ranges 145 kV and 220–800 kV, respectively.

10.3.1   GIS Enclosure Configuration

The GIS enclosure forms an electrically integrated, grounded casing for the entire switchgear. The GIS enclosure can be either of the three-phase type, as in Figure 10.2, or of the single-phase type, as in Figure 10.3. The advantages of the three-phase common enclosure design are:

1. A smaller number of enclosures is required per feeder (one-third).
2. In the case of a three-phase common enclosure, an arc between phase and ground will, within a few milliseconds, evolve into a phase-to-phase fault between conductors, owing to ionization of the gap, and at the same time the phase-to-ground arc will extinguish. Consequently, an enclosure burn-through is not possible.
3. For the same parameters (voltage level, conductor size, clearances between phases and phase-to-ground) the resultant field stress in a common three-phase enclosure is approximately 30% less than that in a single-phase enclosure and hence less likely to cause failure.
4. The absence of complicated tie rods and linkages between poles for the circuit breaker, isolator, and grounding switch drives simplifies the drive system.

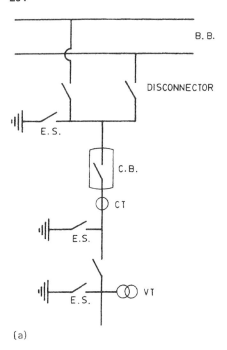

(a)

Figure 10.1   (a) Single-line diagram of a feeder bay of a double-busbar system.

However, three-phase enclosures are used only for voltages below about 200 kV. Above that level, insulation requirements necessitate the use of single-phase enclosure types.

10.3.2   Enclosure Material

Both aluminum and steel are used for SF_6 GIS enclosures. Both materials fulfill specific purposes. However, the use of steel for single-phase enclosure GIS is, for example, limited to lower current densities because of the problem of electrical losses and heating in the enclosure resulting from induced circulating and eddy currents. In the case of a three-phase or single-phase enclosure high-pressure GIS, steel has the major disadvantage that it cannot be shaped and welded into the same homogeneous form as can be achieved with aluminum castings to ensure optimum field stress distribution for a given gas pressure and conductor configuration.

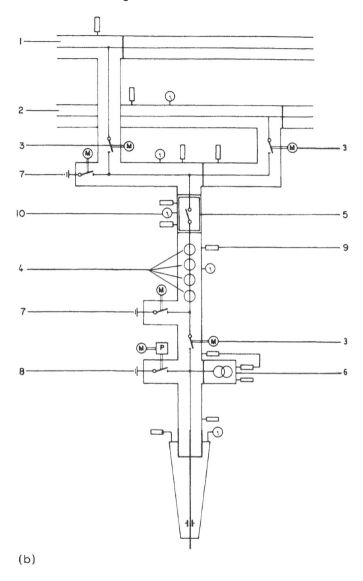

(b)

Figure 10.1 (continued)   (b) diagram of a typical GIS feeder bay showing gas-filled compartments. 1, busbar i; 2, busbar ii; 3, disconnector; 4, current transformer; 5, circuit breaker; 6, voltage transformer; 7, maintenance grounding switch; 8, high-speed grounding switch; 9, gas connector; 10, density monitor.

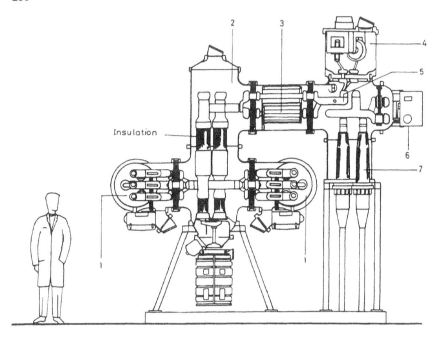

Figure 10.2 Sectional view of a typical SF_6 gas-insulated, 145 kV circuit breaker bay using one enclosure for the three phases. 1, busbar with combined disconnector grounding switch; 2, circuit breaker; 3, current transformer; 4, voltage transformer; 5, combined disconnector/grounding switch; 6, high-speed grounding switch; 7, cable-end unit.

10.3.3 SF_6 Insulating Gas Pressure

With respect to the SF_6 insulating gas pressure, there are two principal GIS designs: high-pressure GIS operating at about 4 bar (405 kPa), and low-pressure GIS, which operates at 1.2 bar absolute. While for a rate voltage of 72.5 kV and below the low-pressure GIS is the more suitable design, at higher voltage levels the advantages of high-pressure GIS become dominant.

A higher SF_6 gas pressure results in higher dielectric strength. Consequently, for a given voltage level the required conductor spacing is reduced, resulting in a more compact design. On the other hand, the dielectric strengh increases at a lower rate than the gas pressure. The reason behind this is the increasing sensitivity of SF_6 insulating gas to the roughness of conductor and enclosure surfaces and to contamination at higher gas pressures, which in turn can reduce the reliability and increase the require-

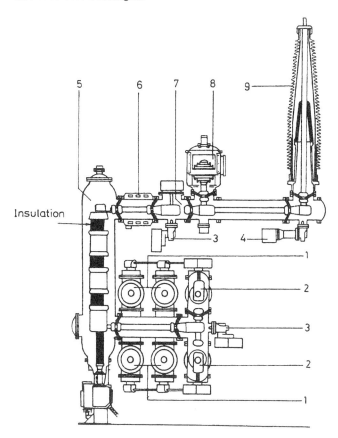

Figure 10.3 Sectional view of a typical SF_6 gas-insulated EHV (220–800 kV) bay using single-phase enclosures. 1, busbar; 2, busbar disconnector; 3, maintenance grounding; 4, high-speed grounding switch; 5, circuit breaker; 6, current transformer; 7, disconnector; 8, voltage transformer; 9, SF_6/air brushing.

ment for servicing and maintenance. As the gas pressure is increased, so are the requirements for a homogeneous electric field distribution if the higher pressure is to be fully utilized. Consequently, for higher SF_6 gas pressures it becomes increasingly necessary to ensure careful homogeneous shaping of conductors, components, and enclosures.

From the above it can be deduced that there is a limit to the SF_6 insulating gas pressure above which the economies of a design decrease as the disadvantages of the higher gas pressure dominate. By the same reasoning, for voltages of 110 kV and above there is also a limit to the SF_6 insulating gas pressure below which the increasing production cost for larger enclosures and greater material expenditure as well as larger buildings make the design uneconomical.

Another important consideration when selecting the SF_6 insulating gas pressure is the behavior of the GIS when a leak occurs. A leakage resulting in a drop in the SF_6 insulating gas pressure automatically means a reduction in the dielectric strength and therefore the integrity of the switchgear. In the case of high-pressure GIS with a rated insulating gas pressure or approximately 4 bar, a severe gas leakage with a pressure drop down to 1 bar will result in a reduction of the dielectric strength, and therefore of the BIL, by about 75% of the rated value. On the other hand, in the case of low-pressure GIS with a rated insulating gas pressure of 1.2 bar, the same leak will result in only a 15% deterioration in dielectric strength.

10.3.4   Conductor System

The conductors of a GIS normally consist of aluminum tubes, the diameter and wall thickness of which depend on voltage and rated current (Sections 10.6 and 10.6.3). Spring-loaded copper contact fingers constitute the female contacts and copper plugs the male contacts. The contact surfaces are silver plated and the contacts are welded to the aluminum conductors. The conductor system, together with supporting insulators, must be properly designed to withstand the electrical, thermal, and mechanical stresses that arise during normal service and during short-circuit conditions.

10.3.5   Solid Spacers

Solid insulators are used in gas-insulated apparatus for physical support of high-voltage conductors and for mechanical operations of the switchgear. They take various shapes, such as annular disks, truncated cones, and post supports. The presence of a solid insulating material in the compressed SF_6 medium distorts the field distribution in the GIS duct (Stone et al., 1987). Several problems are introduced by the use of spacers in GIS:

1. Spacers can limit the operating gradient (in kV/m) of the system, due to aging of the solid material.
2. Spacers can affect the short-term behavior of the GIS (i.e., its dielectric strength) because of a number of factors, as detailed in Section 10.7.3.

10.4   COMPONENTS OF GIS

In addition to circuit breakers (discussed in Chapter 11), a GIS bay includes disconnectors, voltage and current transformers, surge arresters, bushings and cable-end boxes, and gas density monitors.

10.4.1   Disconnectors

Disconnectors are made up from insulators, enclosures, and conductors of different geometrical shapes to give an optimum layout, as shown in Figures 10.2 and 10.3 and in detail in Figure 10.4. They are equipped with copper contacts that are spring loaded to give the disconnector high electrical efficiency and high mechanical reliability. Disconnectors must be carefully designed and tested to be able to break small charging currents without generating too-high overvoltages, otherwise a flashover to earth may occur. The operating mechanisms of the disconnectors and earthing switches are of the same design for most GIS. The main features are motorized or manual operation, electrical interlocking against incorrect operation, and mechanically lockable end positions.

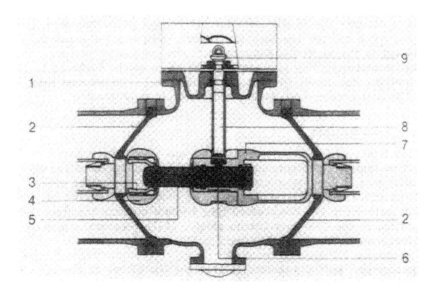

Figure 10.4   GIS disconnector: (1) enclosure, (2) barrier insulator, (3) fixed contact, (4) shielding for fixed contact, (5) moving contact, (6) rack-and-pinion drive, (7) contact support, (8) insulated drive shaft, (9) driving motor.

10.4.2 Earthing Switches

Two different types of earthing switch are normally used, the slow-operating earthing switch and the fast-closing (high-speed) earthing switch (see Figs. 10.2 and 10.3). Slow-operating earthing switches are used for protection purposes when work is being done in the substation, but are operated only when it is certain that the high-voltage system is not energized. The fast-closing earthing switch can close against full voltage and short-circuit power. The fast-closing operation is achieved by means of a spring-closing device.

10.4.3 Voltage Transformers

The most commonly used voltage transformer is of the inductive type (Fig. 10.5). In three-phase enclosed GIS designs, three voltage transformers are placed in one enclosure. It is also possible to design a voltage transformer consisting of a low-capacitive voltage divider connected to an electronic amplifier. The capacitance between the inner conductor and a concentric

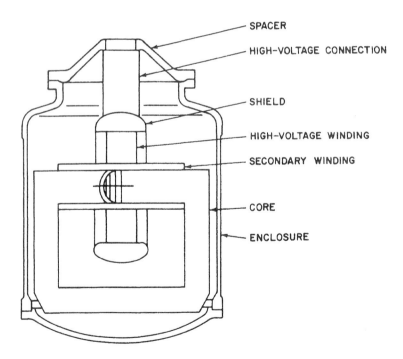

Figure 10.5 GIS voltage transformer.

measuring electrode near the enclosure is then used as the high-voltage capacitor. This design is suitable only for the highest system voltages (Reisinger et al., 1987).

10.4.4 Current Transformers

In the single-phase enclosed GIS, the core of a current transformer is located outside the enclosure, thus ensuring a completely undisturbed electrical field between the enclosure and the conductor. The return current in the enclosure is broken by an insulating layer. Figure 10.6 shows the single-phase enclosed arrangement. In the three-phase enclosed GIS design the cores of the current transformers are normally located inside the enclosure (Fig. 10.2). They are preferably placed outside on the SF_6 bushings or on cables.

10.4.5 Surge Arresters

SF_6 gas-insulated arresters are based on the same active parts as conventional arresters (i.e., varistors and spark gaps) but have very compact designs. The spark gap elements are sealed off from the atmosphere, and the entire arrester is insulated with dry compressed gas, which creates a highly consistent performance within tolerances. It is therefore possible to use arresters with lower sparkover voltages, which provides a closer margin

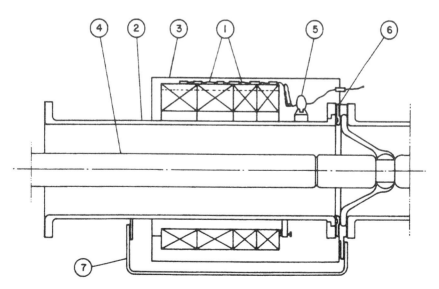

Figure 10.6   GIS current transformer: (1) core unit, (2) enclosure, (3) cover, (4) HV conductor, (5) secondary connection, (6) insulation layer, (7) short-circuiting bar.

for protecting the system insulation. The reduction can reach 10% as compared to conventional arresters.

In SF_6 gas-insulated arresters the metallic earthed parts are much closer to live parts, exploiting the high insulation strength of compressed gases. Thus capacitances to earth of live parts are far greater than in conventional arresters. Therefore, extra precaution is needed to compensate the resulting nonlinearities in voltage distribution along the components of gas-insulated arresters. This is achieved in their design (e.g., by including a metallic hood) (Fig. 10.7). The shape and dimensions of such a hood, necessary for linearizing the field distribution along the arrester elements, can be determined by field computations (Chapter 2). Figure 10.7 shows an arrester with a rated voltage 120 kV. In arresters with higher rated voltages the hood

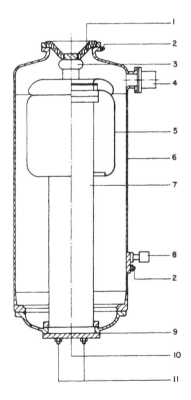

Figure 10.7 GIS-enclosed 120 kV arrester: (1) support and barrier insulator, (2) SF_6 gas connections, (3) HV connection, (4) bursting plate connection, (5) hood for field control, (6) metal cladding, (7) insulating tube, (8) manometer, (9) insulation, (10) ground connection, (11) N_2 gas connection.

is supplemented with a number of grading metal rings fitted at certain points along the arrester's active parts. SF_6 gas-insulated arresters can, of course, be integrated into the GIS in any desired position, depending on the protection requirements.

10.4.6 Bushings

Overhead lines and all-air insulated components are connected to the SF_6 GIS by air/SF_6 gas-filled bushings. These bushings (Fig. 10.3) employ capacitive grading and are divided into two independent gas compartments by a barrier insulator.

The space surrounded by the porcelain insulator is filled with SF_6 gas to slightly above atmospheric pressure. If the porcelain is damaged, this reduces the risk to a minimum. The gas space on the switchgear side of the barrier insulator has the same SF_6 gas pressure as the switchgear. Oil-filled condenser bushings can also be used for high voltages. Bushings used for the direct connection of a GIS to a transformer may also be of the oil-filled type.

10.4.7 Cable-End Boxes

High-voltage cables of various types are connected to the SF_6 switchgear via cable-end boxes (Fig. 10.2), which consist of the cable-end bushing with connecting flange, the enclosure, and the barrier insulator with female plug contacts.The pressure-tight bushing separates the SF_6 gas compartment from the insulating medium of the cable. A completely dry cable termination for connecting XLPE cables to GIS has the advantages of smaller dimensions and better thermal properties.

10.4.8 Gas Density Monitors

The dielectric strength of the switchgear insulated with SF_6 gas and the breaking capacity of the SF_6 circuit breaker depend on the density of the gas. Since the pressure varies with temperature, it is the gas density that is monitored. For this, a density monitor is employed. Each of the gas compartments separated by barrier insulators (Fig. 10.1b) is monitored by its own density monitor.

10.4.9 New Trends in GIS Design

During the past years some improvements have been introduced in the design and structure of GIS. Following are some of those features.

10.4.9.1 Gas-Insulated Transformers

As shown earlier, GIS substations normally do not encompass power transformers in the compressed gas complex. This has been mainly because the
cooling capacity of the gas is lower than that of oil in oil-immersed transformers. Recently, however, gas-insulated transformers have been manufactured (Toda et al., 1995). They fall into two broad categories:

1. Gas-insulated transformers in the capacity range below 60 MVA
   which do not generate much heat and are therefore cooled by SF_6
   gas circulating. They are referred to as the "gas cooling type."
   Transformers of this type normally serve medium-voltage systems
   (up to 77 kV).
2. Large-capacity transformers with a capacity above 60 MVA.
   These are normally liquid-cooled gas-insulated transformers
   using perfluorocarbons ($C_8F_{16}O$ or C_8F_{18}), which are incombus-
   tible and have a high cooling capacity as a cooling medium. Such
   transformers with a rating of up to 300 MVA at 275 kV are now
   available (Toda et al., 1995). Measures taken to improve the cooling capability include raising the SF_6 gas pressure, optimizing the
   layout of gas ducts in the windings to produce large gas flow, and
   employing highly thermal-resistant insulating materials to raise
   the temperature-rise limit of the windings.

Gas-insulated transformers enjoy all the advantages of GIS such as
safety, cleanliness, non-inflammability, and compactness. Figure 10.8 shows
the layout of a GIS housing a gas-insulated transformer and gas-insulated
shunt reactors.

10.4.9.2 Voltage and Current Transformers

Inductive instrument transformers which are heavy and large, are being
replaced by optical or electrical sensors. The function of the traditional
high-voltage current and voltage instrument transformers is performed by
an advanced generation of current and voltage sensors combining both the
functions in a single component. For the current measurement, a Rogowski
coil (an air-core ring coil) is used (Högg et al., 1998). This measures the
current with the highest precision and without saturation across the entire
operating range. The voltage is measured by means of a metal-enclosed
capacitor voltage divider, which avoids any ferromagnetic resonance. The
measured signals are processed after being digitized and then sent, in serial
form, via the optical process bus (IEC 1375) to the bay protection and
control level. The sensors are normally located in the outgoing circuit of

Gas-insulated shunt reactor

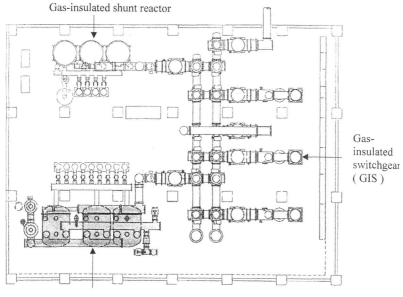

Gas-
insulated
switchgear
(GIS)

Gas-insulated transformer

Figure 10.8 Layout of a gas-insulated substation housing a gas-insulated transformer. (Courtesy of Toshiba.)

the breaker and meet all the demands made on control and measurement as well as on state-of-the-art protection and revenue metering.

10.4.9.3 Drive Motors

The trend now is for drive motors for disconnectors and earthing switches to be controlled electronically. The function of the auxiliary switch is performed by a sensor mounted directly on the operating rod.

10.9.4.4 Control and Protection

Control and protection signals are transmitted in digital form over fiber optic links. Drive control, density measurement, and various measurements for monitoring are also realized with modern sensor technology. The measured analog signals are digitized before serial transmission over the optical process bus to the bay level (Högg et al., 1998). For this connection to the control and protection sytem (control cabinet), there is a plug-ended cable which incorporates the fiber optic (bus) cable for analog and binary signal exchange as well as the auxiliary power supply for the drive mechanisms. This version can be connected to numerical or digital protection systems.

Conventional toroidal core current or voltage transformers are required for the electromechanical and solid-state protection relays. The wide use of sensors makes it possible to implement a large range of monitoring features, such as self-checking, trend analysis of the gas density, and circuit-breaker conditions (pump running time, operations, energy needs, contact displacement curve, remaining lifetime, etc.).

10.4.9.5 Integrated Compartments

A metal-enclosed, gas-insulated combination of circuit breaker, disconnectors, earthing switch, current and voltage sensors, and bushing may be located in a common gas compartment (Högg et al., 1998). The high-voltage switchgear is limited to the minimum amount of equipment really necessary to guarantee the functionality of the bay or the substation for all typical configurations.

10.4.9.6 Resistor-fitted Disconnectors

Upon switching a charging current, the GIS disconnector causes repeated restriking, thus generating very fast transient (VFT) overvoltages. Such overvoltages can cause ground faults between disconnector contacts or at a bus contaminated by metallic particles (Ziomek and Kuffel, 1997). Interference with a low-voltage circuit control system may also develop. With UHV GIS systems underway the trend is towards reducing the lightning-impulse withstand level, e.g., to only 2250 kV for 1000 kV apparatus. Therefore, there is a need to suppress VFTs in such systems by employing resistor-fitted disconnectors. A disconnector catering for the above needs was manufactured as shown in Figure 10.9 (Yamagata et al., 1996). The disconnector connects a resistor in series with the circuit in the event of restriking, and should be able to to so using movable electrodes, i.e., without needing mechanical contacts to connect the resistor. The disconnector was said to withstand the duty of suppressing very fast transient overvoltages due to restrikings at the time of switching charging currents. It met the limiting requirements of the rising speed of the voltage on the resistor, the energy consumed by the resistor, and the damping time constant of the resistor voltage.

10.5 COMPRESSED GAS-INSULATED CABLES

Although overhead transmission still represents the most economical solution to bulk power transmission, it is becoming more difficult to construct overhead lines in densely populated areas, and underground cables are then required (Chapter 12). SF_6 compressed gas-insulated cables present possibilities for underground transmission of high power with rated voltage above

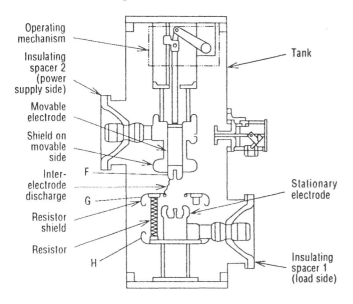

Operating mechanism

Insulating spacer 2 (power supply side)

Movable electrode

Shield on movable side

Inter-electrode discharge

Resistor shield

Resistor

F

G

H

Tank

Stationary electrode

Insulating spacer 1 (load side)

Figure 10.9 Typical resistor-filled disconnector in a GIS.

123 kV. The range of application of compressed gas-insulated cables extends from short links to transmission of energy from cavern-type power stations over long distances. Conventional cables are adequate for rated voltages up to about 150 kV over short distances and without extremely high rated currents. The application range of these cables can be expanded only by forced cooling.

For higher voltages it will become increasingly more difficult to dissipate the heat losses since the insulation thickness, and hence the internal thermal resistance, will be necessarily increased. Also, the dielectric loss increase with voltage—raised to an exponent of about 1.4—can hardly be contained within acceptable limits. The compressed-gas cable is an attractive alternative in this case.

10.5.1 Comparative Conductor Size

The active losses of a transmission line depend primarily on the cross-section of the conductor. Additional active losses occur in cables and compressed gas-insulated cable systems due to loss currents in the metallic enclosure and to dielectric losses (Chapter 12). The cross-sections of individual transmission media that can be utilized in practice are determined largely by their active resistance, which also influences their range of application.

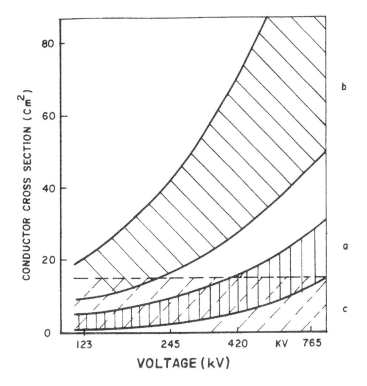

Figure 10.10 Comparative conductor cross-sections as functions of rated voltage: (a) overhead lines, (b) compressed-gas cables, (c) conventional cables.

Figure 10.10 shows a comparision of cross-sections that can be produced in practice. Overhead lines are generally aluminum-stranded conductors with cross-sections of 240–500 mm². This relatively small range is, however, increased by using bundled conductors for higher voltages. Furthermore, the total of reactive losses is also superior to those of compressed gas-insulated cables, which is particularly important when the transmission distance is long.

Whereas conventional cables show heavy capacitive losses that only slightly change with loading, a compressed gas-insulated cable changes from capacitive to inductive losses as loading is increased. A point is passed where the ideal operating condition is reached (i.e., at zero reactive losses). The surge impedance load of compressed gas-insulated cables is three to four times smaller than that of overhead lines. Therefore, the transmission properties of a compressed gas-insulated cable correspond to those of three to four overhead lines operating in parallel (Chapter 14).

10.6 DIMENSIONING OF COMPRESSED GAS ENCLOSURE

The dimensions of the enclosure are generally decided by considering the duct to be a coaxial cylinder arrangement with a conductor of outer radius r and an enclosure of inner radius R. As explained in Chapter 2, the electric field at the surface of the conductor is

$$E_r = \frac{V}{r \ln(R/r)} \tag{10.1}$$

where V is the applied phase voltage.

As shown in Chapters 2 and 12, E_r is minimum (i.e., ionization is least likely to develop) when

$$\frac{R}{r} \approx e \approx 2.72 \tag{10.2}$$

If this relation is combined with knowledge of the breakdown threshold of the gas (i.e., the maximum allowable value for E_r), the enclosure dimensions will be determined. For coaxial cylinders, several formulas for the breakdown threshold have been published. One formula sets the threshold under positive impulse voltages at

$$E_{th} = 81.2p + 10 \qquad \text{kV/cm}, \qquad 0.1 < p < 0.4 \, \text{MPa} \tag{10.3}$$

For a 420 kV GIS whose basic impulse insulation level (BIL) is 1425 kV and which operates at 0.35 MPa, equations (10.1)–(10.3) give $r = 48$ cm and $R = 13.2$ cm. Values obtained in this way are somewhat less than what should be used in practice, where particle contamination and surface roughness have adverse effects.

10.6.1 Condensation Threshold

Another important factor in the dimensioning of GIS ducts is that condensation of the gas should be avoided, typically at the lowest ambient temperature. According to Figure 10.11, for a given minimum temperature there exists an upper limit of gas density above which condensation is sure to occur (e.g., a gas density of 60 g/l at $-20°$C). This maximum allowable gas density, in turn, corresponds to a maximum allowable withstand electric field, as shown in Figure 10.11 (e.g., 430 kV/cm at 60 g/l). Finally, the electric field in the gas can be ensured not to exceed this maximum only if a lower limit is imposed on the duct dimensions according to equation (10.1). For the current example, the field is maintained below 430 kV/cm for a typical system BIL of 1425 kV if the conductor and inner duct radii are larger than 3.3 and 9.0 cm, respectively.

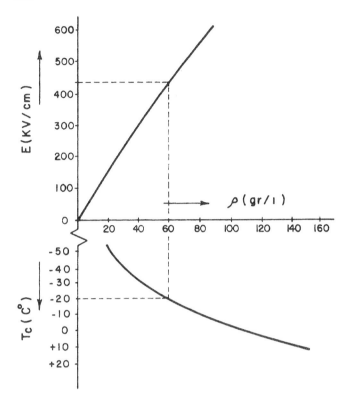

Figure 10.11   Relations between the impulse breakdown strength E of SF_6, the gas density ρ, and condensation temperature T_c.

10.6.2   Spacer Flashover

Solid support insulators and spacers represent a parallel insulation to the compressed gas insulation, and hence breakdown may, in general, materialize in either insulation. Under low gas pressures a flashover across a solid insulator is highly unlikely to occur prior to a breakdown in the gas and therefore poses no design problem. At higher gas pressures, in contrast, a flashover across an insulator is likely since the dielectric strength of the gas is much increased while that of a solid insulator is nearly unaffected by increasing gas pressure.

The surface flashover strength of cast resin, the material of which solid supports are usually made, is about 30 kV/cm (rms). The relation between the applied phase voltage and the maximum field on a supporting insulator depends largely on the insulator shape. Efforts have been made to reduce the maximum field to only 1.2 of its average value along the insulator (Stone et al., 1987). Therefore, the maximum field on the insulator is related to the phase voltage by

$$E_m = \frac{1.2V_{ph}}{R - r} \qquad (10.4)$$

If E_m is equated to the reported dielectric strength of 30 kV/cm, then, for a 420 kV GIS,

$$R - r = 9.7 \text{ cm}$$

If the optimal relation between the inner and outer radii, equation (10.2), is considered, the minimum dimensions of the system can be determined as

$$\left. \begin{array}{l} r = 5.6 \text{ cm} \\ R = 15.3 \text{ cm} \end{array} \right\} \qquad (10.5)$$

10.6.3   Heat Dissipation Considerations

While the GIS duct diameter computed on an insulation basis increases more or less linearly with rated system voltage, the rated current increases much less than linearly with voltage. At lower system voltages, therefore, the duct diameter may be required to be larger than that determined on an insulation basis to limit the temperature rise in the duct.

The heat generated by the "copper" losses in both the conductor and the duct wall will be dissipated to the atmosphere by radiation and convection. By increasing the thicknesses of both the hollow conductor and the duct wall, more current can be passed in the system without excessive temperature rise.

10.7   FACTORS AFFECTING INSULATION STRENGTH

The level of dielectric strength of compressed SF_6 outlined in the foregoing sections may not be fully attainable in reality. Several factors are known to have an effect on the dielectric strength of SF_6 when used in an actual GIS. In the following sections the effects of those factors are discussed.

10.7.1   Effect of Electrode Material

Under very high pressures and when the breakdown strength exceeds about 200 kV/cm, the material of the electrodes in an SF_6 gap begins to have an influence over breakdown. The uniform field breakdown strength is larger with steel electrodes than with copper. The dependence of breakdown strength on electrode material is especially evident in uniform and quasi-uniform field gaps.

10.7.2 Conductor Conditioning and Surface Roughness

Particularly at high SF_6 gas pressures, repeated breakdowns affecting the electrode surfaces have their influence on the strength of the gap. The dielectric strength increases with the number of previous breakdowns until it levels off at an ultimate value. This phenomenon is explained by the conditioning of the electrodes. The effect of conditioning on breakdown in SF_6 was reported to depend on the gas pressure and the electrode area (Anis and Ward, 1988).

The breakdown strength of an SF_6 gap no longer has a single value at a given gas pressure. The fluctuation in the breakdown strength is thus better expressed by its statistical distribution. The dispersion is larger at higher pressures. The breakdown strength is also reduced by the roughness of the electrode surface. It has been found that the reduction in the breakdown voltage due to electrode roughness depends on the type of gas insulation as well as the product of the protrusion height R_{max} and gas pressure p. When this product exceeds 0.8 kPa·cm, the protrusion causes a decrease in the breakdown voltage. However, the effect is negligible if the product is less than 0.4 kPa·cm. The reduction in the maximum withstand stress E related to the gas pressure p as a function of the product pR_{max} is shown in Figure 10.12.

10.7.3 Problems Associated with Solid Spacers

Several factors contribute to the reduction of dielectric strength of spacers:

1. *Imperfect solid conductor adhesion.* Narrow gaps may exist between electrodes and spacers due to imperfect casting and/or mechanical stresses at the solid/gas interface. Studies of the effect of small gas gaps between the spacer and conductors on breakdown performance showed that the reduction in breakdown strength is more pronounced at higher gas pressures. As explained in Chapters 2 and 8, a small gas gap stressed in series with insulation of a dielectric constant ε_r would undergo an electric stress ε_r times higher than the average. Thus a microdischarge is then sure to occur in the gap.

2. *Moisture content.* The presence of some degree of moisture adversely affects the dielectric strength of the spacer/gas interface. It was reported that the breakdown voltage dropped by more than

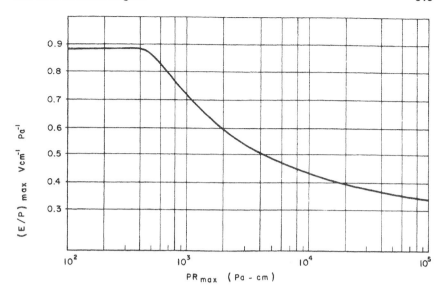

Figure 10.12 Effect of surface protrusions on the strength of SF_6.

50% when the humidity was increased from zero to 4000 ppm (Masetti et al., 1982).

3. *Contaminating particles.* Contaminating particles may eventually adhere to the solid spacer by electrostatic forces, causing local field enhancement. In the following section the subject of particle contamination in GIS is discussed in more detail.

As mentioned in Section 10.6.2, the breakdown in a high-pressure GIS is more likely along the gas/spacer interface, unlike low-pressure GIS where breakdown is purely gaseous. Efforts have been made to optimize the design of post-type spacers based on minimizing the field enhancement in the gas at the spacer boundary while accounting for resultant electrodynamic forces on the conductors (Stone et al., 1987; Trinh et al., 1984).

10.7.4 Particle Contamination in GIS

The presence of particle contaminants in gas-insulated systems is by far the most significant factor responsible for the deterioration of insulation integrity. The effect of metallic particles on the SF_6 breakdown voltage is more pronounced at high gas pressures, as shown in Figure 10.13 for a 150/250 mm coaxial system and 0.4 mm diameter wire contaminants.

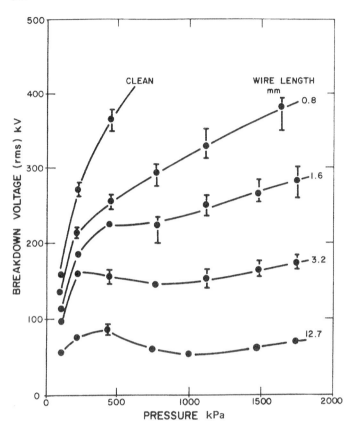

Figure 10.13 Effects of conducting wire particles of different lengths on the strength of an SF$_6$ gap at different pressures. [From Cookson and Farish (1973).]

In the case of free particles, the particle motion depends largely on the type of applied voltage. Under AC voltage, for a wire particle of given radius, the activity increases with particle length since the particle charge-to-mass ratio at lifting increases with length. In addition, lateral particle movement along the central conductor is possible. Under DC voltage, the particles oscillate between the electrodes, and wire particles may exhibit intense "firefly" activity.

10.7.5 Particle-initiated Breakdown

Under the influence of the applied field, free conducting particles become charged and oscillate in the interelectrode gap. As a charge particle approaches either electrode, it may lose its charge to the electrode through a microdischarge in the gas. This microdischarge may well trigger the breakdown of the main gap (Cooke, 1978).

Particle-initiated breakdown in compressed gas insulation generally depends, among other parameters, on the position of the particle in the gap. It has been reported that, for a given particle-contaminated GIS system at some particle positions in the gap, the breakdown voltage is lower than for others (Cooke et al., 1977). An attempt was made to establish an analytical "breakdown voltage profile" that relates the instantaneous breakdown voltage magnitudes to the instantaneous particle position in the interelectrode gap (Anis and Srivastava, 1981a). For a 1 mm radius spherical particle in a parallel-plane SF_6 gap at a gas pressure of 5 atm, the breakdown voltage profile is as illustrated in Figure 10.14. The profile explains the experimentally discovered existence of a critical particle position at which the insulation displays minimum dielectric withstand. This "critical" breakdown voltage may be taken as the insulation withstand limit under transient as well as steady-state applied voltages. The breakdown voltage profile helped to explain the significant difference between the breakdown voltages of fixed and free particles, and the general effects of particle size, gas pressure, and electrode configuration.

10.7.6 Particle Control Techniques

For systems to be reliable and economical, the problem of particle contamination should be overcome. Contamination control in GIS can be achieved either by designing the system with a certain degree of immunity against the presence of particles, or by providing designated low-field areas in the system in the form of "particle traps" where the particles can be safely trapped and contained. An electrostatic particle trap can be designed such that a region of low or zero electric field is provided at the outer enclosure by a slightly elevated metal shield. Particles can enter the low-field region created by the trap through slots in the shield. There are several important factors to be considered for particle control with such traps:

1. Particle contaminants have to be moved through the GIS system to the location of the traps. This must be achieved without risking damage to the system from flashovers.
2. The trap must provide rapid capture of the particles without any restriction on particle size and shape.
3. The trap must have positive retention characteristics under AC, DC, and transient voltages as well as mechanical vibrations.

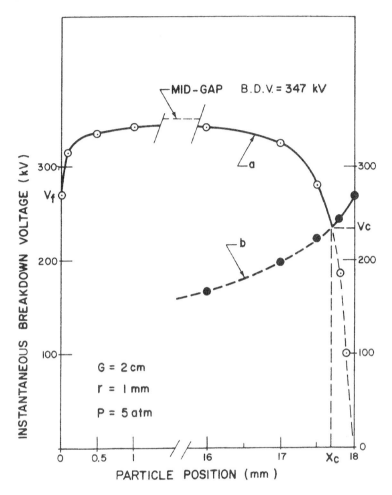

Figure 10.14   Breakdown voltage profile of a particle-contaminated SF_6 2 cm gap, showing the effect of a 2 mm spherical particle position along the gap: (a) locus of breakdown voltage between a particle and the right-hand elecrode; (b) locus of breakdown voltage between a particle-caused protrusion on the right-hand electrode and the left-hand electrode.

10.7.7   Conductor Coating

Conductors in a gas-insulated system may be coated with a dielectric material to restore some of the dielectric strength of the compressed gas that is lost as a result of surface roughness and contamination with conducting particles. The improvement in dielectric strength of the system, due to coating, can be attributed to the following effects:

1. Coating reduces the degree of surface roughness on conductors, thus decreasing the high local electric fields (Endo et al., 1983).
2. The high resistance of the coating dielectric impedes the development of pre-discharges in the gas, thus increasing the breakdown voltage.
3. In the presence of metallic contaminating particles, dielectric coating can be of further benefit in two main ways:
   a. The electric field necessary to lift a particle resting on the bottom of a GIS enclosure is much increased, due to the coating (Parekh et al., 1979).
   b. Once a particle begins to move in the gas gap under the applied voltage, it may collide with either conductor. If the conductor is coated, the particle will acquire a drastically reduced charge, if any. Thus the risk of a breakdown initiated by a discharge is reduced significantly.

A particle resting on the bottom of a GIS duct acquires the charge necessary for lifting either by conduction through the dielectric coating or by means of microdischarges in the gas initiated at the particle surface.

10.7.7.1 Particle Charging through Coating Layer Conductance

The dielectric coating layer has a finite nonzero conductance, and therefore a conductive charging current will pass through. The charge accumulated on the particle should reach a value at which the electrostatic lifting force on the particle overcomes the gravitational force (Anis and Ward, 1988). Using this mechanism, it can be shown that the lifting field E_1 and the corresponding lift-off charge q_1 of a filamentary particle in a coaxial GIS under AC configurations are

$$E_1 = \left[\frac{mg}{R \ln(R/r) A \alpha(\phi)} \right]^{1/2} \tag{10.6}$$

and

$$q_1 = \left[\frac{(1 + C_d/C_g)^2}{G^2} + \frac{1}{\omega^2 C_g^2} \right]^{-1/4} \frac{[mgR \ln(R/r)]^{1/2}}{\omega k} \psi(\phi) \tag{10.7}$$

where

$$\psi(\phi) = \left\{ \frac{\cos\phi - \cos[(2\pi + 2\phi)/3]}{\sin[(2\pi - \phi)/3]} \right\}^{1/2}$$

$$A = \frac{k/\omega}{\sqrt{(1 + C_d/C_g)^2/G^2 + 1/\omega^2 C_g^2}}$$

$$\alpha(\phi) = \left(\cos\phi - \cos\frac{2\pi + 2\phi}{3} \right) \sin\frac{2\pi - \phi}{3}$$

$$\phi = \cot^{-1}\left(\frac{\omega C_g(1 + C_d/C_g)}{G} \right)$$

where C_g and C_d are, respectively, the capacitance between the particle and the inner and outer electrodes; G is the conductance of the particle-to-outer-electrode path; and ω is the angular frq
uency. k is a factor, less than unity, to account for the effect of image charges on the electrostatic force; k is approximately 0.82 for spherical and horizontal wire particles, and nearly unity for long vertical wire particles (Anis and Srivastava, 1981b). r and R are the inner and outer radii of the GIS system, and g is the gravitational acceleration.

10.7.7.2 Particle Charging through Microdischarges

The other possible mechanism through which charges can be transferred to a contaminating particle from a coated conductor in a GIS is that of micro-discharges. If the field near the particle is high enough, discharges may occur between the particle and the electrode through the dielectric coating, thus charging the particle. This mechanism was used to compute lifting fields of spherical particles (Parekh et al., 1979).

When the particle is in contact with the dielectric coating and the field is sufficiently high, electron avalanches will be initiated at a point on the particle surface and propagate along a field line toward the coating layer. The point on the particle surface at which avalanches will develop is that which ensures maximum avalanche size, which is a function of both field and space. The size of an electron avalanche started by one electron is given in Chapter 4 as

$$n_e = \exp\left[\int_{x_1}^{x_2} (\alpha - \eta)\,dx \right] \tag{10.8}$$

where x_1 and x_2 are the avalanche limits in space, α the ionization coefficient, and η the electron attachment coefficient. Having n_e, the charge transferred from the conductor surface to the particle is en_e, e being the electronic

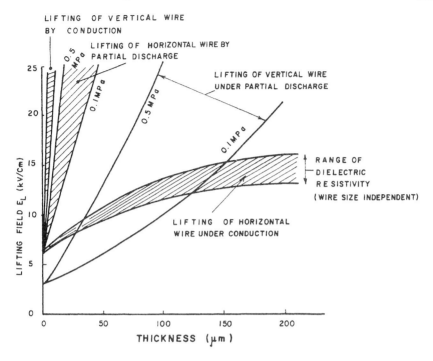

Figure 10.15 Comparative lifting fields for conducting-wire particles off coated electrodes in compressed-gas ducts.

charge. Equation (10.8) can be applied by first computing the field distribution in space, which is then related to $(\alpha - \eta)$ (Chapter 2).

The controlling mechanism causing the particle to lift off at a lower applied field will be either conduction or microdischarges. A comparison between the possible effectiveness of the two mechanisms for lift-off is shown in Figure 10.15. Two initial positions for wire particles are shown in each case, the horizontal and the vertical positions. It is clear from the figure that particle lift-off for all coating layer thicknesses beyond 100 μm is not possible through a microdischarge mechanism. Conduction through the coating or some other mechanism has to be responsible for lift-off. For thick coatings, a particle will lift off from a horizontal position.

10.7.8   Particle Contamination under Very Fast Transients

Very fast transient overvoltages (VFT) are generated in GIS during the operation of disconnectors. These surges are steep-front oscillatory over-voltages with a maximum magnitude of the order of 2.9 times the operating voltage and a fundamental frequency of several megahertz. The insulation characteristics of spacer surfaces for VFT under particle contamination conditions were investigated by Okabe et al. (1996). For particle lengths in the range 3–10 mm it was reported that the flashover voltage decreased as the particle length increased. Figure 10.16 depicts the resulting voltage-time characteristics. However, the V–t characteristics for VFT and lightning impulse were nearly identical for each of the particle lengths considered and, for a VFT frequency in the range 1.7–18 MHz, no dependence on frequency was observed. It was concluded that the dielectric strength along a spacer surface for VFT is at least equivalent to that for lightning impulse.

10.7.9   Breakdown of GIS at Low Temperature

Outdoor equipment may be exposed to temperatures dipping down to −50°C, a temperature sufficiently low to cause liquefaction of the insulating medium (SF_6), given the appropriate pressure. The electric strength of GIS was believed to be determined by the gas density, independent of the environmental temperature. To verify this, the low-temperature DC breakdown of an SF_6 gas-insulated system was investigated experimentally for temperatures ranging from −50 to 24°C (Fréchette et al., 1995). The experiment was conducted for molecular densities ranging from 2.596 to 16.43 × 10^{19} cm$^{−3}$ (equivalent to pressures of 105 and 624 kPa, respectively, at 24°C) and for gap lengths starting at 0.5 mm and extending to 7 mm. Despite a constant molecular density and under preferably uniform-field conditions, it was found that lowering the environmental temperature brought about a 10% reduction in the withstand capability of the system. The decrease appeared at a temperature threshold between −25 and −30°C, and remained constant down to −50°C. No definite trend with regard to a temperature dependence was traced under high field conditions, which suggested that the variance

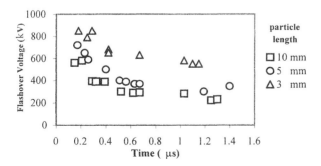

Figure 10.16 Particle flashover under very fast transients.

observed under highly uniform field conditions may be related to a triggered disturbance (field related), linked to a modification in the state of the cathode surface.

Since the breakdown behavior of GIS was shown to depend on the environmental temperature, the conclusion may be that the equipment design should take this into account because of the implicit reduction in the operational safety factor. In this regard, the findings set the order of magnitude for the withstand drop below $-25°C$, i.e., 10%, which provides a basis for deciding what measures to take.

10.8 ELECTROMAGNETIC COMPATIBILITY IN GIS SUBSTATIONS

As the complexity of the electronics grows in a GIS substation, so does its susceptibility to increasing electromagnetic interference in the environment. To ensure high reliability for the installation, it is therefore necessary to make the electronic equipment compatible with its electromagnetic surroundings. The electromagnetic environment in GIS substations depends on the noise sources, which are essentially switching operations carried out with disconnect switches, earthing switches, and circuit breakers. Other noise sources include lightning discharges, earth faults, and short circuits (Meppeling and Remde, 1986). Also to be taken into account are external sources such as local communication systems and power transmitter stations. The noise sources themselves are characterized by such variables as noise voltage, noise current, electromagnetic noise fields, and noise energy.

The disturbance variables are transmitted over a coupling path to the noise sink (electronic equipment or secondary units), where they are reduced by the coupling mechanism. Typical coupling paths are secondary cables, voltage and current transformers, earthing systems, and electromagnetic field coupling into control cubicles. Electromagnetic compatibility is achieved when the disturbance variables active at the noise sink are smaller than the immunity to disturbance of the sink.

Several means may be used to inhibit the coupling paths, thus approaching EMC compatibility. Some of those means are:

Meshed earthing system
Use of control cables and connectors with low transfer impedance
Cable shields earthed at both ends
Secondary circuits of instrument transformers earthed only once
Coaxial entry of secondary cable shields in cubicles
Filtering of the mains supply

Limitation of transient overvoltages
Use of fiber-optic cables for communication and data transmission

10.9   ON-SITE TESTING

In contrast to conventional switchgear installations, where the switchgear need not be installed according to any given sequence, GIS installations require systematic assembly according to a plan laid down at the manufacturing stage. As soon as the components forming a common gas chamber are assembled, they are evacuated to about 100 Pa to remove any moisture that may have entered during the assembly phase. The high-voltage equipment is assembled at the works to form large units, tested, sealed in a gastight housing, evacuated, and filled with dry nitrogen at 150 kPa, and then transported.

At the site of installation the following tests are performed during commissioning:

1. Every flanged joint made on site is tested for leaks. The maximum permissible values per joint are set so that the entire installation cannot lose more than 1% of its gas per year.
2. The moisture content of the SF_6 gas is measured between two and three weeks after the first filling. Experience shows that a final value is reached after this period.
3. Functional checks on the breakers, isolators, and earthing switches, and also in the settings of the supervisory instruments, are carried out as soon as the appropriate stage of completion is reached.
4. If the design of the voltage source and ambient conditions permit, measurement of partial discharges may be carried out.
5. The amount of air in the SF_6 is measured.
6. The earthing system is checked.
7. High-voltage tests are performed on the installation to detect damage or contamination that may have occurred in transit or during assembly.

10.9.1   High-Voltage On-Site Testing

10.9.1.1   Power-Frequency Voltage

Because it corresponds to the continuous stress met with in service, a power-frequency test should be carried out wherever possible. Of disadvantage in such cases is the high reactive power required for extensive installations, and the high test voltages. The state of the insulation can be judged more accurately if frequent partial discharge tests are carried out together with the AC test. In most cases, however, the test results obtained on site cannot be conclusive, owing to the abnormally high ambient noise level. Certain drawbacks in conventional testing equipment can be avoided by the use of resonant circuits. In a few cases a power-frequency test can be carried out by energizing the voltage transformers on their low-voltage side. This distorts the shape of the wave, but although the distortion does not much affect the test result it must be taken into consideration when measuring the voltage level. When the manufacturing tests have been carefully supervised, the test on the completed installation may be considered a repetition. The voltage amplitude can be limited to 80% of the rated test voltage without overlooking assembly errors.

10.9.1.2 Impulse Voltage

Applying impulse voltages to complete installations is not an entirely suitable way of checking the quality of erection because it indicates only shortcomings in the geometric arrangement whereas contamination remains undetected. However, testing with oscillatory switching surges, which correspond to the standard switching surge when flashover occurs, is sometimes used.

10.9.1.3 DC Voltage

In cases where AC and impulse testing cannot be performed, it is possible to test the switchgear with DC voltage instead, although this stress rarely occurs in operation. A DC voltage test can still be carried out at 80% of the rms value of the rated power-frequency test voltage.

Since the switchgear contact gaps are already factory tested, testing of the installation can be limited to checking the dielectric strength between the live parts and the enclosure. It is, however, expedient to establish single testing sections in order to localize faults during testing. These sections are determined by the position of the isolators and breakers and are normally tested with AC or DC voltage for a period of 10 s, or with switching surges up to a maximum of three shots.

The final test, in which the largest possible number of installation sections are coupled together, is performed for a maximum of 1 minute for AC or DC voltage or up to five shots with oscillating switching surges. For stresses varying with polarity, it may be sufficient for the test to be carried out with the more critical polarity.

10.10 MAINTENANCE

The GIS is virtually maintenance free and is designed to avoid any opening of the enclosures during its lifetime of at least 30 years. The circuit breaker can normally withstand 20 interruptions at its rated short-circuit breaking current and 2000 interruptions at full-load current. Therefore, even the circuit breakers need not normally be opened for maintenance. The operating devices require minor maintenance a few times during the GIS lifetime.

10.10.1 Gas Handling

Charging the GIS with compressed SF_6 gas is effected by evacuating each gas compartment (Fig. 10.1b), and the section is filled with SF_6 gas to the required density. During this process the gas moisture content is checked. Nitrogen is used to dry out the internal parts of the GIS. Finally, moisture absorbers are mounted in circuit breakers and each section is filled with SF_6 gas to the working pressure. A special gas-handling chart is normally used for this purpose.

10.11 DIAGNOSTICS OF MICRODISCHARGES IN GIS

The main objectives of GIS diagnostics are to detect whether there is any defect in the GIS, to identify it as a particle, floating shield, and so on, and to locate it so that it can be corrected. Various diagnostic techniques have been developed, both in the laboratory and on site during commissioning (Pearson et al., 1995). A partial discharge (PD) is the localized breakdown of gas over a distance of usually less than 1 mm. For surface defects such as small protrusions, the discharge takes the form of corona streamers which give rise to current pulses with very short rise times (< 1 ns). Discharges in voids, and microdischarges associated with poor contacts or with the transport of conducting particles, are also characterized by high rates of change of current.

In all cases, the very fast rise time of the PD pulse causes electromagnetic energy to be coupled into the GIS chamber, and the energy dissipated in the discharge is replaced through a pulse of current in the EHV supply circuit. In microdischarges and intense coronas the discharge is followed by rapid expansion of the ionized gas channel, and an acoustic pressure wave is generated. Partial discharges are also accompanied by the emission of light from excited molecules, and by the creation of chemical breakdown products. The particle-based discharges in a GIS and the subsequent acoustic and high-frequency emission have been widely investigated (Runde et al., 1997; Sellars et al., 1994; Ziomek et al., 1997). The PD, therefore, has many effects, physical, chemical, and electrical, and in principle any of them could be used to detect the presence of the discharge.

10.11.1 Light Output

Detecting the light output from a discharge is probably the most sensitive of all diagnostic techniques, because a photomultiplier can detect the emission of even a single photon. The radiation is primarily in the UV band, and since this is absorbed strongly by both glass and SF_6 it is necessary to use quartz lenses and a reasonably short path length. Although this is a powerful laboratory tool for finding the onset of activity from a known corona point, there are many difficulties in using it to detect a discharge that might be anywhere in a GIS.

10.11.2 Chemical Products

This is based on the fact that the chemical decomposition is not affected by the electrical interference which is inevitably present in the GIS, and with any steady discharge the concentration of the diagnostic gas should rise in time to a level where it can be detected. The main decomposition product of SF_6 is SF_4, which is a highly reactive gas. It reacts further, typically with traces of water vapor, to form the more stable compounds thionyl fluoride (SOF_2) and sulfuryl fluoride (SO_2F_2). These are the two most common diagnostic gases, and by using a gas chromatograph and mass spectrometer they may be detected with sensitivities down to 1 ppmv. In small-volume laboratory tests a reasonably small discharge of 10 to 15 pC can be detected, typically after about 50 h. However, in a GIS the diagnostic gases would be greatly diluted by the large volume of SF_6 in which they occur, and much longer times would be needed. It therefore appears that the chemical approach is too insensitive to be considered for PD monitoring.

10.11.3 Acoustic Signals

Acoustic signals are generated both from pressure waves caused by partial discharges (PD) and from free particles bouncing on the GIS enclosure. The latter is the only instance of a diagnostic signal not coming from a PD (although of course the particle generates PD as well). An example of an acoustic signal is shown in Figure 10.17 (Runde et al., 1997).

The signals in GIS have a broad bandwidth, and travel from the source to the detector by multiple paths (Lungaard et al., 1992). Those originating at the chamber wall propagate as flexural waves, at velocities which increase with the square of the signal frequency to a maximum of 3000 m/s. Propagation through the gas is at the much lower velocity of

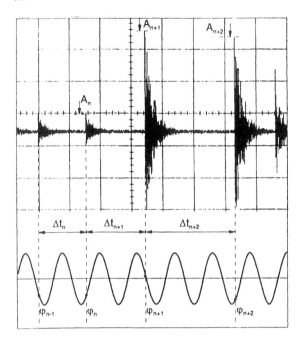

Figure 10.17 Acoustic signal from a bouncing particle and the corresponding 50 Hz voltage trace.

150 m/s, and the higher frequencies in the signal are absorbed quite strongly. Filled epoxide barriers also attenuate the signal markedly.

The different propagation velocities of the wave as it passes through various materials, and the reflections it experiences at boundaries between them, give rise to a complex signal pattern. This can be picked up by accelerometers or acoustic emission sensors attached to the outside of the chamber. The acoustic signal from a particle bouncing on the chamber floor is characterized by a signal not correlated with the power frequency cycle. It also has other features, such as the crest factor (ratio of the peak rms value), the impact rate, and the ratio of the lift-off/fall-down voltages, from which the particle shape and its movement pattern can be inferred. Other sources of discharge may be identified in a similar way from their own characteristics.

One advantage of acoustic measurements is that they are made nonintrusively, using external sensors which may be moved from place to place on the GIS. Because of the rather high attenuation of the signals, the sensors should preferably be on the chamber containing the source. This in itself gives the approximate location of the defect, but a more accurate position can be found within 1 cm by using a second sensor and a time-of-flight method.

The acoustic technique is not suited to a permanently installed monitor, because too many sensors would be needed.

10.11.4 Conventional Electrical Method

A GIS partial discharge test circuit is described in IEC Publication 270. In that circuit the charge flowing through a coupling capacitor fitted in parallel with the GIS is measured using a quadrupole and a detector. The PD current pulse at the defect probably has a duration of < 10 ns, and propagates as a traveling wave in each direction along the chambers. The pulses are attenuated and undergo multiple reflections, but do not appear in the external circuit. After 1 µs the pulses die away, and the GIS appears to the external circuit as a lumped capacitor with a depleted charge. From then a replacement charge flows into the GIS, and is measured by the detector.

To obtain the maximum sensitivity of measurement a completely shielded test arrangement is required, which is possible for a test assembly but may be inconvenient when testing a complete GIS. Also, the total capacitance of a GIS is high, and it must be divided into sections for test. In addition, there is no means of locating the discharge, and since a coupling capacitor is not normally provided in a GIS the technique cannot be used for in-service measurements.

10.11.5 Ultra-High-Frequency (UHF) Technique

The current along the PD path rises in < 1 ns and contains frequency components which extend from DC to more than 1000 MHz. They excite the GIS chambers into various modes of electrical resonance, which because of the low losses in these chambers may persist for $\sim$ 3 µs (Hampton et al., 1990; Judd et al., 1996). The resonances are indicative of PD activity, and if they are picked up by couplers installed in the GIS they may be displayed on a spectrum analyzer.

Over the past years much experience has been gained using UHF at frequencies from 300 to 1500 MHz. UHF couplers may be fitted to the inside of hatch cover plates on GIS equipment and UHF monitoring systems are permanently installed. Since the UHF signals propagate throughout the GIS with relatively little attenuation and even low levels of corona can be detected readily, it is sufficient to fit couplers at intervals of 20 m along the chambers. This can be undertaken at VHF frequencies from 30 to 300 MHz and at UHF. UHF couplers can be either internal or external.

It has been found that the background noise level in cable-fed GIS usually is very low, but when the entry is by overhead line then interference from air corona can be fed into the chambers. Since the intensity of air corona falls rapidly with increasing frequency, it is preferable to make measurements in the UHF range, say at more than 500 MHz. Even then communications interference may be picked up, for example from portable telephones, but this can be easily be recognized.

The UHF components of the resonant signal give it a very fast rise-time, and the source of the PD can be located to within 10 cm from the time interval between the PD pulse wave front arriving at couplers on either side of the defect. The coupler signals can be phase resolved to show the point on the AC wave where the discharges occur, as seen in Figure 10.18 (Sellars et al., 1994).

The principal advantages of the UHF method are its high sensitivity, the ability to locate discharges accurately by time-of-flight measurements, and that it can be used readily in a continuous and remotely operated monitoring system.

A study compared experimentally the above different detection techniques (CIGRE WG 15.03, 1992). For a needle attached to the busbar, Figure 10.19 depicts the results from the various techniques expressed as signal-to-noise ratios. It was reported that although the acoustic, conventional PD (IEC 270), and UHF techniques all showed good sensitivity, the acoustic measurements were nonintrusive and thus could be made on any GIS, but the attenuation of the signal across barriers and along the chambers was high. Also, conventional PD measurements needed an external

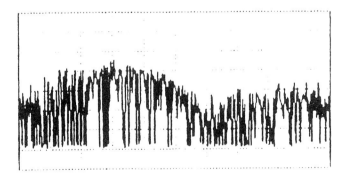

Figure 10.18   Point-on-AC wave for a particle in GIS.

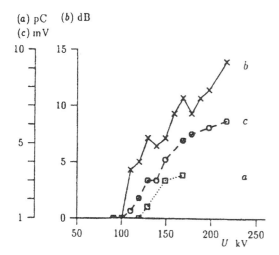

(a) pC   (b) dB
(c) mV

Figure 10.19 Comparative partial-discharge signals from different detection methods; (a) conventional, (b) UHF, (c) acoustic.

coupling capacitor, and could not be used on GIS in service. The UHF technique was suitable for in-service monitoring, but for maximum sensitivity couplers needed to be fitted inside the GIS.

In a more recent study no significant difference between patterns obtained from the different methods was observed (Meijer et al., 1998).

10.12 ENVIRONMENTAL CONSIDERATIONS

The world production of SF_6 has been steadily increasing, resulting in increased concentration of SF_6 in the atmosphere (Christophorou et al., 1997). The amount of SF_6 in the atmosphere has been increasing at an annual rate of 8.7%. The atmospheric concentration of SF_6 could reach 10 parts in 10^{12} by volume by the year 2010. In many industrial applications SF_6 is not recoverable, and its release into the environment by the electric power industry comes from normal equipment leakage, maintenance, reclaiming, handling, testing, etc. Without disposal methods that actually destroy SF_6, it can be expected that all of the SF_6 that has ever been or will ever be produced will eventually enter the atmosphere (within the next few centuries). This is so even if the present SF_6 leak rate from enclosed power system equipment has improved to only 0.5% per year. This release of SF_6 into the environment cannot be reduced significantly since there are no currently accepted and economically feasible methods for controlling or destroying SF_6 as it leaks from enclosures. It has been suggested that impure used SF_6 in storage containers can be destroyed by thermal decomposition in industrial waste treatment furnaces at elevated temperatures ($T > 1100°C$).

However, decreasing the rate of SF_6 leakage and increasing the level of recycling are high priorities since they both will curtail production needs of SF_6 and thus will reduce the quantities of SF_6 that are eventually released into the environment. Indeed, efforts have recently been undertaken by the electric power industry to better monitor the gas pressure in SF_6-insulated equipment and the amount of SF_6 released into the environment. These efforts include improved methods to quantify and stop leakages, better pumping and storage procedures, setting of standards for recycling, manufacturing tighter and more compact equipment, development of sealed-for-life electrical apparatus, gradual replacement of older equipment which normally leaks at higher rates, and implementation of a sound overall policy of using, handling, and tracing SF_6.

10.12.1 SF_6 Greenhouse Effect

Sulphur hexafluoride is a potent greenhouse gas, i.e., it absorbs a portion of the infrared radiation emitted by the earth and returns it to earth by emitting it back. The effective trapping of infrared radiation by the greenhouse gases and its re-radiation back to earth results in an increase in the average temperature of the earth's atmosphere in what is called the "greenhouse effect." The greenhouse effect shifts the balance between incoming and outgoing radiation, causing "global warming."

Sulfur hexafluoride absorbs infrared radiation at wavelengths near $10.5\,\mu m$. It is also largely immune to chemical and photolytic degradation and therefore its contribution to global warming is virtually permanent.

While the potency of SF_6 as a greenhouse gas is extremely high, the amount of SF_6 in the atmosphere should, in the near term, be too small to have significant environmental consequences. The relative contribution of SF_6 to nonnatural global warming is nearly 0.2%. Government and environmental protection agencies, electrical, chemical, and other industries have expressed concerns over the possible long-term environmental impact of SF_6. Because this gas is already widely used, there are obvious economic implications in any attempts to regulate or control its production, use, and eventual disposal.

One considered solution is the alternative use of gases that are benign and environmentally ideal, such as N_2 which is nonelectronegative and has lower dielectric strength (three times lower than SF_6) and lacks the fundamental requirements for use by itself in circuit breakers. Nonetheless, such environmentally friendly gases might be used by themselves at higher pressures, or at comparatively lower pressures, as the main component in mixtures with electronegative gases, including SF_6. Considerations are made for the use of high-pressure N_2 and mixtures of N_2 with SF_6 for insulation, arc quenching, and current interruption, as well as the use of high-pressure nitrogen for gas-insulated transmission. Mixtures of N_2 and SF_6 are being used in circuit breakers under severe weather conditions ($T < -40°C$) where SF_6 used under pressure in circuit breakers may liquefy and thus lose part of its current interruption properties. It was found that for such uses the SF_6/N_2 mixtures with 50% SF_6 are efficient arc interrupting media. Besides SF_6/N_2, other mixtures in use include SF_6/CF_4 and SF_6/He.

10.12.2 Radio Interference and Transit Field from GIS

A study of radio interference from a GIS revealed that they were either below the background levels (i.e., levels due to all RI sources including power lines but excluding the substation), or contributed little to the background (Harvey et al., 1995). Measured broadcasting signal levels from AM, FM and TV stations were higher than the RI levels in these frequency bands. Peak levels were higher than average levels by about 13 dB at 0.05 MHz, while peak levels were higher than quasi-peak levels by 4 dB at 0.5 and 5 MHz, and by 26 dB at 73.5 MHz for both horizontal and vertical polarizations.

In the same study, transient peak fields of up to 580 V/m were measured inside the station building and dropped to background values of 10 V/m at 120 m from the station. Dominant frequencies were between 20 and 25 MHz for sites within the station, and between 10 and 15 MHz for sites outside the station, which suggested that the station building attenuated the higher-frequency components more than the lower-frequency components. Oscillatory transients rose to their peak values in an interval between 0.2 and 0.5 µs, and decayed to below 50% of those values in less than 1 µs and to below 10% in less than 3 µs. In brief, transient field levels diminish rapidly with distance from GIS stations.

10.13 PROBLEMS

(1) A 380 kV, 0.4 MPa compressed gas-insulated feeder is to be designed, i.e., its coaxial dimensions evaluated. The basic insulation level (BIL) of such a system is 1050 kV. The positive breakdown field of SF_6 is nearly a linear function of the gas pressure and approximately equals 81.3 (kV/m)/kPa. The surface discharge field of GIS spacers is 3.2 MV/m.

Design the duct system on both gaseous and solid insulation withstand bases, and comment on the results. Prove any relation you may use.

(2) Discuss the factors that may affect the withstand strength of compressed-gas-insulated switchgear.

(3) A compressed gas-insulated coaxial system operates at 380 kV at a pressure of 3 atm. The system's solid spacers are designed in such a way as to have the maximum field 25% higher than its average field; the surface discharge withstand is 2.5 kV/mm. The BIL peak AC design voltage ratio is 2.1. Choose acceptable dimensions for the system, based on gas and solid breakdown considerations, showing your method of calculation.

(4) Compare, without calculation, the modes of breakdown if the GIS is operated at pressures of 1 atm and 6 atm.

(5) Describe the "breakdown voltage profile" in GIS switchgear.

(6) What are the main advantages and disadvantages of SF_6 when used in GIS systems?

(7) Why are GIS usually sectionalized?

(8) Compare low- and high-pressure GIS systems.

(9) Why are conductors in GIS systems sometimes coated with dielectric material?

(10) Write down the equations governing the movement of solid conducting particles in GIS systems under DC, AC, and impulse voltages, showing the various forces acting on them.

(11) What are the factors governing the conductor and enclosure diameters in GIS?

(12) Explain why, in monitoring SF_6 in GIS systems, the gas density rather than the gas pressure is measured.

(13) What are the main troubles caused by solid spacers used in GIS?

(14) For a 240 kV GIS system whose basic insulation level is 1050 kV and which operates at 0.35 MPa, calculate the conductor and enclosure radii. Assume minimum conductor surface field. The equation governing the impulse breakdown strength and pressure for SF_6 is given by

$$E_{BD} = 81.2p + 10 \text{ kV/cm}$$

where p is the gas pressure (multiples of atmospheric pressure). Comment on the results.

REFERENCES

Anis H, Srivastava KD. IEEE Trans PAS-100:3694–3702, 1981a.

Anis H, Srivastava KD. IEEE Trans EI-16:327–338, 1981b.

Anis H, Ward S. Proceedings of a Conference on Electrical Insulation and Dielectric Phenomena, Ottawa, 1988, pp 312–317.

Christophorou LG, Olthofe JK, Van Brunt RJ. IEEE Electr Insul Mag 13:20–23, 1997.

CIGRE Working Group 15-03 CIGRE paper 15/23-01, 1992.

Cooke CM. Proceedings of 1st International Symposium on Gaseous Dielectrics, Knoxville, TN, 1978, pp 162–189.

Cooke CM, Wootton RE, Cookson AH. Trans IEEE PAS-96:768–775, 1977.

Cookson, AH, Farish O. IEEE Trans PAS-92:871–876, 1973.

Endo F, Ishikawa T, Yamagiwa T, Ozawa J. Proceedings of 4th International Symposium on High Voltage Engineering, Athens, 1983, Paper 32-05.

Fréchette MF, Roberge D, Larocque RY. IEEE Trans Dielectr Electr Insul 2:925–951, 1995.

Hampton B, Irwin T, Lightle D. CIGRE paper 15/33-01, 1990.

Harvey SM, Wong PS, Balma PM. IEEE Trans Power Deliv 10:357–363, 1995.

Högg P, Füchsle D, Kara A. ABB Rev 2:12–20, 1998.

IEC. Partial Discharge Measurements. Publication 270. Geneva: International Electrotechnical Commission, 1981.

Judd MD, Farish O, Hampton BF. IEEE Trans Dielectr Electr Insul 3:213–228, 1996.

Maller VN, Naidu MS. Advances in High Voltage Insulation and Arc Interruption in SF_6 and Vacuum. Oxford: Pergamon Press, 1981.

Masetti C, Pigini A, Bargigia A, Brambilla R. Proceedings of 4th BEAMA International Conference on Electrical Insulation, Brighton, 1982, pp 119–123.

Meijer S, Gulski E, Smit J. IEEE Trans Dielectr Electr Insul 5:830–842, 1998.

Meppeling J, Remde H. Brown Boveri Rev 73(9):498–502, 1986.

Lungaard LE, Tangen G, Skyberg D, Faugstad K. IEEE Trans Power Deliv 5:1751–1759, 1992.

Okabe S, Koto M, Endo F, Kobayashi K. IEEE Trans Power Deliv 11:210–216, 1996.

Parekh H, Srivastava KD, Van Heeswijk RG. IEEE Trans PAS-98:748–755, 1979.

Pearson JS, Farish O, Hampton BF, Judd MD, Templeton D, Pryor BM, Welch IM. IEEE Trans Dielectr Electr Insul 2:895–905, 1995.

Reisinger F, Muhr M, Schenner H, Diessner A. Proceedings of CIGRE Symposium on New and Improved Materials for Electrotechnology, Vienna, 1987, Report 1010-02.

Runde M, Aurud T, Ljøkelsøy K, Nøkleby JE, Skyberg B. IEEE Trans Power Deliv 12:714–721, 1997.

Sellars AG, Farish O, Hampton BF. IEEE Trans Dielectr Electr Insul 1:323–331, 1994.

Stone G, Boggs S, Braun J, Kurtz M. Proceedings of CIGRE Symposium on New and Improved Materials for Electrotechnology, Vienna, 1987, Report 400.05.

Toda K, Ikeda M, Teranishi T, Inoue T, Yanari T. Ninth International Symposium on High Voltage Engineering (ISH'95), Graz, Austria, 1995, Paper 2296.

Trinh NG, Mitchel GR, Vincent C. Proceedings of 4th International Symposium on Gaseous Dielectrics, Knoxville TN, 1984, pp 335–341.

Yamagata Y, Tanaka K, Nishiwaki S, Takahashi N, Kokumai T, Miwa I, Komukai
    T, Imai K. IEEE Trans Power Deliv 11:872–879, 1996.
Ziomek W, Kuffel E. IEEE Trans Dielectr Electr Insul 4:39–43, 1997.

11

Circuit Breaking

R. RADWAN *Cairo University, Giza, Egypt*

11.1 INTRODUCTION

Circuit breakers differ from switches in that they not only manually make and break the circuit while carrying their normal currents, but are also capable of making and breaking the circuit under the severest system conditions. Breaking or making the circuit under load conditions represents no real problem for a circuit breaker since the interrupted current is relatively low and the power factor is high. Under short-circuit conditions, however, the current may reach tens of thousands of amperes at a power factor as low as 0.1. It is the duty of a circuit breaker to interrupt such currents as soon as possible to avoid equipment damage. Loss of system stability is a consequence of slow fault clearance. Fault clearance time has been immensely reduced during the last 50 years due to the high technology adopted in circuit breaker design and the use of static relays. Fault clearing time on the order of two to three cycles have been achieved, of which the circuit breaker arcing occupies one-half to one cycle.

When switching short-circuit currents, small inductive currents, and capacitive currents, high overvoltages appear in the switched circuit. These voltages may reach 5 times, or even greater, the normal circuit voltage and damage of circuit insulation or restrike in the circuit breaker may occur. To reduce these voltages, external elements are added to the breaker such as switching resistors, inductors, arresters, and electronic circuits for controlled switching.

11.2   ARC INTERRUPTION

The circuit breaker contacts, fixed and moving, are usually made of highly conducting material, with adequate contact area ensured by a suitable contact pressure. When the contacts part to break the circuit, the contact area decreases to a very small value just before contact separation. The heat produced within such a small area causes the metal to melt and even evaporate (Chapter 6).

In air circuit breakers, metal vapor and hot air bridge the gap between the contacts, forming an electric arc. For oil circuit breakers, the heat generated within the arc decomposes some oil and generates gases. The amount of gases generated is a function of the arc energy and the gases are composed of about 66% hydrogen, 17% acetylene, 9% methane, and 8% others. At such extreme temperatures, these gases and the metal vapor will be highly ionized. They thus provide a conducting path between the circuit breaker's contacts and the circuit current will continue to flow through the arc.

All arc interruption methods are aimed at disturbing the energy balance of the arc. These methods are to cool the arc, increase its length, and split it into a number of arcs in series, as happens in multi-break circuit breakers and breakers equipped with splitter plates. Before describing in detail the different types of circuit breaker, it may be well to define the basic rated quantities of circuit breakers and discuss the currents they are called upon to interrupt in AC circuits.

11.3   CIRCUIT BREAKER RATED QUANTITIES

In addition to the breaker's rated voltage and frequency, there are other rated quantities that are important for its operation and selection. These quantities are discussed briefly in the following paragraphs.

11.3.1   Rated Current

The rated circuit breaker current is the rms value of the current that it can carry continuously without the temperature rise of its components exceeding the specified limits.

11.3.2   Rated Breaking Current

The rated breaking current is the rms value of the current that it can break under specified conditions of recovery voltage. Figure 11.1 shows the short-

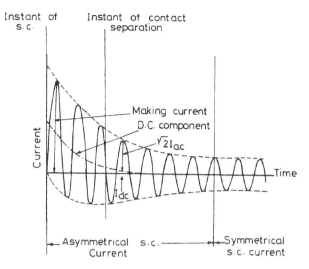

Figure 11.1 Symmetrical and asymmetrical short-circuit currents.

circuit current flowing in an inductive circuit with negligible resistance containing an AC source and a circuit breaker with a fault applied to its terminals. The value of the rated breaking current may include the DC current component, in which case it is termed the "asymmetrical" breaking current. This value can be calculated as follows. The maximum value of the DC current component is

$$I_{dc} = \frac{\sqrt{2}V_p}{X}$$

where V_p is the source phase voltage and X is the total circuit inductive reactance per phase. The rms AC component is

$$I_{ac} = \frac{V_p}{X}$$

The rms asymmetrical breaking current is thus

$$I_b = \sqrt{2\left(\frac{V_p}{X}\right)^2 + \left(\frac{V_p}{X}\right)^2}$$

$$= \sqrt{3}\,\frac{V_p}{X} \tag{11.1}$$

The breaking current value above is the value of current at the instant of short circuit. However, the circuit breaker contacts part a few milliseconds later and the interrupted current thus drops below this value (Fig. 11.1). Normally, the asymmetrical breaking current is taken by designers as 1.6 (V_p/X).

11.3.3 Rated Making Current

The rated making current is defined as the peak value of the current that a circuit breaker can make when closed onto a short circuit. Obviously, the duty imposed on the circuit breaker during this process is very severe, due to the high mechanical stresses produced within the circuit breaker. The rated making current is given by

$$I_m = \text{maximum value of DC component} + \text{peak value of AC}$$
$$\text{component}$$
$$= \frac{\sqrt{2}V_p}{X} + \frac{\sqrt{2}V_p}{X}$$
$$= 2\sqrt{2}\frac{V_p}{X}$$

$$(11.2)$$

The first current peak occurs after about one-fourth of a cycle from the instant of short circuit. To allow for the slight drop in current, the factor 2 (doubling-effect factor) is replaced by 1.8.

11.3.4 Rated Short-Time Current

The rated short-time current is the maximum current that the circuit breaker can carry for 1 s without damage to its conductors, insulation, operating mechanism, or tank.

11.3.5 Circuit Breaker Breaking Capacity

The breaking capacity of a three-phase circuit breaker (in MVA) is given by

$$\text{MVA}_b = \sqrt{3}VI_b \qquad (11.3)$$

where V is the rated voltage in kV and I_b is the rated rms breaking current in kA. According to British practice, the rated breaking current is the rms value of the alternating component only at the instant of contact separation and in this case is termed the symmetrical breaking capacity. In American practice the rated breaking current includes the DC component, which increases the breaking capacity by a factor of 1.6 (Section 11.3.2) and in this case is termed the "asymmetrical" breaking capacity.

11.3.6   Rated Insulation Level

The rated insulation level of a circuit breaker is the withstand power frequency and impulse voltage of its insulation (see Tables 18.1 and 18.2 in Chapter 18).

11.3.7   Rated Operating Sequence (Duty Cycle)

Circuit breakers, particularly those equipped with auto-reclosures, are subjected to frequent and successive operations. Under such circumstances the breakers and dielectric medium suffer severe mechanical and electrical stresses respectively. Circuit breakers should satisfy one of the following tests

$$O\text{–}t\text{–}CO\text{–}T\text{–}CO$$

or

$$CO\text{–}t'\text{–}CO$$

where O is an opening operation, C is a closing operation, CO is a closing operation immediately followed by an opening operation, t is the time between operations (3 min and 0.3 s for circuit breakers without and with auto-reclosures respectively), T is 3 min, and t' is 15 s for circuit breakers not to be used with rapid auto-reclosures.

11.4   SWITCHED CURRENTS AND CIRCUITS

In addition to normal load currents, circuit breakers are called upon to interrupt short-circuit currents, small inductive currents, and capacitive currents. Examples for small inductive and capacitive currents are the no-load current of a power transformer and the charging current of an extensive unloaded cable network, respectively. The duty imposed on a circuit breaker is strongly affected by the magnitude of the interrupted current and also by the circuit parameters and how far the fault is from the breaker terminals. The interruption of each of these currents and circuits is presented below.

11.4.1   Three-Phase Short Circuit

Symmetrical three-phase-to-ground faults at the circuit breaker terminals can be represented by an equivalent single-phase circuit as shown in Figure 11.2a, although interphase reactions cannot always be neglected (Guile and Paterson, 1980). The stray capacitance of the circuit breaker bushing and other connections, represented by C, determines the shape of the transient part of the recovery voltage (Fig. 11.2b). This is called the restriking voltage and builds up across the circuit breaker contacts. The severity of the circuit

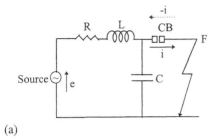

(a)

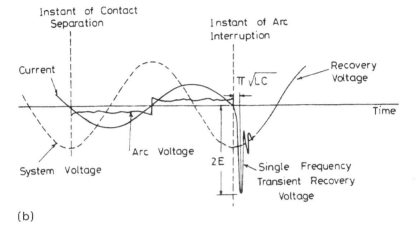

(b)

Figure 11.2 (a) Single-phase equivalent circuit of a symmetrical three-phase-to-ground short circuit; (b) single-frequency transient recovery voltage.

breaker duty is determined by the value of the short-circuit current together with the shape and amplitude of the restriking voltage.

In case of a fault at point F (Fig. 11.2) and the breaker interrupting it at a normal current zero, the restriking voltage can be obtained as follows. Let the supply voltage be in the form $e = E \cos \omega t$. The circuit voltage equation at the instant of contact separation is given by

$$Ri + L\frac{di}{dt} + \frac{1}{C}\int i \, dt = E \cos \omega t \tag{11.4}$$

The solution of equation (11.4) gives an expression for the recovery voltage across the circuit breaker contacts in the form

$$e_r = E\left[\cos \omega t - \exp\frac{-Rt}{2L}\cos \omega_0 t\right] \tag{11.5}$$

The transient recovery voltage will oscillate at a natural angular frequency $\omega_0 = 1/\sqrt{LC}$. If the effect of circuit resistance R is neglected and the natural frequency of oscillation is much greater than the supply frequency, the recovery voltage can be approximated to the form

$$e_r = E(1 - \cos\omega_0 t) \tag{11.6}$$

The recovery voltage as expressed by equation (11.5) is a single-frequency transient where a high-frequency oscillatory voltage is super-imposed on the supply normal-frequency voltage during the transient period (Fig. 11.2b). Neglecting damping in the circuit, the maximum value of the transient recovery voltage will be $2E$ and it occurs after a time $\pi\sqrt{LC}$ from the instant of arc interruption. In actual circuits the effects of resistance and system losses are considered and the value of the maximum transient voltage is lower than $2E$. The maximum rate of rise of the transient recovery voltage $(de_r/dt)_{max}$ can be easily derived from equation (11.6) and it is equal to $E/\sqrt{LC}$ at a time $(\pi/2)\sqrt{LC}$.

Since the currents in a three-phase system are displaced by 120° from each other, the arc in one of the phases will be extinguished at a normal current zero, while the other two phases are still arcing. After 90° from the first phase to clear, simultaneous interruption in the other two phases occurs. If the system neutral is isolated or the fault is not to ground, the recovery voltage across the first phase to clear will reach three times the maximum phase voltage, provided that the three phases are balanced. The first-phase-to-clear factor is 1.5. It may be less than this value for an earth fault in a system with earthed neutral, where its value depends on the ratio between the zero- and positive-sequence impedances.

When a three-phase fault occurs away from the circuit breaker term-inals, the transient recovery voltage will have more than one frequency component. The equivalent single-phase circuit representing this case is shown in Figure 11.3. This circuit comprises two parts: the source side (1) and the line side (2). Each part will produce an oscillatory voltage. The transient recovery voltage across the breaker terminals is the vector differ-ence of both voltages and is of a double-frequency nature (Fig. 11.4). The source-side frequency f_s is equal to $1/2\pi\sqrt{L_1 C_1}$ while the line-side frequency f_l is $1/2\pi\sqrt{L_2 C_2}$.

Under certain conditions, such as relatively low currents and effective interrupting means, interruption of the arc may take place not at current zero but near to it at an angle θ. The recovery voltage for such case can be obtained as follows.

Consider the circuit given in Figure 11.2a with a source voltage $e = \sqrt{2}E\cos\omega t$. The short circuit current is given by

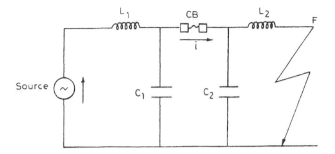

Figure 11.3 Equivalent single-phase circuit representation of a three-phase fault away from the circuit breaker.

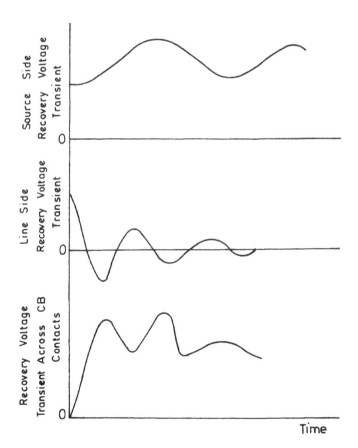

Figure 11.4 Double-frequency transient recovery voltage.

$$i_{s.c.} = \frac{\sqrt{2}E}{j\omega L}\cos\omega t$$

$$= \frac{\sqrt{2}E}{\omega L}\sin\omega t \tag{11.7}$$

The process of current interruption in a circuit can be simulated by driving an equal current in the opposite direction $(-i)$, between the circuit breaker contacts, to the original current (i). This current is called the cancellation current and is driven by the voltage e_r which is the voltage between the open contacts when the source voltage is short circuited (Fig. 11.5a). The recovery voltage in the Laplace form is given by

$$e_r(s) = -i(s).Z(s) \tag{11.8}$$

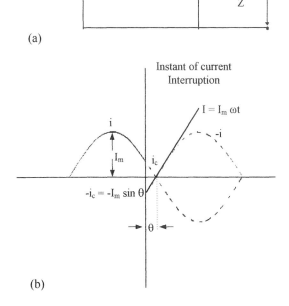

(a)

(b)

Figure 11.5(a) Equivalent electric circuit for determining the cancellation current; (b) main current and cancellation current waveforms.

where $-i(s)$ is the fictitious driven current at the instant of current interruption and $Z(s)$ is the circuit impedance. The sine-wave current near zero can be represented by a straight line (Fig. 11.5b), and hence the current at the instant of interruption is

$$i \approx -i_c + I_m \omega t$$

$$\approx -i_c + \frac{\sqrt{2}E}{\omega L} \omega t \qquad (11.9)$$

$$\approx -i_c + \frac{\sqrt{2}E}{L} t$$

The recovery voltage is given by

$$e_r(s) = \left(-\frac{i_c}{s} + \frac{\sqrt{2}E}{Ls^2}\right)\left(\frac{Ls \cdot \dfrac{1}{Cs}}{Ls + \dfrac{1}{Cs}}\right)$$

$$= -i_c \frac{1}{s} \frac{L/C}{L/s\left(s^2 + \dfrac{1}{LC}\right)} + \frac{\sqrt{2}E}{L} \frac{L/C}{s^2 L/s\left(s^2 + \dfrac{1}{LC}\right)}$$

Put $\omega_0^2 = 1/LC$.

$$e_r(s) = -i_c \sqrt{\frac{L}{C}} \frac{\omega_0^2}{\left(s^2 + \omega_0^2\right)} + \sqrt{2}E \frac{\omega_0^2}{s\left(s^2 + \omega_0^2\right)}$$

and the recovery voltage in the time domain will be

$$e_r = i_c \sqrt{\frac{L}{C}} \sin \omega_0 t + \sqrt{2}E(1 - \cos \omega_0 t) \qquad (11.10)$$

Equation (11.10) shows that interruption at non-current zero introduces the first term $(-i_c\sqrt{L/C}\sin\omega_0 t)$; otherwise the equation is the same as for interruption at current zero.

11.4.2 Asymmetrical Short-Circuit Switching

Symmetrical short circuits on power systems are not very common. Single line-to-ground faults represent more than 90% of the total number of faults on high-voltage and extra-high-voltage systems. Double line faults are rare but they do occur on medium-voltage and low-voltage cables.

The power-frequency recovery voltage under asymmetrical faults to ground is equal to the maximum phase voltage when the power system is solidly earthed. Line-to-line faults produce relatively lower power-frequency voltages on the faulted phases ($\sqrt{3}/2 \times$ maximum phase voltage), since the two phases break the circuit simultaneously. Overvoltages produced by asymmetrical faults have been covered in the literature (e.g., Fakheri et al., 1983; Flurscheim, 1975).

Circuit breakers with rated capacity based on a symmetrical short circuit are not necessarily capable of interrupting asymmetrical faults. If the zero-sequence impedance is less than the positive-sequence impedance ($X_0 < X_1$), a circiut breaker rated on the symmetrical fault level will not be capable of interrupting a single line-to-ground fault, and vice versa. However, in the case of double line-to-ground faults the duty on a circuit breaker is only 50%, on two poles, and thus the rating of the breaker is 75% of the symmetrical duty and rating. This can be proved by the theory of symmetrical components to determine the short circuit levels for different system faults.

11.4.3   Small Inductive Current

When the circuit breaker is interrupting a short-circuit current, the arc energy is adequate to keep the arc path highly ionized and conducting up to the instant of natural current zero when the arc gets interrupted. However, in the case of small inductive currents such as no-load currents of transformers, and with powerful circuit breakers, the ionized gases get blown off violently and rapidly. The arc current thus gets forcibly extinguished before the natural current zero (Fig. 11.6). This "current chopping," with a very high rate of change, induces very high voltage transients $L(di/dt)$. Having excessive rates of rise, these high voltages cause arc restrikes and, more importantly, they have serious effects on the system insulation, especially the terminal parts of the transformer winding. Protection schemes against such voltage transients are usually provided, which may be in the form of surge arresters, capacitive and resistive shunts, or by the use of switching resistors.

The value of the voltage appearing across the circuit breaker terminals, when chopping inductive currents, can be estimated by considering the energy balance just before and after the instant of arc interruption. Referring to the circuit in Figure 11.3 with an arc current i flowing between the breaker contacts, the electromagnetic energy stored in the inductance L_2 is $(1/2)i^2 L_2$. When the arc is extinguished there will be a successive transfer of energy between the electromagnetic and electrostatic fields. Assuming that the arc current is chopped at a magnitude i, the energy balance equation will be in the form

$$\frac{1}{2}i^2 L_2 + \frac{1}{2}C_2 v_2^2 = \frac{1}{2}C_2 v_b^2 \tag{11.11}$$

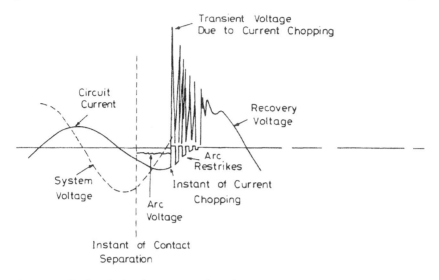

Figure 11.6 Low inductive current chopping.

The terms $(1/2)i^2L_2$ and $(1/2)C_2v_2^2$ are the energies stored in the inductance L_2 and capacitance C_2 at the instant just before arc interruption, respectively; v_2 is the voltage across C_2 just before the instant of arc interruption. The maximum voltage across the breaker contacts immediately after arc interruption is given by

$$v_b = \left(v_2^2 + i^2 \frac{L_2}{C_2} \right)^{\frac{1}{2}} \tag{11.12}$$

It is thus clear that the value of the restriking voltage is a function of the chopped current magnitude and $\sqrt{L_2/C_2}$. The value of $\sqrt{L_2/C_2}$ depends on the type of load connected to the circuit breaker and may reach $10^5\,\Omega$ for power transformers.

11.4.4 Capacitive Current

When a circuit breaker interrupts a purely capacitive current such as that of a capacitor bank or a large network of unloaded cables, a high recovery voltage appears across its contacts. This can lead to several restrikes before complete arc interruption. The current flowing during restriking has a high frequency and its interruption may lead to voltage escalation.

The process of interrupting capacitive currents can be explained by considering the equivalent circuit shown in Figure 11.7b, where C_1 repre-

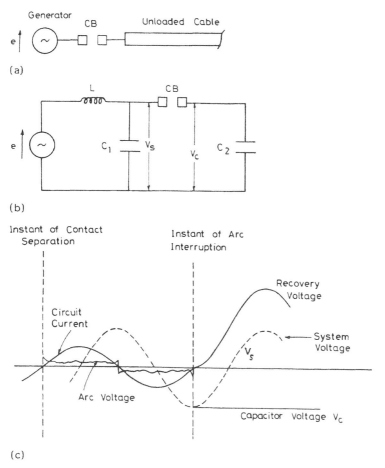

Figure 11.7 (a) Representation of unloaded cable; (b) its equivalent circuit; (c) recovery voltage across circuit breaker contacts when interrupting its capacitive current.

sents the stray capacitance of the system on the source side, L its inductance, and C_2 the cable network capacitance. When the breaker interrupts the capacitive current at its normal zero, the voltage on the source side V_s is at its peak value. The capacitance C_2 is charged to the same value and will remain constant if there is no leakage. The voltage across the circuit breaker contacts half a cycle later will thus reach twice the supply voltage peak value (Fig. 11.7c). The rate of rise of the recovery voltage is relatively low and the arc may be interrupted at the voltage peak. This is reasonably true for air-blast circuit breakers, where the gas blast powerfully deionizes the arc column within half a cycle. In other types, such as oil-circuit breakers, deionization of the arc column is relatively slow when interrupting small currents and the arc may strike. At this instant an oscillatory current flows through the circuit with a frequency

$$f_n = \frac{1}{2\pi\sqrt{LC_2}} \qquad \text{if } C_2 \gg C_1 \tag{11.13}$$

If this current is interrupted at its first zero, the voltage across the circuit breaker contacts will jump to three times the supply voltage. Again, if restriking occurs once more with arc interruption at the first current zero, the recovery voltage will reach four times the supply voltage. Theoretically, this process may continue indefinitely and the recovery voltage becomes escalated. In practical circuits, damping limits voltage escalation and the arc may become finally extinguished after the first restrike.

11.4.5   Short-Line Fault Switching

The fault current is inversely proportional to the impedance between the source and the fault. Consequently, terminal faults produce the maximum fault currents that a breaker has to interrupt. However, the severity of this duty is less than if the fault is a few kilometers away. When a fault occurs at a distance of 1–8 km from the breaker terminals, the restriking voltage will be a double frequency transient (Fig. 11.8). The source-side voltage is a relatively slow rising voltage with a frequency in the range 500–5000 Hz. The line-side voltage is a high-frequency sawtoothed transient. The restriking transient voltage across the circuit breaker contacts is the vector difference between the two voltage components (Fig. 11.8). The frequency of the line-side voltage is in the range 30–100 kHz and is given approximately by

$$f_l = \frac{1}{2\pi\sqrt{L_2 C_2}} \tag{11.14}$$

where L_2 and C_2 are the equivalent lumped series inductance and shunt capacitance of the line from the breaker to the fault. If the inductance and capacitance of the line are considered as distributed parameters, the period of the line-side transient voltage is four times the time of travel between the breaker and the fault (Chapter 14) and is given by

$$T = \frac{4l}{v}$$

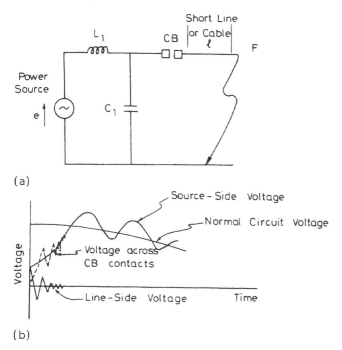

Figure 11.8 Representation of (a) short-line and (b) voltage waveforms.

and

$$f_l = \frac{v}{4l} \tag{11.15}$$

where v is the surge velocity and l the length of the line or cable from the breaker to the fault. The surge velocity is equal to $1/\sqrt{LC}$, where L and C are the line inductance and capacitance per unit length. In the case of a cable the surge velocity is reduced by the factor ε_r, the relative permittivity of its insulation.

11.5 RATE OF RISE OF RESTRIKING VOLTAGE

The previous analysis of the various currents and circuits has shown the nature of the transient recovery voltage across the breaker contacts immediately after arc interruption. Both the magnitude and the rate of rise of this voltage will affect the arc interruption process. At the instant of arc interruption the dielectric medium starts to regain its strength while the voltage across the contacts builds up. If the voltage buildup is slow, there will be no risk of arc reignition. If the dielectric recovery is too slow, the arc will restrike. If, on the other hand, the dielectric regains its strength very quickly, such as in vacuum or air-blast interrupters, current chopping may occur with a consequent abnormal voltage rise, as mentioned in Section 11.4.3.

The rate of rise of restriking voltage (RRRV) depends on the interrupted current and its power factor, the circuit natural frequency, neutral grounding of the system, and the presence of any leaking across the interrupted circuit. In gas-blast breakers, the maximum RRV is a function of the gas pressure and properties and the design features of the breaker. In HV and EHV air-blast circuit breakers, the high-frequency transient recovery voltage can be considerably damped by resistance switching. The basic idea of resistance switching is to interrupt the current in two steps. First, a resistance is connected across the breaker contacts as soon as they part; thus the arc between them gets relatively easily interrupted and the circuit current drops to a magnitude decided by the resistance value. The second step is to interrupt the current completely. This can be carried out by an external isolator since the current is relatively low and its power factor is high. If the resistance is properly selected, the high-frequency voltage oscillation can be critically damped or overdamped. The value of the switching resistance for damping the voltage oscillations is given by the relation

$$R \leq \frac{1}{2}\sqrt{\frac{L}{C}} \qquad\qquad (11.16)$$

11.6 CONTROLLED SWITCHING

Switching operations produce overvoltages which may damage the circuit breaker and associated equipment insulation. These voltages can be reduced to acceptable levels by the use of conventional methods such as opening and closing resistors, closing reactors, and surge arresters. These methods have proved to be effective and reliable since their first use. However, modern technology always seeks more and more effective alternatives. Controlled switching of electrical equipment such as capacitors, transmission lines, shunt reactors, and transformers has drawn the attention of many researchers and manufacturers (Holm et al., 1990; Rajotte et al., 1996).

In controlled switching, electronic circuits are used to control the instants of circuit breakers' poles opening or closing with respect to reference signals. In uncontrolled switching the poles of a circuit breaker open or close simultaneously at random. However, due to the mechanical system a tolerance of ± 2 ms is usually observed. In controlled switching the poles open or close at predetermined instants for optimum reduction of overvoltage or inrush current. In controlled opening the arcing time for each pole is controlled by setting the instant of contact separation with respect to the current waveform. For controlled closing, the source voltage is monitored by the switching controller and the closing command is issued randomly with respect to the phase angle of a reference signal (CIGRE Working Group 13.07, 1999).

It was found that, when energizing a capacitor bank with controlled and uncontrolled closing operation, the overvoltage produced is reduced from 2.95 to 2.39 p.u. under certain conditions (CIGRE Study Committee 13, 1996). The same reference showed that, for opening reactors, the overvoltage can be reduced from 2.4 to 1.1 p.u. while the arcing time is reduced from 10.8 to 4.5 ms. Also, the maximum value of the inrush current of a transformer is reduced from 3.3 to 1.5 p.u. of its rated peak current.

Field tests on circuit breakers equipped for controlled switching showed good results, as mentioned before. However, there are many conservations on their reliability and economy (CIGRE Study Committee 13, 1995).

11.7 TYPES OF CIRCUIT BREAKER

Circuit breakers are classified according to the switching medium in which their contacts part. The main switching media currently in common use are air, oil, sulfur hexafluoride (SF_6), and vacuum. With the rapid advancement in the field of solid-state devices, solid-state breakers have a promising future. The different types of circuit breaker are (a) air, (b) oil, (c) air-blast, (d) SF_6, (e) vacuum, and (f) solid-state. The construction and principle of operation of each type are outlined in the following sections.

11.7.1 Air Circuit Breakers

The arc established between the contacts of air breakers is interrupted in atmospheric air. They are commonly used in low-voltage systems with a normal current up to 3000 A. Due to their simple construction and maintenance, they may replace oil circuit breakers of the same rating in systems where fire hazards exist. In heavy industries having large electric motors with frequent starting, air circuit breakers are superseding oil breakers, due to oil contamination. They are also used extensively with electric furnaces.

To help interrupt the arc, deion champers, air chutes, magnetic blasts, and splitter plates are used. When air circuit breakers open, the arc is drawn between the main contacts and then moves upward, by thermal buoyancy and magnetic effects, to the arcing contacts and then to the arc runners (Fig. 11.9). To speed the arc movement, special magnetic coils are fitted. These

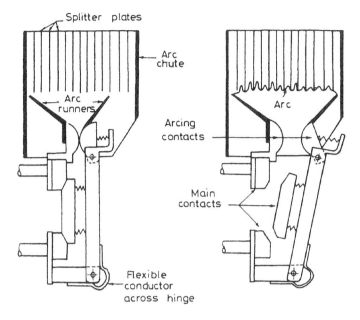

Figure 11.9 Main parts of an air circuit breaker with closed and open contacts. [From Lythall (1986).]

coils carry the arc current only during the arcing period and are connected in such a way that their magnetic field will force the arc upward.

11.7.2 Oil Circuit Breakers

Here oil provides an excellent medium for arc interruption. In oil breakers the arc can be described as "self-extinguishing." By its excessive heat it evaporates and decomposes some of its surrounding oil. The greater the arc energy, the higher will be the pressure of the gas bubble produced. The proper design of oil breakers fully exploits these high pressures in fast arc interruption. Oil current breakers are classified as the bulk oil or minimum oil type. The methods of arc control and interruption are different from one type to the other.

11.7.2.1 Bulk Oil Circuit Breakers

Here oil serves both to insulate the live parts and to interrupt the arc. The contacts of bulk oil breakers may be of the plain-break type, where the arc is freely interrupted in oil, or enclosed within arc controllers. The latter have higher ratings.

Plain-break circuit breakers are evidently limited to the low-voltage range. They consist mainly of a large volume of oil contained in a metallic tank. Arc interruption here depends on the head of oil above the contacts and the speed of contact separation. The head of oil above the arc should be sufficient to cool the gases, mainly hydrogen, produced by oil decomposition.

Bulk-oil circuit breakers with arc control devices are used in the medium- and high-voltage ranges. In this type of breaker the arc is confined and interrupted in "explosion pots" rather than in the open volume of oil. The explosion pot encloses the fixed and moving contacts. When they part, the arc is drawn between them inside the pot. The gases so produced in the confined space will cause high turbulence of the oil and rush outside the pot. The rushing gases and oil will disturb the arc, and interruption may occur even before the moving contact has left the pot (Figs. 11.10, 11.11).

11.7.2.2   Minimum Oil Circuit Breakers

Bulk oil circuit breakers have the disadvantage of using large quantities of oil, with their associated handling and storage problems. With frequent breaking and making of heavy currents, the oil will deteriorate and may lead to circuit breaker failure. Minimum oil circuit breakers work on the same principles of arc control as those used in bulk-oil breakers. Their containers, however, are made of porcelain or other insulating material.

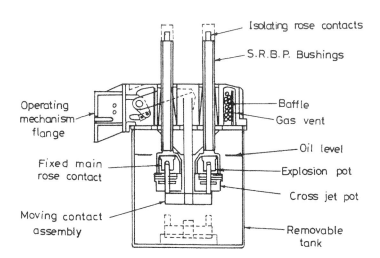

Figure 11.10   Three-phase dead-tank bulk-oil 13.8 kV circuit breaker in its closed position. (Courtesy of South Wales Switchgear.)

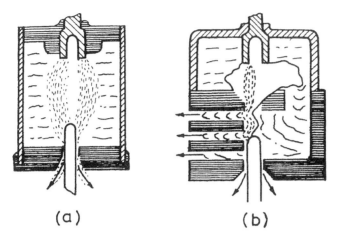

(a)   (b)

Figure 11.11 Explosion pot arc controllers: (a) simple type; (b) cross jet.

In this type of breaker, arcing is confined to a much reduced volume of oil inside an explosion pot. Figure 11.12 shows one pole of a minimum oil circuit breaker. The lower chamber contains the operating mechanism and the upper chamber contains the moving and fixed contacts together with the arc-control device.

Single-break minimum oil breakers are available in the voltage range 33–132 kV with breaking capacities from 1500 to 5000 MVA. For higher voltages and ratings, multibreak breakers are constructed from a number of modules in series. Equalizing resistors or capacitors are connected in shunt with each interrupter unit to ensure uniform voltage distribution across the breaks (Soderberg, 1978).

Oil circuit breakers are characterized by their long times of interrupting small currents and short times for large currents, which is contrary to other circuit breakers. The reason for this is the high turbulence caused during large current interruption, and vice versa.

11.7.3 Air-Blast Circuit Breakers

The principle of arc interruption in air-blast circuit breakers is to direct a high-pressure blast of air longitudinally or perpendicularly at the arc. Fresh and dry air will thus rapidly replace the ionized hot gases within the arc zone and the arc length is considerably increased. Consequently, the arc may be interrupted at the first current zero.

The merits of air-blast breakers are (a) cheapness and availability of the interrupting medium, (b) chemical stability and inertness of air, (c) great

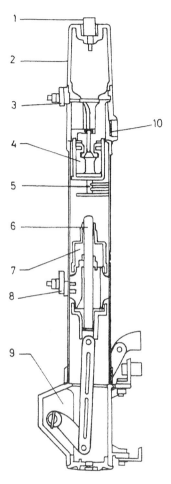

Figure 11.12 One pole of a 12 kV minimum-oil circuit breaker. 1, vent; 2, air chamber; 3, upper main terminal; 4, tulip contact; 5, arc control chamber; 6, contact rod; 7, contact roller; 8, lower main terminal; 9, crank housing; 10, oil level observation glass. (Courtesy of Asea Brown Boveri AG, Switzerland.)

reduction in the erosion of contacts from frequent switching operations, (d) high-speed opreation, (e) short arcing time, (f) operation in fire hazard locations, (g) reduction of maintenance frequency, and (h) consistent breaking time resulting from use of the interrupting medium pressure to open the contacts. However, the use of air compressors and high-pressure vessels increases the production cost. Upon arc interruption, air-blast breakers produce high-level noise when discharging to open atmosphere. In residential areas, air-blast breakers should be equipped with silencers to reduce the noise to an acceptable level.

In many designs, contact separation is achieved by admitting high-pressure air from the air receiver. Prior to contact separation the air pressure in the interrupter head is atmospheric, but it rapidly builds up to separate the contacts by piston action. The contact separation should be enough to interrupt the arc and withstand the recovery voltage under high pressure. Designs with voltages up to 110 kV are equipped with series isolators to open the circuit while the interrupter head is pressurized. When the air supply is stopped, the contacts reclose under spring action.

Modern designs use permanently pressurized interrupter heads and airflow pipes (Fig. 11.13). The relative positions of the interrupter heads and the admission valve ensure simultaneous operation of the breaks. The number of breaks per pole or phase depends on the voltage level and it is

Figure 11.13 Air-blast circuit breaker rated 10,000 MVA at 300 kV, having eight interrupter heads per phase and equipped with switching resistors and equalizing capacitors. (Courtesy of Asea Brown Boveri AG, Switzerland.)

customary to use four and eight breaks for the voltage levels 230 and 750 kV, respectively.

As with EHV minimum oil breakers, switching resistors and equalizing capacitors are usually fitted across the interrupters. The resistors help to interrupt the arc and limit transient overvoltages. The capacitors are used to equalize the voltage across the open breaks.

The velocity and pressure of the air blast are independent of the interrupted current. Thus, when it interrupts a small inductive current, there is a chance of its chopping, as explained in Section 11.4.3. The arcing time of air-blast breakers does not vary considerably with the interrupted current, which is different from the case of oil circuit breakers, where arc interruption relies on the turbulences produced in the oil.

11.7.4 SF$_6$ Circuit Breakers

Sulfur hexafluoride is an excellent insulating and arc-quenching medium. The physical, chemical, and electrical properties of SF$_6$ are superior to many of the other media. It has been used extensively during the past 30 years in circuit breakers, gas-insulated switchgear (GIS), high-voltage capacitors, bushings, and gas-insulated cables (Chapter 10).

11.7.4.1 Properties of SF$_6$

Sulfur hexafluoride has a high thermal conductivity and its thermal time constant is about 1000 times shorter than that of air, a great advantage in arc quenching. The velocity of sound in SF$_6$ is about 40% of that in air, and therefore the required gas flow for arc interruption can be produced during an opening operation without the need for a gas compression plant (Schumann and Evans, 1981).

SF$_6$ is chemically inert and does not attack metals or glass under normal conditions. However, at temperatures of the order of 1000°C, as in electric arcs, it decomposes to SF$_4$, SF$_2$, S$_2$, F$_2$, S, and F. The decomposition products recombine shortly after arc extinction (within about 1µs). The highly toxic S$_2$F$_{10}$ has never been traced in the arcing products. In SF$_6$ breakers the decomposition products attack the contacts, metal parts, and rubber sealings in the presence of moisture. The gas should therefore be meticulously dried. Most of the decomposition products can be absorbed by a mixture of soda lime (NaOH + CaO) and activated alumina placed in the arcing chamber. The high dielectric strength and good arc-quenching properties of SF$_6$ are ascribed primarily to its high affinity for electron attachment (Chapter 3).

One of the major problems associated with the use of SF$_6$ is its condensation at high pressures and low temperatures. At a pressure as high as 14 bar, SF_6 liquefies at $0°C$. SF_6 breakers to be used in areas where such low ambient temperatures are encountered may be equipped with special heaters for the gas. SF_6 is also highly sensitive to strong localized fields, moisture, and foreign solid particles (Chapter 10). Mixtures of SF_6 with other gases, such as nitrogen, are used to overcome some of these problems.

SF_6 circuit breakers cover voltage levels in the range 6.6–765 kV. The construction and principle of operation of each type are presented in the following sections.

11.7.4.2 Double-Pressure SF_6 Circuit Breakers

This is the early design of the SF_6 circuit breaker. Its operation is similar in principle to that of air-blast circuit breakers. It comprises mainly a high-pressure metal reservoir, where most of the SF_6 is kept, and an interrupter compartment containing the breaker contacts. For the current-breaking operation, the circuit breaker contacts part while the high-pressure gas is released from its reservoir to the interrupter compartment, where it blows out the arc. After the current interruption, the gas is pumped back to its reservoir.

In medium-voltage circuit breakers the interrupter compartment is a grounded metal "dead tank," whereas in EHV breakers the "live tank" type is more suitable. As with air-blast circuit breakers, EHV SF_6 breakers comprise a number of "interrupter compartment" modules in each phase. Because of its need for various auxiliaries, such as gas compressors, filters, monitors, control devices, and their complicated design and construction, breakers of this type have been outmoded by simple designs of the self-extinguishing and puffer types.

11.7.4.3 Self-Extinguishing SF_6 Circuit Breakers

The interrupting chamber is divided into two main compartments (Fig. 11.14). Both have the same gas pressure, about 5 atm, while the breaker is closed. When the breaker is being opened, the contacts separate and an arc is drawn between them. The heat generated in the arc heats the gas in the arc compartment (6) and rapidly increases its pressure. The gas blasts from the arc compartment to the other compartment (7). This rapid expansion cools the arc column and extinguishes it at a current zero. A third compartment is incorporated to augment the gas pressure, by piston (10) action, while interrupting smaller currents. By this arrangement the arcing time is independent of the current magnitude, and the currents, large and small, get interrupted at their natural zero (Schumann and Evans, 1981). A coil (2) surrounding the main contacts speeds up arc interruption by exerting a magnetic force on the arc. This force makes the arc rotate around the periphery of the contacts, thus cooling it and avoiding spitting of the contacts. The coil is dis-

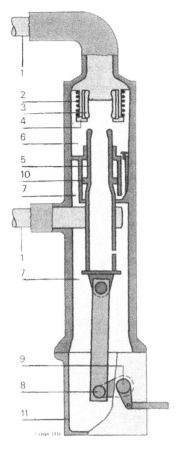

1 Main connections
2 Cylindrical coil
3 Load current contact
4 Fixed arcing contact
5 Moving contact
6 Breaking chamber
7 Pressure equalizing chamber
8 Operating lever
9 Operating shaft with rotary seal
10 Auxiliary compression piston
11 Transmission casing

Figure 11.14   Section through a self-extinguishing SF_6 circuit breaker. (Courtesy of Asea Brown Boveri AG, Switzerland.)

connected from the main circuit during normal operation and is in series with it immediately when the arc is established between the contacts (Fig. 11.14).

Breakers of this type are normally used in medium-voltage (MV) networks with rated voltage up to 24 kV, rated current up to 2350 A, and rated breaking current of 31.5 kA.

11.7.4.4   Puffer-Type SF$_6$ Circuit Breakers

These are also sometimes called "single-pressure" or "impulse"-type breakers. Their principle of arc interruption is by compressing the SF$_6$ gas during contact separation, with the moving contact acting as a piston. Thus the gas pressure in the interrupter compartment builds up rapidly to levels high above its steady value of about 3–5 atm. Its operation can easily be visualized by reference to Figure 11.15.

While the breaker is closed, the current flows between the current terminals (1) via the main contacts: the fixed (2) and the moving contact (3). When the breaker is being opened upon contact separation, the arc is drawn between the arcing contacts (4) and (5), through the insulating nozzle (7). During the contact travel, the moving cylinders of (3) and (4) act as pistons and pump compressed SF$_6$ gas, which flows axially and blows out the arc.

For the EHV range, up to 765 kV, a number of modules are arranged in series on insulating supports as shown in Figure 11.16, where two modules per phase are sufficient in a 362 kV breaker.

11.7.5   Vacuum Circuit Breakers

The advantages of vacuum as an insulant and arc interrupting medium have been known for many years. A hard vacuum, on the order of 10^{-4} Pa, has a dielectric strength and an arc-interrupting ability superior to those of other media, including compressed gases and oil. In vacuum breakers a contact separation of about 1 cm is adequate. Since it is so compact, the power needed to close or open it is much less than for other types of breaker. The rate of dielectric recovery of vaccum after arc interruption is about one order of magnitude faster than in air-blast breakers. Modern vacuum breakers successfully interrupt capacitive currents and small inductive currents and short line faults without producing excessive transient overvoltages. Being compact, with their simple mechanism, they do not need much maintenance. Their main shortcoming is that a failure in their hard vacuum cannot be easily detected while in service.

In a vacuum circuit breaker each phase consists of an evacuated interrupter compartment and an external operating mechanism (Fig. 11.17a).

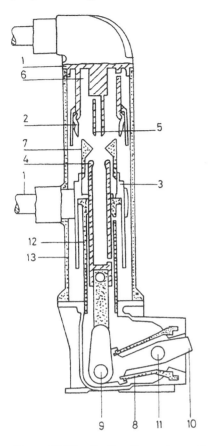

Figure 11.15 Section through a puffer-type SF$_6$ circuit breaker. 1, current terminals; 2, main contact, fixed; 3, main contact, moving as a piston; 4, arcing contact, moving; 5, arcing contact, fixed; 6, exhaust compartment; 7, insulating nozzle; 8, sheath seal; 9, transmission link; 10, operating level; 11, lever fulcrum; 12, compression cylinder; 13, insulating cylindrical enclosure. (Courtesy of Asea Brown Boveri AG, Switzerland.)

The contacts are of large surface area, with spiral segments so that the arc current produces an axial magnetic field (Fig. 11.17b) to help move the arc over the contact surfaces and interrupt it rapidly (Yanabo et al., 1983). Moving the arc spots over the contacts minimizes metal evaporation and thus arc erosion (Chapter 6). The wall of the interrupter compartment is made either totally or partly of an insulating material (e.g., glass). The metal bellows (Fig. 11.17a) make possible the movement of the moving contact while maintaining the hard vacuum (Sunada et al., 1982).

Figure 11.16 362 kV two-break three-pole puffer SF$_6$ circuit breaker. (Courtesy of GEC Switchgear.)

In a vacuum breaker the process of arc extinction differs from that in other types of breaker. When its contacts part, their last points to separate get heated up to the metal's boiling point. The ionized metal vapor thus evolved provides the arc medium (Chapter 6). Here the arc is burning almost exclusively in metal vapor. Therefore, the metal or alloy of which the contacts are made directly affects the characteristics of the vacuum breaker.

The metal vapor swept radially outward from the arc column is bound to condense on the walls of the interrupter compartment. Therefore, insulating parts of the wall facing the arc region should be shielded from the metal vapor to keep their insulation strength. In the breaker illustrated in Figure 11.17a, this part of the wall is already metallic; other parts take care of the insulation strength.

At low currents the arc is characterized by a diffuse appearance (Lafferty, 1980), whereas, with large currents of kiloamperes, intense ionization confines the arc within an intense core, causing excessive heating and local melting of its spots on the contacts. Therefore, the arc is whirled magnetically (Fig. 11.17b) in order to minimize contact erosion.

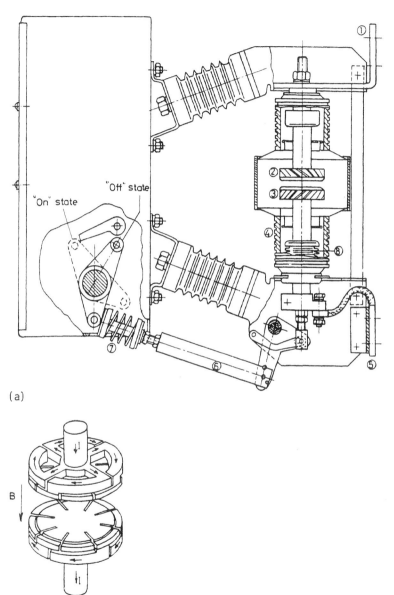

(a)

(b)

Figure 11.17 (a) Construction of 24 kV vacuum breaker. 1, upper breaker terminal; 2, fixed contact; 3, moving contact; 4, interrupter body; 5, lower breaker terminal; 6, insulating coupler; 7, contact pressure spring; 8, metal bellows. (Courtesy of Siemens.) (b) Details of the contacts for the current *I* to produce an axial magnetic field *B* in the vacuum.

11.7.6 Solid-State Circuit Breakers

This type of circuit breaker uses solid-state devices such as thyristors, triacs, or power transistors. They do not suffer from the shortcomings of electro-mechanical devices. Solid-state breakers can clear a fault within only about one half-cycle (Fig. 11.18a). If the breaker closes a circuit that happens to be faulty, it will still interrupt the fault current within only about one half-cycle

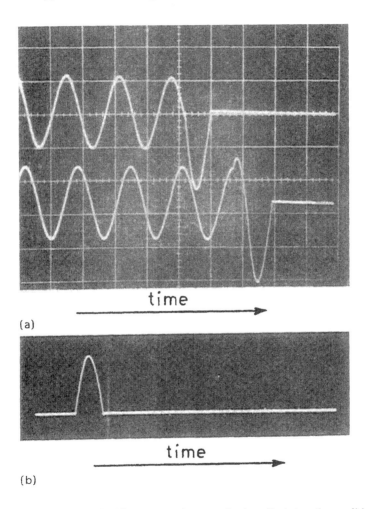

Figure 11.18 Oscillograms of currents handled by the solid-state circuit breaker: (a, top waveform) current interrupted within about half a cycle from the instant it exceeds the overcurrent setting. (b) current flowing when the breaker closes a faulty circuit.

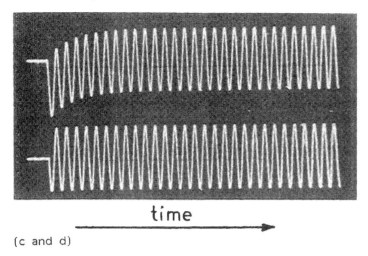

time

(c and d)

Figure 11.18 continued, (c) input current in a circuit where the current has a very pronounced DC component; (d) input current when the instant of circuit closing is properly synchronized.

(Fig. 11.18b) (Khalifa et al., 1979a,b). It has no moving contacts that get eroded by the electric arc, and thus does not need to be maintained or replaced. The instant of making a circuit with a solid-state breaker can be synchronized such that the doubling of inrush currents into inductive circuits and the accompanying electromechanical overstresses on the circuit conductors and supports can be eliminated. Switching transient overvoltages can also be eliminated.

The principle of operation of solid-state breakers is illustrated by the simple circuit shown in Figure 11.19. In this figure two power thyristors are connected antiparallel. Each thyristor will allow the load current to flow during one complete half-cycle when properly triggered. As the instantaneous circuit current exceeds the set level, the current sensor and the gate

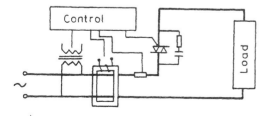

Figure 11.19 Simplified circuit with solid-state circuit breaker.

control device will cut off the gate signals. This will turn off the thyristors and the circuit current will be broken at the ensuing zero crossing (Fig. 11.18a). In some low-voltage and low-current power systems applications, a triac may replace the power thyristors.

Up until now, solid-state breakers have been used in few power system applications in the low-voltage range. By series–parallel connections of the thyristors, solid-state breakers for higher voltages and currents can be constructed. The main disadvantage of solid-state breakers is their relatively high power losses. This situation may be much improved with the development of new semiconductor materials.

11.7.7   Direct-Current Circuit Breakers

High-voltage DC lines are now commonly used to transmit large bulks of electrical energy for long distances. They are point-to-point transmission systems. They cannot be tapped or paralleled because of the lack of HV DC breakers with high interrupting capacity.

Air-break circuit breakers are usually used to interrupt low DC currents at low voltages. They are installed in some traction systems, electrochemical plants, and similar applications.

Not all types of circuit breaker used in HV AC systems can be installed in HV DC systems unless equipped with additional circuits to bring the full DC current smoothly to zero for arc interruption (Yanabo et al., 1982). The basic principle of such circuits is shown in Figure 11.20. Before the main circuit breaker is opened, switch S is closed to discharge the precharged capacitor C_1 in the opposite direction of the circuit current. This will force it to zero with a few oscillations and the arc will thus be interrupted at a current zero. The self-saturated reactor L_2 and capacitance C help to reduce the rate of decrease of the circuit current and the rate of increase of the transient recovery voltage, respectively.

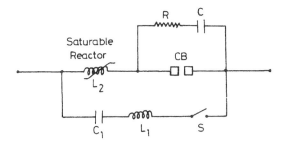

Figure 11.20   Principle of commutation circuit for DC interruption.

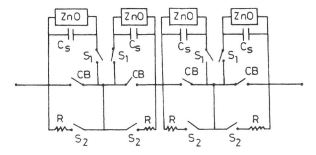

Figure 11.21 Schematic of a 500 kV HV DC SF₆ puffer-type circuit breaker. [From Lee et al. (1985).]

Lee et al. (1985) described a 500 kV DC circuit breaker to be used for switching load and fault currents up to 2200 A. The breaker is a modified SF₆ puffer type. Four breaker modules are connected in series (Fig. 11.21). When the breaker contacts start to open to break the circuit current, an arc is drawn between them. As the contacts continue to open, the arc voltage increases, due to arc lengthening and cooling. After the buildup of enough arc voltage, switch S_1 is closed and this causes a current diversion into C_s. The current diversion results in a current oscillation that grows. When its magnitude is large enough, a current zero is produced in the interrupter. As soon as this current zereo is produced, the arc is interrupted and the full circuit current is diverted to C_s. When the voltage across C_s reaches the clipping voltage of the ZnO arrester, it will conduct and stop any further increase of voltage across the contacts. The circuit energy is absorbed by the arresters and the fault current is cleared in a few milliseconds.

Air-blast and SF₆ puffer 400 kV DC circuit breakers have been developed by Vithayathile et al. (1985). Both breakers included commutating circuits and energy-absorbing elements. The speed of operation of these breakers is comparable to that of AC breakers of similar types.

11.8 PROBLEMS

(1) A power system with 1000 MVA generating capacity feeds a heavy industrial area at 66 kV. The p.u. impedance of the system up to the 66 kV busbar is 3 p.u. If a three-phase fault occurs at the 66 kV bus, calculate the symmetrical and asymmetrical breaking and making currents of circuit breakers located at the point.

(2) A three-phase synchronous generator has a line voltage of 13.8 kV. The generator is connected to a circuit breaker. The circuit inductive

reactance and capacitance up to the circuit breaker are $4\,\Omega$ and $0.015\,\mu F$ respectively. Determine the following:

1. Peak restriking voltage across the circuit breaker contacts
2. Frequency of restriking voltage transient
3. Average rate of rise of restriking voltage
4. Maximum rate of rise of restriking voltage
5. Time to peak restriking voltage and to maximum rate of rise of restriking voltage

(3) A generator/transformer unit of 200 MVA, 0.45 per unit impedance, is connected to a capacitor bank through a circuit breaker. The voltage at the capacitor bank side is 220 kV. Calculate the value of a resistance to be connected across the circuit breaker contacts to prevent voltage transient when switching the charging current of the bank. The rating of the condenser bank is 20 MVA.

(4) A 60 Hz circuit breaker interrupts a current in the form: $i = I_m \sin \omega t$. The circuit breaker's contacts open at a current of 0.1 I_m, after the peak value. If arc interruption occurs at the second current zero, calculate the time taken for arc interruption from the instant of contacts opening.

(5) Prove that the first phase to clear in a three-pole circuit breaker interrupts its current 90° ahead of the other two phases. Also prove that the first-phase-to-clear factor is 1.5.

(6) In a system of 220 kV, the circuit inductance is 5 H and its capacitance to ground is 0.01 μF. Calculate the voltage appearing across the pole of a circuit breaker if a current of 8 A is chopped at its peak value. Calculate also the value of resistance to be connected across its contacts for critical damping of the restriking voltage.

(7) An oil circuit breaker is installed to protect an 11 kV feeder with an overcurrent relay. The main particulars of the relay and breaker are as follows:

Relay operating time	60 ms
CB contact stroke	10 cm
CB contact speed	10 m/s
Arcing time	0.5 cycle

If a fault occurs on the feeder, calculate the total time needed for fault clearing in seconds and cyles. Assume the system frequency is 60 Hz.

(8)   In a 60 Hz circuit breaker, the speed of the moving contact is 3 m/s, and the contact's stroke is 6 cm. The first phase to clear interrupts the arc 10 ms after reaching full stroke. Calculate, in seconds and cycles, the time taken by the first phase to clear as well as the other two phases to interrupt the arc.

(9)   If a short circuit occurs 1.5 km away from a circuit breaker on a transmission line having an inductance of 1.2 mH/km and a capacitance of 0.012 μF/km, deduce the lineside natural frequency and show by what factor its exceeds the value obtained by considering the line inductance and capacitance to be lumped.

(10) In a short circuit test on a circuit breaker, the following readings were obtained:

Time to reach the peak restriking voltage   90 μs
Peak restriking voltage   130 kV

Calculate:

1.   The average rate of rise of the restriking voltage
2.   The natural frequency of the circuit
3.   The time at which maximum RRRV occurs
4.   The value of the maximum RRRV
5.   The rms voltage of the testing source

(11) A 66 kV, 60 Hz transmission line has an inductance and capacitance per phase of 0.933 mH/km and 0.007 78 μF/km respectively. If a symmetrical 3-phase fault occurs 6 km away from the circuit breaker protecting the line, calculate:

1.   The steady-state interrupted current
2.   The interrupted current, assuming full asymmetry
3.   The maximum rate of rise of the restriking voltage

If a cable of the same length, having an inductance and capacitance per phase of 0.155 mH/km and 0.187 μF/km respectively, is used instead of the transmission line, calculate the same values in 1, 2, and 3 above. Comment on the results.

(12) A three-phase, star-connected, 60 Hz, 13.8 kV synchronous alternator has its neutral solidly earthed. The alternator is connected to a power system through a circuit breaker. The alternator series inductance is 15 mH and its stray capacitance is 6 μF. If a terminal three-phase fault occurs on the circuit breaker, obtain an expression for the restriking

voltage across its contacts if arc interruption occurs $8°$ ahead of current zero.

(13) A circuit breaker is installed to protect a generating source as shown in the single-phase equivalent circuit (Fig. 11.22). The source inductance, resistance, stray capacitance, and switching resistance are L, R, C, and r respectively. If a short circuit occurs on the circuit breaker terminals, prove that the frequency of oscillation of the transient recovery voltage is given by

$$\frac{1}{2\pi}\sqrt{\frac{1}{LC} - \left(\frac{R}{2L} - \frac{1}{2Cr}\right)} \quad \text{Hz}$$

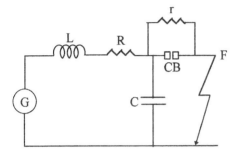

Figure 11.22 Equivalent circuit for Problem 13.

(14) A 60 Hz generator is connected to a transmission line 1.5 km long having an inductance of 1.2 mH/km and a capacitance of 0.015 µF/km. The generator side circuit has an inductive reactance of 5 Ω./phase and a stray capacitance to ground of 0.01 µF. Calculate the source-side and transmission lineside frequencies of the recovery voltage appearing on the circuit breaker poles when clearing a three-phase fault occurring at the far end of the transmission line. If the short circuit current at the far end of the line is 3000 A, calculate the initial rate of rise of the line voltage side.

(15) A 3-phase, star-connected, 11 kV, 60 Hz alternator is protected by a circuit breaker. The alternator neutral is earthed through a 1 Ω reactance and at the instant of fault it has a positive, negative, and zero sequence impedance of 10, 7, and 3 Ω respectively. Obtain and compare the ratings of the circuit breaker based on three-phase, double line, and single line to ground faults. Obtain also the voltage of the neutral point above ground under a single line to ground fault.

(16) A circuit breaker is connected to the high voltage side of a 30 MVA, 132/11 kV transformer. The stray capacitance to ground is 8 nF and the magnetizing inductance is 50 H. Calculate the voltage appearing across the circuit breaker contacts due to chopping the magnetizing current of the transformer at its peak value. What is the value of a switching resistance required for critical damping of the transient recovery voltage?

(17) What is the difference between the symmetrical and asymmetrical rated breaking and making currents of a circuit breaker?

(18) Why does the arcing time in oil circuit breakers decrease with increasing arc current? Does this occur in air-blast breakers?

(19) Why are oil circuit breakers not recommended for installation in power systems with a high frequency of load interruption?

(20) Explain the principle of magnetic blowout in air-break circuit breakers.

REFERENCES

CIGRE Study Committee 13. Electra 164, 1996.

CIGRE Working Group 13.07. Electra 183, 1999.

Fakheri A, Bhatt N, Ware B, Sybille G, Belanger, J. IEEE Trans PAS-102:3315–3328, 1983.

Flurscheim CH. Power Circuit Breakers Theory and Design. IEE Monograph 17. Stevenage, UK: Peter Peregrinus, 1975.

Guile AE, Paterson W. Electric Power Systems, Vol. 2. Oxford, UK: Pergamon, 1980, p 49.

Holm A, Alvinsson R, Akesson U, Karlen O. CIGRE Session 1990, Report 13–201.

Khalifa M, Arifur-Rahman S, Enamul-Haque S. Proc IEE 126:75–76, 1979a.

Khalifa M, Arifur-Rahman S, Enamul-Haque S. Proceedings of IEEE Industrial Commercial Power System Technical Conference, Seattle, WA, 1979b, pp 103–106.

Lafferty JM. Vacuum Arcs, Theory and Applications. New York: John Wiley, 1980.

Lee A, Slade PG, Yoon KH, Porter J, Vithayathile J. IEEE Power Eng Rev 10:32, 1985.

Lythall RT. The J & P Switchgear Book. London: Butterworth, 1986, p 32.

Rajotte J, Brault S, Desmarais J, Charpentier C, Hung H. CIGRE Session 1996, Paper 13-301.

Schumann R, Evans GJ. Proceedings of Conference of the Electricity Supply Engineers Association of New South Wales, Sydney, 1981, pp. 12-1–12-12.

Soderberg G. ASEA J, 51:3–5, 1978.

Sunada Y, Ito N, Yanabo S, Awaji H, Okumura H, Kanai Y. Proceedings of CIGRE Conference on Large High Voltage Electric Systems, 1982, Report 13-04.

Vithayathile JJ, Courts AL, Peterson WG, Hingorani NG, Nilsson S, Poryer JW. IEEE Power Eng Rev, 10:29, 1985.

Yanabo S, Tohru T, Shoichi I, Tsuneo H, Shozo T. IEEE Trans PAS-101:1958–1965, 1982.

Yanabo S, Kaneko H, Koide T, Tamagawa T. IEEE Trans PAS-102:1395–1402, 1983.

12

High-Voltage Cables

M. ABDEL-SALAM *Assiut University, Assiut, Egypt*

12.1 INTRODUCTION

Since Ferranti's first cable in 1881, underground cables have been used in power distribution networks in cities and densely populated areas. The last few decades have witnessed great development in cable technology. Underground cables can now supersede overhead lines for electric power transmission over short distances.

Cables have the following advantages compared with overhead lines:

1. No interruption of supply, even under severe weather conditions such as thunderstorms (on overhead lines, insulator flashover, short circuits, etc., result in supply interruptions)
2. No liability of accident to the public
3. No wayleave troubles that may occur with short circuits of overhead lines crossing main roads
4. No objectionable effect on the aesthetics of the environment

Therefore, underground cables are installed in densely populated regions even when their cost is much higher than that of overhead lines. Cables are usually more expensive than overhead lines at all supply voltages, with a cost ratio of about 20, 8, and 2 at 400, 132, and 11 kV, respectively (King and Halfter, 1982).

Under certain circumstances, DC transmission has many advantages over AC. DC cables have their recommended applications, as in the case of long submarine transmission. In this chapter we present the many aspects of AC high-voltage cables, with particular emphasis on fundamentals. Readers interested in DC cables are invited to consult specialized books on cables (Weedy, 1988; King and Halfter, 1982).

12.2 CABLE CONSTRUCTION

Any cable used in the power industry usually comprises one, three, or four cores. Each core is a metallic conductor surrounded by insulation. The cable has an overall sheath. For underground cables, armoring is added for mechanical protection. In HV and EHV multicore cables, each core has a metallic screen.

12.2.1 Conductors

For all types of power cable, copper and aluminum are in common use. The metal purity is very important (> 99.95%). Impurities seriously reduce the conductivity (Fig. 12.1). The conductors are usually stranded to secure flexibility. Aluminum was considerably cheaper than copper, but now the difference has diminished. The conductivity of aluminum is only 60% that of copper. Therefore, a larger aluminum cross-sectional area is required for the same current-carrying capacity. Both solid and stranded aluminum conductors are presently used in cable construction. When there is more than one layer of strands, alternate layers are spiraled in opposite directions.

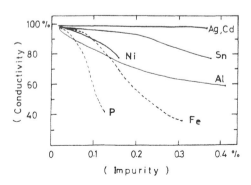

Figure 12.1   Effect of impurities on the conductivity of copper.

Flexibility obtained by spiraling of the strands is at the expense of a small increase in cable resistance, due to increased strand length. The increase in conductor resistance with stranding is significant only in the case of aluminum, where the oxide layer prevents effective contact between the strands.

Sodium has been proposed for use as a cable conductor (Humphrey et al., 1966). Sodium is abundant in seawater and is easily and cheaply manufactured by electrolysis of NaCl. Sodium conductors would be considerably larger in diameter than equivalent conventional ones, as the conductivity of sodium is about 30% of that of copper. Sodium possesses larger values of specific heat and latent heat of fusion. Thus sodium cables have good stable performance under short-circuit conditions. On the other hand, when sodium comes in contact with moisture a violent chemical reaction takes place, evolving hydrogen, which is highly flammable. Also, sodium has virtually no mechanical strength.

12.2.2 Insulation

Insulating materials for power cables are generally classified as (a) impregnated paper, (b) synthetic materials, and (c) compressed gases. The main characteristics that cable insulating material should possess are:

1. High dielectric strength
2. High insulation resistance
3. Sufficiently low thermal resistivity
4. Reasonably long life
5. Low relative permittivity (ε_r) and loss tangent ($\tan \delta$) when used in AC cables
6. Chemical stability over a wide range of temperature
7. Easy handling, manufacture, and installation
8. Economy

Cable insulation used to be oil-impregnated paper, with oil under high pressure in HV and EHV cables. Now polymers are in common use. Cross-linked polyethylene (XLPE) is used in LV and HV cables; poly(vinyl chloride) (PVC) and ethylene propylene rubber (EPR) are limited to LV cables. Attempts to use EPR in HV cables have been carried out (Ross, 1987). In Table 12.1 are given the ranges of voltage ratings in use for cables insulated with oil-impregnated paper and synthetic materials of the oil-filled, gas-filled, and compressed-gas types.

12.2.2.1 Impregnated Paper

Paper is well known to be hygroscopic. It is therefore impregnated with oil to improve its electrical properties. Typical values of 50 Hz breakdown

Table 12.1 Insulation and Voltage Rating of Various Cable Types

Cable type	Impregnated-paper cables	Plastic-insulated cables	Oil-filled cables	Gas-filled cables	Compressed-gas cables
Long-term electric strength (AC working stress)	2–4 MV/m (particularly for XLPE)	6–8 MV/m	15 MV/m	10–15 MV/m	20–25 MV/m
Voltage ratings	6.6, 11, and 22 kV (with screening of each core in three-core cables)	PVC cables: 11, 22, and 33 kV PE cables: up to 22 kV XLPE cables: up to 345 kV EPR cables: up to 66 kV (132 kV under development)	33–525 kV (oil pressure up to 500 kPa)	Up to 220 kV (gas pressure = 1380 kPa)	400–500 kV (SF_6 pressure = 300–500 kPa)

stress are 6–8 MV peak per meter when tested in thin layers before impregnation, and 50–60 MV peak per meter after impregnation.

12.2.2.2 Synthetic Materials

Polymers such as PVC, XLPE, EPR, and polyethylene (PE) show some technical advantages over impregnated paper when used to insulate underground cables in domestic and power distribution networks (Olsson and Hjalmarsson, 1986). More detail is given in Chapter 8.

12.2.2.3 Compressed Gases

In some EHV cable installations, compressed gases such as sulfur hexafluoride (SF_6), Freon (CCl_2F_2), and nitrogen (N_2) are used for insulation. As explained in Chapters 4 and 10, SF_6 has several advantages as an insulating medium. Cables insulated with compressed SF_6 are discussed in Chapter 10.

12.2.3 Screening

Experimental and theoretical investigations have revealed that conductor stranding can increase the maximum electric field in the cable insulation by about 20–30%. To mitigate this effect, conductor screens in the form of aluminum foil or semiconductor carbon paper tape are lapped over the stranded conductor. Screening the insulation around each core of a multi-core cable confines the electric field and makes it symmetrically radial, thereby minimizing the possibility of surface discharges due to tangential stresses. Laminated insulation usually has higher dielectric strength under normal than under tangential field stresses.

For paper-insulated cables, a metallic sheath is essential for mechanical protection against handling and to prevent ingress of moisture to cable insulation. In the past, metallic sheaths used to be manufactured from lead. High-purity lead is not only expensive but is mechanically weak and would crack under vibration. Therefore, lead alloys containing small amounts of tin, antimony, or cadmium were commonly used for cable sheaths. Recently, aluminum sheaths have been introduced. The main advantage claimed for aluminum-sheathed cables are about a 50% saving in weight, improved mechanical properties, and better economy.

The drawbacks of aluminum sheaths are that they are less chemically inert and have a higher sheath loss than that of lead. Sheath losses are discussed in Section 12.5.

12.2.4 Armoring

For cables subject to mechanical stresses, armoring is required. A bedding of hessian is wound or a PVC sheath is extruded over the metal sheath to provide a mechanical cushion and chemical insulation between it and the armoring. The armoring consists of steel tapes or wires. Sometimes, non-magnetic wires (bronze) are used to increase the magnetic reluctance and thus minimize losses. An outer serving of hessian or PVC sheathing is then applied over the armoring to protect it against corrosion (Ross, 1987).

12.3   TYPES OF CABLES

Power cables can be classified into two distinct categories:

1. Solid-type cables in which the pressure within their insulation is atmospheric
2. Pressurized cables in which the pressure is always maintained above atmospheric, either by oil in oil-filled cables or by gas in gas-filled and compressed-gas-insulated cables, to help raise the insulation strength, as explained in Sections 12.3.4 and 12.3.5.

12.3.1   Paper-Insulated Cables

Three-core belted cables use impregnated paper wherein each of the conductors is insulated for half of the line voltage. Then extra insulation is applied as a circumferential belt over the three cores to provide sufficient insulation to withstand the phase voltage between each conductor and the sheath. A serious difficulty that occurs with belted cables is due to the electric field distribution throughout the insulation. The electric field is no longer radial to the conductor as in the case of single-core cables. Paper insulation is weaker under a tangential electric field than under a radial field. Therefore, in belted cables there is a tendency for leakage currents to flow along the layers under the tangential component of the field. Subsequently, heat is generated which may eventually lead to breakdown (Chapter 8). For this reason, belted cables are restricted to voltages of less than 33 kV where the tangential-field component throughout the insulation is fairly insignificant (Weedy, 1988; King and Halfter, 1982).

Tangential field components in belted cables can be eliminated by screening each core separately, as in H-type cables. The conductor screens are in electrical contact with each other and with the overall metal sheath. The cable is therefore equivalent to three single-core cables with a radial field in each.

For voltages exceeding 220 kV, three-phase cable insulation is more efficiently employed if the cable design takes the form of three single cores rather than a three-core belted cable. The cable size bcomes less bulky and more economical.

Another serious problem with paper-insulated cables results from the inelastic property of the lead or aluminum sheath. With cyclic loading, the insulation and sheath expand with different rates and do not return to their original dimensions upon cooling. With time, voids are formed within the volume of insulation. These voids cause losses and ultimate breakdown of the cable insulation, as explained in Chapter 8.

Void formation seems to be a main cause of service failures of cables rated at 66 kV and above. For this reason the gas in the voids is replaced by oil under pressure in "oil-filled" cables, as explained in Section 12.3.3, or the gas is compressed by applying static pressure on the paper insulation in "gas-filled" cables, as explained in Section 12.3.4.

12.3.2 Synthetic-Insulation Cables

The reliability of paper-insulated cables has been so good that the introduction of thermoplastic materials in this field has not been rapid. Nevertheless, two advantages of plastic-insulated cables have accelerated their use: the resistance of plastic sheath to the ingress of corrosive moisture, and the omission of the lead or aluminum sheath which is essential for paper-insulated cables (Weedy, 1988; King and Halfter, 1982). Other advantages of plastic-insulated cables include lighter weight, reduced liability to damage during installation, and colored cores for instant identification. Figure 12.2 shows the construction of a typical plastic-insulated cable.

Following the successful application of PVC cable designs and with the advent of thermosetting materials, XLPE-insulated single- and three-core cables were introduced for 11 kV and higher voltages (Brown, 1983).

Reports on the performance of PE- and XLPE-insulated cables installed in wet environments indicate the deleterious effects of moisture. This includes a decrease in breakdown strength and insulation resistance and an increase in dielectric loss due to water treeing (Bottger et al., 1987). This is discussed in more detail in Chapter 8.

12.3.3 Oil-Filled Cables

Oil-filled cables are paper insulated and always completely filled with oil under pressure (3–4 atm) from oil reservoirs along the route. The oil pressure inhibits the formation of voids. Their contained gases, being compressed, will stand higher electric stresses. For this reason, the long-time breakdown strength of an oil-filled cable is approximately twice that of the equivalent paper-insulated solid cable. Moreover, the oil acts as a coolant (Fig. 12.3).

The design of the hydraulic system for oil-filled cables depends mainly on the cable route and its profile. A gradient in the profile will produce a rise

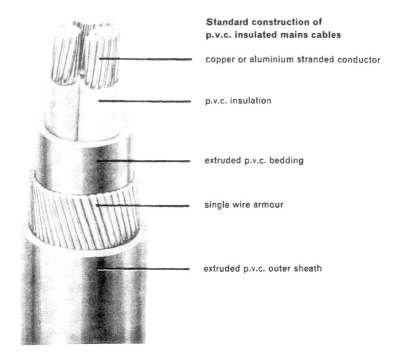

**Standard construction of
p.v.c. insulated mains cables**

copper or aluminium stranded conductor

p.v.c. insulation

extruded p.v.c. bedding

single wire armour

extruded p.v.c. outer sheath

Figure 12.2   Construction of a low-voltage PVC insulated cable.

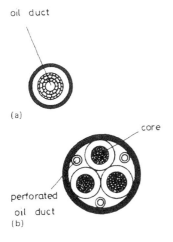

oil duct

(a)

core

perforated
oil duct
(b)

Figure 12.3   Cross-sectional views of typical oil-filled cables.

in static oil pressure. Different oil-filling methods have been used in practice, their choice being dependent on the length of the cable.

A pressure reservoir at one cable end is applied to short lengths of cable routes. The reservoir consists of a number of atmospheric gas-filled sealed cells within a tank completely filled with cable oil and connected to the cable. During cable heating, the excess oil flows along the oil ducts in the cable to the pressure reservoir, where it compresses the sealed cells. When the cable cools at light loads, the oil in the cable contracts but the cells in the pressure reservoir reexpand and force oil back into the cable along the cable ducts and into the insulation. The formation of voids in the insulation is thus mitigated.

For comparatively long cable routes a feeding reservoir is added. In contrast to conventional pressure reservoirs, balanced pressure reservoirs are used to make it possible to adjust the pressure–volume characteristics of the reservoir to suit the profile of the cable. The disadvantages of oil filling are the higher initial cost and complication at joints. Oil leaks can be a considerable nuisance and maintenance costs are high (Ross, 1987).

12.3.4 Gas-Filled Cables

Gas-filled cables are paper-insulated cables but are placed in steel pipes containing compressed gas, usually nitrogen, thus putting the insulation under pressure. The voids that may form in the cable insulation are filled with the gas at a pressure sufficient to suppress ionization. Gas-filling approximately doubles the working voltage of a cable of the same size. Moreover, the gas acts as a coolant and the steel pipe provides mechanical protection for the cable. There are, in some EHV substations, short lengths of cables insulated with compressed SF_6. These are described in Chapter 10.

12.4 CABLE INSULATION CHARACTERISTICS

12.4.1. Breakdown Stress

As explained in Chapter 8, the breakdown strength of the cable dielectric is greatly influenced by the time span during which the voltage is applied. The breakdown stress of impregnated paper reaches a value of about 50 MV/m if the stress duration is a fraction of a second, but drops to less than 20 MV/m for durations of 50–100 h. Therefore, short-term values of breakdown stresses have to exceed the long-term operating stress by a suitable safety margin.

12.4.2 Dielectric Loss

The losses occurring in a cable dielectric include leakage loss and hysteresis loss. The former occurs since the cable insulation has a finite volume resistivity. It therefore draws a leakage current, however small, under both DC and AC voltages. The additional dielectric loss under AC is that caused by the hysteresis involved in the process of dielectric polarization, as explained in Chapter 8. The total dielectric loss under AC is expressed by the loss angle $\delta = 90° - \phi_d$, where ϕ_d is the dielectric power-factor angle. The dielectric loss $= \omega C V^2 \tan \delta$, where C is the capacitance to neutral per metre and V is the phase voltage. For paper–oil cables $\tan \delta$ lies between 0.002 and 0.003, but this value increases rapidly with temperature above 80°C in oil-filled cables. In low-voltage cables this loss is negligible, but it is very appreciable in cables rated at 275 kV and above. For instance, in a 400 kV oil-filled cable, the dielectric losses reach about 80% of the three-phase conductor losses (King and Halfter, 1982).

Several tests have shown a marked dependence on temperature of the loss factor ($\tan \delta$) of oil-impregnated paper insulation of cables, whether they are of the solid, oil-filled, or gas-filled type. Tan δ assumes values of 0.002–0.01, depending on the temperature and the type of cable insulation. A minimum value for the loss factor occurs at about 40–80°C. This characteristic results from the combined effects of temperature on both the resistivity of paper and the viscosity of oil. For cable insulation in general the loss factor increases with applied electric stress. If the insulation contains gas voids, the rate of increase of the loss factor with applied voltage becomes rather high above the onset level for ionization of these gas voids. This is evident in the case of solid-type cables, much more than those with pressurized insulation. This point is explained in Chapter 8.

12.4.3 Internal Partial Discharges

In extruded insulation small cavities or voids exist along with impurities as a result of the manufacturing process. Breakdown of the gas occurs regularly in the AC cycle and randomly on DC, giving discharges known as partial discharges. These erode the surrounding solid insulation and can cause breakdown. In extruded insulation the erosion is often in the form of a propagation channel and discharge paths take the general shape of trees. In paper–oil cables the working AC stress is made considerably below the inception stress for discharges. With extruded polyethylene cables the working stress is made low, typically an average value of about 3 kV/mm, to avoid discharges and "treeing."

12.4.4 Life

Experience over many years on samples and real cables has indicated that the life of a cable at constant temperature is governed by the empirical equation $t\,E^n = $ constant. This law is tested by maintaining constant stress on the dielectric and measuring time to failure. Life under service conditions is obtained by extrapolating the straight line resulting from the plot of $\log E$ against $\log t$. This assumes that the law which has shown to be valid over relatively short test times holds for a value of, say, 30 years or more.

12.5 SHEATH PHENOMENA

AC currents flowing along cable conductors produce alternating magnetic fields, as is well known. These link the metallic sheaths of the cable and induce electromotive forces in them. The EMF induced along the sheath of every core of single-core cables drives a current if these sheaths are bonded. Their magnitudes depend on the core currents, the power frequency, the arrangement and spacing of the cores, and the sheath resistance (El-Kadi, 1984).

In addition to these circulating-current losses in single-core cables, eddy currents are induced by the alternating magnetic fields linking every part of the metallic sheaths. These eddy currents flow in both single-core and multi-core cables. Both types of sheath losses are discussed in the following paragraphs.

12.5.1 Circulating-Current Losses

For three-phase power transmission by three single-core cables, the cables are symmetrically disposed at the vertices of an equilateral triangle. Let $I_1 = I\angle 0°$, $I_2 = I\angle 120°$, and $I_3 = I\angle 240°$ be the line currents in phases 1, 2, and 3 and $I_{sh1} = I_{sh}\angle - \psi$, $I_{sh2} = I_{sh}\angle 120 - \psi$, and $I_{sh3} = I_{sh}\angle 240 - \psi$ be the respective sheath currents, where ψ is the phase shift between the conductor and sheath currents. Let R_c and R_{sh} be the resistances of each cable conductor and sheath, respectively. Due to the symmetrical arrangement of the cables, the mutual inductances (M_{12}, M_{23}, M_{31}) between the conductor of a cable and the sheath of another cable are equal:

$$M_{12} = M_{23} = M_{31} = M = \frac{\mu}{2\pi}\ln\frac{D}{r_{sh}} \tag{12.1}$$

where D is the cable-to-cable spacing, r_{sh} the mean sheath radius, μ the permeability of free space, and ω the angular frequency. The sheath voltage drop per meter of cable length E_{sh} is

$$E_{sh} = I_{sh1}R_{sh} - j\omega M(I_2 + I_{sh2}) - j\omega M(I_3 + I_{sh3})$$
$$= I_{sh1}R_{sh} + j\omega M(I_1 + I_{sh1}) \tag{12.2}$$

The leakage reactance that accounts for the flux between the sheath and conductor of any cable is so small that its corresponding voltage drop is neglected.

When all cable sheaths are open-circuited at one or both ends of the cable system,

$$I_{sh} = 0$$

and

$$E_{sh} = \omega MI = \frac{\mu \omega I}{2\pi} \ln \frac{D}{r_{sh}} \tag{12.3}$$

It is obvious that, with long runs of heavily loaded cables, the sheath voltage to neutral may reach dangerous values, particularly under short-circuit conditions. This may result in arcing between sheaths, with considerable damage. Equation (12.3) shows that the voltage between sheaths increases with an increase in the spacing between cables.

When all cable sheaths are bonded at each end, equation (12.2) gives

$$I_{sh} = I \frac{\omega M}{\sqrt{R_{sh}^2 + \omega^2 M^2}} = I \frac{\omega M}{Z_{sh}} \tag{12.4}$$

and

$$\psi = 90° - \tan^{-1} \frac{\omega M}{R_{sh}} \tag{12.5}$$

It is obvious from equation (12.4) that cross-bonding of the sheaths along the route of the cable does not affect the value of the sheath current (I_{sh}), which is directly proportional to the conductor current (I). The sheath loss as a percentage of the core loss per phase can be easily evaluated, thus:

$$\frac{W_{sh}}{W_c} = \frac{R_{sh}}{R_c} \frac{\omega^2 M^2}{R_{sh}^2 + \omega^2 M^2} \tag{12.6}$$

Usually, the sheath loss is small with respect to the ohmic conductor loss (Table 12.2).

The $I^2 R_{ac}$ loss of the conductor respresents the largest heat source. The alternating-current resistance of the conductor R_{ac} is the direct-current resistance R_{dc}, modified to account for the skin and proximity effects. Skin effect, even at power frequency, is significant and increases with conductor

Table 12.2 Dielectric and Sheath Losses Relative to the Conductor Losses in Three-Phase Cables of the Oil-Filled Type with Different Rated Voltages[a]

Voltage (kV)	Relative losses (%)		
	Conductor	Dielectric	Sheath
66	1.0	0.04	0.13
132	1.0	0.12	0.15
275	1.0	0.36	0.29
400	1.0	0.80	0.40

[a]Conductor area = 2000 mm². The cables were buried at a depth of 1 m in a soil of thermal resistivity = 1 K/mW.
Source: King and Halfter (1982).

cross-section. With large conductors, e.g., 2000 mm², the increase in R_{ac} due to this cause is of the order of 20%.

A three-phase cable system with sheaths bonded at both ends is equivalent to a three-phase 1:1 air-cored transformer. Its primary reactance per phase X_c is quite small. The secondary reactance per phase is practically equal to ωM.

It can easily be shown that bonding the three sheaths of the cable effectively changes the impedance per phase of the three-phase balanced cable from Z_c to Z_{ef}. This is illustrated by the equivalent circuit of Figure 12.4, where the conductor and sheath form a coupled circuit.

Consider a metre length of the circuit. The voltage drop per metre

$$\Delta V = I(R_c + j\omega L_c) + I_{sh}j\omega M$$

and in the sheath

$$0 = I_{sh}(R_s + j\omega L_c) + I j\omega M$$

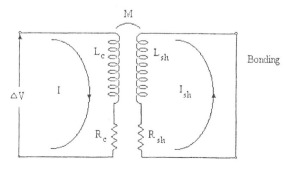

Figure 12.4 Equivalent circuit formed by conductor and bonded sheaths.

From these two equations

$$Z_{ef} = \frac{\Delta V}{I} = \left[R_c + R_{sh}\left(\frac{\omega^2 M^2}{R_{sh}^2 + \omega^2 M^2} \right) \right] + j\left(\omega L_c - \frac{\omega^3 M^3}{R_{sh}^2 + \omega^2 M^2} \right)$$

$$= R_{ef} + jx_{ef}$$

$$= (R_c + j\omega L_c) + \left[R_s\left(\frac{\omega^2 M^2}{R_{sh}^2 + \omega^2 M^2} \right) - j\frac{\omega^3 M^3}{R_{sh}^2 + \omega^2 M^2} \right]$$

$$= Z_c + \left[R_s\left(\frac{\omega^2 M^2}{R_{sh}^2 + \omega^2 M^2} \right) - j\frac{\omega^3 M^3}{R_{sh}^2 + \omega^2 M^2} \right]$$

$$\text{(12.7)}$$

Thus the phase conductor resistance is effectively increased while its inductance is reduced.

When the three single-core cables are disposed in a flat arrangement, each of the outer phases will suffer a higher induced EMF than that of the middle phase. In case of bonding, the outer phases will also have higher sheath losses.

12.5.2 Eddy-Current Losses

In three-phase cable sytems, no point on any sheath is equidistant from all three current-carrying conductors, and hence the EMF induced in the sheath will vary from point to point on its circumference. Subsequently, an eddy current will flow circumferentially, resulting in a so-called sheath eddy loss. These seldom exceed 2% of the conductor loss and are usually neglected.

12.5.3 Cross-Bonding in Three-Phase Single-Core Cable Systems

The simplest form of sheath bonding for three-phase single-core cables is to bond and earth the three sheaths only at one end, or at an intermediate point, and to accept the total induced voltages at the remote end. These voltages to ground, however, should be limited to a safe value (i.e., 100 V in protected positions and about 25 V in exposed positions). Therefore, with cable lengths exceeding a few hundred meters, "cross-bonding" is resorted to for limiting sheath voltages (Kuffel and Poltz, 1981). This cross-bonding of individual core sheaths is carried out in addition to their transposition (Fig. 12.5).

Consider a system consisting of three equal lengths of cable per phase, laid in trefoil, in which the continuity of the sheaths and insulation screens is

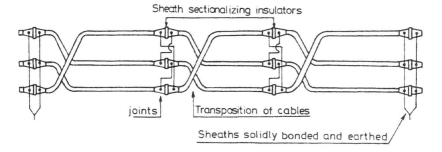

Figure 12.5   Cross-bonding of cable sheaths. (Courtesy of Pirelli.)

broken at each joint. If the sheaths are cross-bonded, the total voltage at the end of each sheath circuit will be zero for a balanced loading of the cables. At the ends, therefore, the three sheaths can be bonded and earthed without any sheath current flowing (Pirelli, 1986) (Fig. 12.5).

12.6   ARMOR LOSS

The armoring of a cable may be considered as a supplementary metallic sheath, with corresponding losses. If the cable armoring is magnetic, additional losses due to magnetic hysteresis will take place. In single-core cables, the hysteresis losses are often of such a magnitude that they preclude the use of ferrous armoring materials. The derivation of armor losses has been dealt with thoroughly in the literature (e.g., Ametani, 1979).

In three-phase single-core cable systems, the armor losses are usually divided into:

1.   Losses due to circulating and eddy currents in the armoring.
2.   Losses due to the magnetic field produced by the cable conductor current itself and by currents in neighboring cables if unscreened. These combined magnetic fields can produce significant hysteresis losses.

12.7   CABLE CONSTANTS

12.7.1.   Conductor Resistance

With DC, the current gets uniformly distributed over the conductor cross-section. With AC, however, the current's own magnetic field effectively drives it toward the skin of the conductor. This "skin effect" causes the AC resistance of a conductor to exceed its DC resistance. At 50 Hz, for

example, the increase in resistance is 2.5% and 7.5% for conductor diameters of 2.5 and 3.8 cm, respectively. The effect is much more significant
at higher frequencies. A similar effect is that of neighboring conductors
carrying AC currents, known as the "proximity effect." The skin effect is
evidently lower for tubular conductors than for solid cylindrical conductors;
it is therefore a favored design for conductors in EHV cables. The duct
inside the tubular conductor is very useful for cooling in oil-filled cables
(Fig. 12.3).

12.7.2 Cable Capacitance

Consider the cross-section of a single-core cable with a metallic sheath (Fig.
12.3a). The capacitance per unit length of such a cable can easily be shown
to have the value

$$C = \frac{2\pi\varepsilon_0\varepsilon_r}{\ln(b/a)} \tag{12.8}$$

where a is the conductor radius, b the inner radius of the sheath, ε_0 the
permittivity of free space, and ε_r the relative permittivity of the cable insulation. The determination of capacitance is important for determining the
cable charging current.

   As the charging current flows in the cable conductor, a severe decrease
in the value of load current transmittable (derating) occurs if the thermal
rating is not to be exceeded: in the higher voltage range, lengths of the order
of 30 km create a need for drastic derating. A further current reduction is
caused by the appreciable magnitude of dielectric losses at higher voltages.
When the charging current becomes equal to the rated current, the cable
length is termed critical. A 345 kV cable with approximately 25.3 mm thickness of insulation and a relative permittivity of 3.5 has a critical length of
42 km; the corresponding MVAr requirement is about 10.6 MVAr per km.

   For three-core belted cables, accurate calculation of the capacitance
per phase is very difficult. Resort is made to numerical computation (Malik
and Al-Arainy, 1988). It is much easier to determine it by bridge measurements. There are capacitances between conductors and a capacitance of
each conductor to the sheath.

12.7.3 Cable Inductance

Compared to overhead lines, the cable inductance is much smaller, and its
accurate calculation is much more difficult. It is usually determined by
bridge measurements. The inductance per phase of a cable is decided by
the magnetic flux linkage, which depends on:

1. The amount of screening afforded by the metal sheaths
2. The presence of armoring and whether or not it is ferrous
3. The proximity of the cable to other conductors and ferrous objects

12.8   ELECTRIC FIELDS IN CABLE INSULATION

As explained in Chapter 2, the electric field intensity in single-core coaxial cables can be calculated. Its maximum value occurs at the surface of the inner conductor (Table 2.1) and is

$$E_{max} = \frac{V}{a \ln(b/a)} \tag{12.9}$$

where V is the applied voltage, a the conductor radius, and b the inner radius of the sheath. The stress, which is a maximum at the surface of the conductor, is further increased by the effects of conductor stranding by an amount dependent on the strand diameter, but in the range 15–25%. This increase is overcome by the use of taped screens of metallized paper or carbon-black paper, or, in the case of extruded cables, an extruded screen of semiconducting plastic. Thus, for a given sheath size, the maximum electric stress varies with the conductor size and has a minimum value when $a = b/2.718$ (see Example (2)). In practice such a ratio is far from economical. However, an enlarged conductor diameter for the same net cross-section is achieved with tubular conductors. This is a suitable design for oil-filled cables (Fig. 12.3).

In three-core cables with unscreened cores, the electric fields have tangential components at some points. The field distributions in such cases are evaluated by measurements on models or by numerical computations (Chapter 2). Some empirical formulas have been developed for estimating the field intensities at some points on the surface of the cable conductors (King and Halfter, 1982; Malik and Al-Arainy, 1987).

The big difference between the maximum and average fields (i.e., E_{max} and E_{av}) in any cable means that the insulation is being inefficiently utilized; the inner parts are much more highly stressed than those near the sheath (Fig. 12.6). Some suggestions have been made for reducing the ratio E_{max}/E_{av}. One is to use multilayer insulation, the outer layer having a lower value of ε_r (see Section 12.12.2). Another suggestion is to subdivide the cable's homogeneous insulation by intermediate metallic sheaths to be energized at suitable fractions of the cable voltage (see Section 12.12.1). Both suggestions have been faced with numerous practical difficulties in both manufacture and operation.

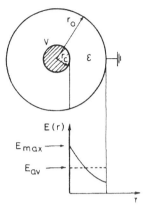

Figure 12.6 Electric field distribution in a single-core cable with one uniform insulation.

12.9 CURRENT-CARRYING CAPACITY AND STEADY-STATE HEATING OF CABLES

The current-carrying capacity of a cable is defined as the maximum current it can carry continuously without the temperature at any point in its insulation exceeding the limit prescribed for it according to its thermal class (Section 8.3). Thus the rated current for a cable depends on the rates of heat generation within it and dissipation to its surrounding. This, in turn, depends on the thermal resistances of the different parts of the cable and of the soil in which it is buried. It depends also on whether the cable is buried directly in the soil or run through troughs or ducts; whether it is buried by itself, being one of a group of cables; or is laid near gas or water pipes. All these factors contribute to heating or cooling the cable under study, and should be considered (King and Halfter, 1982; El-Kadi et al., 1984).

Recommended steady-state operating temperatures of the conductors of paper/oil cables are as follows:

Armored cables buried direct in soil: 65°C
Oil-filled or gas pressure cables direct in soil: 85°C

The above values assume an earth temperature of 15°C; the range of earth temperature to be expected depends upon the depth and geographical location.

The rated current of a cable can be evaluated by setting up the equivlant thermal circuit of the cable–soil system. Take the simple case of the single-core cable represented in Figure 12.7. There the maximum temperature of the insulation θ_c occurs at the conductor surface. The heat developed by the conductor power loss $W_c = I^2 R$ flows through the thermal resistance G_d of the insulation, those of the bedding G_b, serving G_{ser}, and the soil G_s to the ambient, of which the temperature is θ_a. The dielectric loss W_d is assumed concentrated halfway through the insulation. The losses W_{sh} and W_{ar} in the sheath and armoring also contribute to the cable temperature rise, so they are included at the appropriate nodes in the equivalent circuit (Fig. 12.7).

Thus the thermal form of Ohm's law gives the relation

$$\theta_c - \theta_a = I^2 R_c(G_d + G_b + G_{ser} + G_s) + W_d\left(\frac{G_d}{2} + G_b + G_{ser} + G_s\right)$$
$$+ W_{sh}(G_b + G_{ser} + G_s) + W_{ar}(G_{ser} + G_s)$$

$$(12.10)$$

As the dielectric loss flows through one-half of the internal thermal resistance of the cable, the temperature rise $\Delta\theta_d$ above θ_a is due only to the dielectric loss W_d and

$$\Delta\theta_d = W_d\left(\frac{G_d}{2} + G_b + G_{ser} + G_s\right) \qquad (12.11)$$

Substitution yields

$$W_{sh} = \lambda_{sh}(I^2 R) \qquad \text{and} \qquad W_{ar} = \lambda_{ar}(I^2 R)$$

where λ_{sh} and λ_{ar} are, respectively, the ratio of the sheath loss and armor loss to the conductor loss.

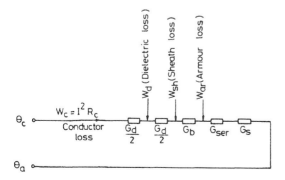

Figure 12.7 Equivalent thermal circuit of a single-core buried cable.

After simplification,

$$I = \left[\frac{\theta_c - \theta_a - \Delta\theta_d}{R_c[G_d + G_b(1 + \lambda_{sh}) + (G_{ser} + G_s)(1 + \lambda_{sh} + \lambda_{ar})]} \right]^{1/2} \qquad (12.12)$$

Equation (12.12) does not take into account correction factors for the skin and proximity effects (Section 12.7). Both effects, together with typical ratios λ_{sh} and λ_{ar}, can be evaluated (IEC, 1974).

What is left for calculating the current-carrying capacity of the cable is the determination of its own thermal resistance and the maximum permissible temperature. Calculation of the thermal resistance of a single-core cable is easy and is determined in terms of the cable radii and the thermal resistivity of the insulation. The thermal resistance of a single-core cable G_d is determined by the thermal resistivity of the dielectric material $g(°C·m/W)$, the conductor radius a, and the sheath radius b. It is expressed as

$$G_d = \frac{g \ln b/a}{2\pi} °C/W \qquad (12.13)$$

For a three-core belt-type cable, however, the calculation of thermal resistance is more complex, due to the arrangement of the thermal field, and computer programs are best used (Glicksman et al., 1978).

The external thermal resistance depends on the type of soil, the amount of moisture present, and how the cable cores are disposed in the earth. Theoretical and experimental studies for evaluating the internal and external thermal resistances and the temperature rise for different cooling systems of general buried cables have been reported in the literature (Abdel-Aziz and Riege, 1980; Burghardt et al., 1983; Prime et al., 1984).

12.10   TRANSIENT HEATING OF CABLES

The following regimes of transient heating occur (Weedy, 1988):

1. *Short-circuit.* All the heat evolved is assumed to be stored in the cable. A temperature limit of 120°C is often assumed.
2. *Cyclic loading.* Calculations are often based on an idealized daily load curve of full-load current for 8 h and no-load for 16 h. This results in a lower temperature rise than when permanently on full load, which is allowed for in practice by the use of cyclic rating factors. For directly buried cables of conductor area less than 64.5 mm^2 this factor is 1.09, and for conductor areas from 64.5 to 645 mm^2 it is taken as 1.13; i.e., 1.13 times the rated full-load current can be passed on cyclic loading.

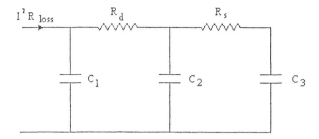

Figure 12.8 Simple model for transient thermal response of buried cable: C_1 = thermal capacitance of conductor; C_2 = thermal capacitance of sheath; C_3 = effective thermal capacitance of soil; R_d = thermal resistance of insulation; R_s = effective thermal resistance of soil.

3. *Short-time emergency loading.* An overload may be sustained for perhaps several hours without excessive temperatures due to the long thermal time-constant of the cable.

A lumped-constant thermal network may be used for transient calculations by the connection of appropriate thermal capacitances between the nodes of the steady-state network and the ambient or reference line. For a cable a uniformly radial heat flow is assumed; i.e., the sheath is isothermal and the thermal circuit is an *R–C* ladder network. The number of nodes depends on the number of annular cylinders into which the dielectric is divided. The conductor and sheath temperatures may be obtained from a simple model (Fig. 12.8). Reasonable results over a few hours of transient are obtained for pipe-type cables by making C_3 in Figure 12.8 the thermal capacity of the pipe oil.

The simplest analysis results from a single lumped constant representation, i.e., a single *R–C* network. The response of such a network to a time-varying load is illustrated in Figure 12.9, in which τ is the thermal time constant ($R \times C$) and θ_{m1}, θ_{m2}, and θ_{m3} are steady-state temperatures.

12.11 JOINTING AND TERMINATING HIGH-VOLTAGE CABLES

Because of manufacturing, shipping, and installation limitations, all HV cables are produced and laid in a number of limited lengths that have to be joined together on site and terminated at required positions.

There are four main methods of installing cables, as follows:

1. Direct in the soil: the cable is laid in a trench which is refilled with a backfill consisting of either the original soil or imported material of lower or more stable thermal resistivity.

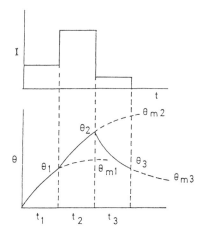

Figure 12.9 Temperature rise–time curves under varying load.

2. In ducts or troughs, usually of earthenware or concrete.
3. In circular ducts or pipes through which cables are drawn. This has the advantage of further cables being installed without excavation.
4. In air; e.g., installed in tunnels built for other purposes.

Although cables are manufactured under carefully controlled conditions, the joints and terminations cannot always be made under similar circumstances. Nevertheless, when completed, they should be as reliable as the rest of the cable, to eliminate any risk of supply interruptions (King and Halfter, 1982; Pirelli, 1986).

In general, cable joints are classified as straight-through joints, branch or T-joints, trifurcating joints, stop joints of oil-filled cables, and outdoor sealing ends for terminating cables outdoors. The techniques of jointing and terminating cables depend on long experience with the specific type of cable, its conductors, and insulation materials. Extreme care is taken to ensure the high current-carrying capacity and high insulation strength of the joints and sealing ends.

12.12   STRESS CONTROL IN CABLES

There are two main methods by which a more uniform distribution of stress in the cable insulation may be achieved: by the introduction of intersheaths, and by the use of layers of insulating material with different permittivities.

12.12.1 Intersheath Grading

Suppose that intersheaths of radii r_1 and r_2 are inserted into the dielectric and maintained at voltages V_1 and V_2. The conductor and the sheath have radii a and b. The stress between any two metallic cylinders varies inversely as the distance from the axis (Table 2.1). Thus, the stress at a point distant r from the conductor $(a < r < r_1)$

$$E_1 = A_1/r$$

where A_1 is a constant which is found by integrating E_1 from the conductor to the intersheath as follows:

$$V - V_1 = \int_a^{r_1} E_1 dr = A_1 \ln\frac{r_1}{a}$$

where V is the conductor voltage (applied voltage). Thus,

$$A_1 = \frac{V - V_1}{\ln(r_1/a)}$$

and

$$E_1 = \frac{V - V_1}{r \ln(r_1/a)}$$

The maximum stress is

$$E_{1\,\text{max}} = \frac{V - V_1}{a \ln(r_1/a)}$$

Similarly, the maximum stress between the first and second intersheaths is

$$E_{2\,\text{max}} = \frac{V_1 - V_2}{r_1 \ln(r_2/r_1)}$$

whilst the maximum stress between the second intersheath and the sheath is

$$E_{3\,\text{max}} = \frac{V_2}{r_2 \ln(b/r_2)}$$

By choice of V_1 and V_2 the maximum stresses can be made equal, and the stress distribution is like that shown by curve C_1 of Figure 12.10, instead of curve C_2 which represents the stress without intersheaths.

It is possible to choose r_1 and r_2 so that the stress varies between the same maximum and minimum in the three layers, by taking

$$r_1/a = r_2/r_1 = b/r_2 = \alpha$$

So, the minimum stresses are the maximum stresses divided by α. Equating the maximum stresses, one obtains

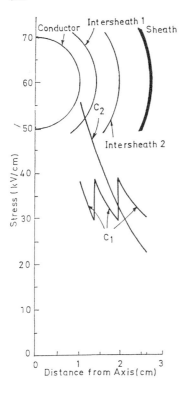

Figure 12.10 Effect of intersheaths on stress distribution in single-core cables.

$$V_2 = \frac{V}{1 + 1/\alpha + 1/\alpha^2}, \quad V_1 = \frac{V(1 + 1/\alpha)}{1 + 1/\alpha + 1/\alpha^2}$$

The maximum stress is then

$$\frac{V - V_1}{a \ln \alpha} = \frac{V}{(1 + \alpha + \alpha^2) a \ln \alpha} = \frac{3V}{(1 + \alpha + \alpha^2) a \ln(b/a)}$$

since $\ln(b/a) = \ln \alpha^3 = 3 \ln \alpha$. Without the use of intersheaths, the maximum stress is

$$(E_{\max})_0 = \frac{V}{a \ln(b/a)}$$

so that the maximum stress has been reduced due to the use of intersheaths in the ratio

$$1 : \frac{1}{3}(1 + \alpha + \alpha^2)$$

The use of intersheaths has not been general practice because of the complications involved. The sheaths must be supplied with the requisite potentials and must carry quite large charging currents. Jointing is made very difficult.

12.12.2 Capacitance Grading

Suppose that the cable dielectric consists of two layers with a dividing radius r_1, the permittivities being ε_1 and ε_2. The stress in the inner layer is

$$E_1 = A_2/(\varepsilon_1 r)$$

whilst in the outer layer

$$E_2 = A_2/(\varepsilon_2 r)$$

Then

$$V = \int_a^{r_1} E_1 \mathrm{d}r + \int_{r_1}^b E_2 \mathrm{d}r$$

$$= A_2 \left[\frac{1}{\varepsilon_1} \ln(r_1/a) + \frac{1}{\varepsilon_2} \ln(b/r_1) \right]$$

$$\therefore A_2 = \frac{V}{\dfrac{1}{\varepsilon_1} \ln(r_1/a) + \dfrac{1}{\varepsilon_2} \ln(b/r_1)}$$

The maximum value of E_1 is

$$E_{1\,\mathrm{max}} = \frac{V}{a \left[\ln(r_1/a) + \dfrac{\varepsilon_1}{\varepsilon_2} \ln\left(\dfrac{b}{r_1}\right) \right]}$$

and

$$E_{2\,\mathrm{max}} = \frac{V}{r_1 \left[\dfrac{\varepsilon_2}{\varepsilon_1} \ln(r_1/a) + \ln\left(\dfrac{b}{r_1}\right) \right]}$$

12.13   ARTIFICIAL COOLING OF UNDERGROUND CABLES

In recent years progress has been made in schemes for reducing the external thermal resistance of cables by artificial cooling. These methods take two main forms:

1. *Internal cooling*: the pumping of water or insulating oil along a central duct formed in the cable conductor.
2. *External cooling*: water is pumped through plastic pipes laid beside the cables. From an engineering standpoint this scheme is much simpler than the first. Greatly increased currents are achievable by the use of large pipes of diameter, say, 150 mm.

12.14   LOCATING FAULTS IN CABLES

Faults in multiconductor cables may be divided according to their nature as follows:

1. High- or low-resistance earth faults involving one or more of the conductors
2. Open-circuit faults with low or high resistance at the break
3. Flashing (self-healing) faults

The faulty conductor and the type of fault can easily be identified by simple tests at both ends of the cable. These include measuring the insulation resistances by a megger of suitably high voltage. Normal practice is to identify the zone in which the fault has occurred by an approximate measurement of its distance X from either end of the cable, and then to locate the fault more accurately. The distance X can be measured either by the "pulse" method or by using a bridge. A DC bridge is suitable for low-resistance faults, whereas AC bridges are appropriate for faults of broken conductors, as noted below.

12.14.1   Pulse Method

A voltage pulse from a suitable source is applied between the faulty conductor and a sound conductor or the metal sheath. This pulse will travel along the cable up to the fault, where it gets reflected. The amplitude and polarity of the reflected pulse depend on the resistance of the fault (Fig. 12.11; see also Chapter 14). Using such a trace of the pulses on a cathode-ray oscilloscope, the time t_x corresponds to the pulse traveling from the sending end of the cable to the fault and back. Knowing v, the velocity of propagation of electromagnetic waves along the cable, the distance X to the fault can be calculated. The error in the result depends on errors in v and the measured t_x.

12.14.2   Bridge Method

Low-resistance faults can be located by connecting a DC bridge to the fault and a sound conductor at each end of the cable. These two conductors are

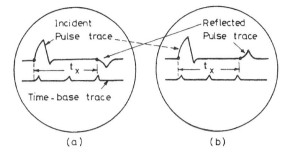

Figure 12.11 Pulse method for locating the faulty zone of a cable: (a) low-resistance fault; (b) high-resistance fault.

connected at the far end by a jumper of sizable cross-section forming a loop. Thus the balance conditions of the bridge are used in calculating the fault distance. Inaccuracy in measuring the distance X by this method results from errors such as insensitivity of the galvanometer, tolerances in the bridge elements, and the contact resistances at both ends and at the jumper. Too high a resistance of the fault itself would no doubt affect the sensitivity of the bridge. A self-healing fault, for instance, would need to be burned further in order to be located by the bridge measurement with fair accuracy. The further burning of the fault is usually carried out using a special powerful energy source.

If the fault is a break in one of the cable conductors, it can be located using an AC bridge (Fig. 12.12). The distance X of the fault is proportional

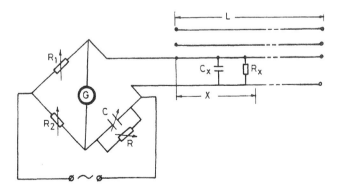

Figure 12.12 AC bridge method for locating a break in one of the cable conductors. G is either earphones, meter, or oscilloscope.

to the capacitance C_x. The proportionality can be evaluated by measuring the entire length of the same cable by the bridge. Errors in measuring the distance X by such a bridge result from insensitivity of the detector G, tolerances in the bridge elements, and impedance between the two parts of the broken conductor.

For a more precise location of the fault, a train of high-frequency pulses is sent along the cable while their magnetic field is detected (e.g., by headphones and a search coil) along the cable route near the suspected fault location. There will be a detectable signal all along the cable up to the point of the fault, beyond which it ceases.

12.15 SUPERCONDUCTING CABLES

Superconductors are materials whose electrical resistivities become immeasurably small or actually zero below critical temperatures T_c unless exposed to magnetic fields exceeding critical strengths H_c. Over the years intensive research has been carried out to discover more superconducting metals and alloys with the highest possible critical temperatures. Even as late as 1987 some ceramics were discovered, including oxides of copper, barium, yttrium, and lanthanum. These ceramic superconductors have critical temperatures approaching 90 K. For such materials the cost of refrigeration would be about four orders of magnitude lower than the case of conventional metallic superconductors (Anon., 1987).

If employed in cables, superconductors would mean loss-free transmission of very large powers. In machines, a drastic reduction in their sizes for large ratings would be possible. The possibility of producing very strong magnetic fields by very high currents without resistance losses and with no EMF for continuously driving it would make possible electromagnetic levitation of fast trains, magnetohydrodynamic propulsion of marine vessels, and many other applications. In electronics and computers it would mean an increase in switching rates and memory sizes.

12.15.1 Superconductivity

12.15.1.1 Experimental Evidence

A superconducting material manifests zero electrical resistivity and zero magnetic permeability as long as it is kept at a temperature lower than T_c. Its zero permeability results in its repulsion to magnetic fields, which accounts for the great possibility of magnetic levitation.

The critical temperature T_c and the critical magnetic field H_c of strength lower than H_0 are correlated as follows:

$$H_c(T) = H_0 \left(1 - g \frac{T^2}{T_c^2} \right) \tag{12.14}$$

where g is a constant; g was considered equal to unity by King and Halfter (1982). Equation (12.14) states that the critical field strength $H_c(T)$ at the absolute temperature T is lower than the critical field strength H_0 above which the material's superconductivity is destroyed at absolute zero temperature. This relation is diagrammatically shown in Figure 12.13 for different metals.

Superconducting coils have actually been manufactured for producing very strong magnetic fields in cyclotrons and other particle accelerators (Maix et al., 1987). To ensure stability of superconductor performance, it is drawn in the form of several filaments of the superconductive alloy 10–60 μm in diameter. The group of filaments is embedded inside a conducting wire of copper or copper–nickel with an overall diameter of about 1 mm. The stable rated current of such a wire (Fig. 12.14) under magnetic fields of about 2 T is about 300 A.

12.15.1.2 Theory

A theory to explain superconductivity was proposed in 1957 by Bardeen, Cooper, and Schrieffer, based on the concept of a perfect pair of free electrons (Hummel, 1985). Different from normal conductors, in which the current is carried by a stream of individual elecrons, the current is carried through a superconductor by pairs of electrons, as explained very briefly below.

According to this theory, positive ions located in the crystal lattice along the path of the electrons through the superconductor help to cancel out some of the repulsive forces between pairs of electrons having mutually suitable energies. They have equal and opposite movements. Their energy

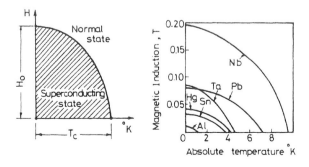

Figure 12.13 Graphs illustrating the states of superconducting metals.

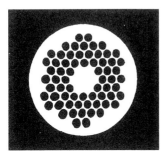

Figure 12.14   Superconducting filamentary wires clad in a copper wire. Filament diameter = 57 μm, overall wire diameter = 0.7 mm. (Courtesy of BBC.)

levels lie near free states. Thus, as each electron pair drifts through the superconductor, one of the two electrons perturbs its atomic lattice. A displaced ion reverts back into its original position with a momentary oscillation. A phonon acts as an elastic coupling between the two electrons as they pass together through the atomic lattice with no net perturbation. The perturbation of the atomic lattice of normal conductors caused by the flowing electrons accounts for their electrical resistivities.

Cryogenic cooling of a suitable metal or alloy to near 0 K damps out vibrations of its lattice atoms and furnishes the condition suitable for free electron pairs. This theory no doubt would need modification to accommodate superconducting ceramics, discovered recently.

12.15.2   Prospects for Superconducting Cables

The dielectric loss in superconducting cables should be minimal as it imposes an additional load on the refrigeration system. Insulating materials, including paper, cannot be used because of the unacceptable loss tangent. Synthetic materials are not permitted for use because of their differential contraction, which causes cracking. In the early days of superconducting cables, a relatively low-loss dielectric such as helium or vacuum was used for cable installation, together with solid dielectric spacers. Later, liquid nitrogen and liquid hydrogen have been introduced.

Until 1987, active studies were progressing for developing the least expensive design for superconducting cables with suitable core material and size. Some researchers have also studied the suitability of liquid air-impregnated paper as an insulating material. Other materials, such as glass–epoxy tapes and enamels, have been considered (Kofler and Telser, 1987; Bobo et al., 1987).

Now, with the discovery of ceramics that exhibit superconductivity at temperatures approaching normal room temperature, many problems have vanished. On the other hand, the ceramic superconductors available so far have unsuitable mechanical and physical properties. They are brittle and chemically unstable (Bogner et al., 1987). However, the next millennium may well witness the solution of these various technical problems and the commercial availability of normal-temperature superconducting cables and wires for numerous power and electronics applications.

12.16 CABLE TESTING

Detailed descriptions of the procedures for the various type, routine, and acceptance tests on cables are laid down in the appropriate standards (Chapter 18).

12.17. SOLVED EXAMPLES

(1) The inner conductor of a coaxial cable has a diameter of 4 cm, the sheath diameter over the insulation being 10 cm. The cable is insulated with two materials having permittivities of 6 and 4 respectively with corresponding safe operating stresses of 40 kV/cm and 25 kV/cm. Calculate the radial thickness of each insulating layer and the safe operating voltage of the cable.

Solution: Let r be the outer radius of the inner insulating material, and the charge on the conductor be q/cm length. Then, if ε_1 is the permittivity of the inner dielectric and ε_2 the permittivity of the outer dielectric, the electric stress at any radius r is

$$E = \frac{q}{2\pi\varepsilon_0\varepsilon_1 r} \tag{12.15}$$

where ε_0 is the permittivity of free space, and ε_r is the relative permittivity of the dielectric.

Therefore, the maximum stress in the inner dielectric (i.e., at the surface of the conductor) is

$$E_1 = \frac{q}{2\pi\varepsilon_0 \times 6 \times 2}$$

The maximum stress in the outer dielectric (at radius r) is

$$E_2 = \frac{q}{2\pi\varepsilon_0 \times 4 \times r}$$

Therefore,

$$\frac{E_1}{E_2} = \frac{4r}{12} = \frac{r}{3}$$

As $E_1 = 40\,\text{kV/cm}$ and $E_2 = 25\,\text{kV/cm}$,

$$r = \frac{40}{25} \times 3 = 4.8 \text{ cm}$$

Therefore, the radial thicknesses of the dielectric are

inner $= 4.8 - 2 = 2.4\,\text{cm}$
outer $= 5 - 4.8 = 0.2\,\text{cm}$

The voltage drop across a dielectric between radii r_1 and r_2 is given by the expression

$$V = E_m r \ln \frac{r_2}{r_1} \tag{12.16}$$

where E_m is the maximum stress in the insulation for the inner dielectric,

$$E_m = 40 \text{ kV/cm}, \qquad r_1 = 2 \text{ cm}, \qquad r_2 = 4.8 \text{ cm}$$

Therefore,

$$V_1(\text{peak}) = 40 \times 2 \ln \frac{4.8}{2} = 70.04 \text{ kV}$$

For the outer dielectric,

$$E_m = 25 \text{ kV/cm}, \qquad r_1 = 4.8 \text{ cm}, \qquad r_2 = 5 \text{ cm}$$

Therefore,

$$V_2(\text{peak}) = 25 \times 4.8 \ln \frac{5}{4.8} = 4.9 \text{ kV}$$

$$\begin{aligned}
\text{Peak voltage of cable} &= V_1(\text{peak}) + V_2(\text{peak}) \\
&= 70.04 + 4.9 = 74.94 \text{ kV}
\end{aligned}$$

$$\text{Safe operating voltage (rms)} = \frac{74.94}{\sqrt{2}} = 53 \text{ kV}$$

(2) For a single-core cable, determine the value of conductor radius r which gives the minimum voltage gradient for fixed values of applied voltage V and sheath radius R. Calculate the optimum value of r if V is $100\,\text{kV}$ and the maximum permissible gradient is $55\,\text{kV/cm}$.

Solution: The highest value of the voltage gradient occurs at the surface of the conductor and is given by

$$E_{\mathrm{m}} = \frac{V}{r \ln R/r} \tag{12.17}$$

If V and R are fixed and r is variable, the value of r which gives the minimum value of E_{m} is that which makes $\dfrac{\mathrm{d}E_{\mathrm{m}}}{\mathrm{d}r}$ equal to zero.

$$\frac{\mathrm{d}E_{\mathrm{m}}}{\mathrm{d}r} = \frac{-V\left(\ln\dfrac{R}{r} + r\dfrac{r}{R}\left(\dfrac{-R}{r^2}\right)\right)}{\left(r\ln\dfrac{R}{r}\right)^2}$$

For E_{m} to be a minimum

$$\ln\frac{R}{r} + r\frac{r}{R}\left(\frac{-R}{r^2}\right) = 0$$

i.e.,

$$\ln\frac{R}{r} - 1 = 0, \qquad \ln\frac{R}{r} = 1, \qquad \frac{R}{r} = \mathrm{e}$$

$$r = R/\mathrm{e}$$

If the value of r is substituted in equation (12.17), the minimum value of E_{m} becomes

$$E_{\mathrm{m}} = V/r$$

The significance of these results is that if the overall and core radii are in the ratio of e, then for a given core radius the voltage gradient at the surface of the conductor, which is always where the insulation is most heavily stressed, is a minimum. Under these circumstances the voltage gradient at the conductor surface is inversely proportional to the core radius.

If $V = 100\,\mathrm{kV}$ and the voltage gradient at the conductor surface must not exceed $55\,\mathrm{kV/cm}$, then the optimum value of r is given by

$$55 = \frac{100\sqrt{2}}{r}$$

$$r = 2.57\,\mathrm{cm}$$

It is assumed here that the value given for V is rms and that for the voltage gradient is peak value.

(3) A single-core cable 10 km long has an insulation resistance of 0.5 MΩ. The core diameter is 3 cm and the sheath diameter of the cable is 6 cm. Calculate the resistivity of the insulating material. Prove the formula used.

Solution: Consider an element of the insulating material at a radius r from the conductor center and having a thickness dr in a length l of the cable. Let ρ be the resistivity of the material in MΩ·m.

$$\text{Insulation resistance through the element} = \frac{\rho dr}{2\pi rl}$$

$$\text{Total insulation resistance} = \int_r^R \frac{\rho dr}{2\pi rl} = \frac{\rho}{2\pi l}\ln\frac{R}{r} \tag{12.18}$$

In the problem given, the insulation resistance $= 0.5 \times 10^6\,\Omega$

$$l = 10 \times 10^3 = 10^4\,\text{m}, \qquad R/r = 3/1.5 = 2$$
$$\rho = 2\pi \times 10^4 \times 0.5 \times 10^6 / \ln 2$$
$$= 4.53 \times 10^4\,\text{MΩ·m}$$

(4) Calculate the KVA_r taken by 10 km of such a cable which has a capacitance of 0.5 µF/km measured between two of the cores, when it is connected to a 10,000 V, 50 Hz frequency supply.

Solution: Figure 12.15 shows the arrangement of capacitors which is equivalent to the capacitances in a 3-phase, lead-sheathed cable, C_1 representing the intercore capacitances and C_2 the capacitance of each core to the sheath.

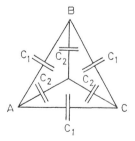

Figure 12.15 In-core and core-to-sheath capacitances in three-phase, three-core cables.

Figure 12.15 may be replaced by the equivalent diagram in which the star capacitances C_3 take the place of the delta capacitances C_1. For the two circuits to be equivalent, $C_3 = 3C_1$. Now it is obvious that the star point N and the sheath will be equipotential. Hence, the cable may also be represented by the system of capacitors shown in Figure 12.16.

Line charging current for the star-connected system of Fig. 12.16

$$= V_{\text{ph}}(C_2 + C_3)\omega$$

If C_4 is the capacitance measured between A and B with C insulated, then

$$C_4 = \tfrac{1}{2}(C_2 + C_3)\omega$$

Therefore,

$$C_2 + C_3 = 2C_4$$

Line charging current $= V_{\text{ph}}\omega 2C_4$

If V is the line voltage,

$$\text{Line charging current} = \frac{2V\omega c_4}{\sqrt{3}} \tag{12.19}$$

$$= 7.25 fVC_4$$

In this case

$$C_4 = 0.5\,\mu\text{F/km} \times 10\text{ km} = 5\,\mu\text{F}$$

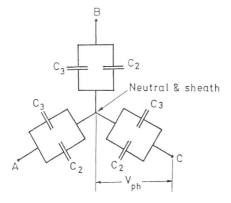

Figure 12.16 Equivalent core-to-neutral capacitances in three-phase, three-core cables.

$$\text{Line charging current } I_c = \frac{2 \times 10^4 \times 2\pi \times 50 \times 5 \times 10^{-6}}{\sqrt{3}} \text{ A}$$

$$\begin{aligned}
\text{Charging kVA} &= \sqrt{3}VI_c \times 10^{-3} \\
&= \frac{\sqrt{3} \times 10{,}000 \times 2 \times 10{,}000 \times 2\pi \times 50 \times 5}{\sqrt{3} \times 10^9} \\
&= 314.16 \, \text{kVA}_r
\end{aligned}$$

(5) In a 3-core, 3-phase, lead-sheathed cable, the capacitance between the three cores bunched together and the lead sheath is $0.75 \, \mu\text{F/km}$ and that between two of the cores connected together and the third core is $0.6 \, \mu\text{F/km}$. Calculate:

(a) the capacitance per km measured between any two of the cores,
(b) the kVA_r required to keep 10 km of this cable charged when connected to a 33 kV, 3-phase, 50 Hz supply.

Solution: (a) Using the notation of Figs. 12.15 and 12.16:

$$3C_2 = 0.75 \, \mu\text{F/km}$$

$$C_2 = 0.25 \, \mu\text{F/km}$$

If B and C are joined together the capacitance between A and B consists of two capacitances of $2(C_2 + C_3)$ and $(C_2 + C_3)$ in series. Therefore, capacitance between A and B

$$= \frac{2}{3}(C_2 + C_3)$$
$$= 0.6 \, \mu\text{F/km}$$

$$C_2 + C_3 = 0.9 \, \mu\text{F/km}, \qquad C_3 = 0.65 \, \mu\text{F/km}$$

Hence,

$$\therefore C_4 = \frac{1}{2}(C_2 + C_3)$$
$$= 0.45 \, \mu\text{F/km}$$
$$= \text{capacitance per km between any two cores}$$

(b) Capacitance of 10 km between 2 cores $= 4.5\,\mu F$

Charging kVA_r of 10 km of cable

$$= \sqrt{3}VI_c \times 10^{-3}$$

$$= \sqrt{3}V\frac{2V\omega C_4}{\sqrt{3}}$$

$$= \frac{2(33 \times 10^3)^2 2\pi \times 50 \times 4.5}{10^9} = 3079.1\,kVA_r$$

Therefore, to keep 10 km of the cable charged requires $3079.1\,kVA_r$.

(6) Determine the effective electrical parameters of a three-phase, 66 kV, 50 Hz underground circuit, 85 km in length and comprising three single-core cables each of conductor radius 1 cm. The internal and external radii of the lead sheath are 2.5 and 3.0 cm respectively. The cables are located at vertices of an equilateral triangle, 8 cm each side. The sheaths are bonded to ground at several points. The conductor AC resistance per kilometer at 15°C is $0.0875\,\Omega/km$ and the resistivity of lead at the operating temperature may be assumed to be $23.2 \times 10^{-6}\,\Omega\,cm$.

Solution: Conductor resistance $= 0.0875(1 + 0.004 \times 50) = 0.105\,\Omega/$ km (conductor operating temperture assumed to be 65°C and temperature coefficient 0.004).

For the whole length,

$$R_c = 0.105 \times 85 = 8.9\,\Omega$$

Resistance of sheath

$$R_{sh} = \frac{23.2 \times 10^{-6} \times 85 \times 10^5}{\pi(3^2 - 2.5^2)} = 22.83\,\Omega$$

In a three-phase system the equivalent phase-to-neutral reactance is considered, and in this case the conductor-to-sheath mutual inductive reactance

$$X_m = \omega M = 2\pi \times 50 \times 2\ln\left(\frac{D}{r_{sh}}\right) \times 10^{-7}\,\Omega/m \qquad (12.20)$$

where $r_{sh} =$ mean radius of sheath and $D =$ cable-to-cable spacing ($= 8\,cm$);

$$X_m = 2\pi \times 50 \times 2\ln\left(\frac{8}{\frac{1}{2}(2.5+3)}\right) \times 10^{-7} \times 85\,000$$

$$= 5.7\,\Omega \text{ for 85 km length}$$

Effective AC resistance of conductor

$$R_{ef} = R_c + \frac{X_m^2 R_{sh}}{R_{sh}^2 + X_m^2} = 8.9 + \frac{5.7^2 \times 22.83}{22.83^2 + 5.7^2}$$

$$= 10.24\,\Omega$$

Effective reactance per cable

$$X_{ef} = X_c - \left(\frac{X_m^2}{R_{sh}^2 + X_m^2}\right)$$

where X_c = reactance with sheaths open-circuit

$$= \omega L_c = \omega \times 2\ln\left(\frac{D}{r_c}\right) \times 10^{-7} \times 85,000 = 11.1\,\Omega$$

Effective reactance

$$X_{ef} = 11.1 - \frac{5.7^2}{22.83^2 + 5.7^2} = 11.04\,\Omega$$

$$\frac{\text{Sheath loss}}{\text{Conductor loss}} = \frac{R_{sh}}{R_c} \cdot \frac{X_m^2}{R_{sh}^2 + X_m^2}$$

$$= \frac{22.83}{8.9} \frac{5.7^2}{22.83^2 + 5.7^2} = 0.15$$

For a current of 400 A, the emf induced without bonding (per sheath)

$$= IX_m = 2280\,V = 2.28\,kV$$

(7) Find the induced sheath voltage per km of a symmetrical three-phase cable system with conductor spacing 15 cm and sheath diameters 5.5 cm. The current is 250 A at 50 Hz.

Solution:

$$E = IX_{\mathrm m} = I\omega M = I\omega 2 \ln \frac{D}{r_{\mathrm{sh}}} \times 10^{-7}$$

$$= \frac{\mu\omega I}{2\pi} \ln \frac{D}{r_{\mathrm{sh}}}$$

$$= \frac{4\pi \times 10^{-7} \times 314 \times 250}{2\pi} \ln \frac{15}{2.75}\ \mathrm{V/m}$$

$$= 26.6\ \mathrm{V/km}$$

If the sheaths are bonded at one end, the voltage between them at the other end is $\sqrt{3} \times 26.6 = 46\ \mathrm{V/km}$

(8) A single-core 66 kV cable has a conductor radius of 1 cm and a sheath of inside radius 2.65 cm. Find the maximum stress. If two intersheaths are used, find the best positions, the maximum stress, and the voltage on the intersheaths.

Solution:

$$b/a = 5.3/2 = \alpha^3, \qquad \alpha = 1.384$$

Thus, $r_1 = 1.385$ cm and $r_2 = 1.92$ cm are the radii of the intersheaths.
The peak voltage on the conductor

$$V = \frac{66}{\sqrt{3}}\sqrt{2} = 53.8\ \mathrm{kV}$$

$$V_2 = \frac{53.8}{1 + 1/1.384 + 1/1.384^2} = 23.9\ \mathrm{kV}$$

$$V_1 = (1 + 1/1.384)23.9 = 41.1\ \mathrm{kV}$$

Without sheaths, the maximum stress is

$$(E_{\max})_0 = \frac{53.8}{1\ln(2.65/1)} = 55.3\ \mathrm{kV/cm}$$

and the minimum stress is

$$(E_{\min})_0 = \frac{53.8}{2.65\ln(2.65/1)} = 20.8\ \mathrm{kV/cm}$$

With the intersheaths, the maximum stress is

$$E_{\max} = \frac{3 \times 55.3}{(1 + 1.384 + 1.384^2)} = 38.7\ \mathrm{kV/cm}$$

while the minimum stress is

$$38.7/\alpha = 27.96 \text{ kV/cm}$$

Figure 12.10 shows the stress distribution in both cases. The maximum stress has been reduced by the ratio 1:1.43.

(9) Suppose that in the previous example the intersheaths are spaced at equal distances from each other, the conductor and the sheath. Find their voltage for the same maximum stresses in the layers, and find the maximum stress.

Solution:

$$b = 2.65, \qquad a = 1 \text{ cm}, \qquad \frac{1}{3}(b - a) = 0.55 \text{ cm}$$
$$\therefore r_1 = 1.55 \text{ cm and } r_2 = 2.1 \text{ cm}$$

$$E_{1\,\text{max}} = \frac{V - V_1}{a \ln(r_1/a)} = 2.28(V - V_1)$$

$$E_{2\,\text{max}} = \frac{V_1 - V_2}{r_1 \ln(r_2/r_1)} = 2.12(V_1 - V_2)$$

$$E_{3\,\text{max}} = \frac{V_2}{r_2 \ln(b/r_2)} = 2.06 V_2$$

As $E_{1\,\text{max}} = E_{2\,\text{max}} = E_{3\,\text{max}} = E_{\text{max}}$,

$$V_1 = 45.2 \text{ kV and } V_2 = 23 \text{ kV}$$

The maximum stress

$$E_{\text{max}} = 47.5 \text{ kV}$$

(10) Suppose that the cable of the last two examples has an inner layer 1 cm thick of rubber of relative permittivity 4.5 and the rest is impregnated paper of relative permittivity 3.6. Find the maximum stress in the rubber and in the paper.

Solution:

$$a = 1 \text{ cm}, \qquad r_1 = 2 \text{ cm}, \qquad b = 2.65 \text{ cm},$$
$$\varepsilon_{r1} = 4.5, \qquad \varepsilon_{r2} = 3.6, \qquad V = 53.8 \text{ kV}$$

$$E_{1\,\text{max}} = \frac{53.8}{1[\ln 2 + (4.5/3.6) \ln 1.325]} = 51.5 \text{ kV/cm}$$

and

$$E_{2\,\text{max}} = \frac{53.8}{2[(3.6/4.5)\ln 2 + \ln 1.325]} = 32.18\,\text{kV/cm}$$

Thus, the maximum stress has been reduced from 55.3 to 51.5 kV/cm. The reduction is hardly worthwhile, and in practice the only grading used is for strength; i.e., a better quality paper is put near the conductor than near the sheath. This method of grading is quite practical.

REFERENCES

Abdel-Aziz M, Riege H. IEEE Trans PAS-99:2386–2392, 1980.
Ametani A. IEEE Trans PAS-98:902–910, 1979.
Anon. High-Tech Mater 4(9):3, 1987.
Bobo J, Poitevin J, Nithart H. Proceedings of CIGRE Symposium on New and Improved Materials for Electrotechnology, Vienna, 1987, Report 100-04.
Bogner G, Lambrecht D, Sabrié J. Electra 114:96–107, 1987.
Bottger O, Golz W, Saure M. Proceedings of CIGRE Symposium on New and Improved Materials for Electrotechnology, Vienna, 1987, Report 620-07.
Brown M. IEEE Trans PAS-102:373–381, 1983.
Burghardt R, Mathews H, Purnhagen D, Engelhardt J. IEEE Trans PAS-102:2133–2144, 1983.
El-Kadi M. IEEE Trans PAS-103:2043–2050, 1984.
El-Kadi M, Chu F, Radhakrishna H. IEEE Trans PAS-103:2735–2740, 1984.
Glicksman L, Sanders J, Robsenow W. IEEE Trans PAS-97:134–139, 1978.
Hummel R. Electronic Properties of Materials, Berlin: Springer-Verlag, 1985.
Humphrey L, Hess R, Addis G, Ruprecht A, Ware P, Steeve E, Schneider J, Matthysse I, Scoran E. IEEE Spectrum 11:73, 1966.
IEC. Calculation of the Continuous Current Rating of Cables (100% Load Factor). Publication 287-2. Geneva: International Electrotechnical Commission, 1974.
King SY, Halfter NA. Underground Power Cables. Harlow, UK: Longman, 1982.
Kofler H, Telser E. Proceedings of CIGRE Symposium on New and Improved Materials for Electrotechnology, Vienna, 1987, Report 100-02.
Kuffel E, and Poltz J. IEEE Trans PAS-100:369–374, 1981.
Maix R, Rauch J, Benz H, Vecsey G, Jakob B, Zichy J. Brown Boveri Rev 74:40–48, 1987.
Malik N, Al-Arainy A. IEEE Trans PWRD-2:589–595, 1987.
Malik N, Al-Arainy A. Int J Elect Eng Educ 25(1):27–32, 1988.
Olsson J, Hjalmarsson G. ASEA J, 59:8, 1988.
Pirelli. Self-Contained Oil-Filled Cables. Milan: Pirelli, 1986, p. 55.
Prime J, Valdes J, Macias C, Porven A. IEEE Trans PAS-103:2794–2798, 1984.
Ross A. IEE Power Eng J 1:51, 1987.
Weedy BM. Thermal Design of Underground Systems. New York: John Wiley, 1988.

13

Grounding Systems

A. EL-MORSHEDY *Cairo University, Giza, Egypt*

13.1   INTRODUCTION

The use of electricity brings with it an electric shock hazard for humans and animals, particularly in the case of defective electrical apparatus. In electricity supply systems it is therefore a common practice to connect the system to ground at suitable points. Thus, in the event of fault, sufficient current will flow through and operate the protective system, which rapidly isolates the faulty circuit. It is therefore required that the connection to ground be of sufficiently low resistance.

It is essential to mention here that the terms "ground" and "earth" cannot quite be used interchangeably. The proper reference ground may sometimes be the earth itself, but most often, with small apparatus, it is its metallic frame or grounding conductor. The potential of this ground conductor may be quite different from zero (i.e., that of the earth itself) (IEEE, 1972). The term "ground" is therefore used throughout this chapter.

Grounding is of major importance in our efforts to increase the reliability of the supply service, as it helps to provide stability of voltage conditions, preventing excessive voltage peaks during disturbances. Grounding is also a measure of protection against lightning. For protection of power and substations from lightning strokes, surge arresters are often used which are provided with a low earth resistance connection to enable the large currents encountered to be effectively discharged to the general mass of the earth. Depending on its main purpose, the ground is termed either a power system ground or a safety ground.

13.2   RESISTANCE OF GROUNDING SYSTEMS

The value of resistance to ground of an electrode system is the resistance between the electrode system and another "infinitely large" electrode in the ground at infinite spacing. The soil resistivity is a deterministic factor in evaluating the ground resistance. It is an electrophysical property. The soil resistivity depends on the type of soil (Table 13.1), its moisture content, and dissolved salts. There are effects of grain size and its distribution, and effects of temperature and pressure.

Homogeneous soil is seldom met, particularly when large areas are involved. In most cases there are several layers of different soils. For non-homogeneous soils, an apparent resistivity is defined for an equivalent homogeneous soil, representing the prevailing resistivity values from a certain depth downward (Sverak et al., 1982; Nahman and Salamon, 1984; Komaragiri and Mukhedkar, 1981; Sunde, 1968).

The moisture content of the soil reduces its resistivity (Fig. 13.1). As the moisture content varies with the seasons, the resistivity varies accordingly. The grounding system should therefore be installed nearest to the permanent water level, if possible, to minimize the effect of seasonal variations on soil resistivity. As water resistivity has a large temperature coefficient, the soil resistivity increases as the temperature is decreased, with a discontinuity at the freezing point.

The resistivity of soil depends on the amount of salts dissolved in its moisture. A small quantity of dissolved salts can reduce the resistivity very remarkably. Different salts have different effects on the soil resistivity, which explains why the resistivities of apparently similar soils from different locations vary considerably (Fig. 13.2).

Table 13.1   Typical Values of Resistivity of Some Soils

Type of soil	Resistivity ($\Omega \cdot$ m)
Loam, garden soil	5–50
Clay	8–50
Sand and gravel	60–100
Sandstone	10–500
Rocks	200–10,000

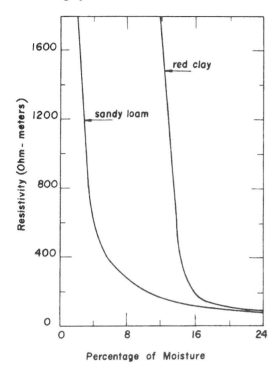

Figure 13.1   Variation of soil resistivity with its moisture content.

The distribution of grain size has an effect on the manner in which the moisture is held. The finer the grading, the lower will be the resistivity. There is not much experimental work on the effect of pressure, but it is reasonable to assume that higher pressures resulting in a more compact body of earth will result in lower values of resistivity.

13.2.1   Resistance of a Grounding Point Electrode

The simplest possible electrode is the hemisphere (Fig. 13.3). The ground resistance of this electrode is made up of the sum of the resistances of an infinite number of thin hemispheric shells of soil. If a current I flows into the ground through this hemispherical electrode, it will flow away uniformly in all directions through a series of concentric hemispherical shells. Considering each individual shell with a radius x and a thickness dx, the total resistance R up to a large radius r_1 would be

$$R = \int_{r}^{r_1} \frac{\rho dx}{2\pi x^2} = \frac{\rho}{2\pi}\left(\frac{1}{r} - \frac{1}{r_1}\right) \tag{13.1}$$

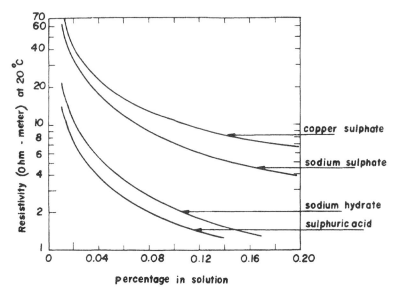

Figure 13.2   Resistivities of different solutions.

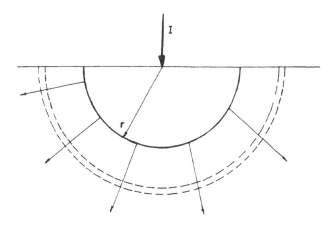

Figure 13.3   Current entering ground through a hemispherical electrode.

where ρ is the earth resistivity. As $r_1 = \infty$

$$R_\infty = \frac{\rho}{2\pi r} \tag{13.2}$$

The general equation for electrode resistance is

$$R = \frac{\rho}{2\pi c} \tag{13.3}$$

where c is the electrostatic capacitance of the electrode combined with its image above the surface of the earth. This relation is applicable to any shape of electrode.

13.2.2   Resistance of Driven Rods

The driven rod is one of the simplest and most economical forms of electrode. Its ground resistance can be calculated if its shape is approximated to that of an ellipsoid of revolution having a major axis equal to twice the rod's length and a minor axis equal to its diameter d; then

$$R = \frac{\rho}{2\pi l}\ln\frac{4l}{d} \tag{13.4}$$

If the rod is taken as cylindrical with a hemispherical end, the analytical relation for R takes the form

$$R = \frac{\rho}{2\pi l}\ln\frac{2l}{d} \tag{13.5}$$

If, however, the rod is assumed to be carrying current uniformly along its length, the formula becomes

$$R = \frac{\rho}{2\pi l}\left[\ln\left(\frac{8l}{d}\right) - 1\right] \tag{13.6}$$

Table 13.2 gives approximate formulas for the resistance of electrodes with various shapes.

The resistance of a single rod is, in general, not sufficiently low, and it is necessary to use a number of rods connected in parallel. They should be driven as far apart as possible so as to minimize the overlap among their areas of influence. In practice, this is very difficult, so it becomes necessary to determine the net reduction in the total resistance by connecting rods in parallel. One of the approximate methods is to replace a rod by a hemispherical electrode having the same resistance. The method assumes that each equivalent hemisphere carries the same charge. Evaluating their average potential and total charge, the capacitance and hence the resistance of the system can be calculated.

Table 13.2 Approximate Formulas for Resistance of Various Electrodes

Ground rod	$R = \dfrac{\rho}{2\pi\ell}\left[\ln\left(\dfrac{8\ell}{d}\right) - 1\right]$	
Two ground rods $S > \ell$ $S < \ell$	$R = \dfrac{\rho}{4\pi\ell}\left[\ln\left(\dfrac{8\ell}{d}\right) - 1\right] + \dfrac{\rho}{4\pi S}\left(1 - \dfrac{\ell^2}{3S^2}\right)$ $R = \dfrac{\rho}{4\pi\ell}\left(\ln\dfrac{32\ell^2}{dS} - 2 + \dfrac{S}{2\ell} - \dfrac{S^2}{16\ell^2}\right)$	
Horizontal wire	$R = \dfrac{\rho}{4\pi\ell}\left(\ln\dfrac{16\ell^2}{dh} - 2 + \dfrac{h}{\ell} - \dfrac{h^2}{4\ell^2}\right)$	
Horizontal strip (section $a \times b$)	$R = \dfrac{\rho}{4\pi\ell}\left[\ln\dfrac{8\ell^2}{ah} + \dfrac{a^2 - \pi ab}{2(a+b)^2} - 1 + \dfrac{h}{\ell} - \dfrac{h^2}{4\ell^2}\right]$	

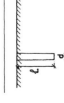

Table 13.2 (Continued)

Four-point star		$R = \dfrac{\rho}{8\pi\ell}\ln\dfrac{4\ell^2}{dh} + 2.9 - 2.14\dfrac{h}{\ell} + 2.6\dfrac{h^2}{\ell^2}$
Six-point star		$R = \dfrac{\rho}{12\pi\ell}\left(\ln\dfrac{4\ell^2}{dh} + 6.85 - 6.26\dfrac{h}{\ell} + 7\dfrac{h^2}{\ell^2}\right)$
Ring of wire		$R = \dfrac{\rho}{2\pi^2 D}\ln\dfrac{16D^2}{dh}$
Horizontal round plate		$R = \dfrac{\rho}{4D} + \dfrac{\rho}{8\pi h}\left(1 - 0.036\dfrac{D^2}{h^2}\right)$
Vertical round plate		$R = \dfrac{\rho}{4D} + \dfrac{\rho}{8\pi h}\left(1 + 0.018\dfrac{D^2}{h^2}\right)$

The resistance of n rods in parallel is thus found to exceed $(1/n)$ of that of a single rod because of their mutual screening. The screening coefficient η for n electrodes in parallel is defined as

$$\eta = \frac{\text{resistance of one electrode}}{(\text{resistance of } n \text{ electrodes in parallel})}$$

It is difficult to determine the value of η for complicated systems and usually it is listed in tables obtained by calculations and measurements (Dawalibi and Blattner, 1984).

13.2.3 Grounding Grids

A common method for obtaining a low ground resistance at high-voltage substations is to use interconnected ground grids. A typical grid system for a substantion would comprise 4/0 bare solid copper conductors buried at a depth of 30–60 cm, spaced in a grid pattern of about 3–10 m. At each junction, the conductors are securely bonded together.

The side of grid conductors required to avoid fusing under the fault current I is estimated as (IEEE, 1987)

$$a = I\sqrt{\frac{76t}{\ln\{(234 + T_\mathrm{m})/(234 + T_\mathrm{a})\}}} \tag{13.7}$$

where a is the copper cross-section (circular mils), t is the fault duration (s), T_m is the maximum allowable temperature, and T_a is the ambient temperature. Such a grid not only effectively grounds the equipment, but has the added advantage of controlling the voltage gradients at the surface of the earth to values safe for human contact. Ground rods may be connected to the grid for further reduction in the ground resistance when the upper layer of soil is of much higher resistivity than that of the soil underneath (Garrett and Holley, 1980).

13.2.3.1 Resistance to Ground and Mesh Voltages of Grounding Grids

The resistance to ground determines the maximum potential rise of the grounding system during a ground fault. The following equation for grid resistance may be used:

$$R = \frac{\rho}{L}\left(\ln\frac{2L}{\sqrt{dh}} + K_1\frac{L}{\sqrt{A}} - K_2\right) \tag{13.8}$$

where L is the total length of all conductors, A the total area of the grid, d the grid conductor diameter, h is the depth of burial of the grid, and K_1 and K_2 the factors presented graphically as functions of length-to-width ratio of the area. For practical design purposes, various approximate formulas based on the similarity of a grid and a round plate of equal area have been proposed (Table 13.2).

The mesh voltage represents the maximum touch voltage to which a person can be exposed at the substation. It is the potential difference between the grid conductor and a point at the ground surface above the center of the grid mesh. Mesh voltages of ground-grounding grids are calculated using a relation of the form (IEEE Committee, 1982)

$$E_{mesh} = K_m K_i \rho \frac{I}{L} \tag{13.9}$$

where I is the current flowing into the ground and K_m is a coefficient that takes into account the effect of number n of the grid conductors, their spacing S, diameter d, and depth of burial h,

$$K_m = \frac{1}{2\pi} \ln \frac{S^2}{16hd} + \frac{1}{\pi} \ln \prod_{j=3}^{n} \left(\frac{2j-3}{2j-2} \right) \tag{13.10}$$

K_i is an irregularity correction factor to allow for nonuniformity of ground current flow from different parts of the grid:

$$K_i = 0.65 + 0.172n \tag{13.11}$$

13.2.3.2 Scale Models of Grounding Grids

Scale model tests with an electrolytic tank are very useful for determining the ground resistance and surface potential distributions during ground faults in complex grounding arrangements where accurate analytical calculations are hardly possible (El-Morshedy et al., 1986). By measuring the voltage applied to the model and the current flowing through the electrolyte between the model grid and the return electrode, the effective grid resistance can be evaluated.

Surface potential profiles for a 4×4 mesh grid $10\,cm \times 10\,cm$, $1\,mm$ conductor diameter, at a depth of $1\,cm$ are shown in Figure 13.4 for the normal profile. The surface potential is given as a percentage of applied grid voltage and the horizontal axis is in centimeters measured from the center of the grid. The maximum and minimum potentials throughout the grid can be determined. Using such a model, the effect of changing the grounding mesh parameters and adding grounding rods of different depths at different locations can easily be estimated. Sample results are given in Figure 13.5 (El-Morshedy et al., 1986).

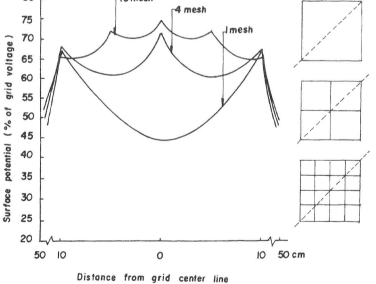

Figure 13.4 Profiles of surface potential for a 4 × 4 mesh, 10 cm ×10 cm grid. Conductor diameter = 1 mm, its depth below electrolyte surface = 1 cm.

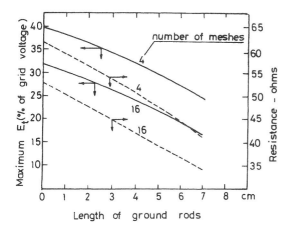

Figure 13.5 Maximum touch voltage and ground resistance for a model of grounding grids of different mesh as functions of the length of ground rods. Grid size = 20 × 20 cm, conductor diameter = 1 mm, depth below electrolyte surface = 1 cm.

13.2.4 Unequally Spaced Grounding Grids

Equally spaced grounding grids have some disadvantages, such as fringing effect. This leads to more currents emanating from the peripheral conductor of a grid, resulting in touch voltages on the corner of the grid much higher than those in the center. An unequal spacing technique has been proposed to overcome the drawbacks of equally spaced grids (Huang et al., 1995). This reference gives a computer program based on the surface element method which calculates the emanating current density, potential distribution, and resistance of any configuration of grounding grids buried in uniform soil. The program gives the optimum spacing between grid conductors. Figure 13.6 shows two rectangular grounding grids with the same grid area and number of buried conductors. The numerical calculation results are given in Figures 13.7 and 13.8.

Unequally spaced grids have some advantages in uniform soils over equally spaced grids. Current density emanating from grid conductors can be more uniform using unequal spacing. Unequal spacing can also decrease potential gradients on the earth's surface and enhance the level of safety for people and equipment. In addition, unequal spacing can greatly save grounding grid material, resulting in a more than 30% decrease in grid cost (Huang et al., 1995).

In a recent work (Lee et al., 1998), the effects of rectangular ground grids—with and without ground rods—on the performance of a grounding

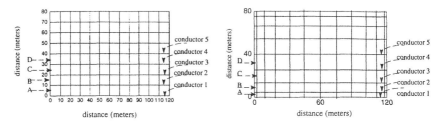

Figure 13.6 Grounding grids with area of 120 m × 80 m and 22 conductors. (A, B, C, D) refer to the potential curve positions of Figures 13.7 and 13.8. Soil resistivity is 250 Ω · m, burial depth is 0.6 m.

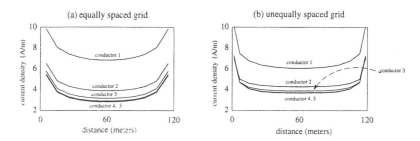

Figure 13.7 Current density distribution.

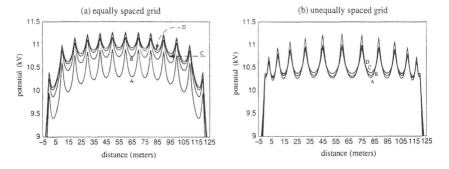

Figure 13.8 Potential distribution.

system have been investigated. The study involves uniform, two-layer, and multilayer soils. The study reveals that:

• In uniform or two-layer soils with a low-resistivity top layer, unequally spaced grids having denser conductors at the edges provide the most efficient design.

- Equally spaced grids provide the most efficient designs in soil structures with a high-resistivity top layer.
- Ground rods are effective only if they penetrate a low-resistivity bottom layer to a significant depth.
- Multilayer soils appear to behave like uniform or two-layer soils derived from the multilayer soils by averaging the resistivities of adjoining layers which are not in contact with the grid.

13.3 IMPULSE IMPEDANCE OF GROUNDING SYSTEMS

Knowledge of the impulse impedance of a grounding system is necessary for determining its performance while discharging impulse currents to ground, as in the case of lightning and transient ground faults.

13.3.1 Performance of Driven Rods

In Figure 13.9a the current I travels down the rod electrode and from there diffuses into the ground. In addition to its resistivity, the soil has a dielectric constant ε_r. When the electrode voltage changes with time, there will be a conductive current in addition to a capacitive current. The capacitance of the ground electrode is (IEEE, 1987)

$$C = \frac{\varepsilon_r l}{18 \ln(4l/d)} \times 10^{-9} \quad \text{F} \tag{13.12}$$

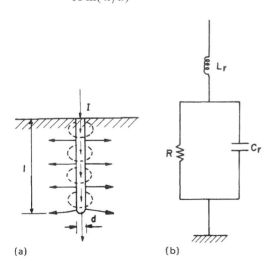

(a) (b)

Figure 13.9 (a) Impulse current spreading from a driven rod. Its magnetic field is also represented. (b) Equivalent circuit for a rod under impulse.

The current in the electrode and ground produces a magnetic field. This is highest where the current is most concentrated (i.e., at the top). The inductance of such a rod is

$$L = 2l \, \ln\frac{4l}{d} \times 10^{-7} \quad \text{H} \tag{13.13}$$

At high frequencies of about 1 MHz the ground will draw a considerable capacitive current in addition to its conductive current. The total current flows through the self-inductance of the electrode. Figure 13.9b represents the equivalent circuit (Gupta and Thapar, 1980). At low frequencies the inductance and capacitance can be safely neglected, whereas at extremely high frequencies the ground would be represented by a distributed network.

13.3.2   Performance of Grounding Grids

Ground wires of considerable length are usually buried as counterpoises along transmission lines or near high-voltage substations, for lightning protection. A buried wire can be represented by the usual transmission-line circuit of distributed constants. Its distributed resistance to ground is:
For a deeply buried wire:

$$R = \frac{\rho}{2\pi}\ln\frac{4l}{d} \quad \Omega \cdot m \tag{13.14a}$$

For a wire buried very near the surface:

$$R = \frac{\rho}{\pi}\ln\frac{4l}{d} \quad \Omega \cdot m \tag{13.14b}$$

The impulse impedance of a buried ground wire in operational form (Verma and Mukhedkar, 1981) is expressed as

$$Z = \sqrt{R(r + SL_d)} \tag{13.15}$$

where r is the metallic resistance of the wire per meter (Ω) and L_d is the distributed inductance per meter (H). In the case of a grounding grid, its equivalent circuit is analyzed in response to the applied impulse current. The circuit parameters can be estimated, knowing the dimensions of the grid. To account for the impulse current, however, flowing off each conductor along its length, its effective inductance is reduced to one-third the steady-state value.

Naturally, the impulse impedance is initially higher than the power-frequency impedance, but decreases with time to the steady-state value at a rate depending on the circuit and wave parameters. The inductance of the grid is the governing factor contributing to its impulse impedance.

13.4 PRINCIPLES OF DESIGN OF A SUBSTATION GROUNDING SYSTEM

The ground of a substation is very important, as it provides the ground connection for the system neutral, the discharge path for surge arresters, and ensures safety to operating personnel. It also provides a low-resistance path to ground to minimize the rise in ground potential. The ground-potential rise depends on fault-current magnitude and the resistance of the grounding system.

Low-resistance substation grounds are difficult to obtain in desert and rocky areas. In such cases, the use of grids will provide the most convenient means of obtaining a suitable ground connection. Many utilities add ground rods for further reduction of the resistance. The size of the grid and the number and length of driven rods depend on the substation size, the nature of the soil, and the ground resistance desired.

The practical design of a grid requires inspection of the layout plan of equipment and structures. The grid system usually extends over the entire substation yard and sometimes several meters beyond. To equalize all ground potentials around the station, the various ground cables or buses in the yard and in the substation building should be bonded together by heavy multiple connections and tied into the main station ground. It is also necessary to adjust the total length of buried conductors, including cross-connections and rods, to be at least equal to those required to keep local potential differences within acceptable limits.

13.4.1 Ground Conductor Size

The ground conductor should have low impedance and should carry prospective fault currents without fusing or getting damaged, taking into account future expansion of the connected power system. The size of ground conductor is given by equation (13.7).

13.4.2 Conductor Length Required for Gradient Control

Equation (13.9) gives the value of the mesh voltage; the value of the step voltage, E_{step}, is given by the formula (Fig. 13.10) (IEEE, 1987)

$$E_{\text{step}} = K_s K_i \rho \frac{I}{L} \tag{13.16}$$

where K_s is a coefficient that takes into account the effect of number, spacing S, and depth h of burial of the ground conductors.

$$K_s = \frac{1}{\pi}\left(\frac{1}{2h} + \frac{1}{S+h} + \frac{1}{2S} + \frac{1}{3S} + \cdots\right) \tag{13.17}$$

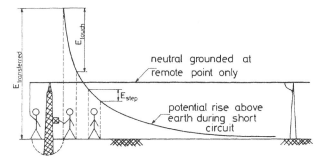

Figure 13.10 Step, touch, and transferred voltages near a grounded structure.

The number of terms within the parentheses is equal to the number of parallel conductors in the basic grid, excluding cross-connections.

The tolerable step voltage with duration t, E_{step}, which is the voltage between any two points on the ground surface that can be touched simultaneously by the feet, is (IEEE, 1987)

$$E_{step} = \frac{165 + \rho_s}{\sqrt{t}} \tag{13.18}$$

where ρ_s is the resistivity of ground beneath the feet, in ohm-meters, taking its surface treatment into account.

The tolerable touch voltage, E_{touch}, which is the voltage between any point on the ground where a person may stand and any point that can be touched simultaneously by either hand, is (IEEE, 1987)

$$E_{touch} = \frac{165 + 0.25\rho_s}{\sqrt{t}} \tag{13.19}$$

Equating the value of E_{mesh} to the maximum value of E_{touch} yields

$$\frac{K_m K_i \rho I}{L} = \frac{165 + 0.25\rho_s}{\sqrt{t}} \tag{13.20}$$

The approximate length of buried conductor required to keep voltage within safe limits is thus

$$L = \frac{K_m K_i \rho I \sqrt{t}}{165 + 0.25\rho_s} \tag{13.21}$$

13.5   NEUTRAL GROUNDING

Grounding of the neutral points of generators, transformers, and transmission schemes is an important item in the design of power systems, as it has a considerable bearing on the levels of transient and dynamic overvoltages stressing the equipment insulation. It also directly affects the levels of short-circuit currents in the power network and, accordingly, the ratings of switchgear needed to cope with them.

The methods of system neutral grounding include resistance and low reactance for effective grounding. They also include tuned reactance, solid grounding, and grounding through a high-impedance such as that of a potential transformer (Fig. 13.11) (Brown et al., 1978; Gulachenski and Courville, 1984). Each of these methods has advantages and limitations. For example, with isolated or high-impedance grounding, excessive overvoltages appear on the system in the case of line-to-ground (L–G) faults. The "healthy" phases acquire transient overvoltages several times higher than the normal peak phase voltage (Chapter 14). Also, some contingencies may develop ferroresonance, causing high power-frequency overvoltages. In both cases, the equipment insulation, if not suitably designed, would be very vulnerable. Further, with the same method of neutral grounding, the magnitudes of L–G fault currents would be so low that only special protective gear could detect them. This type of neutral grounding, however, is favored only in some LV and MV isolated networks where the need of supply continuity is extremely pressing and the equipment insulation is adequate.

Neutral grounding through a reactor tuned to match the system capacitance C to ground $(= 3\omega C)$ neutralizes the system L–G fault currents

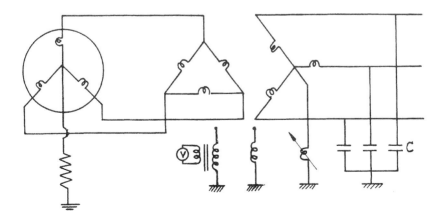

Figure 13.11   Examples of power system neutral grounding.

almost completely. Thus the fault arc becomes unstable and easily gets extinguished. This method has been in common use in some high-voltage networks in Europe. It helps maintain the continuity of the supply without endangering the system insulation.

For resistance grounding, a resistance of a suitable design is connected between the system neutral point and the grounding electrode group as an addition to the system ground resistance. This technique is suitable for generators, as it helps to maintain their stability by the power consumed during L–G faults; otherwise, the generators might race out of step (Ershevich et al., 1982).

By grounding solidly or through a small reactance, the system over-voltages are limited to their possible minimum (Chapter 14). On the other hand, the L–G short circuit currents will be excessive unless the grounding reactor is designed with a suitable magnitude.

13.6 GROUNDING OF LOW-VOLTAGE (LV) SYSTEMS

The different methods of grounding low voltage systems are (IEC, 1980):

1. Exposed conductive parts connected to neutral (TN)
2. Grounded neutral (TT)
3. Ungrounded neutral (IT)

The purpose of these three systems is to protect persons and property. They are not identical in dependability of the LV electrical installation with respect to electrical power availability and installation maintenance.

The LV grounding system characterizes the grounding mode of the secondary of the MV/LV transformer and the means of grounding the installation frames. Identification of the system types is thus defined by means of two letters:

- the first one for transformer neutral connection, T for neutral con-nected to ground and I for neutral isolated from ground (Fig. 13.12a);
- the second one for the type of application frame connection, T for directly connected to the ground and N for connection to the neu-tral at the origin of the installation, which is connected to the ground (Fig. 13.12b).

Combination of these letters gives three possible configurations: TT, TN, and IT.

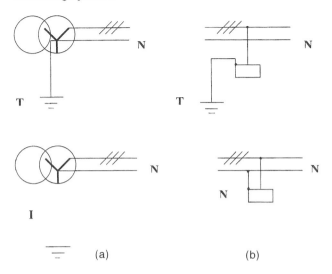

Figure 13.12 Connection modes of the neutral of the installation and of the frames of the electrical loads.

The TN system includes several subsystems:

• TN-C: if the neutral conductor (N) and protective conductor (PE) are one and the same (PEN).
• TN-S: if the N and PE conductors are separate. The use of this system is compulsory for networks with conductors of a cross-section less than or equal to 10 mm^2 Cu.
• TN-C-S: use of a TN-S downstream from a TN-C. The opposite is forbidden.

Each grounding system can be applied to an entire LV electrical installation. However, several grounding systems may be included in the same installation, as shown in Fig. 13.13.

13.7 GROUNDING OF GAS-INSULATED SUBSTATIONS

The main difference between gas-insulated substations (GIS) and conventional substations is their metallic enclosures. Under faults, these enclosures carry induced currents of significant magnitude. These currents must be confined to specific paths (Chapter 10).

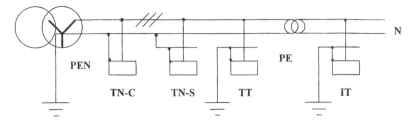

Figure 13.13 Example of the various grounding systems included in the same installation.

13.7.1   Grounding of Enclosures

The following requirements should be met to minimize the undesirable effects caused by circulating currents:

1.   All metallic enclosures should normally operate at ground voltage level.
2.   No significant voltage differences should exist between individual enclosure sections.
3.   The supporting structures and any part of the grounding system must not be influenced by the flow of induced currents.
4.   Precautions should be taken to prevent excessive currents being induced into adjacent frames and structures.
5.   As GIS substations have limited space, reinforced-concrete foundations may cause irregularities in a current discharge path. The use of a simple monolithic slab reinforced by steel serves as an auxiliary grounding device. The reinforcing bars in the foundations can act as additional ground electrodes.

13.7.2   Touch Voltages for GIS

The enclosures of GIS should be properly designed and adequately grounded so as to limit the potential difference between individual sections within the allowable limit of 65–130 V during faults (Sverak et al., 1982). The analysis of GIS grounding includes estimation of the permissible touch voltage. For this, equation (13.19) can be used with $\rho_s = 0$ for a metal-to-metal contact. Dangerous touch and step voltages within the GIS area are drastically reduced by complete bonding and grounding of the GIS enclosures, and by using grounded conductive platforms connected to the GIS structures.

13.7.3 Transient Impedance of GIS Grounding Grid

The transient impedance of a grounding grid for a commercial 550 kV GIS was measured on site using steep front currents with rise time from 100 ns to 2 μs (Karaki et al., 1995). From the measured results, it was found that the transient impedance of the grid for those currents is simulated by a series circuit with an inductance and a resistance of 1 μH and 3 Ω respectively.

Overvoltages in the GIS due to lightning surges were analyzed using this grounding impedance. It was noticed that:

- The radius of the grid area in which the grounding potential is raised by the injected steep current is very small (1.5 m).
- The grounding impedance of the GIS enclosure can be simulated by attaching the grounding impedance obtained to each grounding point of the GIS enclosure.
- The influence of the grounding impedance connected to the bushing enclosure and the gas circuit breaker enclosure is negligible.
- The isolation of an arrester enclosure end and the grounding impedance of the surge arrester reduce the protecting performance of the surge arrester in the same manner.

13.8 TRANSMISSION-LINE TOWER GROUNDING

To design and operate a transmission system with a low-outage rating and safety to the maintenance personnel as well as the public, it is necessary to have a suitable grounding system. This can be furnished by overhead ground wires, counterpoises, and by earthing tower bodies and foundations.

The degree to which low tower footing resistance can be met depends on local soil conditions. The method used to reduce the equivalent footing resistance and the degree to which this is carried out is a matter of economics. Experience indicates that some means of reducing the footing resistance to an equivalent of 10 Ω, as measured with the ground wires removed, is more economical than adding extra insulation.

Unfortunately, improved grounding cannot be economically effective in the event of direct lightning strokes to phase conductors. The importance of having low tower footing resistance in this case is in avoiding a high rate of back flashovers and thereby improving the conditions for successful fault suppression by ground-fault neutralizers. Ground wires are so located as to shield the line conductors adequately from direct lightning strokes. Their design is discussed in detail elsewhere (IEEE Committee, 1975, 1978).

With underground cables the situation is different. Dangerous voltages to earth may result from insulation failure, charges due to electrostatic induction, flow of currents through the sheath, or from the voltage rise during discharging of faults to the station ground system to which the sheaths are connected.

13.9   SAFETY GROUND

Safety ground is meant to ensure that persons working with electrical equipment will not be exposed to the danger of electric shocks. Safety of operating personnel requires grounding of all exposed metal parts of power equipment. There is no simple relation between the resistance of the ground system as a whole and the maximum shock current to which a person might be exposed. Thus a low station ground resistance is not in itself a guarantee of safety (Sverak et al., 1981; El-Kady and Vainberg, 1983).

13.9.1   Range of Tolerable Currents

The effects of an electric current passing through the vital parts of a human body depend on its duration, magnitude, and frequency. Human beings are most vulnerable to currents with frequencies of 50–60 Hz. The human body can tolerate slightly larger currents at 25 Hz and approximately five times larger direct currents and still larger currents at 3–10 kHz (IEEE Committee, 1982).

 Currents of 1–6 mA, often termed "let-go currents," although unpleasant to sustain, do not impair a person's ability to control his or her muscles. Currents of 9–25 mA may be quite painful and can make it hard to release energized objects grasped by the hand. For still higher currents, uncontrollable muscular contractions can make breathing difficult. In the range of 100 mA ventricular fibrillation, stoppage of the heart, or inhibition of respiration may occur and cause severe injury or death. It has been observed (Sverak et al., 1981) that for the same effect the current magnitude I varies with its duration t according to a relation of the form $I \approx t^{-1/2}$. High-speed clearing of ground faults is evidently crucial, as the risk of electric shock will thus be greatly mitigated.

 To avoid danger from electric shock, one's body should under no circumstances be a part of an electric circuit. There are several means of reducing the hazard of electric shock, including grounding, isolation, guarding, insulation and double insulation, shock limitation, isolation transformers, and employing high-frequency and direct current. The first four methods are normally adopted in high-voltage systems.

13.9.2   Tolerable Step, Touch, and Transferred Voltages

Using the magnitude of the tolerable body current and the appropriate circuit constants, it is possible to calculate the tolerable potential difference between possible points of contact. The step voltage (Fig. 13.10) increases the closer one gets to the site of a ground fault or to the grounding point. If the enclosure of grounded equipment in which an earth fault has occurred is touched by a person, he or she will be subjected to a potential termed a "touch voltage" (Fig. 13.10).

When a person standing within the station area touches a conductor grounded at a remote point, or when a person standing at a remote point touches a conductor connected to the station ground grid, he or she is subjected to a transferred potential. The shock voltage in this case may be equal to the full voltage rise of the ground grid under fault conditions (El-Kady and Vainberg, 1983). The tolerable step and touch voltages with durations t for persons weighing 50–70 kg can be estimated as

$$E_{\text{step}} = [1000 + 6C(\rho_s)]\frac{0.116}{\sqrt{t}} \tag{13.22}$$

$$E_{\text{touch}} = [1000 + 1.5C(\rho_s)]\frac{0.16}{\sqrt{t}} \tag{13.23}$$

where C is a factor depending on the soil homogeneity and equals 1 for uniform soil.

13.9.3   Electric Shock Hazard in Hospitals

Increased application of electric instrumentation has greatly increased the risk of electric shock, especially when a patient in a hospital is actually connected into the circuit. The hazard increases when probes or needles are inserted into the body. Sometimes these are placed directly on or inside a heart chamber. Some medical authorities believe that a current as small as $20\,\mu\text{A}$ at 50–60 Hz applied directly to the heart can produce ventricular fibrillation. The hazard is increased if pacemakers are used.

13.10   METHODS OF DECREASING GROUND RESISTANCE

Decreasing the ground resistance of a grounding system in high-resistivity soil is often a formidable task. Recently, some new methods have been proposed to decrease ground resistance. Three methods are listed in the following section.

13.10.1   Cracks with Low-Resistivity Materials (LRM)

This method requires 3 steps: drilling deep holes in the ground, developing cracks in the soil by means of explosions in the holes, filling the holes with low-resistivity materials (LRM) under pressure (Meng et al., 1999). Most of the cracks around the vertical conductors will be filled with LRM, and a complex network of low-resistivity tree-like cracks linked to the substation grid is formed (Fig. 13.14). Field tests show that the optimum span between vertical conductors is in the range of 1.5–2 times the length of the vertical conductor. This method is effective in reducing ground resistances in rocky areas.

13.10.2   Chemical Rods

Chemical rods are electrodes with holes along their length, filled with mineral salts. The specially formulated mineral salts are evenly distributed along the entire length of the electrode (Fig. 13.15). The rod absorbs moisture from both air and soil. Continuous conditioning of a large area insures an ultra-low-resistance ground which is more effective than a conventional electrode. If the conductive salts are running low, the rod can be recharged with a refill kit. These rods are available in vertical and horizontal configurations. They may be used in rocky soils, freezing climates, dry deserts, or tropical rain forests. They provide stable protection for many years.

13.10.3   Grounding Augmentation Fill (GAF)

About 95% of the grounding resistance of a given electrode is determined by the character of the soil within a hemisphere whose radius is 1.1 times the length of the rod. It is obvious that replacing all or part of that soil with a highly conductive backfill will facilitate the achievement of a low-resistance ground connection. The greater the percentage of soil replaced, the lower the ultimate grounding resistance.

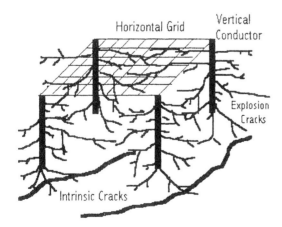

Figure 13.14 Grounding system with explosion and intrinsic cracks.

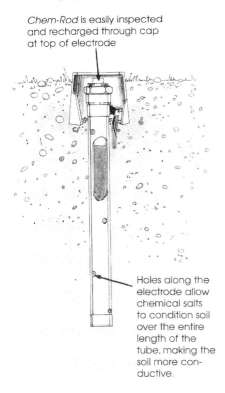

Chem-Rod is easily inspected and recharged through cap at top of electrode

Holes along the electrode allow chemical salts to condition soil over the entire length of the tube, making the soil more con- ductive.

Figure 13.15   Chemical rod. (Courtesy of LEC, Inc.)

13.11   MEASUREMENT OF GROUND RESISTANCE

Ground resistance measurement consists of measuring the resistance of a body of earth surrounding a grounding electrode system. One end of the resistance is definitely available at the grounding system itself, while the other end is not practically available and is called "remote soil". In practice, however, about 98% of the total resistance is contained within a finite distance from the grounding system. The power-frequency reactances of the grounding system can safely be neglected unless its ohmic resistance is extremely low. The resistance is measured using alternating current to avoid possible polarization effects under direct current (IEEE, 1983; Dawalibi and Blattner, 1984). The practical and reliable method for measuring the resistance of a ground is that of "fall of potential."

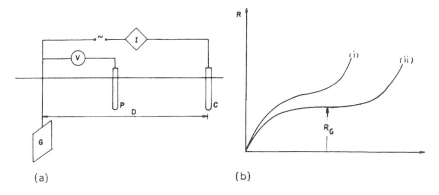

Figure 13.16 Ground resistance measurement by the fall-of-potential method: (a) circuit connection; (b) resistance plots for the electrode separation D, being too short (i) and adequate (ii).

13.11.1 Fall-of-Potential Method

This method has several variations and is applicable to all types of ground resistance measurements. It involves passing through the grounding system G a current I to return from another electrode C (Fig. 13.16a). The passage of this current produces at a distance x from G a voltage drop V_x in the soil. V_x is measured by a potential probe P. The simplest form of the fall-of-potential method is obtained when G, P, and C are on a straight line and P is located between G and C. When V_x/I is plotted as a function of the potential probe distance x, curves similar to those shown in Figure 13.16b are produced. If the distance D is large enough with respect to the grounding system dimensions, the center part of the fall-of-potential curve tends to be flat. It is usually accepted that the flat section of the curve gives the correct magnitude of the resistance R_G.

13.12 MEASUREMENT OF EARTH RESISTIVITY

Soil resistivity tests should be carried out before designing a grounding system. Since the performance and cost of any grounding electrode or ground grid is directly related to soil resistivity, accurate evaluation is a matter of considerable importance.

13.12.1 Four-Point Method

The most common method of measuring soil resistivity employs four electrodes driven into the soil. The theoretical basis of this method has been derived. Small electrodes are inserted into four small holes in the earth all to a depth of b meters and spaced along a straight line at intervals of a meters, and making electrical contact with the earth only at the bottom. A test current I is injected through the earth between the two outer electrodes and the potential E between the two inner electrodes is measured with a potentiometer or high-impedance voltmeter. The ratio between the observed potential and the injected current is referred to as the apparent resistance, which is a function of soil resistivity and the electrode geometry. The apparent soil resistivity is computed by multiplying the apparent resistance by a geometric multiplier. The apparent soil resistivity down to a depth equal to a is

$$\rho = \frac{4\pi a R}{1 + \frac{2a}{\sqrt{a^2 + 4b^2}} - \frac{a}{\sqrt{a^2 + b^2}}} \tag{13.24}$$

where R is the apparent resistance (Ω) and ρ is the soil resistivity ($\Omega \cdot m$). For spacing a much greater than depth b, it can be easily shown that equation (13.24) reduces to

$$\rho = 2\pi a R \tag{13.25}$$

13.12.2 Four-Long Cylindrical Electrodes

It is important to note that in the relation (13.25) the electrodes are considered as "points" buried at a depth b. The following new formula based on the field theory relating soil resistivity to measured resistance has been developed (Baishiki et al., 1987). It relaxes the requirement that rod electrodes have a large spacing compared to their buried length.

$$\rho = \frac{4\pi R b}{2 \ln[(2 + E)/(1 + F)] + 2F - E - a/b} \tag{13.26}$$

with

$$E = \sqrt{4 + \left(\frac{a}{b}\right)^2}$$

$$F = \sqrt{1 + \left(\frac{a}{b}\right)^2}$$

where b is the electrode driven length in meters. When $a \gg b$, equations (13.24) and (13.26) reduce to (13.25).

13.13   GROUND RESISTANCE METERS

Any instrument for measuring the resistance of a grounding point and the resistivity of soil mainly comprises an AC source of an adjustable frequency and a measuring circuit. DC current would not be suitable for such tests because of the entailed electrolysis and polarization, introducing serious errors. The frequency to be used should differ from that of any neighboring network; otherwise, any pickup would cause errors in the measurements.

Figure 13.17 shows a schematic of the circuit in common use, where a hand-driven DC generator G is coupled to a mechanical inverter CR and to a mechanical rectifier PR. The inverter inverts the DC current I to AC (of a frequency adjusted by the speed of cranking the generator) before it is fed into the ground. The rectifier PR converts the AC potential picked up between the electrodes P1 and P2 back to DC before it is fed to its respective coil in the ohmmeter CC. This "ohmmeter" compares the measured potential with the main circuit current I.

The circuit shown in Figure 13.17 is for measuring the soil resistivity according to the four-point method. For measuring the resistance of a

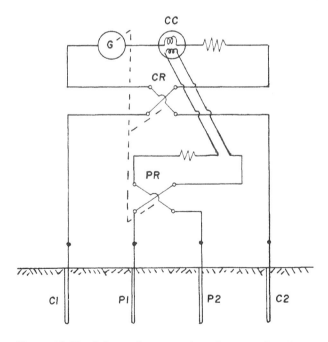

Figure 13.17   Schematic connection of a ground resistance meter used for measuring the ground resistivity by the four-electrode method.

grounding electrode already installed, it would be connected to the ground resistance meter at the two electrodes C1 and P1 (Fig. 13.17) combined. Some variations of the ground resistance meter use a battery instead of the generator G, and include electromechanical vibrators or solid-state devices for the current inversion and rectification.

13.14 PROBLEMS

(1) Calculate the ground resistance for the grounding system shown in Figure 13.18, given that the earth resistive $\rho = 100\,\Omega \cdot m$, the length of the driven rod is 8 m, and its diameter is 6 cm. Discuss the results.

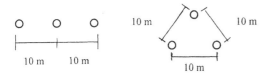

Figure 13.8

(2) Repeat the above problem for the grounding system shown in Figure 13.19.

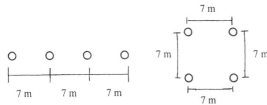

Figure 13.19

(3) A rod of radius 5 cm is driven in a clay soil of resistivity $90\,\Omega \cdot m$. Calculate its required length to have a ground resistance of $4\,\Omega$. If the short circuit current is 1500 A, calculate the step voltage for a man standing 7 m from the rod whose step length is 80 cm. Calculate also the touch voltage for a man with arm length 60 cm.

(4) In the above problem calculate the current passing through a man in both positions, if his resistance is $10,000\,\Omega$ and $500\,\Omega$ respectively. Comment on your results.

(5) A grounding system in a sandy clay soil is composed of five similar driven electrodes of 8 m length and 5 cm diameter, driven vertically at the points shown in Figure 13.20. If the earth resistivity is $80\,\Omega \cdot m$, calculate the actual earth resistance and the overlapping coefficient.

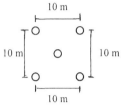

Figure 13.20

(6)   The earth resistivity was measured by the four-electrode method. The
      resistance R in ohms varied with the probe separation a as follows:

a (m)	12	24	40	100
R (Ω)	1.4	1.1	0.5	0.08

Calculate the average resistivity down to depths of 30, 50, and 80 m.

REFERENCES

Baishiki RS, Osterberg CK, Dawalibi F. IEEE Trans PWRD-2:64–71, 1987.

Brown P, Johnson I, Stevenson J. IEEE Trans PAS -97:683–694, 1978.

Dawalibi F, Blattner C. IEEE Trans PAS-103:374–382, 1984.

El-Kady M, Vainberg M. IEEE Trans PAS-102:3080–3087, 1983.

El-Morshedy A, Zeitoun AG, Ghourab MM. Proc IEE 133C:287–292, 1986.

Ershevich VV, Krivwshkin LF, Neklepaev BN, Sheimovich VD, Slavin GA. CIGRE
      paper 32-09, 1982.

Garett D, Holley H. IEEE Trans PAS-99:2008–2011, 1980.

Gulachenski EM, Courville EW. IEEE Trans PAS-103:2572–2578, 1983.

Gupta B, Thapar B. IEEE Trans PAS-99:2357–2362, 1980.

Huang L, Chen X, Yan H. IEEE Trans PWRD-10:716–722, 1995.

IEC. Publication 364, 1980.

IEEE. IEEE Recommended Practice for Grounding of Industrial and Commercial
      Power Systems. Publication 142. New York: Institute of Electrical & Electronics
      Engineers, 1972.

IEEE. IEEE Guide for Measuring Ground Resistance and Potential Gradients in the
      Earth. Publication 81. New York: Institute of Electrical & Electronics
      Engineers, 1983.

IEEE. IEEE Guide for Safety in Alternating Current Substation Grounding.
      Publication 80. New York: Institute of Electrical & Electronics Engineers, 1987.

IEEE Committee. IEEE Trans PAS-94:1241–1247, 1975.

IEEE Committee. IEEE Trans PAS-97:2243–2252, 1978.

IEEE Committee. IEEE Trans PAS-101:4006–4023, 1982.

Karaki S, Yamazaki Y, Nojima K, Yokota T, Murase H, Takahashi H, Kojima S.
      IEEE Trans PWRD-10:723–731, 1995.

Komaragiri K, Mukhedkar D. IEEE Trans PAS-100:2993–3001, 1981.

Lee HS, Kim JH, Dawalibi FP, Ma J. IEEE Trans PWRD-13:745–751, 1998.

Meng Q, He J, Dawalibi FP, Ma J. IEEE Trans PWRD-14:911–916, 1999.

Nahman J, Salamon D. IEEE Trans PAS-103:880–885, 1984.

Sunde ED. Earth Conduction Effects in Transmission Systems. New York: Macmillan, 1968.

Sverak JG, Dick WK, Dodds TH, Heppe RH. IEEE Trans PAS-100;4281–4290, 1981.

Sverak JG, Benson RU, Dick WK, Dodds TH, Garret DL, Idzkowski JE, Keil RP, Patel SG, Ragan ME, Smith GE, Verma R, Zukerman LG. IEEE Trans PAS-101:4006–4023, 1982.

Verma R, Mukhedkar D. IEEE Trans PAS-100:1023–1030, 1981.

14

Overvoltages on Power Systems

H. ANIS *Cairo University, Giza, Egypt*

14.1 INTRODUCTION

The examination of overvoltages on the power system includes a study of their magnitudes, shapes, durations, and frequency of occurrence. This study should be performed not only at the point where an overvoltage originates but also at all other points along the transmission network to which the surges may travel.

14.2 TYPES OF OVERVOLTAGE

The voltage stresses on transmission network insulation are found to have a variety of origins. In normal operation AC (or DC) voltages do not stress the insulation severely. However, they remain the initial factor that determines its dimensions. Overvoltages stressing a power system can generally be classified into two main types:

1. *External overvoltages*: generated by atmospheric disturbances. Of these disturbances, lightning is the most common and the most severe.

2. *Internal overvoltages*: generated by changes in the operating conditions of the network. Internal overvoltages can be divided into (a) switching overvoltages and (b) temporary overvoltages.

14.3   LIGHTNING OVERVOLTAGES

According to theories generally accepted, lightning is produced in an attempt by nature to maintain a dynamic balance between the positively charged ionosphere and the negatively charged earth (Marshall, 1973). Over fair-weather areas there is a downward transfer of positive charges through the global air–earth current. This is then counteracted by thunderstorms, during which positive charges are transferred upward in the form of lightning (Lewis, 1965).

During thunderstorms, positive and negative charges are separated by the movements of air currents forming ice crystals in the upper layer of a cloud and rain in the lower part. The cloud becomes negatively charged and has a larger layer of positive charge at its top. As the separation of charge proceeds in the cloud, the potential difference between the concentrations of charges increases and the vertical electric field along the cloud also increases. The total potential difference between the two main charge centers may vary from 100 to 1000 MV. Only a part of the total charge—several hundred coulombs—is released to earth by lightning; the rest is consumed in intercloud discharges. The height of the thundercloud dipole above earth may reach 5 km in tropical regions.

14.3.1   The Lightning Discharge

The lightning discharge through air occurs as one of the forms of streamer breakdown of long air gaps explained in Chapter 4. The channel to earth is first established by a stepped discharge called a leader stroke. The leader is generally initiated by a breakdown between polarized water droplets at the cloud base caused by the high electric field, or a discharge between the negative charge mass in the lower cloud and the positive charge pocket below it. In Figure 14.1 the development stages of a lightning flash are depicted.

As the downward leader approaches the earth, an upward leader begins to proceed from earth before the former reaches earth. The upward leader joins the downward one at a point referred to as the striking point. This is the start of the return stroke, which progresses upward like a traveling wave on a transmission line. At the earthing point a heavy impulse current reaching the order of tens of kiloamperes occurs, which is responsible for the known damage of lightning. The velocity of progression of the

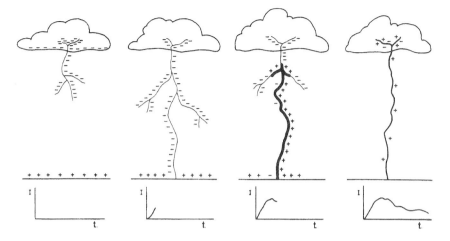

Figure 14.1 Developmental stages of a lightning flash and the corresponding current surge.

return stoke is very high and may reach half the speed of light. The corresponding current heats its path to temperatures up to 20,000°C, causing the explosive air expansion that is heard as thunder. The current pulse rises to its crest in a few microseconds and decays over a period of tens or hundreds of microseconds (Ragaller, 1980).

14.3.2 Lightning Voltage Surges

The most severe lightning stroke is that which strikes a phase conductor on the transmission line as it produces the highest overvoltage for a given stroke current. The lightning stroke injects its current into a termination impedance Z, which in this case is half the line surge impedance Z_0 since the current will flow in both directions as shown in Figure 14.2. Therefore, the voltage surge magnitude at the striking point is

$$V = \frac{1}{2}IZ_0 \tag{14.1}$$

The lightning current magnitude is rarely less than 10 kA (Berger et al., 1975) and thus, for a typical overhead line surge impedance Z_0 of 300 Ω, the lightning surge voltage will probably have a magnitude in excess of 1500 kV. Equation (14.1) assumes that the impedance of the lightning channel itself is much larger than $\frac{1}{2}Z_0$; indeed, it is believed to range from 100 to 3000 Ω. Equation (14.1) also indicates that the lightning voltage surge will have approximately the same shape characteristics. In practice, however, the

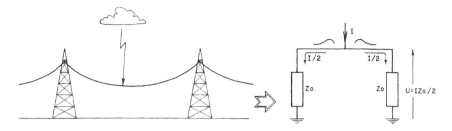

Figure 14.2   Development of lightning overvoltage.

shapes and magnitudes of lightning surge waves get modified by their reflections at points of discontinuity as they travel along transmission lines (Section 14.6).

Lightning strokes represent true danger to life, structures, power systems, and communication networks. Lightning is always a major source of damage to power systems where equipment insulation may break down under the resulting overvoltage and the subsequent high-energy discharge.

14.4   SWITCHING OVERVOLTAGES

With the steady increase in transmission voltages needed to fulfill the required increase in transmitted powers, switching surges have become the governing factor in the design of insulation for EHV and UHV systems. In the meantime, lightning overvoltages come as a secondary factor in these networks. There are two fundamental reasons for this shift in relative importance from lightning to switching surges as higher transmission voltages are called for:

1.   Overvoltages produced on transmission lines by lightning strokes are only slightly dependent on the power system voltages. As a result, their magnitudes relative to the system peak voltage decrease as the latter is increased.
2.   As shown in Chapter 15, external insulation has its lowest breakdown strength under surges whose fronts fall in the range 50–500 μs, which is typical for switching surges.

According to the International Electrotechnical Commission (IEC) recommendations, all equipment designed for operating voltages above 300 kV should be tested under switching impulses (i.e., laboratory-simulated switching surges) (IEC, 1973).

14.4.1 Origin of Switching Overvoltages

There is a great variety of events that would initiate a switching surge in a power network. The switching operations of greatest relevance to insulation design can be classified as follows:

1. *Energization of transmission lines and cables.* The following specific switching operations are some of the most common in this category:
   a. Energization of a line that is open circuited at the far end
   b. Energization of a line that is terminated by an unloaded transformer
   c. Energization of a line through the low-voltage side of a transformer
2. *Reenergization of a line.* This means the energization of a transmission line carrying charges trapped by previous line interruptions when high-speed reclosures are used.
3. *Load rejection.* This is effected by a circuit breaker opening at the far end of the line. This may also be followed by opening the line at the sending end in what is called a line dropping operation.
4. *Switching on and off of equipment.* All switching operations involving an element of the transmission network will produce a switching surge. Of particular importance, however, are the following operations:
   a. Switching of high-voltage reactors
   b. Switching of transformers that are loaded by a reactor on their tertiary winding
   c. Switching of a transformer at no load
5. *Fault initiation and clearing.*

14.4.2 Energization of an Unloaded Transmission Line

When an unloaded transmission line is switched on, the sinusoidal supply voltage is suddenly applied to it as represented by the single-phase circuit of Figure 14.3. The transformer is represented here by its leakage inductance and the line by its inductance and capacitance to ground. The switching operation is effected at an instant T seconds beyond that of zero voltage. The voltage across the capacitor C is the one under study here, as it represents the voltage at the open-circuit end of the line. The resistance R includes all series resistances of the line and transformer.

The circuit performance after switching may be expressed by the following differential equation:

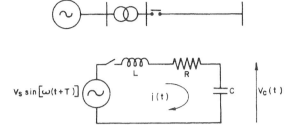

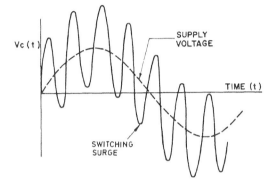

Figure 14.3 Energization switching transient.

$$v_s(t) = Ri(t) + L\frac{di(t)}{dt} + \frac{1}{C}\int i(t)\, dt \tag{14.2}$$

The supply voltage $v_s(t)$ beyond the switching instant is given by

$$v_s(t) = V_s \sin(\omega t + \omega T) \tag{14.3}$$

Equations (14.2) and (14.3) can easily be manipulated (e.g., by using operational calculus) and the expression for the voltage across the line capacitance takes the form

$$v_c(t) = V_c \sin(\omega t + \omega T - \theta) + A_e^{-\alpha t} \sin(\omega_1 t + \beta) \tag{14.4}$$

where

$$\theta = \tan^{-1}\frac{-R}{\omega L - 1/\omega C}$$

$$V_c = \frac{V_s}{\omega C\sqrt{R^2 + (\omega L - 1/\omega C)^2}}$$

$$A = -V_c \frac{\sin(\omega T - \theta)}{\sin \beta}$$

$$\beta = \tan^{-1} \frac{\omega_1 \sin(\omega T - \theta)}{\omega \cos(\omega T - \theta) + \alpha \sin(\omega T - \theta)}$$

$$\alpha = \frac{R}{2L} \qquad\qquad (14.5)$$

$$\omega_0 = \frac{1}{\sqrt{LC}}$$

$$\omega_1 = \sqrt{\omega_0^2 - \alpha^2} = 2\pi f_1$$

14.4.2.1 Example

A typical 132/500 kV, 200 MVA transformer would have an inductance (referred to its 500 kV side) of about 400 mH; a 500 kV transmission line may have a series inductance of about 1 mH/km and a capacitance to ground of about 0.015 μF/km. For a transmission line of 100 km, the total series inductance L of Figure 14.3 would be 500 mH and the capacitance C would be 1.5 μF. Neglecting all series resistances (i.e., letting $\alpha = 0$), the frequency of oscillation of the transient voltage component as given in equation (14.4) would be

$$f_1 = \frac{1}{2\pi\sqrt{LC}}$$

$$= 184 \text{ Hz}$$

This, in effect, means that the time to crest of the switching surge, which is about $1/4f_1$, is

$$t_{crest} = \frac{10^6}{4 \times 400}$$

$$= 1360 \text{ μs}$$

In Figure 14.3 the expected switching surge is sketched for the previous case when energization is made at $T = 0$.

14.5 TEMPORARY OVERVOLTAGES

Temporary overvoltages (i.e., sustained overvoltages) differ from transient switching overvoltages in that they last for longer durations, typically from a few cycles to a few seconds. They take the form of undamped or slightly damped oscillations at a frequency equal or close to the power frequency.

The classification of temporary overvoltages as distinct from transient switching overvoltages is due mainly to the fact that the responses of power network insulation and surge arresters to their wave shapes are different. Some of the most important events leading to the generation of temporary overvoltages are discussed briefly below.

14.5.1   Load Rejection

When a transmission line or a large inductive load that is fed from a power station is suddenly switched off, the generator will speed up and the busbar voltage will rise. The amplitude of the overvoltage can be evaluated approximately, as illustrated in Figure 14.4, by

$$V = E \frac{X_c}{X_c - X_s} \tag{14.6}$$

where E is the voltage behind the transient reactance, which is assumed to be constant over the subtransient period and equal to its value before the incident, X_s the transient reactance of the generator in series with the transformer reactance, and X_c the equivalent capacitive input reactance of the system.

14.5.2   Ferranti Effect

The Ferranti effect of an uncompensated transmission line is given by

$$\frac{V_r}{V_s} = \frac{1}{\cos \beta_0 l} \tag{14.7}$$

where V_r and V_s are the receiving-end and sending-end voltages, respectively, and ℓ is the line length (km). β_0 is the phase shift constant of the

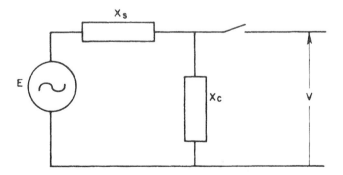

Figure 14.4   Equivalent circuit during load rejection.

line per unit length. It is equal to the imaginary part of $\sqrt{ZY}$, where Z and Y are the impedance and admittance of the line per unit length. For a lossless line $\beta_0 = \omega\sqrt{LC}$ where L and C are the inductance and capacitance of the line per unit length. β_0 has a value of about 6° per 100 km at normal power frequency (Diesendorf, 1974).

14.5.3 Ground Fault

A single line-to-ground fault will cause the voltages to ground of the healthy phases to rise. In the case of a line-to-ground fault, systems with neutrals isolated or grounded through a high impedance may develop overvoltages on healthy phases higher than normal line-to-line voltages. Solidly grounded systems, on the other hand, will only permit phase-to-ground overvoltages well below the line-to-line value. An earth fault factor is defined as the ratio of the higher of the two sound phase voltages to the line-to-neutral voltage at the same point in the system with the fault removed.

14.5.4 Harmonic Overvoltages Due to Magnetic Saturation

Harmonic oscillations in power systems are initiated by system nonlinearities whose primary source is that of the saturated magnetizing characteristics of transformers and shunt reactors. The magnetizing current of these components increases rapidly and contains a high percentage of harmonics for voltages above the rated voltage. Therefore, saturated transformers inject large harmonic currents into the system.

14.6 TRAVELING WAVES

Any overvoltage surge appearing on a transmission line due to internal or external disturbances will propagate in the form of a traveling wave toward the ends of the line. During its travel the overvoltage surge will, in general, experience attenuation and distortion inflicted upon it by earth resistance, line losses, and corona discharges.

To be able to develop the time equation of the traveling wave, a transmission line may be looked upon as being made up of a chain of lumped components such as that shown in Figure 14.5, in which

$R =$ line resistance (Ω/m)

$L =$ line inductance (H/m)

$C =$ line capacitance-to-ground (F/m)

$G =$ line leakage conductance (S/m)

$dx =$ length of a line element (m)

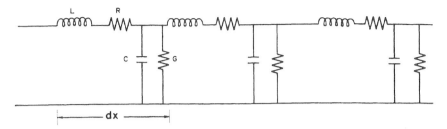

Figure 14.5 Transmission-line equivalent circuit.

The magnetic flux linkage produced by the wave's current is given by

$$\phi = IL\,dx \tag{14.8}$$

Consequently, the total voltage drop over an element dx is given by

$$-dv = \left(R + L\frac{\partial}{\partial t}\right)i\,dx \tag{14.9}$$

The total change in wave current is caused by the leakage component $vG\,dx$ and the current used to charge the line capacitance to ground. Therefore, the total current change takes the form

$$-di = \left(G + C\frac{\partial}{\partial t}\right)v\,dx \tag{14.10}$$

By differentiation, we easily obtain

$$\frac{\partial^2 v}{\partial x^2} = [(R + Ls)(G + Cs)]v \tag{14.11}$$

and

$$\frac{\partial^2 i}{\partial x^2} = [(R + Ls)(G + Cs)]i \tag{14.12}$$

s being the differential operator.

Equations (14.11) and (14.12) are the known transmission-line traveling-wave equations, in which the term $(R + Ls)$ is the impedance operator Z, and $(G + Cs)$ is the admittance operator Y. The solution to equations (14.11) and (14.12) has the general form

$$v = e^{x\Gamma}F_1(t) + e^{-x\Gamma}F_2(t) \tag{14.13}$$

and

$$i = -\sqrt{\frac{Y}{Z}}[e^{x\Gamma}F_1(t) - e^{-x\Gamma}F_2(t)] \tag{14.14}$$

where the operator Γ is defined as

$$\Gamma = \sqrt{ZY} \tag{14.15}$$

The time functions $F_1(t)$ and $F_2(t)$ are integration constants with respect to x only. The exact solution to equations (14.11) and (14.12) depends on the nature of Γ (i.e., on the line constants) and on the boundary conditions of the wave. The operator Γ can easily be shown to take the form

$$\Gamma = \frac{1}{\gamma}\sqrt{(s+\varsigma)^2 - \eta^2} \tag{14.16}$$

in which γ has the dimensions of velocity (m/s) and

$\varsigma = \frac{1}{2}\left(\frac{R}{L} + \frac{G}{C}\right)$ is the attenuation constant

$\eta = \frac{1}{2}\left(\frac{R}{L} - \frac{G}{C}\right)$ is the wavelength constant

14.6.1 Lossless Line Equation

An ideal transmission line with no losses should have both $R = 0$ and $G = 0$. Then the operator Γ reduces to

$$\Gamma = s\sqrt{LC} = \frac{s}{\gamma} \tag{14.17}$$

while both the attenuation and wavelength constants vanish. Remembering that, according to Taylor's expansion,

$$e^{\pm xs/\gamma}F(t) = F\left(t + \frac{x}{\gamma}\right)$$

the voltage and current waves along a lossless line take the general forms

$$v = F_1\left(t + \frac{x}{\gamma}\right) + F_2\left(t - \frac{x}{\gamma}\right) \tag{14.18}$$

$$i = -\sqrt{\frac{C}{L}}\left[F_1\left(t + \frac{x}{\gamma}\right) - F_2\left(t - \frac{x}{\gamma}\right)\right] \tag{14.19}$$

This solution, as it is expressed in terms of $(t \pm x/\gamma)$, indicates that the wave is traveling along x at a velocity γ while maintaining its shape such that for a given value of $(t \pm x/\gamma)$ the wave's instantaneous value is unchanged.

The voltage (or current) wave is also seen in equations (14.18) and (14.19) to be made up of a component F_2 traveling forward along x and a component F_1 traveling backward (Fig. 14.6). The magnitudes of the voltage and current waves are related as follows:

For the forward component

$$\frac{v}{i} = \sqrt{\frac{L}{C}}$$

For the backward component

$$\frac{v}{i} = -\sqrt{\frac{L}{C}}$$

This means that the voltage and current waves traveling along the line have a fixed ratio, termed the line's surge impedance Z_0.

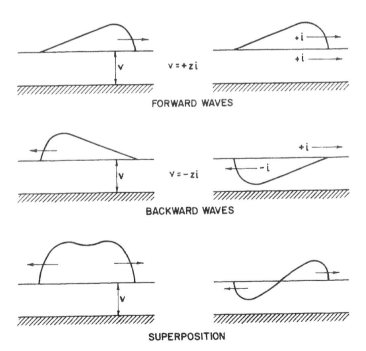

FORWARD WAVES

$v = +zi$

BACKWARD WAVES

$v = -zi$

SUPERPOSITION

Figure 14.6   Voltage and current traveling-wave components.

$$Z_0 = \sqrt{\frac{L}{C}} \tag{14.20}$$

The surge impedance is clearly independent of the line length. In practice, it is about 300–400 Ω for overhead transmission lines and about 30–80 Ω for underground cables.

14.6.2   Velocity of Wave Propagation

For a lossless transmission line, it can easily be shown from equation (14.9) that

$$\frac{\partial v}{\partial x} = -L\frac{\partial i}{\partial t} \tag{14.21}$$

Similarly,

$$\frac{\partial i}{\partial x} = -C\frac{\partial v}{\partial t} \tag{14.22}$$

Realizing that dx/dt is the velocity at which the surge propagates, an expression for that velocity may easily be deduced from equations (14.21) and (14.22) as

$$\gamma = \frac{1}{\sqrt{LC}} \tag{14.23}$$

As we remember, the inductance L per unit length of a conductor with radius r and height h above ground is approximately given by

$$L = \frac{\mu_0}{\pi}\ln\frac{2h}{r} \qquad \text{H/m} \tag{14.24}$$

Also, the capacitance C per unit length of the same conductor arrangement is given by

$$C = \frac{\pi\varepsilon_0}{\ln(2h/r)} \qquad \text{F/m} \tag{14.25}$$

where μ_0 and ε_0 are the permeability and permittivity of free space, respectively, Therefore, the velocity of wave propagation is given by

$$\gamma = \frac{1}{\sqrt{\mu_0\varepsilon_0}} \tag{14.26}$$

which is equal to the speed of light and is independent of line geometry. At that speed a surge will travel 1 km in about 3.3 μs.

In the case of cables, the dielectric constant ε_r of the insulation is larger than unity whereas the permeability remains very much equal to μ_0.

Therefore, the velocity of wave propagation should be expected to be smaller than the speed of light by a factor of $\sqrt{\varepsilon_r}$.

14.6.3 Reflection and Refraction of Traveling Waves

When the traveling wave on a transmission line reaches a point beyond which the line constants are different (e.g., a cable connected to an overhead line or where equipment or more lines are connected), a part of the wave is reflected back along the line (the reflected component) and another part is passed on to the new section (the refracted component). Consider a junction between two lines whose surge impedances are Z_1 and Z_2, such as that shown in Figure 14.7. A surge that is traveling along line 1 has its voltage and current related by

$$\frac{v_1}{i_1} = Z_1 \tag{14.27}$$

At the junction, the reflected and refracted waves will have voltages v_1' and v_2 and currents i_1' and i_2, respectively. These voltages and currents are related by

$$i_1' = \frac{-v_1'}{Z_1} \tag{14.28}$$

and

$$i_2 = \frac{v_2}{Z_2} \tag{14.29}$$

Simply, applying Kirchhoff's laws at the junction yields

$$v_1 + v_1' = v_2 \tag{14.30}$$

and

$$i_1 + i_1' = i_2 \tag{14.31}$$

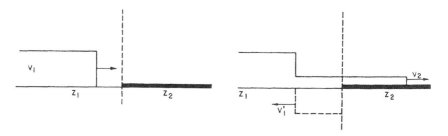

Figure 14.7 Wave reflection at a junction.

Therefore,

$$\frac{v_1}{Z_1} - \frac{v_1'}{Z_1} = \frac{v_2}{Z_2} \tag{14.32}$$

The reflected and refracted wave voltages can finally be put in the form

$$v_1' = \frac{Z_2 - Z_1}{Z_2 + Z_1} v_1 \tag{14.33}$$

$$v_2 = \frac{2Z_2}{Z_2 + Z_1} v_1 \tag{14.34}$$

The quantity $(Z_2 - Z_1)/(Z_2 + Z_1)$ is termed the reflection coefficient, and the quantity $2Z_2/(Z_2 + Z_1)$ is the refraction (transmission) coefficient.

14.6.4 Lattice Diagram

The successive reflection of a traveling wave at the sending and receiving ends of a transmission line may lead to a buildup in the voltage or current at some points along the line. A method—normally called the lattice diagram (Bewley, 1963)—is applied in Figure 14.8 to the simple case of a lossless transmission line with surge impedance Z_0 connected at the sending end to an infinite bus (i.e., one that has a zero internal impedance) and terminated at the receiving end by a load impedance Z. The time–space diagram following the issuing of a voltage surge at the sending end of 1 p.u. is illustrated. The time taken by the surge to cross the entire line is constant and denoted by T. Incident waves at the receiving end are reflected with a reflection coefficient of

$$a = \frac{Z - Z_0}{Z + Z_0}$$

Subsequent reflections at the sending end occur with a reflection coefficient of -1.

The voltage buildup at any chosen point along the line can be estimated by the algebraic summation of the voltage surge magnitudes after each surge reflection. For example, the voltage at the receiving end (R.E.) of the system shown in Figure 14.8 can be evaluated to take the form indicated for the case:

$$\text{R.E. voltage } [rT < t < (r+2)T] = (1-a)(1-a+a^2-a^3+\cdots) \tag{14.35}$$

After a sufficiently long time, the receiving-end voltage in this case settles at a final value of

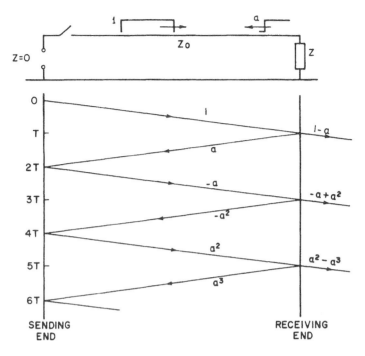

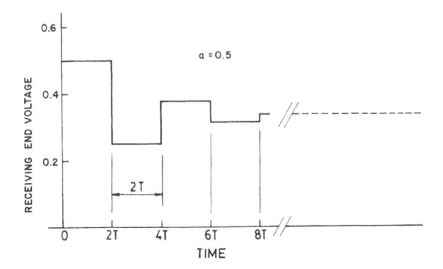

Figure 14.8   Lattice diagram.

$$\text{R.E. voltage (final)} = \frac{1-a}{1+a} \qquad \text{p.u.} \tag{14.36}$$

as shown in Figure 14.8.

14.6.5 Approximations

The discussion above assumed that the line is a lossless single conductor passing over a perfectly conducting ground. In reality, a traveling wave suffers from attenuation and distortion, due to energy loss and corona effects. For steep wavefronts, skin effect in conductors and ground return paths produce some attenuation and distortion (Semlyen, 1981). Corona discharge has the effect of retarding the portion of the wavefront above corona inception voltage, as discussed in Chapter 5 (Abdel-Salam and Stanek, 1987).

Multiconductor lines further distort a traveling wave. The voltage can be looked upon as made up of line components and ground components, each having different velocities of propagation and different rates of attenuation. For ground components, the electrostatic charges induced in the ground are near the surface, whereas the return current is well below the surface. The depth of the current return path depends on frequency and soil resistivity. For a perfectly conducting earth the return current flows at the ground surface at a velocity equal to the speed of light. With finite earth conductivity the conductor's image goes well below the surface and the velocity of propagation is reduced. Distortion in this case results from the different velocities of the two voltage components (Ametani and Schinzinger, 1976; Nakagawa, 1981; Cristina and d'Amore, 1983).

14.7 CONTROL OF OVERVOLTAGES

The adverse effects of overvoltages on power networks can be reduced in two ways: by using protective devices—chiefly surge arresters—or by reducing their magnitudes wherever the surge originates. The latter way is commonly known as overvoltage control. The techniques employed to control switching surges and temporary overvoltages are outlined briefly below.

14.7.1 Control of Switching Surges

Following are some of the techniques currently in use to control the magnitudes of switching surges.

14.7.1.1 Resistor Switching

This is one of the most common methods for reducing energization over-voltages. It is effected by initially applying the supply voltage to the line through a resistor. After a suitable period of time, normally between one-third and one-half of a cycle, the preinserted resistor is short circuited, allowing the full supply voltage to be applied to the line. By the end of the preinsertion period, the magnitude of the energization surge is usually much reduced by the effect of system damping. This effect is evident from equation (14.4), which describes the voltage surge waveform. The initial amplitude of the energization surge when a preinsertion resistor of value R is used will be only $Z_0/(R + Z_0)$ of that reached in the absence of the resistor, where Z_0 is the surge impedance of the line.

When the resistor is shorted at the end of the preinsertion period, another surge will develop. If R is too small, control of the first surge becomes ineffective; if it is too large, the second surge becomes dangerous. An optimal value of R would normally be a fraction of Z_0, and depends on transmission-line length.

14.7.1.2 Phase-Controlled Closure

Referring to equation (14.4), it can be seen that the amplitude of the ener-gization surge depends on the switching phase angle ωT. By properly timing of the closing of the circuit breaker poles, the resulting switching overvol-tage can be greatly reduced. Phase-controlled switching should be carried out successively for the three poles to accomplish a reduction in the initial voltages on all three phases. This is extremely difficult with conventional circuit breakers but is quite possible with solid-state circuit breakers, as explained in Chapter 11.

14.7.1.3 Use of Shunt Reactors

Shunt reactors are used on many high-voltage transmission lines as a means of shunt compensation to improve the performance of the line, which would otherwise draw large capacitive currents from the supply. They have the additional advantage of reducing energization surge magnitudes. This is accomplished mainly by the reduction in temporary overvoltages, as will be seen in the next section.

14.7.1.4 Drainage of Trapped Charges

Charges are trapped on the capacitance to ground of transmission lines after their sudden reenergization. If the line is reenergized soon after, usually by means of automatic reclosures, these charges may cause an increase in the resulting surge. If, in the simple system of Figure 14.3, the capacitance C has an initial voltage $V_c(0) = V_0$ caused by trapped charges, the surge voltage will include an extra component V_0 which, if the same polarity as the surge's peak voltage, will increase the overvoltage on the line (Bickford and Doepel, 1967).

In practice, trapped charges may be partially drained through the switching resistors incorporated in circuit breakers, as mentioned earlier. Magnetic-type potential transformers also drain trapped charges via a low-frequency oscillation which is highly damped by the effect of magnetic saturation.

14.7.2   Control of Temporary Overvoltages

Referring to Figure 14.4 and its related overvoltage equation (14.6), it is evident that by increasing the capacitive input reactance of the transmission line X_c the magnitude of the temporary overvoltage V is reduced. If a shunt reactor of reactance X_r is added to the transmission line, the equivalent input reactance of that line will be increased from X_c to

$$X_c' = \frac{X_c}{1 - X_c/X_r} \tag{14.37}$$

In equation (14.37), the goal of increasing X_c is obviously achieved by means of decreasing X_r (down from its infinite value in the absence of the reactor), thus reducing the overvoltage magnitude according to equation (14.6).

Furthermore, the second harmonic component of temporary overvoltages can be successfully suppressed, or even eliminated, by the use of surge arresters with nonlinear resistor (varistor) characteristics (Ragaller, 1980). A properly designed varistor would conduct in such a way as to provide large losses at the frequency in question.

14.8   STATISTICAL CHARACTERISTICS OF OVERVOLTAGES

All types of overvoltages on a power system are subject to statistical fluctuation in their magnitudes, shapes, and durations. The sources of randomness can generally be grouped into two main categories:

1. Randomness produced simply by the existence of a large variety of overvoltage-producing events
2. Randomness created inherently in the overvoltage-producing process

Examples of the latter category are outlined in the following sections.

14.8.1 Statistical Variations in Lightning Surges

Many factors contribute to fluctuations in the lightning stroke current, which will, in turn, reflect on the consequent overvoltages. The distribution of thunderclouds in the area, the size and charge distribution within each cloud, the weather conditions (e.g., wind, temperature, pressure, etc.), and the distribution of projected objects in the area are among those factors. Figure 14.9 depicts a cumulative frequency distribution of lightning current magnitudes. It has been reported that, in general, higher currents are associated with longer fronts. The marked statistical fluctuation in lightning surges emphasizes the importance of taking this fluctuation into account when dimensioning the insulation and evaluating the risk of insulation failure.

14.8.2 Statistical Variations in Switching Surges

Since all switching surges are products of circuit breaker actions, the random variation in switching surges is attributed primarily to the performance of these devices, as well as to the network circumstances at the time of switching. Random angles of switching are mainly responsible for this variation. The following factors contribute to the randomness in switching angles:

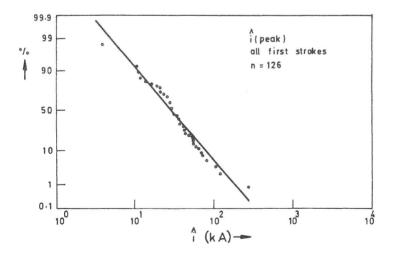

Figure 14.9 Probability distribution of lightning peak current. [From Berger et al. (1975).]

1. The mechanical movement of circuit breaker contacts produces fluctuations about the aiming angle of interruption or closure.
2. Circuit closure may be prematurely effected following a breakdown between the breaker's contacts.
3. Arc interruption, whose timing is relevant to the production of recovery transient voltages, involves a number of physical processes, many of which are inherently random, as described in Chapter 6.

Figure 14.10 shows typical distributions of switching surge magnitudes. Uncontrolled surges may fluctuate in magnitude at a coefficient of variation (standard deviation relative to the mean) of about 25%, while proper control measures can reduce this fluctuation to only about 12%.

14.8.3 Statistical Variation in Temporary Overvoltages

Temporary overvoltages are not subject to randomness in the same sense that switching surges are. Any change in temporary overvoltage magnitude would have to be related to changes in network parameters. If these changes are considered, however, over a long period of time, a statistical distribution of overvoltage magnitudes may be produced. The distributions are highly skewed with very restricted variations—almost deterministic—at lower vol-

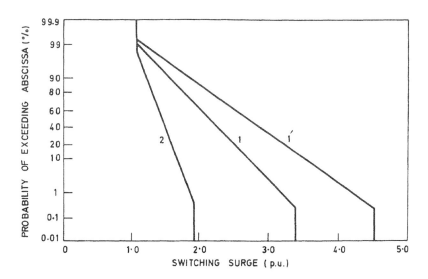

Figure 14.10 Switching surge magnitudes. 1 and 1′, uncontrolled surges; 2, controlled surges. [From Diesendorf (1974).]

tage magnitudes within 1.25 p.u. and a fairly wide variation, with more than a 15% coefficient of variation, at higher voltage magnitudes (Ragaller, 1980).

14.9   THE ELECTROMAGNETIC TRANSIENTS PROGRAM (EMTP)

The Electromagnetic Transients Program (EMTP) is a computer program for simulating electromagnetic, electromechanical, and control system transients on electric power systems (EMTP Development Coordination Group, 1996). The EMTP is used to solve the ordinary differential and/or algebraic equations associated with an "arbitrary" interconnection of different electrical (power system) and control system components. The implicit trapezoidal rule of integration is used in the discretization of the equations of most elements which are modeled by ordinary differential equations. The result is a set of real, simultaneous, algebraic equations which is solved at each time-step using advanced sparsity techniques. These equations are written in nodal-admittance form (with new unknown voltages as variables), and are solved by ordered triangular factorization. Initial conditions for differential equations of the various components can, for most cases of practical interest, be determined automatically by the program. The calculation of initial conditions is normally limited to linear elements. Nonlinear resistances are always ignored during the steady-state solution. Nonlinear reactances can either be linearized during steady state or fully modeled to include harmonic distortion effects. Injections of the electric network may also be specified in terms of power and voltage magnitude, thereby providing multiphase load flow capability.

Large coupled RLC networks, such as the internal transformer representation used by transformer manufacturers, can be manipulated internally without additional approximations or assumptions. The measured response of a power transformer can be used to create frequency-dependent transformer models using the High Frequency Transformer model. Support programs provide additional capabilities such as the calculation of overhead line and cable parameters, as well as the generation of more complex linear and nonlinear models for use in EMTR simulations.

Program output—both printed and plotted—consists of component variables (e.g., branch currents or voltages, machine torques or speeds, etc.) as functions of time.

14.9.1   EMTP Program Features

- Numerous studies can be carried out using the EMTP. They include: switching surges, lightning surges, insulation coordination, shaft torsional stress, high-voltage DC (HVDC), static VAR compensation, carrier frequency propagation, harmonics, ferroresonance, series and shunt resonance, motor starting, out-of-phase synchronization, islanding or other disturbance events, general control systems, grounding, asymmetrical fault current evaluation, phase conductor transposition, ground wire losses, general steady-state analysis of unbalanced systems, capacitor bank switching, and series capacitor protection.
- Series RLC branches are used to represent single-phase resistances, inductances, and capacitances on the basis of their respective voltage–current time relationships.
- Pi-equivalents are used to represent multiphase, coupled RLC circuits.
- A cascaded-pi input is used for untransposed transmission lines.
- Modeling of multiphase, coupled RL circuits is applied for power frequency transformers.
- Frequency-dependent representations of multiphase distributed-parameter transmission lines are used in which propagation time of the line is considered.
- Other representations are dynamic synchronous machines (3-phase balanced design only), unconventional rotating electromechanical energy converters, and control system dynamics.
- A multi-platform graphical user environment, EMTP View, has recently been used whereby data is entered using a circuit schematic diagram and free-format data entry forms. The input file contains the calculation time step, length of time to be simulated, and output requests, as well as the model data. The lumped branches are defined by resistance in Ω, inductance in mH or in Ω at power frequency, and capacitance in μF at power frequency. The simplest traveling-wave models can be defined by surge impedances, resistance per unit length, wave velocity, and line length for positive and zero sequence. Nonlinear elements are specified by current-and-voltage points for resistors and current-and-flux-linkage points for inductors.
- The primary output from a transient simulation takes the form of plotted bus voltages, branch voltages, branch currents, branch energy dissipation, machine variables, and control system variables.

14.9.2   EMTP Example Modeling

Two examples are shown in which the representation model is associated with the governing differential equations.

14.9.2.1   The Pi-Circuit Branch (Fig. 14.11)

This class of branches provides for the representation of lumped-element resistance, inductance, and capacitance matrices. For N conductors, the associated differential equations are

$$V_k - V_m = [L]\frac{di}{dt} + [R]di$$

$$i_m = \tfrac{1}{2}[C]\frac{dv_m}{dt} + i$$

$$i_k = \tfrac{1}{2}[C]\frac{dv_k}{dt} + i$$

All matrices are symmetric; $[C]$ is split in two, with half of the total on each end of the branch.

14.9.2.2   Mutually Coupled RL Branches (Fig. 14.12)

This is to represent lumped-element, mutually coupled RL branches. Associated with these branches are matrices $[L]$ and $[R]$ having performance equations

$$V_k - V_m = [L]\frac{di}{dt} + [R]i$$

Matrices $[R]$ and $[L]$ are assumed to be symmetric.

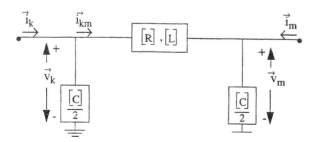

Figure 14.11   Representation of a pi-section.

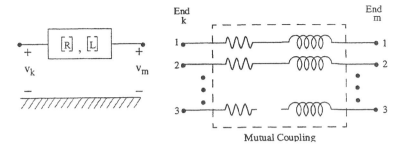

Figure 14.12 Representation of coupled RL elements.

14.9.3 Recent Applications of EMTP

The material given in the following sections has been reported as various extensions of the EMTP program.

14.9.3.1 Cable Switching Transients

Most approaches using EMTP start with steady-state impedance scans of the line or cable of interest and ignore wave reflections at the centers of conductors and at the sheath-to-earth transition. The steady-state starting points for the existing methods were said to fail to recognize the existence of transient conductor surface impedances. It was concluded that most of the complexity and uncertainties of present frequency-dependent modeling techniques would be eliminated simply by recognizing and modeling wave propagation in the conductors as well as along the dielectrics (Meredith, 1997). The approach was termed the method of finite sections to reflect its intermediate position between pi-section modeling and finite-element analysis, as seen in Figure 14.13. It was applied to coaxial cables, submarine and pipe-type cables, parallel pipes, multiphase transformer cores, and transformer tank walls. The finite sections approach, which amounts to a simplification of a finite-element approach where symmetry is obvious, is capable of accurately representing transient electromagnetic wave propagation in conductive materials.

Another study showed that the solid dielectric insulation used in underground distribution cables is subject to failure under the cumulative effect of switching voltage transients which therefore—contrary to standards—should be considered (Walling et al., 1995). The number of transients required to fail the cable increases as the applied surge magnitude is decreased. It was stated that the most accurate assessment of underground distribution system switching transients could be made by making field

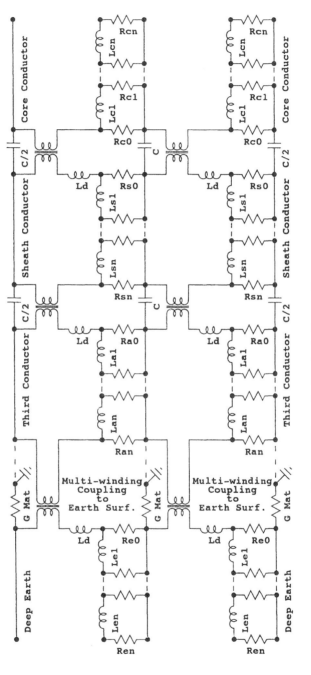

Figure 14.13   Finite section of multiple coaxial conductors including earth.

measurements in actual systems; Figure 14.14 was an example. Although numerical simulation techniques such as the EMTP, can predict the propagation of switching surges in the cable system with reasonable accuracy, the complex behavior of air-break switching devices cannot be predicted by electrical circuit simulation. The work showed that restriking does not lead to voltage escalation when load-break elbows and disconnects are operated because the resulting arc does not extinguish at transient current zeroes, but continues until the transient oscillations are damped. This traps a voltage on the switched cable which is essentially the same as the instantaneous source voltage.

14.9.3.2 Assessment of Very Fast Transients (VFT)

Very fast transients (VFTs) belong to the highest frequency range, i.e., from 100 kHz up to 50 MHz. They are mainly produced in gas-insulated substations (GIS). Fast transients can also accompany switching of vacuum breakers. In GIS, during disconnect switch operation, a number of prestrikes or restrikes occur due to the relatively slow speed of the moving contact. As the contacts close down, the electric field between them will rise until sparking occurs. This first strike will almost inevitably occur at the crest of the power frequency voltage, due to the slow operating. The current will flow through the spark and charge the capacitive load to the source voltage. As it does so, the potential difference across the contacts falls and the spark will eventually extinguish. The rise time of the voltage collapse during one strike is given by (Povh et al., 1996)

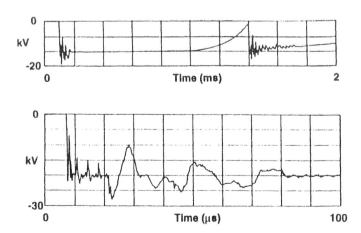

Figure 14.14 Cable end voltage measured during elbow closing energizing a 925 ft cable section.

$$t(\text{ns}) = 13.3 \frac{50}{\left(\dfrac{E}{P}\right)_0 Ph}$$

where $(E/P)_0$ is the dielectric strength of SF_6 which is equal to $860\,\text{kV}/(\text{cm}\cdot\text{MPa})$, P is the pressure in MPa, and h is a field utilization factor. Typically, for $P = 0.5\,\text{MPa}$ and $h = 0.2$, the rise time is about 7.7 ns. As these times are smaller than the transit time of the GIS components, traveling waves will be produced. The VFTs generated depend on the GIS configuration and wave multiple reflections. The main frequencies of the VFTs depend on the length of the GIS section affected by the disconnector switch operation and are in the range of one MHz up to 40 MHz for the basic component and even higher for the superimposed frequencies. The internal damping of the VFT influencing the highest frequency components is determined by the spark resistance; skin effects can be neglected due to the aluminum enclosure. The main portion of the damping of the VFT occurs by outcoupling at the transition to the overhead line. In the case of power transformers feeding the GIS, overall transients with frequencies in the range of 20–100 kHz can be observed, caused by the oscillation of the whole system consisting of the GIS and the transformer.

An example of these transient phenomena, measured in an actual GIS, is given in Figure 14.15 where one prestrike of a disconnector switching is oscillographed showing, for the load and the supply side, the steep voltage transients (a), the basic frequency component of the VFT in the MHz range (b), and the overall transients and the steady-state condition (c,d) (Povh et al., 1996). Normally the highest overvoltage stress is reached at the open end of the load side.

For the calculation of the VFT stress, the trapped charge remaining on the load side of the disconnector must be taken into consideration. For a normal disconnector with a slow speed the maximum trapped charge reaches 0.5 p.u., resulting in a most unfavorable voltage collapse of 1.5 p.u. For these cases the resulting overvoltages are in the range of 1.7 p.u. and reach 2.0 p.u. for very specific cases. In some cases of a high-speed disconnector the maximum trapped charge could be 1.0 p.u. and the highest overvoltages can reach values up to 2.5 p.u. In some cases extreme high values of more than 3 p.u. have been reported. It can be shown, however, that these values have been gained by calculations using unrealistic simplified simulation models.

14.9.3.3 Nonuniform Transmission Line Modeling

It is sometimes required to perform transient analysis of power network elements representing nonuniform transmission lines. An example is the

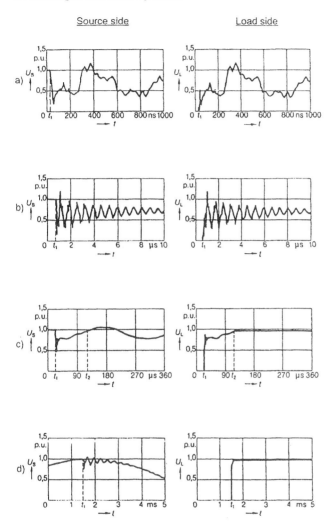

Figure 14.15 Transients on the source and load side of a GIS due to disconnector switching: (a) steep voltage transient; (b) basic frequency component of the VFT in the MHz range; (c) overall transients in the KHz range; (d) low frequency transients and steady-state condition.

lightning surge response of transmission line towers where the surge impedance of a transmission tower varies as the lightning surge travels along it. A nonuniform transmission line model suitable to interface with the Electromagnetic Transients Program (EMTP) was proposed by Correia de Barros et al. (1996). The model was based on a finite-differences algorithm to solve the wave equations for any space variation of the line parameters. The model included the transmission line losses, even when representing them as frequency-constant parameters.

14.9.3.4   EMTP Modeling of Relay Performance

The modeling and testing of relays for transient effects was not a critical issue until the introduction of electronic and microprocessor-based relays. Electromechanical relays are less sensitive to power system transient effects. Digital or sampled-data relays operate in a similar manner to the PC. A method to test the transient performance of a sampled-data relay was reported by Wilson et al. (1993). A model of one measuring unit of a digital relay was created on a personal computer. Computer modeling of the digital relay was done within the Models version of the Transient Analysis of Control System (TACS) subsection of the EMTP Program. The input filter, analog-to-digital converter, Fourier fundamental frequency detector, and the relay measuring principle of one digital microprocessor-based relay were modeled.

14.9.3.5   Transients in a Hybrid AC/DC Transmission System

Adding a DC line to an existing AC transmission circuit, or converting one circuit of a double circuit AC transmission system to DC, has been proposed as an effective method of significantly increasing the transmission capacity of the power corridor up to 200% of the total power-carrying capacity of the corridor. An electromagnetic transient simulation program (EMTDC) was used to investigate the overvoltages on a hybrid AC–DC transmission (Verdolin et al., 1995). The program had the capability of modeling transmission lines for coupling effects as well as the capability to represent HVDC converters in full detail and to represent transformer saturation nonlinearities.

14.9.3.6   Corona Effects on Transient Conditions

Corona is the main factor affecting the attenuation and distortion of travelling surges on transmission lines by modifying the value of the transversal line parameters. Surge propagation on multiphase transmission lines exhibiting corona was studied by Correia de Barros (1995) and Barros et al. (1995). The capacitance coefficients of multiphase transmission lines exhibiting corona were carefully identified. The capacitance coefficients were expressed in terms of space-charge influence factors that took into account the space-charge distribution around each conductor, which was determined by the local electric field distribution.

14.9.3.7 Closing Transients

Control algorithms for point-on-wave closing or auto-reclosing of shunt reactor-compensated transmission lines for reduction of closing overvoltages were reported by Froehlich et al. (1997). The algorithm was independent of the shape of the voltage signal to be synchronized on by means of pattern recognition. For certain system conditions, pre-insertion inductors may be ineffective for mitigating remote overvoltages due to capacitor energization (Bhargava et al., 1993).

14.9.3.8 Reflections on Transformers

The peak values of transient overvoltages may become particularly dangerous to insulation owing to the process of reflection and transmission of waves at discontinuities of surge impedance of transmission lines. Power transformers as terminations of transmission lines with traveling waves were studied by Hasman (1997). The model was based on measuring the input and transmission impedance of the transformer in a suitable frequency range, calculating the reflection and transmission coefficients of the transformer, transforming the incident wave from the time to the frequency domain, modification of every harmonic of the incident wave by the reflection and transmission coefficients, and finally by reverse transformation from the frequency to the time domain.

14.10   LIGHTNING-BASED TRANSIENT MODELING

The expected number of lightning strokes n (per 100 km) to an overhead line, according to the electrogeometrical model, is

$$n = 28 \ h^{0.6} N_g \times 10^{-1}$$

where h is the mean height of the line, and N_g is the density of lightning strokes per km^2 per year; if unknown, N_g may be given by

$$N_g = 0.04 \ T_d^{1.25}$$

where T_d is the isokeraunic level in thunderstorms–days per year.

The density of lightning strokes to ground was related to the record of "lightning flash counters" (LFC) N_c (discharges per year) (Popolanský et al., 1992) by

$$N_g = \frac{N_c}{3.430}$$

14.10.1   Electromagnetic Field Radiation

The measurements and analysis of electromagnetic field radiation from a lightning discharge are important, since the electromagnetic fields depend on the propagation and distribution of the return stroke current along the lightning path. The relationship between electromagnetic field radiation and return stroke current was studied and a calculation model of electromagnetic field radiation proposed by Motoyama et al. (1996). The proposed model consisted of the lightning current propagation and distribution model and the electromagnetic field radiation model. The model realistically assumed that the lightning current in the case of tall structures is initiated near the top of the structure. It also accounted for the propagation and distribution of the lightning current inside the lightning discharge path and the tall structure on which electromagnetic fields depend.

14.10.2   Surge Response Characteristics

To predict the lightning surge overvoltage that occurs in an electric power system using the EMTP, transmission towers need to be simulated on the basis of their surge response characteristics, namely, the surge impedance, the traveling wave propagation velocity, and the attenuation and deformation of the wave. The lightning surge response characteristics of a UHV transmission tower were measured by Yamada et al. (1995). The response waveform corresponding to a $1/70\,\mu s$ ramp-wave current was subsequently evaluated using Laplace transforms. A velocity of propagation of $300\,m/\mu s$ and a tower surge impedance of $120\,\Omega$ were reported for the multistory tower. The surge impedance for a UHV tower was approximately $130\,\Omega$.

14.11   PROBLEMS

(1)   A 3-phase single circuit transmission line is 400 km long. If the line is rated for 220 kV and has the parameters $R = 0.1\,\Omega/km$, $L = 1.26$ mH/km, $C = 0.009\,\mu F/km$, and $G = 0$, find the surge impedance and the velocity of propagation, neglecting the resistance of the line. IF a surge of 150 kV and an infinitely long tail strikes at one end of the line, what is the time taken for the surge to travel to the other end of the line?

(2)   A transmission line of surge impedance $500\,\Omega$ is connected to a cable of surge impedance $60\,\Omega$ at the other end. If a surge of 500 kV travels along the line to the junction point, find the voltage building up at the junction.

(3)   An underground cable of inductance 0.189 mH/km and of capacitance $0.3\,\mu F/km$ is connected to an overhead line having an inductance of 1.26 mH/km and capacitance $0.009\,\mu F/km$. Calculate the transmitted

and reflected voltage and current waves at the junction if a surge of 200 kV travels to the junction, along the cable, and along the overhead line.

(4) A transmission line has the following line constants: $R = 0.1\,\Omega/\text{km}$, $C = 0.009\,\mu\text{F/km}$, and $G = 0$. If the line is a 3-phase line and is charged from one end at a line voltage of 230 kV, find the rise in voltage at the other end if the line length is 400 km.

(5) An overhead power line with an inductanace of 1.0 mH/km and a capacitance of 0.01 μF/km is connected at "j" to a 100 m long underground cable whose total inductance and capacitance are 0.02 mH and 0.03 μF, respectively, and feeds a substation. A 500 kV surge was initiated at a distance of one km from the junction j. Estimate:

1. The transmitted and reflected voltages at j
2. The transmitted and reflected currents at j
3. The time (in μs) which the surge takes to reach the substation.

surge   j
over-head line   UG cable   sub-station

REFERENCES

Abdel-Salam M, Stanek E. IEEE Trans IA-23:481–489, 1987.
Ametani A, Schinzinger R. IEEE Trans PAS-95:773–781, 1976.
Barros H, Carneiro Jr S, Azevedo RM. IEEE Trans Power Deliv 10:1443–1449, 1995.
Berger K, Anderson RB, Kroninger H. Electra 41:23–37, 1975.
Bewley LV. Traveling Waves on Transmission Systems. New York: Dover Publications, 1963.
Bhargava B, Khan A, Imece A, Dipietro J. IEEE Trans Power Deliv 8:1226–1232, 1993.
Bickford JP, Doepel PS. Proc IEE 114:465–477, 1967.
Correia de Barros M. IEEE Trans Power Deliv 10:1642–1648, 1995.
Correia de Barros M, Almeida ME. IEEE Trans Power Deliv 11:1082–1087, 1996.
Cristina S, d'Amore M. IEEE Trans PAS-102:1685–1693, 1983.
Diesendorf W. Insulation Coordination in High Voltage Electric Power Systems. London: Butterworth, 1974.
EMTP Development Coordination Group. Electromagnetic Transients Program (EMTP96). Palo Alto, CA: Electric Power Research Institute (EPRI), 1996.
Froehlich K, Hoelzel C, Stanek M, Carvalho AC, Hofbauer W, Hoegg P, Avent BL, Peelo DF, Sawada JH. IEEE Trans Power Deliv 12:734–740, 1997.

Hasman T. IEEE Trans Power Deliv 12:1684–1689, 1997.

IEC. High Voltage Test Techniques. Publication 60:1, 2, 3. Geneva: International Electrotechnical Commission, 1973.

Lewis WW. The Protection of Transmission Systems against Lightning. New York: Dover Publications, 1965.

Marshall JL. Lightning Protection. New York: Wiley-Interscience, 1973.

Meredith RJ. IEEE Trans Power Deliv 12:489–496, 1997.

Motoyama H, Janischewskyim W, Hussein AM, Rusan RM, Chisholm WA, Chang JS. IEEE Trans Power Deliv 11:1624–1632, 1996.

Nakagawa M. IEEE Trans PAS-100:3626–3633, 1981.

Popolanský F, Špáček Z, Dvořak N. CIGRE report 23/33-04, 1992.

Povh D, Schmitt H, Völcker O, Witzmann R, Chewdhuri P, Imece AF Iravani R, Martinez JA, Keri A, Sarshar A. IEEE Trans Power Deliv 11:2028–2035, 1996.

Ragaller K, ed. Surges in High-Voltage Networks. New York: Plenum Press, 1980.

Semlyen A. IEEE Trans PAS-100:848–856, 1981.

Verdolin R, Gole AM, Kuffel E, Diseko N, Bisewski B. IEEE Trans Power Deliv 10:1514–1522, 1995.

Walling RA, Melchior RD, McDermott BA. IEEE Trans Power Deliv 10:534–539, 1995.

Wilson RE, Nordstrom JM. IEEE Trans Power Deliv 8:984–989, 1993.

Yamada T, Mochizuki A, Sawada J, Zaima E, Amettani A, Ishii M, Kato S. IEEE Trans Power Deliv 1:393–402, 1995.

15

Insulation Coordination

H. ANIS *Cairo University, Giza, Egypt*

15.1 INTRODUCTION

According to the International Electrotechnical Commission (1976), "insulation coordination comprises the selection of the electric strength of equipment and its application in relation to the voltages which can appear on the system for which the equipment is intended, and taking into account the characteristics of available protective devices, so as to reduce to an economically and operationally acceptable level the probability that the resulting voltage stresses will cause damage to equipment insulation or affect the continuity of service."

Two stages of study must precede a successful insulation coordination, namely, an examination of the different types of overvoltage that the power system encounters, a subject dealt with in Chapter 14, and the general assessment of insulation performance under the stress of those overvoltages. In the light of the outcome of those studies, the insulation coordination effort aims at specifying the following:

1. The electrical strength of the insulation of all system components
2. The phase-to-ground and phase-to-phase clearances
3. The leakage distance of external insulators
4. The ratings, type, number, and location of surge arresters and other possible protective spark gaps.

15.1.1 Types of Insulation

When applying insulation coordination, it is useful to classify the insulation of a power network according to location and dielectric performance. Insulation is classified, according to location, as external or internal.

1. *External insulation*: clearances and insulator surfaces in the open air. They are, therefore, influenced by such atmospheric conditions as pollution and humidity. External insulation can be located outdoors or indoors. In the latter case it is less affected by the ambient weather conditions.
2. *Internal insulation*: internal solid, liquid, or gaseous components of equipment insulation and therfore not exposed to atmospheric conditions. Insulating oils, compressed-gas insulation, and solid insulation of cables, transformers, and electric machines belong to this category.

From the dieelectric performance viewpoint, insulation may be either self-restoring or on-self-restoring.

1. *Self-restoring insulation*: regains dielectric integrity following the occurrence of a breakdown. Some insulating films in capacitors produce no permanent leakage path after their breakdown and are therefore called "self-healing" insulation.
2. *Non-self-restoring insulation*: loses its insulating property after a disruptive discharge and must therefore be replaced. Solid insulation, as in cables, transformers, and machines, belongs to this type of insulation.

The coordination of self-restoring insulation, unlike non-self-restoring insulation, lends itself to probabilistic treatment where the evaluation of the insulation's risk of failure is economically justified.

15.2 INSULATION PERFORMANCE UNDER VOLTAGE STRESSES

Power system insulation is subjected to four classes of dielectric stresses: power-frequency normal voltage, temporary overvoltages, switching overvoltages, and lightning overvoltages. The relative durations of these voltages are shown in Figure 15.1. The last three classes of overvoltage are discussed in Chapter 14. Here the response of system insulation to these stresses and the likelihood of failure are outlined.

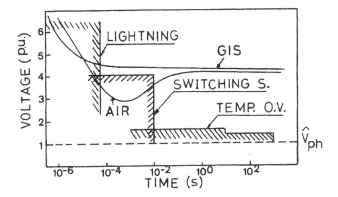

Figure 15.1   Superposition of V–t characteristics on overvoltage distribution pater.

15.2.1   Failure under Power-Frequency Normal Voltage

Following are some of the causes of insulation failure under power-frequency voltages:

1. Contamination on the insulation surface can initiate a mechanism that may lead to total breakdown.
2. Thermal breakdown may develop under the continuous stress and the consequent dielectric losses.
3. Voids and flaws in insulation may develop corona discharges which would lead to premature aging of insulators and their eventual failure.
4. Normal aging of the insulation may be manifested by the gradual reduction in its withstand capability over a long period of time, and a sudden breakdown may then occur under operating voltage.

Atmospheric pollution is the primary cause of breakdown of external insulation under normal voltages (Khalifa et al., 1988). As discussed in Section 9.9, leakage currents on polluted insulators, if heavy enough, may cause eventual flashover at service voltage. The flashover voltage of a polluted insulator decreases considerably as the conductivity of the pollution layer increases.

15.2.2   Failure under Impulse Voltages

Power-line and station insulators are subject to a variet of transient voltages. These voltages—from an insulation flashover point of view—are characterized primarily by their magnitudes and wave shape. The shape of an idealized impulse is made up of its time to crest (or front) and the rate of its decay beyond the crest, which is customarily identified by the impulse tail. Lightning impulses are known to have fronts on the order of a few microseconds, whereas switching surges have fronts on the order of a few hundred microseconds (Chapter 14). In this section the various factors that influence the flashover voltage of external insulation, and are thus relevant to insulation dimensioning, are discussed.

15.2.2.1 Front Duration

Air gaps and insulators shorter than about 1 m have a flashover voltage that is independent of the impulse front. For longer air gaps and insulators the impulse front has a definite effect on the flashover voltage, as discussed in Section 4.8.3. The impulse front corresponding to least dielectric strength is commonly known as the critical impulse front, and, depending on the strike distance and configuration, it takes values between 100 and 500 μs. The IEC standards choose 250 μs to represent the front of a switching surge where the air insulation displays the least strength (IEC, 1976).

15.2.2.2 Gap Geometry and Polarity

The physical process governing breakdown is greatly influenced by the field distribution in the gap (Chapter 4). The breakdown voltage of a uniform field gap can be as high as six times that of a rod–plane gap. Therefore, for practical purposes, the term "gap factor" (K_g) has been introduced to express the breakdown voltage of a given gap relative to a positive rod-to-plane gap of the same length, which is known to have the least breakdown voltage. The gap factor is therefore always larger than, or equal to, unity. For example, rod–rod gaps have a gap factor of about 1.7; a conductor-to-tower clearance normally has $K_g = 1.3$.

Figure 15.2 shows the pronounced influence of ground proximity and that of voltage polarity on the breakdown voltage of an air gap (Electrical Power Research Institute, 1979). Positive breakdown voltages (i.e., those when the more highly curved conductor is positive) are lower than negative voltages, as explained in Chapter 4 in detail. The more nonuniform the field in the gap becomes, the larger will be the difference between positive and negative breakdown voltages.

15.2.2.3 Breakdown Probability Distribution

If an impulse voltage of a certain shape is applied repeatedly to self-restoring insulation, the outcome of all applications may not be the same. Some impulse application may produce flashover, while others, still with the same amplitude, may be withstood by the insulation. This phenomenon is

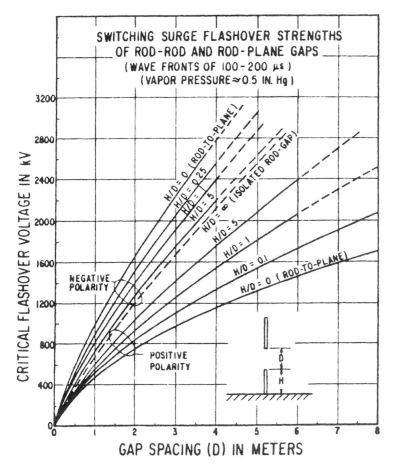

Figure 15.2 Effect of polarity and ground proximity on air gap flashover.

particularly true for sufficiently large gaps (larger than about a 50 cm rod–plane gap in air at atmospheric pressure). Also, it occurs over a finite range of impulse voltage amplitudes, above which all shots produce breakdowns and below which no shots would cause breakdown. This probabilistic property is very essential to the process of insulation dimensioning. In fact, at very high transmission voltages this property is used to set an economical compromise between the insulation cost and that of the risk of insulation failure.

The probability of breakdown (i.e., the number of breakdowns relative to the total number of applied shots) may be plotted against the applied voltage as shown in Figure 15.3a. Experimental investigations have revealed that for most external insulation the breakdown probability function may be expressed by the well-known cumulative Gaussian distribution function

$$P(V) = \frac{1}{\sigma\sqrt{2\pi}} \int_0^V \exp\left[-\frac{1}{2}\left(\frac{V - V_{50}}{\sigma}\right)^2\right] \tag{15.1}$$

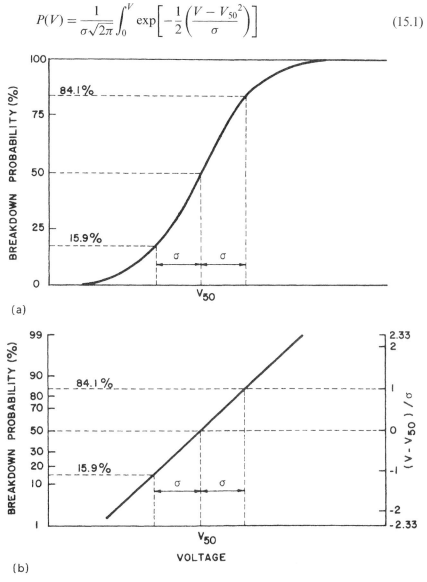

(a)

(b)

Figure 15.3 Breakdown probability of external insulation: (a) on linear coordinates, (b) on a Gaussian scale (σ=standard deviation).

This function is characterized by two parameters:

V_{50} = 50% breakdown voltage; also called the critical flashover voltage, CFO

σ = standard deviation of the representative Gaussian distribution; σ describes the degree of scatter in breakdown voltages about their mean value V_{50}

If the postulate of equation (15.1)—that the breakdown probability fits a Gaussian cumulative function—is accepted, $P(V)$ can alternatively be plotted on a normal probability ordinate scale as seen in Figure 15.3b, in which case $P(V)$ appears linear. The linear units on the ordinate in this case are those of the quantity $(V - V_{50})/\sigma$ rather tha P(V).

15.2.2.4 Time to Breakdown

Upon the application of an impulse voltage, a breakdown, if any, would take place after a certain time lag following the initiation of the impulse. This delay, referred to as the time to breakdown, depends on the gap configuration, the amplitude, and the shape of the voltage impulse. Furthermore, the time to breakdown is a random quantity that may vary markedly from one impulse application to another (Chapter 4). The time to breakdown of insulation is essential for insulation coordination when two or more insulation components are subjected simultaneously to the same impulse stress, particularly when an overvoltage protection device is installed.

15.3 FLASHOVER OF EXTERNAL GAPS IN PARALLEL

A transmission system comprises a large number of insulation components "in parallel" (i.e., subjected to the same normal and abnormal voltages). In a substation, the number of line-to-ground insulation structures may be as high as 100. The number of bus insulators per phase could vary from 100 to 1000, depending on the station voltage and capacity. On transmission lines, the number of insulation strings connecting the same phase to ground can be as large as a few thousands.

In the study of parallel insulation, two physically oriented as assumptions are adopted: first, that upon the incidence of an overvoltage, only one gap at a time is liable to breakdown; second, that whenever more than one gap tends to break down simultaneously, only the one with the shortest "time to breakdown" would actually do so.

15.3.1 The Case of Two Parallel Gaps

When the probabilistic properties of the two gaps are given separately, the behavior of a combination of the two gaps may be predicted under a known stress. If the individual breakdown probabilities of the two gaps before the combination were P_1 and P_2, the new breakdown probability of the first gap during the parallel combination would be

$$P_1' = P_1(1 - P_2) + P_1 P_2 P_{1,2} \tag{15.2}$$

The first term of equation (15.2) is the probability that gap 1 flashes over whereas gap 2 does not. This is complemented by the probability that the two gaps break down simultaneously $(P_1 P_2)$, multiplied by the probability $P_{1,2}$ that in this case gap 1 is faster than gap 2 to develop a breakdown. The breakdown probability of the second gap is, similarly,

$$P_2' = P_2(1 - P_1) + P_1 P_2 P_{2,1} \tag{15.3}$$

Since the time to breakdown of one gap can only be either shorter or longer than that of the other gap, then

$$P_{1,2} + P_{2,1} = 1 \tag{15.4}$$

Obviously, the quantity $P_{1,2}$, hence $P_{2,1}$, is a function of the probability density distributions $f_1(t)$ and $f_2(t)$ of the times to breakdown of the two gaps. Such a function can be derived to be

$$P_{1,2} = \int_0^\infty \int_t^\infty f_1(t_1) f_2(t_2) \qquad dt_1 \, dt_2 \tag{15.5}$$

In the special case when the two density distributions are normal (Gaussian) with mean values T_1 and T_2 and standard deviations S_1 and S_2, respectively, then

$$P_{1,2} = \frac{1}{2} + \frac{1}{\sqrt{2}} \int_0^{(T_2 - T_1)/\sqrt{\sigma_1^2 + \sigma_2^2}} \exp\left(\frac{-t^2}{2}\right) dt \tag{15.6}$$

The overall breakdown probability of the system of two gaps is given by

$$P_c = P_1' + P_2' = P_1 + P_2 - P_2 P_2 \tag{15.7}$$

which is characteristically independent of the times to breakdown.

15.3.2 Flashover of Identical Gaps in Parallel

If the probability of withstand at a given voltage level for a single air gap is W, the probability of withstand for n identical gaps in parallel is then

$$W_n = W^n \tag{15.8}$$

In terms of breakdown probabilities, equation (15.8) can be rewritten as

$$1 - P_n = (1 - P)^n$$

or

$$P_n = 1 - (1 - P)^n \tag{15.9}$$

where P is the breakdown probability of one gap and P_n is that of n gaps. Experimental investigations have shown that the relationships given previously hold accurately. Statistically, if the distribution of the breakdown probability of a single gap P is accepted to be cumulative Gaussian, the distribution of P_n can no longer be Gaussian.

It is concluded from equations (15.8) and (15.9) that the probability of withstand of a transmission system decreases as the number of insulators in parallel increases. For example, with a probability of withstand of a single insulator of 99%, a system of 20 parallel insulators will have a withstand probability of only 82%. If the string of insulators in successive towers are roughly considered to be in parallel, the attenuation of the traveling overvoltage wave should first be considered in evaluating the line's flashover probability.

15.3.3 Minimum Time to Breakdown

The time to breakdown (TBD) of a system of parallel gaps (or insulators) is also a random quantity (Chapter 4). The knowledge of the distribution of this time is vital to insulation coordination and the selection of a protective device. Testing the system of parallel gaps for the determination of its TBD distribution can be impractical in view of the varying number of components of the system. The TBD of a system of parallel gaps can, alternatively, be related to the characteristics of the individual components which make up that system.

The probability density distribution of the collective TBD is composed of the TBD probability densities of the individual components, each being weighted by a corresponding probability of occurrence of a breakdown. To illustrate this relation, the case of two generally unidentical gaps is discussed.

Given two parallel gaps with breakdown probabilities P_1 and P_2 and the TBD distributions $f_1(t)$ and $f_2(t)$ under a given impulse stress, the TBD distribution of the combination is

$$f_c(t) = \frac{\phi[P_1, P_2, f_1(t), f_2(t)]}{\int_0^\infty \phi[P_1, P_2, f_1(t), f_2(t)]\mathrm{d}t} \tag{15.10}$$

where, in view of equation (15.2),

$$\phi = P_1(1 - P_2)f_1(t) + P_2(1 - P_1)f_2(t) + P_1P_2f_{12}(t)$$

and the denominator is introduced to fulfill the constraint

$$\int_0^\infty f_c(t) \, dt = 1$$

Applying the same constraint—that a density distribution integrated over all times equals unity—to the constituents of the denominator gives

$$f_c(t) = \frac{P_1(1 - P_2)f_1(t) + P_2(1 - P_1)f_2(t) + P_1P_2f_{12}(t)}{P_c} \tag{15.11}$$

where $P_c = P_1 + P_2 - P_1P_2$ is the breakdown probability of the combination, as was shown in equation (15.7). The first two terms of $f_c(t)$ are the contributions of the individual TBDs of the two gaps whenever one gap breaks down alone. In the third term, which represents the contribution of a simultaneous breakdown in the two gaps, $f_{12}(t)$ is the density distribution of the minimum TBD of the two gaps since only the gap with the shorter TBD will break down at a time. It can be shown that $f_{12}(t)$ is related to the individual TBD values by

$$f_{12}(t) = [1 - F_1(t)][1 - F_2(t)]\left[\frac{f_1(t)}{1 - F_1(t)} + \frac{f_2(t)}{1 - F_2(t)}\right] \tag{15.12}$$

where F_1 and F_2 are the cumulative distributions of f_1 and f_2, respectively.

To extend the analysis above to a large number of equal gaps n, different combinations of gaps will have to be considered (e.g., singles, twos, threes, etc.). The distribution of the minimum TBD of only r gaps (out of a total of n gaps) can be shown to be given by

$$f_{(r)}(t) = [1 - F(t)]^{r-1}rf(t) \tag{15.13}$$

The overall TBD of the n-gap system can now be deduced in a manner similar to that of two gaps [equation (15.11)]. Considering the binomial combinations of breakdowns in the n gaps, the overall distribution of the TBD of a system of n equal gaps will be

$$f_c(t) = \frac{1}{P_c} \sum_{r=1}^n \frac{n!}{(n-r)!} P^r (1 - P)^{n-r} f_{(r)}(t) \tag{15.14}$$

It is worth mentioning that whenever the individual breakdown probabilities are very small, all combinational contributions to $f_c(t)$ can be neglected and $f_c(t)$ becomes equal to $f(t)$ of any one gap. On the other hand, as the number of gaps in parallel is increased and/or the individual breakdown probability increases, an ultimate distribution of the system's TBD is approached, which can be shown to be

$$f_{\text{ult}}(t) = 2[1 - F(t)]f(t) \tag{15.15}$$

15.4   PRINCIPLES OF INSULATION COORDINATION

The procedure leading to well-coordinated system insulation begins by evaluating the stresses to which insulation is subjected. The insulation dimensions are then decided in view of existing protective devices such that a predetermined level of safety against insulation failure is ensured. Depending on whether the coordinated insulation is self-restoring or non-self-restoring, a statistical or conventional approach is adopted. In the statistical approach a calculated risk of insulation failure is taken, whereas in the conventional approach the coordinator is set to eliminate the possibility of system failure (Diesendorf, 1974).

15.4.1   Statistical Insulation Coordination

With the advent of extra-high-voltage transmission and in anticipation of ultra-high voltage, two phenomena came into existence and justified the use of a statistical approach to insulation coordination. First, overvoltages, particularly those of internal origin, exhibited a distinct random nature wherein they varied markedly in magnitude and form. Second, long insulators and air clearances displayed widely scattered probabilities of breakdown and associated times to breakdown, as explained earlier.

The overvoltages that stress the system insulation may be expressed by a joint statistical distribution of the magnitude V_m and time to create t [i.e., $f(V_m, t)$]. At the same time, the breakdown probability of insulation may also be expressed as a function of the same two quantities, i.e., $P(V_m, t)$. Therefore, overvoltages whose magnitudes fall within an interval dV_m, and with times to crest falling within dt, should produce a risk of insulation failure given by

$$dR = P(V_m, t)[f(V_m, t)dV_m dt] \tag{15.16}$$

The overall risk of insulation failure can thus be written as (Anis et al., 1978)

$$R = \int_0^\infty \int_0^\infty P(V)_m, t)f(V_m, t)dV_m dt \tag{15.17}$$

Due to the complexity of equation (15.17) and the difficulty in acquiring the necessary bivariate data, it is customary to consider only the overvoltage magnitude when calculating the risk of failure, that is

$$R' = \int_0^\infty P(V_m)f(V_m)dV_m \tag{15.18}$$

in which $P(V_m)$ is the breakdown probability distribution, as discussed earlier in this chapter. Figure 15.4 gives a graphical interpretation of equation (15.18).

Insulation can theoretically be made sizable enough to eliminate any risk of insulation failure. This attitude, however, is economically unwise since the excessive cost of insulation may soon exceed the cost of system interruption due to insulation failure. It is therefore necessary to define a reasonable safety factor and design the insulation accordingly. This factor sets a relation between the insulation's withstand threshold as obtained from $P(V_m)$ and a representative overvoltage given by $f(V_m)$. The following quantities are defined for this purpose (IEC, 1976):

Statistical withstand voltage (SWV): peak value of a switching (or lightning) impulse test voltage at which the insulation exhibits, under specified conditions, a probability of withstand equal to 90%.

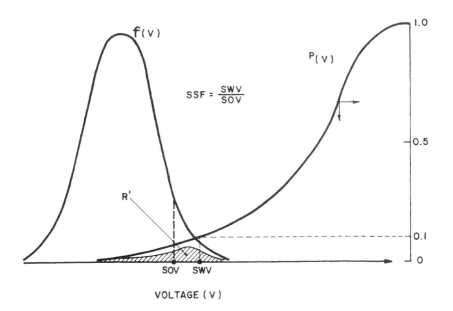

Figure 15.4 Computation of the risk of insulation failure.

Statistical overvoltage (SOV): switching (lightning) overvoltage applied to the equipment as a result of an event of one specific type on the system, the peak value of which has a probability of being exceeded of 2%.

Statistical safety factor (SSF): ratio, for a given type of event, of the appropriate statistical switching (lightning) impulse withstand voltage and the statistical overvoltage, established on the basis of a given risk of failure, taking into account the statistical distributions of withstand voltages and overvoltages.

$$SSF = \frac{SWV}{SOV} \qquad\qquad (15.19)$$

In Figure 15.5 the effect of the statistical safety factor on the consequent risk of insulation failure is demonstrated (IEC, 1976). It appears that the relationship above is almost invariant for all EHV transmission networks—a fact that permits the use of SSF as a reliable design factor.

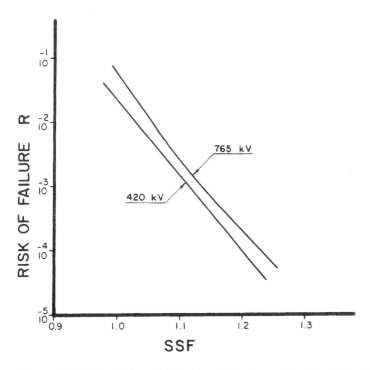

Figure 15.5 Effect of the statistical safety factor on the risk of insulation failure.

15.4.1.1 Example

The following procedural example for a disconnecting switch illustrates the application of statistical insulation coordination:

Highest voltage for equipment (line voltage) = 765 kV (rms)
$$\text{Phase-to-earth voltage } V_{ph} = 442 \text{ kV (rms)}$$
$$= 625 \text{ kV (peak)}$$
$$= 1.0 \text{ p.u.}$$
$$\text{SOV (from system studies)} = 1255 \text{ kV (peak)}$$
$$= 2.0 \text{ p.u.}$$

Maximum acceptable risk of insulation failure to ground $R = 10^{-4}$

Choice of insulation's SWV:

SWV		SSF	R
kV	p.u.		
1300	2.08	1.04	$1 \times 10^{-2} > 10^{-4}$
1425	2.28	1.14	$7 \times 10^{-4} > 10^{-4}$
1550	2.48	1.24	$7 \times 10^{-5} < 10^{-4}$

Conclusion. A rated switching withstand voltage for equipment insulation of 1550 kV is selected. The associated rated lightning withstand voltage is 2400 kV.

15.4.2 Conventional Insulation Coordination

With non-self-restoring insulation the risk of insulation failure should at all times be avoided. This is normally achieved by placing an overvoltage protective device, usually a surge arrester, in the vicinity of the equipment to be protected. The conventional insulation coordination approach seeks the impulse voltage level at which the equipment insulation will not show any disruptive discharge. This procedure is done in view of the characteristics of existing surge arresters.

Figure 15.6 demonstrates the steps leading to the determination of the insulation's basic impulse insulation level (BIL) and basic switching impulse insulation level (BSL). The former expresses the insulation withstand under a standard impulse of 1.2/50 μs, whereas the latter refers to a 250/2500 μs impulse. If the peak value of the conductor-to-earth voltage in a three-phase system is defined as 1 p.u., the most common temporary overvoltage (i.e., the voltage of a healthy phase under single line-to-ground fault conditions) is equal to C_e. This "earth fault factor" is approximately 1.4 for an earthed system. The surge withstand voltage of the system, which is also referred to as the basic impulse insulation level (BIL), is approximately 3.4 p.u. for a transmission system with a rated voltage U_m greater than 200 kV, whereas the switching impulse insulation level (BSL) is about 2.8 p.u. The arrester will limit an overvoltage wave to a certain protection level V_p under normal operating and temporary overvoltage conditions.

The protection level of a protective device is defined as the highest permissible surge voltage that may appear at the terminals of the equipment to be protected; for switching impulses it is taken directly as the maximum sparkover voltage, V_s, in Figure 15.6. For lightning impulses, however, the protection level is taken as the maximum of the following three values:

1. Maximum sparkover voltage
2. Maximum residual voltage at specified discharge current
3. Maximum sparkover voltage in the front, divided by 1.15

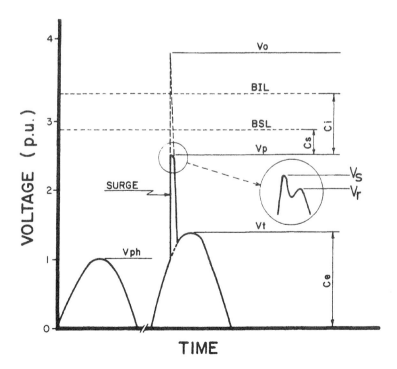

Figure 15.6 Determination of BIL and BSL. The symbols are explained in the text.

The residual voltage V_r is built up in the arrester after sparkover by subsequent flow of current; it can be larger or smaller than V_s, depending on that current. Safety margins (C_i and C_s) exist between the insulation levels BIL and BSL on the one hand and the arrester protection level V_p on the other; in accordance with IEC recommendations, these margins must be at least equal to

$$C_s = 1.15 \qquad C_i = 1.25$$

15.4.2.1 Example

The following procedural example (for a transformer) demonstrates the application of the conventional approach to insulation coordination.

Highest voltage for equipment (line voltage)	= 765 kV (rms)
Phase-to-earth voltage V_{ph}	= 442 kV (rms)
	= 625 kV (peak)
	= 1.0 p.u.

Computed (or measured) temporary overvoltage
to ground
V_t = 605 kV (rms)
= 855 kV (peak)
C_e = 1.37 p.u.

(If temporary overvoltages are not known, take $C_e = 1.4$ p.u.)
Minimum $C_s = 1.15$
Minimum $C_i = 1.25$

Characteristics of surge arrester

Rated voltage (available rating immediately above the temporary overvoltage level of 605 kV)	= 612 kV (rms)
Maximum switching impulse sparkover voltage	= 1230 kV (peak)
Maximum lightning impulse sparkover voltage	= 1400 kV (peak)
Maximum residual voltage at specified discharge current	= 1400 kV (peak)
Maximum sparkover voltage in the front	= 1660 kV (peak)

Protection level V_p:
to switching impulse	= 1230 kV
(i.e., protection factor)	= 1.97 p.u.
to lightning impulse	= 1443 k V
(i.e., protection factor)	= 2.3 p.u.

Recommended insulation

Minimum switching impulse voltage $(C_s V_p)$	= 1415 kV
Rated switching impulse voltage BSL	
(nearest higher standard)	= 1425 kV
	= 2.28 p.u.
Minimum lightning impulse voltage $(C_i V_p)$	= 1800 kV
Rated lightning impulse voltage BIL	
(nearest higher standard)	= 1800 kV
	= 2.88 p.u.

15.5 OVERVOLTAGE PROTECTIVE DEVICES

The duty of overvoltage protection is to ensure that transients arriving at an installation from the power lines or originating in the installation itself are reduced to a level bearing a definite relationship to the rated voltage of the network. One of the oldest known means of providing protection against overvoltages, and one that is still used to some extent, is the spark gap. As a rule it consists of two metal electrodes pointing toward one another, one being connected to the high voltage, the other to ground. Frequently, the spark gap is referred to by a descriptive name (e.g., arcing horns). Evidently, the breakdown voltage depends on the gap width and weather conditions (IEC, 1973).

By allowing the gap to flash over, the propagation of undesired voltages in the network can be prevented. However, the spark gap is not really a very efficient means of protection against overvoltages. The main drawbacks are the time lag that occurs before the gap sparks over, the variation of the sparkover voltage with the polarity and surrounding conditions, and the fact that once an arc has started to burn, it continues even after the overvoltage has disappeared, causing a line-to-ground short circuit on the network.

15.5.1 Magnetically Blown Surge Arresters

The surge arrester does not exhibit the disadvantages described previously. Over the wide range of overvoltages met in practice, the surge arrester possesses an almost unchanging sparkover voltage characteristic, thereby affording high operational safety, consistency, and reliability. Its follow current, resulting from the service voltage, is interrupted automatically.

A major requirement of surge arresters is that they must be able to discharge high energy without changes in their protective levels or damage to themselves or adjacent equipment. This property is decided by the arrester's thermal capacity. Unlike lightning strokes, which have short durations (hundreds of microseconds), switching surges generally have long durations (thousands of microseconds). The arrester is then required to discharge a considerable amount of energy, a situation that simple arresters with "unassisted" spark gaps cannot handle. Therefore, only heavy-duty surge arresters with magnetically blown spark gaps are recommended to operate on switching surges.

Figure 15.7 shows schematically the arrangement of a modern, magnetically blown arrester unit. The main components (i.e., nonlinear resistance elements R_a and the spark gaps E) are mounted in an airtight porcelain housing. The electrodes of the tongue-shaped spark gaps are embedded in the disk-shaped chamber K. The movement of the point of origin of the arc, which is influenced by the shape of the spark gaps, prevents local overheating or serious pitting of the electrodes. In normal operating conditions the control current i_0 flowing through the grading resistors R_s ensures that the voltage is distributed almost evenly among the arrester elements. When the discharge dies down with a rapid change in the current, a follow current flows, initially determined by the system voltage and the value of the discharge resistors.

The blowout coil presents no great impedance to the follow current, which changes only slowly, so the current commutates from the resistor R_b to the blowout coil. The current, now flowing through the coil, creates a strong magnetic field which extends throughout the stack of spark gaps. This tends to draw the arc L and extend it. A high arc voltage u_L thus builds up. This arc voltage helps to extinguish the follow current even before its natural zero. In Figure 15.8 the current/voltage characteristics of the arrester are shown.

15.5.2   Metal-Oxide Surge Arresters

A recently developed ceramic material, based on zinc, bismuth, and cobalt oxides, can be used to make resistors with a much higher degree of non-linearity ($I = cV^\alpha$; where $\alpha > 20$) over a large current range. With such resistors one can design arresters having voltage/current characteristics close to idea. The electrical properties of the material make it possible to dispense with the series-connected spark gaps and thereby to produce solid-state arresters suitable for system protection up to the highest voltages. To enhance the energy absorption capacity it is also possible to connect disks in parallel. The most important property of these resistors is their current–voltage characteristics, which are illustrated in Figure 15.9 for a disk of 80 mm diameter and 32 mm thickness. The voltage arising for a standard lightning current surge of 10 kA (8/20 µs wave) is just over twice the reference DC voltage (Knecht and Menth, 1979).

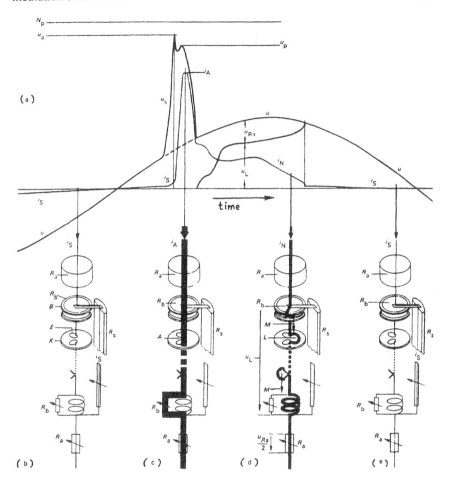

Figure 15.7 Operation of a magnetically blown arrester: (a) arrester voltage and current; (b) and (e) normal operation; (c) passage of surge current; (d) passage of follow current. N_p = guaranteed protection level, U_a = sparkover voltage, U_p = residual voltage during diversion, U_s = surge voltage, U = service voltage at arrester assembly, U_{Ra} = voltage drop across R_a resistors during quenching, U_L = arc voltage during quenching, i_A = surge current, i_N = follow current, i_s = control current, R_b = bypass resistor, and R_s = grading resistor. [From Ragaller (1980).]

The arrester, which is a series-connected stack of such disks, operates in a very simple fashion. It is dimensioned so that the peak value of the phase-to-earth voltage in normal operation never exceeds the sum of the reference voltages of the series-connected disks. The resistive losses in the

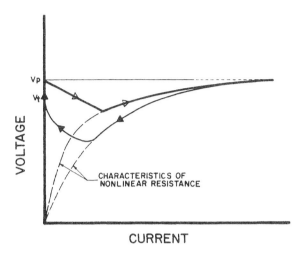

Figure 15.8   Arrester current–voltage characteristics.

arrester in normal operation are therefore very small. When an overvoltage occurs, the current will rise with the wavefront according to the characteristics of Figure 15.9 without delay. No breakdown occurs but, rather, a continuous transition to the conducting state. At the end of the voltage transient the current is reduced, closely following the $I-V$ curve (i.e., in contrast to the conventional arrester, there is no follow current).

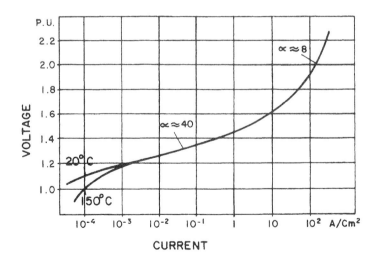

Figure 15.9   Characteristics of metal-oxide arrester disks.

A considerable advantage of the zinc oxide arrester is its very simple construction. Meanwhile, the absence of spark gaps results in a continuous flow of current through the device so that, theoretically, the danger of thermal runaway is present. However, very stable resistors have been developed, so these apprehensions are almost eliminated. Furthermore, the absence of spark gaps makes the voltage grading system unnecessary. The zinc oxide arrester is inherently self-regulating in that the current flow of 0.5–1 mA at normal supply voltage leads to reliable operation even in polluted conditions. An extremely stable I–V characteristic would guarantee constant losses for the entire life of the arrester as well as a constant protection level (Ieda et al., 1987).

15.5.3   Zinc Oxide Varistors

Zinc oxide (ZnO) varistors are semiconducting ceramics having highly nonohmic current–voltage characteristics, which originate at the grain boundaries. In other words, they are voltage-dependent switching devices, as shown in Figure 15.10. These varistors are widely utilized in protecting not only electric power lines but also electronic components against dangerous voltage surges (Eda, 1989).

The resistivity of a ZnO varistor is very high (more than $10^{10}\,\Omega{\cdot}cm$) below a certain threshold voltage (V_{tb}), whereas it is very low (less than

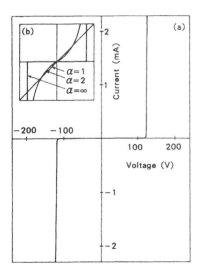

Figure 15.10   (a) Current–voltage characteristics of a typical ZnO varistor; (b) schematic I–V curves for different nonohmic exponents α.

several $\Omega \cdot$cm) above the threshold voltage. Hence, ZnO varistors are very useful devices as elements of surge arresters for protecting electric power lines—as outlined in the preceding section—or as surge absorbers to protect electronic components against voltage surges. ZnO varistors are fabricated by sintering ZnO powders with small amounts of additives such as Bi_2O_3, CoO, MnO, and Sb_2O_3. The nonohmic property comes from grain boundaries between semiconducting ZnO grains. Various additives to improve the electrical characteristics are sometimes used to optimize the processing conditions. The microstructures and the physical properties of the grain boundaries are now well identified, and applications are rapidly found in protecting electrical equipment and electronic components such as transistors and ICs against voltage surges.

ZnO varistors have highly nonohmic current–voltage characteristics above a threshold voltage. Since the range showing the highly nonohmic property is wide, the I–V characteristics are usually expressed logarithmically. The degree of nonohmic property is usually expressed by a nonohmic exponent, α, defined by the following equation:

$$\alpha = (V/I)(dI/dV) \tag{15.20}$$

Empirically, the simple equation

$$I = cV^\alpha \tag{15.21}$$

is used, where c is a constant. When $\alpha = 1$, it is an ohmic resistor, and when $\alpha = \infty$, it is an ideal varistor, as can be seen from Figure 15.10(b). Typical α values of ZnO varistors are from 30 to 100. On the contrary, α values of conventional varistors such as SiC varistors do not exceed 10.

The I–V characteristics of ZnO varistors are classified into three regions. In the region below the threshold voltage (typically corresponding to $1\,\mu A/cm^2$, the nonohmic property is not so prominent and is highly dependent on temperature. In the region between the threshold voltage and a voltage at a current density of about $100\,A/cm^2$, the nonohmic property is very prominent and almost independent of temperature. In the region above $100\,A/cm^2$, the nonohmic property gradually decays. ZnO varistors are characterized by the magnitude of the α values and the width of the range where the highly nonohmic property is exhibited.

The I–V characteristics below $100\,mA/cm^2$ are usually measured using a DC electric source, whereas those above $1\,A/cm^2$ are measured by an impulse current source to avoid heat generation and thermal breakdown. The waveform of the impulse current has an $8\,\mu s$ rise time and $20\,\mu s$ decay time up to one-half the peak value. This waveform is used as a standard impulse current to test surge arresters. The I–V characteristics measured by the impulse currents show voltages higher than those measured using DC.

The discrepancy is usually 10–20% and is caused by the delay of electrical response in the ZnO varistor. The response delay is caused by electron trapping and hole creation at the grain boundaries.

Another important electrical characteristic of ZnO varistors is the dielectric property. Below the threshold voltage, ZnO varistors are highly capacitive. The dielectric constant of ZnO is 8.5, whereas an apparent dielectric constant of a ZnO varistor is typically 1000. The dielectric properties are mainly caused by thin depletion layers ($\sim 1000\,\text{Å}$) at the grain boundaries.

15.6 INSULATION COORDINATION OF GAS-INSULATED SUBSTATIONS

The temporal development of breakdown in compressed SF_6 in a GIS is significantly different from that in open-air insulation. The former case is characterized by a weak field nonuniformity, short clearances, and high gas pressure, as well as the strong electronegativity of the gas. In contrast, open-air insulation may exhibit strong field nonuniformity, long gaps, and it is always at atmospheric pressure. These differences produce sharply distinct voltage–time characteristics, as seen in Figure 15.1. While the withstand of air gaps drops to a minimum with increasing impulse front time before gaining strength under longer-fronted impulses, GIS insulation shows a steadily decreasing strength with increasing impulse time.

In Figure 15.1 the $V-t$ characteristics of insulation are superimposed on the frequency-of-occurrence diagram for the various types of overvoltage. It is concluded that lightning overvoltages are more significant for GIS insulation dimensioning, whereas conductor spacings and other open-air insulation for EHV and UHV systems are basically determined by switching overvoltages and to a lesser extent by lightning overvoltages. Due the rather flat insulation withstand dependence of the GIS on the front time of overvoltage, the withstand of the GIS against temporary overvoltages is satisfactory, as the insulation must be designed for the much higher lightning impulse.

For a conventional substation it is normally deemed sufficient if the transformers, being the most valuable equipment and difficult to replace, are well protected against excessive overvoltage stresses. In contrast, in the case of GIS, even if a part of the GIS could be replaced comparatively quickly and on site, a failure would lead to a great disturbance in the operation of the system. Good overvoltage protection is thus needed for all gas-insulated parts of the substation. This causes the need to protect also the distance from the line entrance to an open circuit breaker, which in turns demands installation of lightning arresters at the line entrance where overhead lines are connected.

15.6.1 Incident Lightning Surges

Design values of stroke current amplitude and steepness are about 70 kA and 70 kA/μs, respectively. The structures comprising towers and overhead lines have, at least during the first microsecond after a stroke has hit the line, a wave impedance of about 100 Ω. It is therefore understood that the steepness of generated surges can be up to 7000 kV/μs at the stroke origin. When the wave travels along the line, its peak and front steepness will be reduced because of corona discharges (see Chapter 5). The wave impedance of an overhead line is considerably greater than that of GIS (typically 75 Ω), and thus reflection will occur at the entrance to the GIS. At an open gap (disconnector or circuit breaker) the surge will be totally reflected, meaning the doubling of amplitude.

As explained in Chapter 14, when a surge enters, one part α is transmitted into the substation and the other part $(1 - \alpha)$ is reflected back along the line. The entered wave is reflected at the open end and, when reaching the GIS entrance, will be reflected again. The growth of the overvoltage will follow the equation (Eriksson and Holmborn, 1978)

$$\frac{U}{U_0} = 2 - (2 - \alpha)(1 - \alpha)^n \tag{15.22}$$

where n is the number of reflections. As n increases, U/U_0 approaches 2.

With finite surge steepness S, however, the voltage at the entrance to the GIS is given as a function of time by

$$V = S\alpha t \tag{15.23}$$

before the first reflection. After n reflections, it will take the form

$$V = S\alpha\left\{t + \sum_{m=2}^{n}(1 - \alpha)^{m-2}(2 - \alpha)[t - (m - 1)T]\right\} \tag{15.24}$$

where

$$2(n - 1) < \frac{t}{T} < 2n$$

The voltage at the open end at time t will be

$$U = S\sum_{m=1}^{n}2(1 - \chi)^{m-1}[t - (2m - 1)T]$$

where

$$2(n - 1) < \frac{t}{T} < (2n - 1)$$

and where S is the steepness of the wave on the overhead line, T the wave traveling time in a GIS, L the length of the GIS from entrance to the open circuit, and n the number of reflections. $T = L/300\,\mu s$. The voltage increase is dependent on three parameters: steepness S, transmission coefficient (Chapter 14) α, and length of substation L.

15.6.2   Effect of Arrester Presence

With the presence of a surge arrester, the sequence of events differs. When the voltage has reached the sparkover voltage of the arrester, it will take a further travel time T between arrester and the open bus end. Meanwhile, a negative wave originates at the arrester and travels to the open bus. Thus the total voltage is reduced. Therefore, in case of surges with steeper fronts and with longer GIS buses, the voltage buildup on the system is greater. On the other hand, for shorter bus lengths, the voltage increase by repeated reflections becomes rapid, resulting in a higher spark-over voltage of the arrester, whereas the voltage buildup reaches a relatively lower level (Fig. 15.11).

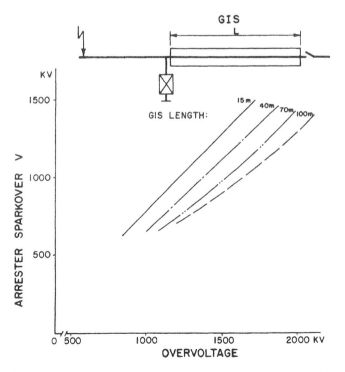

Figure 15.11   Overvoltage versus sparkover voltage of GIS. Enclosed arrester at line entrance; 2000 kV/μs incoming surge.

The arrester sparkover voltage for steep surges is of great interest for the insulation coordination of GIS since the increase of the GIS withstand level is limited for steep surges. Due to the flat SF_6 $V-t$ characteristics, it is necessary also to have a low surge arrester sparkover voltage for front times less than 1 μs.

In Figure 15.12 the average rate of rise of the voltage across an arrester at the line entrance has been plotted versus the distance from entrance to an open gap in the GIS. With a longer bus length, the surge steepness across the arrester will decrease.

15.6.3 Insulation Coordination with Arresters

The first important decision to make is to select the surge arrester rated voltage V_r (reseal voltage). A choice of a reseal voltage equal to the temporary overvoltage results in reseal voltages 1.3 to 1.6 times the normal phase-to-ground voltage (IEC, 1976). Based on the rated voltage (V_r), the following arrester characteristics can be obtained:

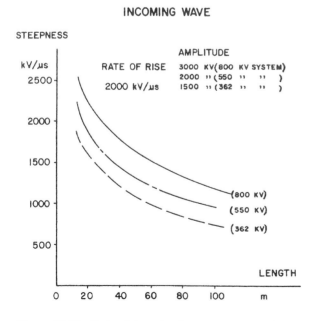

Figure 15.12   Rate of rise of voltage across arrester versus GIS length.

Front of wave sparkover voltage ($\sim 2.3 V_r$ if unknown)
1.2/50 μs sparkover voltage ($\sim 2.1 V_r$ if unknown)
Residual voltage at 10 kA ($\sim 1.9 V_r$ if unknown)

the largest of which determines the protection level of the arrester.

15.6.3.1 Example

The following procedural example demonstrates the determination of the BIL of GIS equipment.

Rated system voltage (line voltage) = 420 kV (rms)
Rated phase voltage V_{ph} = 234 kV (rms)
                                = 344 kV (peak)
Temporary overvoltage factor = 1.4 p.u.

    Arrester specifications

Arrester rated voltage $C_e \times V_{ph}$ = 340 kV (rms)
Protection level V_p (determined as stated above) = 2.2 p.u.
                                = 755 kV (peak)

    System characteristics

Maximum lightning overvoltage in the substation (computed at open end of GIS as explained) V_{ov} = 2.7 p.u.
                                = 929 kV
Safety margin $C_i = BIL/V_{0v}$ = 1.4
(The factor C_i is usually based on experience.)
BIL of GIS = 1300 kV

15.7 CONTAMINATION OF EXTERNAL INSULATORS

If external insulation is designed on the basis of switching surge and lightning requirements, the strength for power frequency voltages and clean insulators, even in wet conditions, is normally very high. However, power frequency flashovers may occur on transmission systems without any switching or lightning surges. These flashovers usually take place in wet weather conditions, such as dew, fog, or drizzle, and are caused by contamination of the insulator surfaces. Contamination flashovers on transmission systems are initiated by the deposition of airborne particles on the insulators. These particles may be of natural origin or generated by artificial pollution which is mostly a result of industrial activities. A common natural deposit is sea-salt which causes a severe contamination of insulators of transmission systems in coastal areas. In industrial areas, there is a great variety of contaminants that can reduce the insulation strength. However, these deposits themselves do not decrease the insulation strength when they are dry. The loss of strength is caused only by the combination of two factors: contaminants and moisture.

15.7.1 Mechanism of Contamination Flashover

The surface impedance of a contaminated insulator unit changes as a function of the wetting rate created by fog and the drying rate created by the leakage current, as seen in Figure 15.13. With the advance of wetting, the impedance changes from capacitive to resistive. The increase of the capacitance is a result of the increase of the conductive area on the insulator surface. The voltage distribution on the unit also changes as a function of the time. With the progress of wetting, most of the applied voltage is impressed across a very narrow dry band formed around the pin. The dry band, whose width can be a few millimeters, is produced by the leakage current. This fact is seen in Figure 15.13.

When this dry band cannot withstand the voltage applied to the unit, scintillation activity begins. The dry band is bridged by partial discharges, so the voltage is distributed over the wetted area. The flashover is determined by the flashover strength of the contamination layer which holds the whole voltage at the final voltage distribution just before flashover. Scintillations may not only depart from the pin but also from any place on the outer skirts, as has been frequently observed.

The mechanism of long string flashover consists of different phases. Initially, the contaminated insulator surface is completely dry, so the voltage distribution on the string can be regarded to be the same as that on a dry and clean insulator string. The equivalent circuit can be represented by a

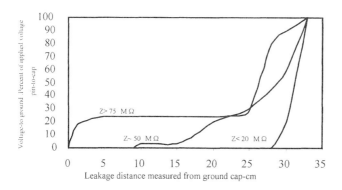

Figure 15.13 Voltage distribution measured from grounded cap before onset of scintillation, for different values of surface impedance.

network of capacitances only. The distribution is usually nonuniform, mainly as a result of the capacitance between insulator units and ground. As wetting progresses, the surface impedance becomes a combined capacitance and resistance. The value of the resistance is influenced by the drying effect created by leakage current, which is a function of the voltage across the unit; therefore, resistances of different units are not equal to each other. The surface resistance of the units under high voltage stress is higher than that of units under low voltage. The high voltage stress usually is concentrated at the bottom section of the insulator string, drying the contamination layer in this region until it forms a dry zone on the string. The surface temperature of these insulators is much higher than those on the remaining section of the string. A wet zone is usually formed at the midsection of the string, where the voltage stress is the lowest. As the wetting condition progresses further, this trend is accelerated. Finally, the units on the bottom section cannot withstand the voltage stress and they flash over. This is observed when the partial discharge at the bottom section bridges several units. Soon this activity develops upward. The bridging action over the bottom section results in an overvoltage for the insulators in the other sections and produces heavier activity along the string. This activity appears as surge currents, usually having peak values ranging from 500 to 700 mA (rms). These surge currents dry up the insulator surfaces in the wet zone, linearize the voltage distribution along the region not bridged by the scintillation arcs, and extinguish automatically due to their own linearization. However, this heavy activity does not make the insulator surfaces in the wet zone as dry as those of the dry zone of the string, so the whole voltage distribution remains nonuniform.

15.7.2 Contaminated Insulator Design

The most common contaminants causing contamination flashover are: seasalt, cement, fertilizer, fly ash, road salt, potash, chemicals, gypsum, limestone, phosphate, sulfate, paint industry, paper mill, acid exhaust, bird droppings, zinc industry, carbon steel works, carbide residue, sulfur, copper and nickel salt, wood fiber, bulldozing dust, aluminum plant, sodium plant, and rock crusher. The insulator electrical design has to have sufficient length to prevent an air breakdown at the appropriate voltage and have sufficient creepage to withstand the degree of pollution prevailing at the site where the insulators will be located in service.

Table 15.1 serves as a guideline for insulator design under contaminating conditions. The insulation size is related to the severity of contamination (General Electric company, 1968, 1979). The table shows that the use of fog-type insulators may permit operation at one higher contamination class.

Table 15.1 EHV Line Insulation Design (All Strings in Vertical Position)

Type of contamination	ESDD (mg/cm^2)	Leakage distance (in/kV rms line-to-ground)	Average kV rms per standard unit	Number of units							
				230 kV		345 kV		500 kV		700 kV	
				Standard	Fog type	Standard	Fog type	Standard	Fog type	Standard	Fog type
Clean atmosphere	0–0.03			Insulation not chosen by contamination requirements							
Slight atmospheric contamination	0.04	1.04	11.5	12	9	18	13	25	19	35	26
Moderate contamination containing soluble salts up to 5%	0.06	1.31	9.1	15	12	22	17	32	25	44	35
Severe contamination, 15% or more of soluble salts	0.12	1.74	6.9	19	15	27	22	42	31	59	44
Salt precipitation–seaside regions, salt marshes	0.3	2.11	5.7	23	18	35	27	50	39	71	54

Figure 15.14 gives the flashover strength of contaminated standard insulators (type A-11) (General Electric Company, 1979). The results are expressed in flashover voltage per unit axial insulator length. As shown in the figure, the performance of insulators is examined for contaminant mixtures having three different insoluble materials, namely kaolin (K-1), tonoko (K-2), and Fuller's earth (K-3). The performance of V-suspension strings is about 25–30% higher than that of tangent strings. As shown in the figure, the influence of the amount of insoluble material is significant but the differences in the physical properties of the insoluble materials (K-1, K-2, K-3) do not affect the flashover performance markedly.

15.7.3 Contamination Severity

Contamination severity for a site can be judged from the behavior of insulators in the area or by measurement of the contamination. Several indices are known to quantify the severity of contamination:

Equivalent salt (NaCl) deposit density (ESDD). This is expressed in mg of equivalent NaCl salt per cm^2 of insulator surface.

Surface conductivity. The surface conductance is defined as the ratio of the power-frequency current flowing over a sample insulator to the applied voltage.

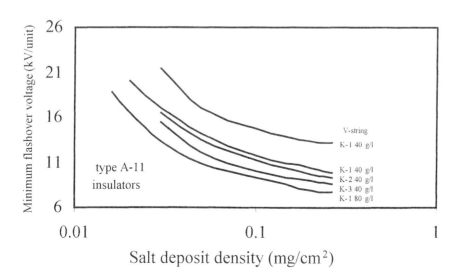

Figure 15.14 Performance of contaminated insulators with four kinds of contaminant mixtures.

Surge count. The number of leakage current pulses above a certain amplitude measured over the insulator at working voltage.

Highest current. The highest peak leakage current recorded during a given period over an insulator energized at the working voltage.

Flashover stress. The power-frequency flashover voltage divided by the overall insulator length.

15.7.4 Contamination Test Methods

Contamination test methods can be roughly classified into three categories.

15.7.4.1 Salt-Fog Test

The contamination and wetting are applied simultaneously by spraying salt water. The degree of contamination is defined by the amount of salt in the sprayed solution. The insulator surface is purely resistive during the test, whose duration is one hour.

15.7.4.2 Wet-Contaminant Tests

The test either uses a "light mixture," in which case contamination is defined by the amount of dry contaminant; or it uses a "heavy mixture," in which case contamination is defined by surface conductivity. In the former case the layer has a resistive–capacitive impedance.

15.7.4.3 Clean-Fog Tests

The test uses either a "slow" or a "quick" wetting process. The insulator surface changes from dry to wet, yet only in the case of slow wetting is the surface impedance capacitive–resistive.

15.7.5 Contamination Countermeasures

The most common ways of combating the problem of insulator contamination are summarized below. A combination of more than one measure sometimes applies.

15.7.5.1 Over-Insulation

This is the most common countermeasure for all contamination conditions, and is achieved by increasing the number of insulators.

15.7.5.2 Use of Fog-type Insulators

In heavily polluted areas fog-type insulators are sometimes employed. Fog-type insulators have a leakage distance up to 150% of that of the standard type of insulator. Figure 15.15 compares a fog-type insulator to a standard

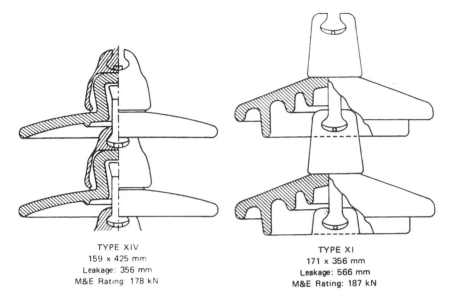

TYPE XIV
159 x 425 mm
Leakage: 356 mm
M&E Rating: 178 kN

TYPE XI
171 x 356 mm
Leakage: 566 mm
M&E Rating: 187 kN

Figure 15.15 (a) Standard insulator unit; (b) fog-type insulator unit.

unit. It should be emphasized that leakage distance is not the only factor affecting contamination flashover of insulators, and there is a variety of fog-type insulators whose performance is not satisfactory. Today there are two basic types of insulator design. One design, considered as more traditional, has a deep skirt at the outer rim of the bottom surface. Another fog-type design is generally labeled as "deep rib type" or "high leakage distance insulator," and is similar to a standard type in appearance.

15.7.5.3 Application of Silicone Grease

Silicone grease limits the ability of the insulator surface to retain moisture and produce leakage currents. This countermeasure is extensively used in many countries, and is particularly effective for "spot contamination" where maintenance is possible. This technique is also applicable to sea-salt contamination. Silicone greases, because of their temperature stability, can be used at higher temperatures. The greases, which can be either of hydrocarbon or of silicone materials, envelop the pollution and thus prevent the occurrence of discharges.

Hydrocarbon greases soften on heating, melt at the site of any discharge, and envelop the pollution. The tendency to soften on heating restricts their use to temperate climates. The grease is applied about 3 mm thick either by hand or by dipping. The removal of the grease at the end of its useful life can be rather time consuming.

Silicone greases are more costly than hydrocarbon greases. It has been shown that if too thin a layer of grease is applied glaze damage can result. Insulators coated with silicone rubber may withstand exposure to severe weather conditions for 3 years without damage.

15.7.5.4   Insulator Washing

Washing and cleaning of deenergized insulators has been used regularly in the past. Live line washing is sometimes employed. Line insulators can be cleaned either manually or by the use of water jets to remove pollution deposits. Washing techniques can be carried out with the insulators energized, using a fixed spray installation or a manually controlled portable jet. The frequency of washing depends on the site condition and should be adequate to maintain the insulator in a clean condition.

15.7.5.5   Semiconductive Glazed Insulators

This provides the insulator surface with a resistive coating, thus limiting leakage currents. The heat produced by the resistive coating keeps the surface dry and, to a certain extent, provides for a linear potential distribution.

15.7.5.6   Room-Temperature Vulcanizing

In order to suppress the leakage current, a low-energy surface is created by coating porcelain and glass with room temperature vulcanizing (RTV) silicone rubber (Deng et al., 1996). The RTV coating initially provides water repellency (hydrophobicity) which prevents the formation of continuous water filming on the surface, even when a layer of contamination has built up on the surface, and thus suppresses the leakage current and flashover. This is attributed to the continuous diffusion of the silicone fluid from the bulk and on to the surface of the pollution deposits.

15.7.6   Polymer Insulators

The most commonly used materials for outdoor insulators are glass and ceramics. Polymers are now being increasingly used as replacement for porcelain and glass for outdoor high-voltage insulation, such as line and station insulators, phase spacers, terminations for underground cables, surge arrester housings, etc. This is due to several significant advantages they offer (Matsuoka et al., 1996; Gorur et al., 1994; Xu et al., 1996). Polymer insulators are lightweight (due to the high strength-to-weight ratio of the central fiber glass rod), possess good vandal resistance, are easy to install, and facilitate line compaction. Polymeric cable terminations have provided users with highly simplified application techniques that have made them more reliable and inexpensive than porcelain. For surge arrester housings, they can contain violent explosions that can occur with porcelain within the housing, thereby minimizing injury to personnel during arrester failure. Due to their hydrophobicity or water-beading property, polymers can also provide higher resistance to leakage current, which improves the performance of insulating systems under contaminated conditions.

In the case of polymer insulators with silicone rubber (SiR) sheds, it is said that hydrophobicity on the surface is maintained for a long period because of migration of low molecular silicone from the bulk of the silicone rubber to the surface. Hydrophobicity, however, may not be fully expected when the hydrophobic surface of silicone rubber insulators is covered with a large amount of contaminant and/or water. Under such contamination and/or wetting conditions, the contamination withstand voltage of silicone rubber insulators may not be significantly different from that of porcelain insulators.

15.7.6.1 Aging in Polymer Insulation

With polymers, owing to their inorganic nature, the material is subject to aging and will degrade with time, with the result that the electrical and mechanical functions are interdependent to a greater degree (Karady et al., 1994). Leakage current leading to dry band arcing, promoted in the presence of moisture and contamination, is mainly responsible for aging of polymers. Other factors such as ultraviolet (UV) in sunlight, heat, chemicals, etc., act to accelerate the aging.

Aging is responsible for two types of failure observed in service, namely, material degradation in the form of tracking and erosion, and flashover at the operating voltage under contaminated conditions. Knowledge of life expectancy and ability to evaluate the long-term performance in the laboratory within a reasonable time are of paramount importance to utilities. But these require a detailed understanding of mechanisms responsible for aging. Presently, the aging mechanisms responsible for field failures are not fully understood.

15.7.7 Insulator Contamination Monitors

Many methods have been established to monitor the accumulation of contamination. Some of those methods are based on monitoring the leakage current (Habib and Khalifa, 1986). Other methods are based on the fact that leakage current monitoring may be an unreliable indicator of contamination level except near pre-flashover levels. This is because leakage current is a function of both the water present on the surface and the amount of contamination. Alternatively, a contamination liquid water sensor (LWS) was shown to be a successful surrogate indicator of the buildup of contamination on insulators (Richards and Renowden, 1997). The sensor measures both the amount of liquid water and the level of contamination on an insulator surrogate. As the amount of liquid water increases, one of the parameters of the LWS changes. The LWS is effective for the determination of contamination levels when the relative humidity is greater than 65%.

A contamination monitor was developed which used ellipsometry to provide valuable information about the optical parameters of the contamination which determine its thickness (Mahmoud and Azzam, 1997). This method provided an alternative to the laborious ESDD measurements for quantifying insulator surface contamination. However, without wavelength scanning it lacked the ability to distinguish the different types of contaminant. A single-wavelength ellipsometer was suggested for anticipating the dominant type of contamination in a certain area. The work proposed the use of spectrophotometry to measure the thickness and refractive index. This would deliver important information about the existence of certain contaminants by showing their characteristic absorption lines.

15.8 PROBLEMS

(1) An external insulation when tested in the laboratory gave a flashover probability of 5% under 120 kV impulse voltage. The probability rose to 75% when the voltage was raised to 135 kV. Estimate the 50% breakdown voltage and the corresponding standard deviation. Estimate the breakdown probability under 145 kV impulses.

(2) Three insulators of the type described above were connected in parallel. Estimate the overall breakdown probability under 130 kV.

(3) Two dissimilar insulators are connected in parallel. Under a certain impulse voltage the breakdown probabilities of the two insulators were 15% and 25%. The distribution of their times-to-flashover could be assumed uniformly distributed between 120 and 140 μs for the first insulator and between 130 and 150 μs for the second. Estimate the overall breakdown probability for the combination. Derive the time-to-breakdown of the compound system.

(4) The insulator system in a substation is characterized by a breakdown probability distribution having a 50% FOV of 220 kV and a standard deviation of 10 kV. If the overvoltages on the substation are statistically distributed by a normal distribution whose mean is 180 kV and standard deviation 40 kV, estimate the statistical withstand voltage, the statistical overvoltage, the statistical safety factor, and the resultant risk of failure.

(5) Repeat the conventional insulation coordination example given in the text for a 550 kV line subjected to temporary overvoltages of magnitude 445 kV (rms). Consult surge arrester manufactureres' specifications and make any necessary assumptions.

REFERENCES

Anis H, Radwan R, El-Morshedy A. Proceedings of IEEE Winter Power Meeting, New York, 1978, Paper A-78-154-7.

Deng H, Hackam R, Cherney EA. IEEE Trans PWD-11:431–443, 1996.

Diesendorf W. Insulation Coordination in High Voltage Electric Power Systems. London: Butterworth, 1974.

Eda K. IEEE Electr Insul Mag 5:28–41, 1989.

Electrical Power Research Institute. Transmission Line Reference Book—345 kV and Above. Project UHV. Palo Alto, CA: Electrical Power Research Institute, 1979.

Eriksson R, Holmborn H. Proceedings of 1st International Symposium on Gaseous Dielectrics, Knoxville, TN. 1978, pp 314–337.

General Electric Company. EHV Transmission Line Reference Book. New York: Edison Electric Institute, 1968, Ch 7.

General Electric Company. Transmission Line Reference Book—345 kV and Above. Palo Alto, CA: Electrical Power Research Institute (EPRI), 1978, Ch 10.

Gorur RS, Bernstein BS, Champion T, Hervig HC, Orbeck T. CIGRE, 15–107, 1994.

Habib S, Khalifa M. IEE Proc 133:105–108, 1986.

IEC. High Voltage Test Techniques. Publication 60. Geneva: International Electrotechnical Commission, 1973.

IEC. Insulation Coordination. Publication 71. Geneva: International Electrotechnical Commission, 1976.

Ieda M, Mizutani T, Suzuki Y, Ohki A. Proceedings of CIGRE Symposium, Vienna, 1987, Report 300.01.

Karady GG, Schneider HM, Rizk FAM. CIGRE, 33–103, 1994.

Khalifa M, El-Morshedy A, Gouda O, Habib S. Proc IEE 135C:24–30, 1988.

Knecht B, Menth A. Brown Boveri Rev 66(11):739–742, 1979.

Mahmoud F, Azzam RMA. IEEE Trans DEI-4:33–38, 1997.

Matsuoka R, Shinokubo H, Kondo K, Mizuno Y, Naito K, Fujimara T, Terada T. IEEE Trans PWD-11:1895–1900, 1996.

Ragaller K, ed. Surges in High-Voltage Networks. New York: Plenum Press, 1980, pp 251–281.

Richards CN, Renowden JD. IEEE Trans PWD-12:389–397, 1997.

Xu G, McGrath PB. IEEE Trans DEI-3:289–298, 1996.

16

High-Voltage Generation

M. ABDEL-SALAM *Assiut University, Assiut, Egypt*

16.1 INTRODUCTION

In the field of electrical engineering and applied physics, high voltages (DC, AC, and impulse) are required for several applications. To name just a few examples:

1. Electron microscopes and x-ray units require high DC voltages on the order of 100 kV or more.
2. Electrostatic precipitators, particle accelerators, and so on, require high DC voltages of several kilovolts and megavolts.
3. Testing the insulation of power apparatus requires high AC voltages of up to millions of volts, depending on their normal operating voltages, as discussed in Chapter 18.
4. Simulation of overvoltages that occur in power systems requires high impulse voltages of very short and longer durations.

In fact, one main concern of high-voltage engineers is for insulation testing of various power-system components under power-frequency AC, DC, switching, and lightning impulse voltages. Normally, in high-voltage testing, the currents are limited to a small value up to about 1 A under AC or DC voltages and a few amperes under impulse voltages. However, during testing of surge arresters or short-circuit testing of switchgear, currents several orders of magnitude higher are required. Methods of generating high voltages and high impulse currents are discussed in this chapter.

16.2   GENERATION OF HIGH ALTERNATING VOLTAGES

For generating AC test voltages of less than a few hundred kV, a single transformer can be used. For higher voltages, construction of a single transformer would entail undue insulation problems. Also, expenses, transportation, and erection problems connected with large testing transformers become prohibitive. These drawbacks are avoided by cascading several transformers of relatively small size with their high-voltage windings effectively connected in series.

16.2.1   Single-Unit Testing Transformers

Single-unit testing transformers do not differ from single-phase power transformers with regard to the design of core and windings in relation to the kVA output. However, particular attention is given to heavy insulation of the high-voltage winding and to low magnetizing currents for minimum distortion of the output voltage waveform. The equivalent impedances of testing transformers are usually within about 5%.

The iron core, as well as one terminal of the low-voltage and high-voltage windings, is usually maintained at earth potential. The other terminal of the high-voltage winding would be insulated to the full output voltage. Considerable economy is achieved, however, if the center point rather than one terminal of the high-voltage winding is earthed. Thus each terminal of the high-voltage winding will be insulated to only half the output voltage. A compensating winding is installed close to the core to reduce the high leakage reactance between the high- and low-voltage windings. The high-voltage winding is arranged in layers that are carefully insulated and the potential distribution over it is carefully controlled.

16.2.2   Cascaded Transformers

Cascaded transformers were first used by Petersen, Dessauer, and Welter (Fig. 16.1) (Kind, 1978). The low-voltage winding (L) of each upper stage is fed from an excitation winding (E) in the stage immediately below it. The L and E windings are therefore rated at currents higher than those of the H windings; the excitation windings at the lower stages evidently carry higher loading than those of the upper stages. If the power carried by the third stage of Figure 16.1 is P, the power carried by the second and first stages are, respectively, $2P$ and $3P$. The total short-circuit impedance of a cascade

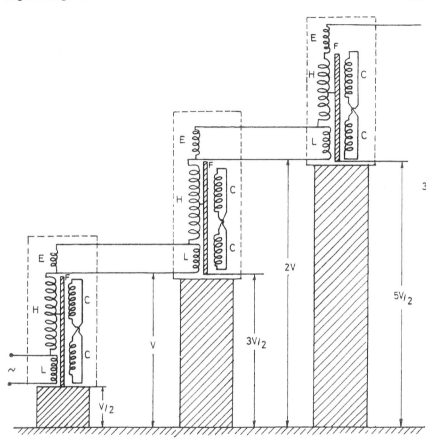

Figure 16.1 Cascaded transformers with compensating windings and extra insulation. H, high-voltage winding; L, low-voltage winding; E, excitation winding; C, compensating winding; F, iron core.

transformer can easily be related to the impedance of the individual stages (Kind, 1978).

If the output voltage of the first stage is V, the full output voltage is its multiple according to the number of stages. The iron core and the container (if metallic) of each stage other than the first are accordingly insulated from earth. This scheme is more expensive and requires more space than that of Fig. 16.1. To reduce the size and cost of the insulation, transformers with center-tapped high-voltage winding are cascaded as shown in Figure 16.1. The tank of the first stage is kept at $V/2$, while those of the second and third stages are at $3V/2$ and $5V/2$ (Fig. 16.1).

Cascade transformers of ratings up to 10 MVA and voltages up to 2.25 MV are available for both indoor and outdoor testing (Fig. 16.2). These transformers, being driven by a full-rated voltage regulator, are referred to as "straight circuits," as distinguished from resonant circuits.

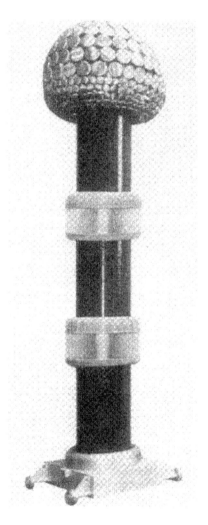

Figure 16.2 1200 kV cascaded transformer with insulating containers. Total height about 12 m. (Courtesy of Messwandler-Bau AG.)

16.2.3 Resonant Transformers

The test circuits discussed in Section 16.2.2 consist of a test transformer (single-unit or cascade) and a voltage regulator, both rated to take full power from the supply. A disadvantage of this arrangement is that harmonics can appear in the test voltage, either by accentuation of harmonics present in the incoming supply or generated in the transformer itself due to its nonlinear magnetic performance. Without filtering out these harmonics, such simple straight test circuits would not be entirely suitable for certain tests, such as those for partial discharges. An alternative method that is more economical and sometimes technically superior is offered by resonant circuits.

16.2.3.1 Parallel Resonance

The addition of parallel reactors either in the primary low-voltage circuit or in the secondary high-voltage circuit may partially or completely neutralize the capacitive load current, thus improving the power factor. If a motor–alternator is used as the supply source, the risk of self-excitation of the alternator is eliminated. Input power reductions of 10:1 are feasible, thus reducing drastically the cost of the regulator, reactors, and filters.

16.2.3.2 Series Resonance

An alternative system is the series resonance circuit. By resonating the circuit through a series reactor L (Fig. 16.3a) at the test frequency (50 or 60 Hz), harmonics are heavily attenuated. The shunt capacitance C usually represents the high-voltage bushing and the test object. Figure 16.3b shows the equivalent circuit of the test transformer. Since $R_e \ll \omega L_e$ and the voltage V_2 is almost in phase with V'_1 (Fig. 16.3c),

$$V_2 = V'_1 \frac{1}{1 - \omega^2 L_e C} \tag{16.1}$$

As resonance is approached, $V_2 \gg V'_1$. Thus there is no longer a fixed ratio of primary to secondary voltage. Therefore, the secondary voltage itself should be measured accurately in the tests.

The main advantages of series resonant transformers are:

1. Pure sinusoidal output waveforms.
2. Less power requirements from the mains (5–10% of straight circuit requirements).
3. No high-power arcing or heavy current surges occur if the test object fails, as the resonance is heavily disturbed by the resulting short circuit.
4. Cascading is also possible for producing very high voltages.

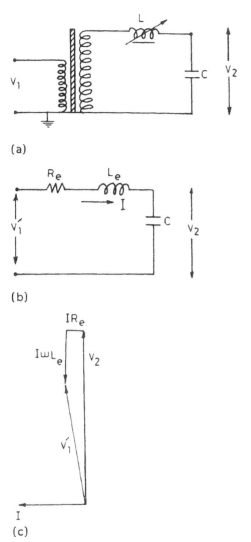

(a)

(b)

(c)

Figure 16.3 (a) Simplified circuit of a resonant transformer; (b) equivalent circuit; (c) phasor diagram.

5.   Simple and compact test setup.

   The main disadvantage is that the additional variable reactors should withstand the full-test voltage and full-current rating. Series resonance sets are usually brought into resonance by mechanical adjustment of an air gap in the iron core of the reactor. These sets are designed and commonly used for partial discharge testing (Fig. 16.4) and for testing installed gas-insulated systems in the field.

16.3 GENERATION OF HIGH DC VOLTAGES

Besides the Van de Graaff generators, with their special applications (Section 16.3.7.1), high-voltage AC and rectifiers are usually used for generating HVDC. The output DC has a trace of AC (ripple), which increases with the current drawn by the load. The "ripple factor," by definition, is equal to

$$\text{RF} = \frac{V(\text{av})_0 - V(\text{av})}{V(\text{av})_0} \tag{16.2}$$

where $V(\text{av})$ is the average value of the rectified output voltage, and $V(\text{av})_0$ is the value of $V(\text{av})$ at no load. Silicon rectifiers are commonly used. The

Figure 16.4 800 kV/8000 kVA series-resonant test system. (Courtesy of Hypotronics.)

AC supply to the rectifiers may be of power frequency or may be of a higher frequency. High frequency is recommended when a ripple of very small magnitude is required without use of expensive filters to suppress it. Different transformer–rectifier circuits for generating high DC voltages are discussed below.

16.3.1 Half- and Full-Wave Rectifier Circuits

In half-wave and full-wave rectifier circuits (e.g., Fig. 16.5), the capacitor C is gradually charged to $V_s(\max)$, the maximum AC secondary voltage of the HV transformer during the conduction periods. During the other periods, the capacitor is discharged into the load with a time constant $\tau = CR$. For reasonably small ripple voltage ΔV, τ should be at least 10 times the period

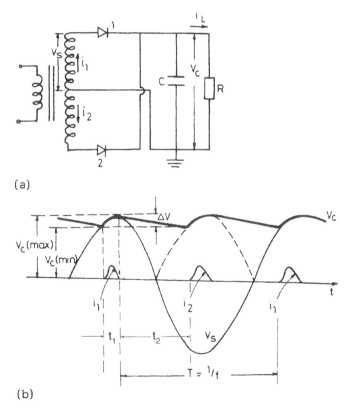

(a)

(b)

Figure 16.5 Single-phase full-wave rectifier with smoothing capacitor: (a) circuit diagram; (b) steady-state voltages and currents with load R.

of the AC supply. The rectifier element should have a peak inverse voltage of at least $2V_s(\max)$. To limit the charging current, an additional resistance is connected in series with the secondary of the transformer.

With half-wave rectification, and during one period $T(= 1/f)$ of the AC voltage, a charge q is supplied from the transformer to the capacitor within the short conduction period t_1 of the rectifier element and is transferred to the load R during the long nonconduction period t_2, so that

$$
\begin{aligned}
q &= \int_{t_1} i(t)\,dt = \int_{t_2} i_L(t)\,dt \\
&= \int_{t_2} \frac{V_c(t)}{R}\,dt = I_L T = \frac{I_L}{f}
\end{aligned}
\tag{16.3}
$$

where $t_1 \ll t_2 \simeq T = 1/f$ and I_L is the average load current.

Neglecting the voltage drops within the transformer and rectifiers during conduction, ΔV is easily found from the charge q transferred to the load:

$$
\Delta V = \frac{q}{C} = \frac{I_L}{fC}
\tag{16.4}
$$

This equation correlates the ripple voltage to the load current and the circuit parameters f and C. The product fC is an important design factor. According to equation (16.2), the voltage ripple factor for half-wave rectification is

$$
\mathrm{RF} = \frac{(\Delta V/2)}{V_s(\max)}
\tag{16.5}
$$

In full-wave rectifier circuits (Fig. 16.5), rectifiers 1 and 2 conduct during alternate half-cycles and charge the capacitor C. The HV transformer requires a center-tapped secondary with a voltage rating of $2V_s$. The ripple voltage according to equation (16.4) is halved.

16.3.2 Voltage-Doubler Circuits

In Figure 16.6a two half-wave rectifier circuits are connected in opposition, thus producing an unsmoothed unidirectional voltage. The source is effectively C_1, C_2, and the transformer in series. The output voltage has a peak value of $3V_s(\max)$ at no load. With the load R connected, the capacitors discharge through it with a corresponding voltage drop. This voltage-doubler circuit may be earthed at any point, provided that the transformer has an adequate insulation.

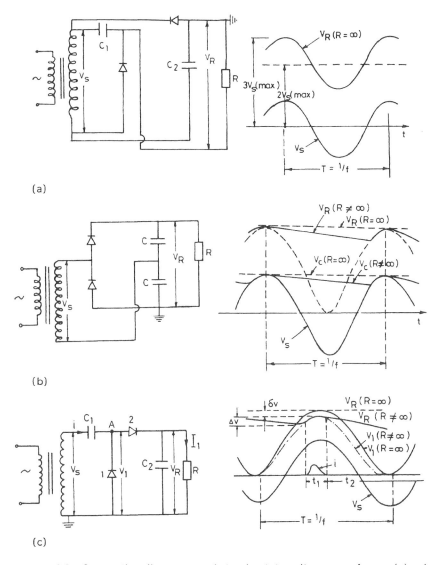

Figure 16.6 Connection diagrams and steady-state voltage waveforms: (a) output voltage contains a considerable AC component; (b and c) output voltage contains only a small ripple.

In the circuit of Figure 16.6b, each of the two capacitors is charged to only $V_s(\text{max})$ in alternate half-cycles and the total output voltage is $2V_s(\text{max})$ at no-load, with a small ripple under load.

As the doubler of Figure 16.6c represents the basic unit of the commonly used Cockcroft–Walton voltage-multiplier circuits, it is discussed in more detail. Rectifier 1 and capacitor C_1 act as a clamp so that the voltage at node A is positive going and is clamped at $0\,\text{V}$ with a peak value of $2V_s(\text{max})$. Rectifier 2 conducts and charges C_2 to $2V_s(\text{max})$ when the voltage at node A is at its peak value. As the voltage at A falls below $2V_s(\text{max})$, rectifier 2 stops conducting as the voltage on C_2 is now greater than the voltage on its anode. The capacitor C_2 provides the load current until the next peak of voltage occurs at node A. Thus, the voltage at the output reaches $2V_s(\text{max})$. The output and input waveforms, obtained from PSpice, are shown in Figure 16.6c (Price, 1997). It is quite clear that the doubler circuit takes a few cycles of the input for the output V_R to reach the maximum value.

Let a charge q be transferred from C_2 to the load per cycle when the average load current is I_1. Then $I_1 \simeq q/t_2$, where t_2 is the time during which rectifier 2 does not conduct; t_2 is much longer than t_1, the conduction time for rectifier 2. Also, let ΔV be the ripple voltage. Since $q = C_2 \Delta V$ and $t_1 \ll t_2 \simeq T = 1/f$, then

$$I_1 = C_2 \Delta V f = qf$$

or

$$\Delta V = \frac{I_1}{C_2 f} \tag{16.6}$$

At the same time, a charge q is transferred from C_1 to C_2 per cycle, so

$$\delta V = \frac{I_1}{C_1 f}$$

Therefore, the main voltage drop below $2V_s(\text{max})$ is $(\delta V + \Delta V/2)$ and the average output voltage

$$
\begin{aligned}
V_{av} &= 2V_s(\text{max}) - \delta V - \frac{\Delta V}{2} \\
&= 2V_s(\text{max}) - \frac{I_1}{f}\left(\frac{1}{C_1} + \frac{1}{2C_2}\right)
\end{aligned}
\tag{16.7}
$$

According to equation (16.2), the voltage ripple factor is

$$\text{RF} = \frac{\delta V + \Delta V/2}{2V_s(\text{max})} \tag{16.8}$$

16.3.3 Voltage-Tripler Circuits

A voltage tripler is an extension to the Cockcroft–Walton voltage-doubler circuit, as shown in Figure 16.7a. The operation is as follows. On the negative half-cycle, C_1 charges through rectifier 1 to $V_s(\text{max})$. On the positive half-cycle, rectifier 2 conducts and the peak voltage is $V_s(\text{max})$ from the source and $V_s(\text{max})$ from C_1; that is, the capacitor C_2 is charged to $2V_s(\text{max})$. During the next negative half-cycle, rectifier 3 conducts and voltage which is applied to C_3 is the sum of $V_s(\text{max})$ from the source and $2V_s(\text{max})$ from C_2 less the voltage on C_1; that is, C_3 is charged to $2V_s(\text{max})$. The output voltage is obtained between nodes A and B and has a value of $3V_s(\text{max})$.

The output and input waveforms obtained from PSpice are shown in Figure 16.7b where the output reaches $3V_s(\text{max})$ after a few cycles of input.

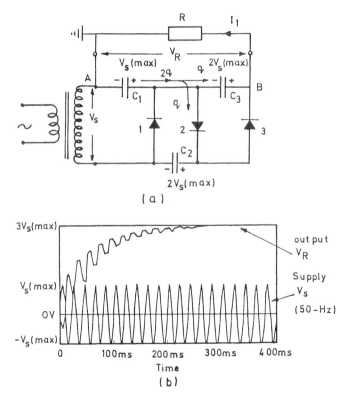

Figure 16.7 Voltage-tripler circuit: (a) circuit diagram; (b) voltage waveform obtained from PSpice.

The same principle can be extended to higher-order voltage multipliers.

Let a charge q be transferred through C_1 and C_3 to the load per cycle when the average load is I_1.

$$q = I_1/f \qquad (16.9)$$

where f is the supply frequency. The ripple voltage at capacitor C_3 is

$$\Delta V_3 = q/C_3 \qquad (16.10)$$

Simultaneously, C_1 transfers a charge q to the load and q to C_2 (Fig. 16.7a), so the ripple voltage at C_1 is

$$\Delta V_1 = 2q/C_1 \qquad (16.11)$$

Hence, the ripple voltage will be

$$\Delta V = \Delta V_1 + \Delta V_3 = \frac{2q}{C_1} + \frac{q}{C_3}$$
$$= q\left(\frac{2}{C_1} + \frac{1}{C_3}\right) = \frac{I_1}{f}\left(\frac{2}{C_1} + \frac{1}{C_3}\right) \qquad (16.12)$$

Because of the charge q transferred from C_1 to C_2 per cycle, the voltage drop

$$\delta V = q/C_2 \qquad (16.13)$$

Therefore, the total voltage drop below $3V_s(\text{max})$ is

$$\delta V + \frac{\Delta V}{2}$$

and the average output voltage

$$V_{av} = 3V_s(\text{max}) - \left(\delta V + \frac{\Delta V}{2}\right)$$
$$= 3V_s(\text{max}) - q\left(\frac{1}{C_2} + \frac{1}{C_1} + \frac{1}{2C_3}\right) \qquad (16.14)$$
$$= 3V_s(\text{max}) - \frac{I_1}{f}\left(\frac{1}{C_2} + \frac{1}{C_1} + \frac{1}{2C_3}\right)$$

According to equation (16.2), the voltage ripple factor is

$$RF = \frac{\delta V + \Delta V/2}{3V_s(\text{max})} \qquad (16.15)$$

16.3.4 Voltage-Multiplier Circuits

The voltage-multiplier circuit using the Cockcroft–Walton principle is shown in Figure 16.8a. The first stage, composed of rectifiers 1 and 2, the

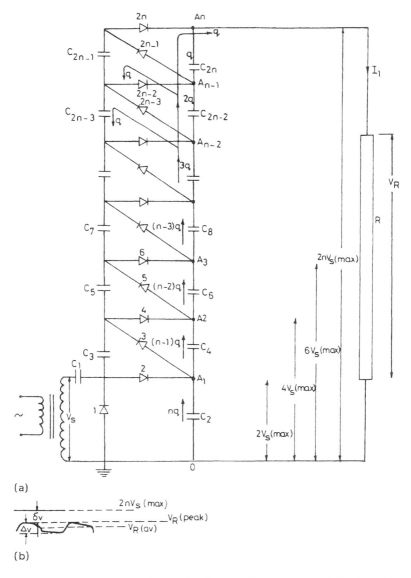

(a)

(b)

Figure 16.8 Voltage-multiplier circuit according to Cockcroft–Walton: (a) circuit diagram; (b) tracing of load voltage—drop δV and ripple ΔV.

capacitors C_1 and C_2, and the HV transformer, is identical to the voltage doubler shown in Figure 16.6c. For higher output voltages of 4, 6, ..., $2n$ of the input voltage $V_s(\text{max})$, the circuit is repeated with a cascade connection. Thus the total output voltage reaches $2nV_s(\text{max})$ above the earth potential (Fig. 16.8a). But the voltage across any individual capacitor or rectifier is $2V_s(\text{max})$, except for C_1, which is charged to $V_s(\text{max})$ only.

16.3.4.1 Ripple in Multiplier Circuits

Referring to Figure 16.8a, let q be the charge transferred from C_{2n} to the load per cycle with a ripple voltage at capacitor C_{2n}:

$$\Delta V_{2n} = \frac{q}{C_{2n}} \tag{16.16}$$

Simultaneously, C_{2n-2} transfers a charge q to the load and q to C_{2n-1}, so the ripple voltage at capacitor C_{2n-2} is

$$\Delta V_{2n-2} = \frac{2q}{C_{2n-2}} \tag{16.17}$$

Similarly, C_{2n-4} transfers a charge q to the load, q to C_{2n-3}, and q to C_{2n-1} with a ripple voltage at C_{2n-4}:

$$\Delta V_{2n-4} = \frac{3q}{C_{2n-4}} \tag{16.18}$$

Proceeding in the same way, the ripple voltage at C_2 is

$$\Delta V_2 = \frac{nq}{C_2} \tag{16.19}$$

Hence, the total ripple voltage will be

$$\Delta V_{\text{total}} = q\left(\frac{1}{C_{2n}} + \frac{2}{C_{2n-2}} + \frac{3}{C_{2n-4}} + \cdots + \frac{n}{C_2}\right) \tag{16.20}$$

It is quite evident that the capacitors near the ground terminal (e.g., C_2) are the major contributors to the ripple voltage If their capacitances are chosen suitably large, the total ripple voltage can be reduced significantly. However, this is not practical, as a transient overvoltage would overstress the smaller capacitors. Therefore, equal capacitances (C) are usually provided, and the corresponding total ripple voltage is

$$\Delta V_{\text{total}} = \frac{q}{C}\frac{n(n+1)}{2} = \frac{I_1}{fC}\frac{n(n+1)}{2} \tag{16.21}$$

16.3.4.2 Voltage Drop on Load

In addition to the ripple voltage ΔV, there is a voltage δV which is the difference between the theoretical no-load and on-load voltages. With reference to Figure 16.8a, capacitor C_2 loses nq during each cycle, so capacitor C_1 has to replenish it. Therefore, C_1 and C_2 charge up to $[V_s(\text{max}) - nq/C]$ and $[2V_s(\text{max}) - nq/C]$ instead of $[V_s(\text{max})]$ and $[2V_s(\text{max})]$, respectively. Similarly, C_3 can be charged only up to a maximum voltage of $[2V_s(\text{max}) - 2nq/C]$ instead of $2V_s(\text{max})$.

Thus one can easily derive the general rule for the total voltage drop at the smoothing column $C_2, C_4, C_6, \ldots, C_{2n}$ (Fig. 16.8a):

$$\delta V_2 = \frac{q}{C} n$$

$$\delta V_4 = \frac{q}{C}[2n + (n-1)]$$

$$\delta V_6 = \frac{q}{C}[2n + 2(n-1) + (n-2)]$$

$$\vdots$$

$$\delta V_{2n} = \frac{q}{C}[2n + 2(n-1) + 2(n-2) + \cdots + 2 \times 3 + 2 \times 2 + 1]$$

Adding all the n voltage drops gives the total voltage drop on load:

$$\delta V_{\text{total}} = \frac{q}{C}\left(\frac{2}{3}n^3 + \frac{1}{2}n^2 - \frac{n}{6}\right) = \frac{I_1}{fC}\left(\frac{2}{3}n^3 + \frac{n^2}{2} - \frac{n}{6}\right) \qquad (16.22)$$

In the case that the generator comprises a large number of stages (i.e., $n > 5$), the first term in equation (16.22) will dominate the others and the load voltage will be approximately (Fig. 16.8b)

$$V_R(\text{peak}) = 2nV_s(\text{max}) - \frac{I_1}{fC} \times \frac{2}{3}n^3 \qquad (16.23)$$

The output voltage $V_R(\text{peak})$ can be increased by increasing either C or the frequency f. An upper limit of 500–1000 Hz is set by the high voltage appearing across the circuit inductances and by the high capacitive currents to be drawn. An increase of supply frequency is, in general, more economical than an increase in the capacitance values. Small values of C provide a DC supply with limited stored energy, which might be an essential design factor for special breakdown investigations of dielectrics. For fast response to load changes and supply variations, high supply frequency and small stored energy are prerequisites.

As seen in Figure 16.8, for a given load, $V_R(\text{peak})$ can be increased with the number of stages n. This reaches an optimum value beyond which it drops if n is chosen too large. For given values of I_1, $V_s(\text{max})$, f, and C, the highest value of $V_R(\text{peak})$ is achieved with the "optimum" number of stages, that is, when

$$\frac{dV_R(\text{peak})}{dn} = 0 \quad \text{or} \quad n_{op} = \sqrt{\frac{V_s(\text{max})fC}{I_1}} \tag{16.24}$$

Thus, for a multiplier circuit with $f = 50\,\text{Hz}$, $C = 0.1\,\mu\text{F}$, $V_s(\text{max}) = 200\,\text{kV}$, and $I_1 = 8\,\text{mA}$, the optimum number of stages $n_{op} \simeq 12$. Using equations (16.21) and (16.22) (Fig. 16.8b), the mean output voltage

$$\begin{aligned} V_R(\text{av}) &= 2nV_s(\text{max}) - \delta V_{\text{total}} - \frac{\Delta V_{\text{total}}}{2} \\ &= 2nV_s(\text{max}) - \frac{I_1}{fC}\left(\frac{2}{3}n^3 + \frac{n^2}{2} - \frac{n}{6}\right) - \frac{I_1}{fC}\frac{n(n+1)}{4} \end{aligned} \tag{16.25}$$

Therefore, the ripple factor is expressed as

$$\text{RF} = \frac{2\delta V_{\text{total}} + \Delta V_{\text{total}}}{4nV_s(\text{max})} \tag{16.26}$$

16.3.5 Cascade Rectifier Circuits with Cascaded Transformers

The multiple charge transfer with the voltage multiplier of Cockcroft–Walton type demonstrated the limitation in the DC output concerning the ripple voltage and voltage drop. This disavantage can be overcome by cascading voltage doublers without a need for additional isolating transformers for every new stage. This can be achieved if the different stages are energized by specially designed cascaded transformers. The transformer in each stage comprises a tertiary low-voltage winding which excites the primary winding of the next upper stage. In this way, low ripple and voltage drops can be achieved when load currents are high. However, there are limitations with regard to the number of stages, as the lower transformers have to supply the energy for the upper ones.

16.3.6 Deltatron Circuits

A very sophisticated cascade transformer HV DC voltage multiplier circuit is shown in Figure 16.9. The very small ripple factor, high stability, fast response, and small stored energies are the main advantages of this circuit. However, the output of the circuit is limited to about 1 MV and a few milliamperes and needs excessive insulation for the higher stages.

The circuit consists primarily of a cascade connection of transformers having no iron cores and fed from a high-frequency source (Fig. 16.9). These

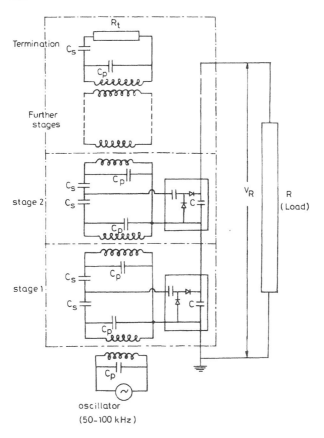

Figure 16.9 Deltatron circuit operating principle.

transformers are coupled by series capacitors C_s to compensate for their leakage inductances. Also, shunt capacitors C_p compensate for the transformers' magnetizing inductances. The string of cascaded transformers is loaded by a terminating resistor R_t (Fig. 16.9). The voltage remains nearly constant over the cascaded transformers but with a small phase shift with respect to the input voltage.

The usual Cockcroft–Walton cascade circuit is connected to each stage (Fig. 16.9). As the frequency is high, the capacitors C of the cascades can be made very small, with a correspondingly low energy stored and fast circuit response to load changes or supply variations. The small ripple voltage characterizing this circuit is caused not only by the small storage capacitors, but also by the phase shift between the input voltages of the different stages provided by their transformers.

16.3.7   Electrostatic HVDC Generators

16.3.7.1   Van de Graaff Generators

Van de Graaff generators have been developed to produce very high voltages of 5–6 MV with output currents of microamperes. They are useful for energizing particle accelerators. The main advantages of these generators are their ripple-free high DC voltages, which can easily be reached with precision and flexibility.

In Van de Graaff generator, the ions of either polarity produced by corona at pointed electrodes are carried mechanically by a very highly insulating belt to the top of the apparatus, where they are conducted to a voluminous high-voltage terminal. As more and more charges accumulate, the generator volltage builds up a shigh as millions of volts. It is limited only by the dielectric strength of the terminal's insulation to ground. Compressed gas and/or grading rings are used to raise the output voltage of the generator. The output current is limited to a few a microamperes (see example 5).

A modification of Van de Graaff's generator was made by Felici (1953) to eliminate its mechanical shortcomings as regards the belt and its vibrations. In Felici's generator, the belt is replaced by an insulating cylinder surrounding a gas discharge chamber and rotating at a high speed. The chamber is subjected to a high DC field and ions of both polarities are drawn from it by a pair of electrodes connected to the load. Such a source can produce an output of a few hundred kilovolts at a few milliamperes.

16.3.7.2   Variable-Capacitance Generators

These generators are designed to convert mechanical energy into electrical energy using the variable-capacitance principle. They consist of a stator with interleaved rotor vanes, forming a variable capacitor that operates in vacuum.

The current through a variable capacitor is given by

$$I = C\frac{dV_1}{dt} + V_1\frac{dC}{dt} \tag{16.27}$$

where C is the capacitance being charged to a voltage V_1. The power intput into the circuit at any instant

$$P = V_1 I = CV_1\frac{dV_1}{dt} + V_1^2\frac{dC}{dt} \tag{16.28}$$

If dC/dt is negative, mechanical energy is converted into electrical energy, and vice versa. With the capacitor charged from a DC voltage, $dV_1/dt = 0$ and the power input will be

$$P = V_1^2\frac{dC}{dt} \tag{16.29}$$

When the rotor-to-stator capacitance is maximum (C_m), the charge between the rotor and the stator is therefore $Q_m = C_m V_1$. As the rotor rotates, the capacitance C decreases and the voltage across $C(= C_m V_1 / C)$ increases, and further rotation of the rotor eventually causes the current to flow from the generator to the load.

A generator of this type, with an output voltage of 1 MV and a field gradient of 1 MV/cm in a hard vacuum and having 16 rotor poles, 50 rotor plates of 1.2 m maximum and 0.6 m minimum diameter, and a speed of 4000 rpm, could develop a maximum power of 7 MW. This vacuum-insulated generator was first discussed by Trump in 1947 and recently investigated by Philp (1977).

16.3.8 Regulation and Stabilization of HVDC Generators

The output of a DC generator changes with the load current as well as with the supply voltage. To maintain a constant voltage at the load terminals, it is essential to have a regulator circuit. It is necessary to keep the change in voltage between ±0.1% and ±0.001%, depending on the applications.

A DC voltage regulator consists of a detector that senses the voltage deviations from the desired value, and a controller actuated by the detector in such a manner as to correct the deviations (Naidu and Kamaraju, 1982). The regulators are generally of a series or a shunt type (Dorf, 1986).

If ΔV is the change in the DC output voltage V as a result of a change ΔV_1 in the AC supply voltage V_1, the stabilization ratio ψ_s is defined as

$$\begin{aligned}
\psi_s &= \frac{\Delta V_1 / V_1}{\Delta V / V} \\
&= \frac{\Delta V_1}{\Delta V} \frac{V}{V_1}
\end{aligned} \tag{16.30}$$

ψ_s is greater than unity and indicates how many times better V is in stability compared wtih V_1.

The "regulation" ψ_r of a DC voltage regulator is defined as the fractional change in V caused by a fractional change in the output current I, that is,

$$\begin{aligned}
\psi_r &= \frac{\Delta V / V}{\Delta I / I} \\
&= \frac{\Delta V / \Delta I}{V / I} = \frac{R_0}{R}
\end{aligned} \tag{16.31}$$

where R_0 is the effective internal resistance of the regulator as seen from its output terminals, and R is the resistance of the load.

16.4   GENERATION OF IMPULSE VOLTAGES

Standardized lightning impulse waves are represented by the general equation

$$e(t) = A[\exp(-\alpha_1 t) - \exp(-\alpha_2 t)] \qquad (16.32)$$

The graph of this expression is shown in Figure 16.10. The tolerances allowed in the front and tail durations are ±30% and ±20%, respectively. However, the tolerance allowed in the peak value is ±3%. The standard lightning impulse wave has a front duration of 1.2 μs and a wavetail duration of 50 μs, and is described as a 1.2/50 μs wave. The standard wave front and tail are defined in Figure 16.21 (IEC, 1973).

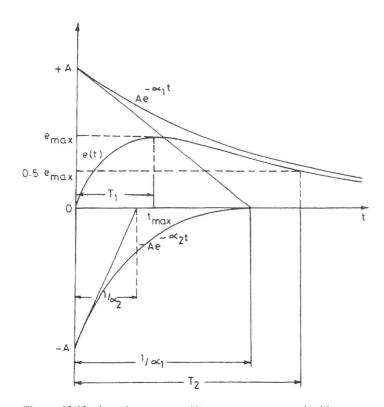

Figure 16.10   Impulse wave and its components as a double exponential.

16.4.1   Circuit for Producing Lightning Impulses

The double-exponential waveform expressed by equation (16.32) may be obtained in the laboratory by any connection of the R–C circuit as exemplified in Figure 16.11a–d or by a series R–L–C circuit under overdamped conditions (Fig. 16.11e).

The circuits commonly used for impulse generators are those of Figure 16.11b and c. The advantages of these circuits are that the wavefront and wavetail are independently controlled by changing R_1 and R_2 separately, which is different from the circuit of Figure 16.11e; second, being capacitive, the test objects admittedly form a part of C_2.

The magnitudes of the different circuit elements together determine the required rapid charging of C_2 to the peak value e_{max}. However, a long wavetail calls for a slow discharge. This is achieved if R_2 is much larger than R_1 (Fig. 16.11b–d). The smaller the time constant of the circuit, the faster is the rate by which the output voltage $e(t)$ approaches its peak value. The peak value e_{max} cannot exceed the value determined by the distribution of the initially available charge VC_1 between C_1 and C_2, so the voltage efficiency

$$\eta = \frac{e_{max}}{V} < \frac{C_1}{C_1 + C_2} \tag{16.33}$$

For high efficiency, C_1 should be chosen much larger than C_2. The exponential decay of the impulse voltage on the tail occurs with a time constant of about $C_1(R_{11} + R_2)$, $C_1(R_1 + R_2)$, $C_1 R_2$, and $C_1(R_1 + R_2)$ in circuits a, b, c, and d of Figure 16.11, respectively. The impulse energy consumed is

$$W = \frac{1}{2} C_1 V^2$$

which is an important characteristic parameter of the impulse generator.

It is sufficient to analyze circuit a of Figure 16.11, being a general one. The other circuits (b–d) are simply special forms of circuit a. Analysis of the inductive circuit e is that of a simple R–L–C circuit.

16.4.1.1   Circuit Analysis

With reference to circuit a of Figure 16.11, on breakdown of the gap G, the capacitor C_1 discharges through the circuit while its voltage decreases from its initial magnitude V_0.

$$i_1 = i_2 + i_0 \tag{16.34}$$

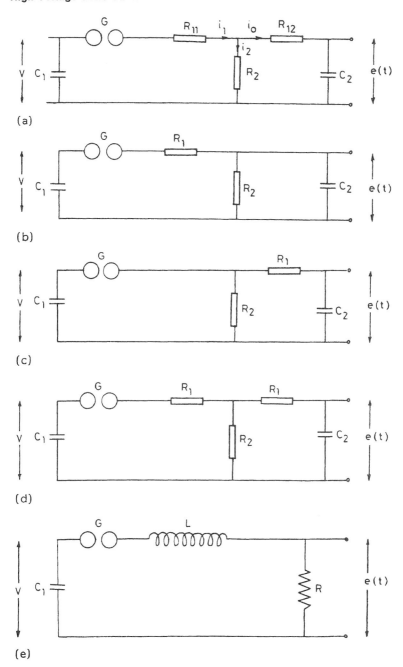

Figure 16.11 Circuits for producing impulse voltage waves.

$$V = \frac{-1}{C_1} \int i_1 \mathrm{d}t + V_0 \tag{16.35}$$

$$e = \frac{1}{C_2} \int i_0 \mathrm{d}t \tag{16.36}$$

Hence

$$i_1 = \left(\frac{R_{12}}{R_2} i_0 + \frac{e}{R_2} \right) + i_0 \tag{16.37}$$

Using simple mathematical manipulation, the following relation for e is reached:

$$A_1 \frac{\mathrm{d}^2 e}{\mathrm{d}t^2} + A_2 \frac{\mathrm{d}e}{\mathrm{d}t} + A_3 e = 0 \tag{16.38}$$

with the initial conditions

$$e \big|_{t=0} = 0$$

and

$$\frac{\mathrm{d}e}{\mathrm{d}t} \bigg|_{t=0} = \frac{C_1 V_0}{A_1}$$

Hence

$$e(t) = \frac{C_1 V_0}{A_1 (\alpha_2 - \alpha_1)} [\exp(-\alpha_1 t) - \exp(-\alpha_2 t)] \tag{16.39}$$

The peak value $e_{\max}$ and the corresponding period $t_{\max}$ can be evaluated by differentiating equation (16.39).

In a normalized form, the impulse wave represented by equation (16.32) can be written as (Auget and Lanovici, 1982)

$$e(t) = \eta \frac{\alpha V_0}{\sqrt{\alpha^2 - 1}} [\exp(-\alpha_1 t) - \exp(-\alpha_2 t)] \tag{16.40}$$

where

$$\alpha_1 = \frac{\alpha - \sqrt{\alpha^2 - 1}}{\theta} \tag{16.41}$$

$$\alpha_2 = \frac{\alpha + \sqrt{\alpha^2 - 1}}{\theta} \tag{16.42}$$

and

$$\theta = \sqrt{C_1 C_2 (R_{11} R_{12} + R_{11} R_2 + R_{12} R_2)} \qquad (16.43)$$

$$\eta = \frac{C_1}{C_1(1 + R_{11}/R_2) + C_2(1 + R_{12}/R_2)} \qquad (16.44)$$

$$\alpha = \frac{1}{2} \frac{R_2 C_1}{\eta \theta} \qquad (16.45)$$

Table 16.1 gives the values of the parameters θ, η, and α and the corresponding elements of the circuits (Kind, 1978; Kuffel and Zaengl, 1984; Auget and Lanovici, 1982). Evidently, in circuit b, $R_{12} = 0$ and $R_{11} = R_1$; in circuit c, $R_{11} = 0$ and $R_{12} = R_1$; and in circuit d, $R_{12} = R_{11} = R_1$.

16.4.1.2 Dimensioning of Circuit Elements

For a given wave shape, the choice of the resistances to control the wavefront and wavetail durations is not entirely independent but depends on the ratio C_1/C_2. However, there is a limitation on this ratio. For circuit b of Figure 16.11 there is an upper limit whereas for circuit c there is a lower limit. Table 16.2 gives the relationship between the time constants $1/\alpha_1$ and $1/\alpha_2$ and their ratios to the durations T_1 and T_2 for typical impulse waves including lightning and switching impulses. Table 16.3 gives the corresponding limiting values of C_1/C_2.

The ratios T_1/T_2 and T_2/θ versus α are shown in the nomograms of Figures 16.12 and 16.13 for lightning and switching impulses (Auget and Lanovici, 1982).

16.4.1.3 Effect of Circuit Inductance

Any residual inductance in the circuit elements would produce some oscillations in the wavefront and wavetail. It is noted that, with increasing series inductance, the wavefront sensitively increases, but the magnitude of the peak value is affected only slightly. The lengthening of the wavefront by increasing the circuit inductance provides a convenient method for generating long-front (switching) impulses. On the other hand, an inductance across the load would drastically change the waveform, as depicted in Figure 16.14.

16.4.2 Multistage Impulse Generators

This is the best way to generate voltage impulses of very high amplitude (millions of volts) using a DC source of a moderate output. A bank of capacitors is charged in parallel and then discharged in series, as originally proposed in 1923 by Marx. Since then, various modifications have been implemented.

Table 16.1   Parameters θ, η, and α and the Wave-Shaping Elements for Different Circuits Producing Impulse Waves

	θ	η	α	X	$R_1(\Omega)$	$R_2(\Omega)$
Circuit (a)	$\sqrt{C_1 C_2 (R_{11}R_2 + R_2 R_{12} + R_{11}R_{22})}$	$\dfrac{1}{(1+R_{11}/R_2 + C_2/C_1(1+R_{12}/R_2))}$	$\dfrac{R_2 C_1}{2\theta\eta}$			
Circuit (b)	$\sqrt{C_1 C_2 R_1 R_2}$	$\dfrac{1}{1+C_2/C_1 + R_1/R_2}$	$\dfrac{R_2 C_1}{2\theta\eta}$	$\dfrac{1}{\alpha^2}\left(1+\dfrac{C_1}{C_2}\right)$	$\dfrac{\alpha\theta}{C_1}(1-\sqrt{1-X})$	$\dfrac{\alpha\theta}{C_1+C_2}(1+\sqrt{1-X})$
Circuit (c)	$\sqrt{C_1 C_2 R_1 R_2}$	$\dfrac{1}{1+C_2/C_1(1+R_1/R_2)}$	$\dfrac{R_2 C_1}{2\theta\eta}$	$\dfrac{1}{\alpha^2}\left(1+\dfrac{C_2}{C_1}\right)$	$\dfrac{\alpha\theta}{C_2}(1-\sqrt{1-X})$	$\dfrac{\alpha\theta}{C_1+C_2}(1+\sqrt{1-X})$
Circuit (d)	$\sqrt{C_1 C_2 R_1 (R_1 + 2R_2)}$	$\dfrac{1}{(1+R_1/R_2)(1+C_2/C_1)}$	$\dfrac{R_2 C_1}{2\theta\eta}$	$\dfrac{1}{4\alpha^2}\left(1+\dfrac{C_1}{C_2}\right)\left(1+\dfrac{C_2}{C_1}\right)$	$\dfrac{2\alpha\theta}{C_1+C_2}(1-\sqrt{1-X})$	$\dfrac{2\alpha\theta}{C_1+C_2}\sqrt{1-X}$
Circuit (e)	$\sqrt{LC_1}$		$\dfrac{R}{2}\sqrt{\dfrac{C}{L}}$		$L(H)=\dfrac{\theta^2}{C_1}$	$R(\Omega)=\dfrac{2\alpha\theta}{C_1}$

Circuit (a)

Circuit (b)

Circuit (c)

Circuit (d)

Circuit (e)

Table 16.2 $1/\alpha_1$, $1/\alpha_2$, and αT for Different Standard Waves

T_1/T_2 (μs)	$1/\alpha_1$ (μs)	$1/\alpha_2$ (μs)	α_2/T_1 (μs)	$\alpha_1 T_2$ (μs)
1.2/5	3.48	0.8	1.49	1.44
1.2/50	68.2	0.405	3.06	0.73
1.2/200	284.0	0.381	3.15	0.7
250/2500	2877.0	104.0	2.4	0.87

Table 16.3 Limiting Values of C_1/C_2 for Different Standard Waves

Circuit	Value measured	T_1/T_2(μs)			
		1.2/5	1.2/50	1.2/200	250/2500
Figure 16.11c	Min (C_1/C_2)	1.57	0.025	0.0054	0.157
Figure 16.11b	Max (C_1/C_2)	–	40.0	185.19	6.37

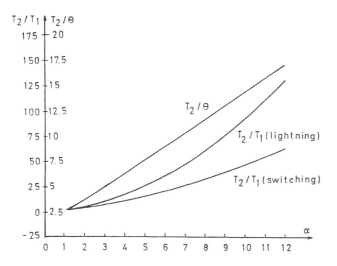

Figure 16.12 T_1/T_2 and T_2/θ versus α for lightning and switching impulses. ($1 \leqslant \alpha < 12$).

A typical modification of a Marx circuit is shown in Figure 16.15. The resistances R_1 and R_2 are incorporated in the individual stages. R_1/n is connected in series with the gap G of each stage. R_2 is also divided into n parts connected across the stage capacitors. These modifications save the space and cost of the generator and result in smaller resistors and higher voltage efficiency.

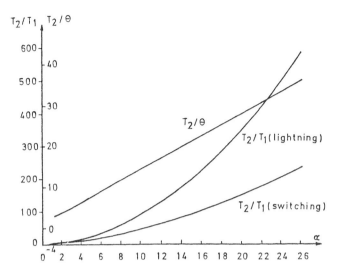

Figure 16.13 T_1/T_2 and T_2/θ versus α for lightning and switching impulses. $(12 \leqslant \alpha \leqslant 26)$.

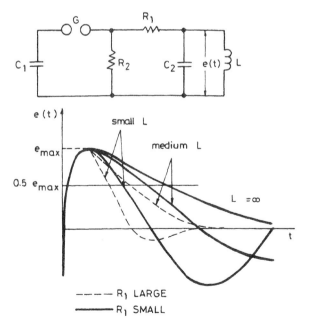

Figure 16.14 Effect of varying the inductance across the output terminals on the generated impulse voltage wave shape $e(t)$.

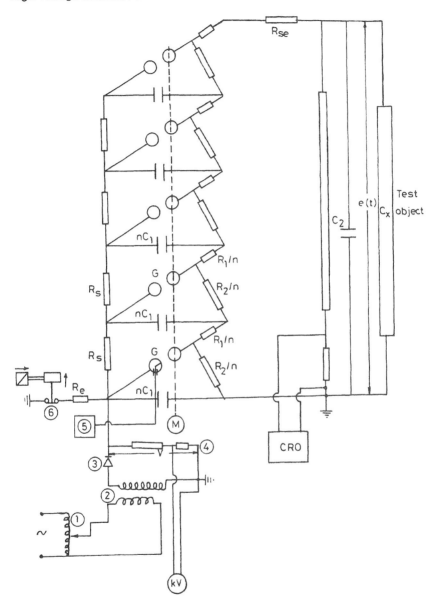

Figure 16.15 Multistage impulse generator incorporating the wave-shaping resistors within the generator proper. 1, regulating transformer; 2, HV transformer; 3, rectifier; 4, measuring resistive divider for the charging voltage; 5, triggering unit; 6, earthing device; M, motor for control of sphere gaps; R_1/n front resistors; R_2/n, tail resistors; R_e, discharging resistor; R_{se}, external part of front resistor.

The gap spacing is chosen such that the breakdown voltage of the gaps G is slightly higher than the charging voltage V. Thus all the capacitors are charged to the voltage V. When the impulse generator is to be discharged, the gaps G are made to spark over simultaneously by some external means. Thus all the capacitors get connected in series and discharge through the wave-shaping resistors R_1 and R_2 into the test object. The discharge time constant $C_1 R_1$ is very small (microseconds) compared to the charging time constant $nC_1 R_s$; thus no significant discharge takes place through the charging resistors R_s. Multistage impulse generators are usually specified by their total output voltage, the number of stages n, and the stored gross energy (Hyltén-Cavallius, 1986).

16.4.2.1   Control of Multistage Impulse Generators

In multistage impulse generators, the spark gaps are arranged such that sparking of the lowest gap induces sparking of the other gaps, the voltage across them being only slightly lower than their breakdown voltage. Thus the impulse is generated at the instant a triggering signal is received by the lowest gap.

Up to the 1950s, the lowest gap was a three-electrode gap, as proposed by White in 1947, but, since then, trigatron gaps have been in common use. The trigatron gap comprises a high-voltage spherical electrode and an earthed main hemispherical electrode, with the main gap between them. A trigger electrode with the shape of a metal rod is located inside and insulated from the main electrode by an annular clearance (Fig. 16.16). The trigatron is included in a pulse circuit such as the one shown in Figure 16.16. Closing the switch S produces, between the trigger and the main electrode, a voltage

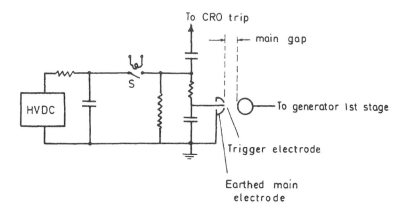

Figure 16.16   Impulse generator tripping circuit using a trigatron spark gap.

pulse sufficient to break down the annular clearance. The ionization of the discharge and the disturbance of the field in the main gap causes its complete breakdown. This breakdown initiates the breakdown of gaps G of the other stages of the multistage impulse generator (Fig. 16.15). Thus the instant of generating the impulse can be controlled. The switch S can be replaced by a synchronizing circuit so as to generate the impulse at the required instant with respect to the power-frequency AC, thus enabling AC and impulse high voltages to be superimposed on a test specimen as required. A trigatron gap connected across the load can be similarly triggered at the appropriate subsequent moment to produce chopped impulse waves.

In a recent work, the tripping pulse for the trigger electrode of the trigatron gap is obtained from an auxiliary impulse generator (Fig. 16.17). The latter is controlled electronically by a triac circuit (Abdel-Salam and El-Mohandes, 1991). Thus, to conrol the firing of the auxiliary generator and therefore the main generator, an electronic circuit was developed. The circuit consisted of: (i) a controlled switch connected in series with the spark gap of the auxiliary generator; (ii) a triac to control the switch; and (iii) a triggering circuit for the triac, including a 555 duty-cycle oscillator, a crossing-detector (Z.C.D.), a 7408 AND gate and a power-transistor amplifier (Fig. 16.17).

A modification of the trigatron gap was suggested in which the triggering spark is produced by breaking the current in a low-voltage inductive circuit at a pair of contacts embedded in the earthed main electrode. When the contacts are remotely separated, the energy stored in the circuit inductance is released in the spark, which initiates the breakdown of the main gap as mentioned before.

16.5   GENERATION OF SWITCHING IMPULSES

The switching waveforms experienced in power systems are not unique. They have front rise times up to several milliseconds and considerably longer tails. They may contain oscillatory components with a frequency ranging from a few hertz to a few kilohertz (Chapter 14). Switching surges are accompanied by energies much larger than those of lightning impulses.

Several circuits have been proposed in the literature for generating switching impulses. They are grouped as follows:

1.   Impulse generator circuits are modified to give wave shapes of longer duration by a proper choice of the front and tail resistors. In order to produce unidirectional damped oscillations, an inductance L is connected in series with the conventional impulse gen-

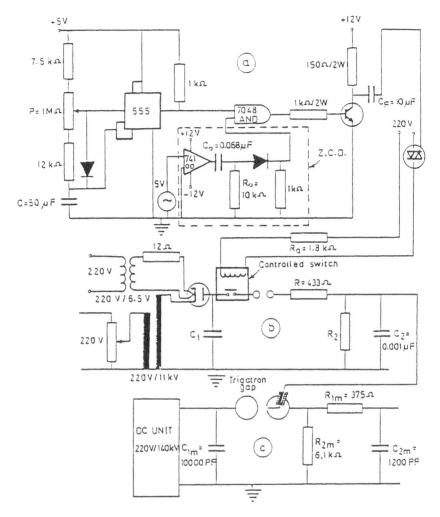

Figure 16.17 Electronic tripping circuit using a trigation spark gap: (a) the triac and its triggering circuit; (b) auxiliary impulse generator; (c) main impulse generator.

erator circuit (Fig. 16.18). These oscillations may have a frequency to 1–10 kHz, depending on the circuit parameters. A sphere gap may be included in parallel with the test object for producing chopped waves, the same as with lightning impulses. The main drawback of these circuits is that the efficiency of the generator gets reduced to 50% or even less.

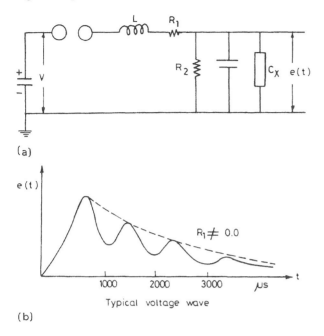

(a)

(b)

Typical voltage wave

Figure 16.18 Circuit for producing a switching impulse voltage and its output waveform. C_X, test object.

2. A testing transformer is excited by a voltage impulse to give long-duration or oscillatory waves in its secondary. The capacitor C_1 of Figure 16.19 is charged to a moderate DC voltage (20–25 kV) and then discharged into the low-voltage winding of a power transformer. The high-voltage winding is connected in parallel to a load capacitance C_2 and the test object. Through autotransformer action, switching impulses of proper waveform can be generated across the test object. The disadvantages of this technique are associated with the considerable amount of high-frequency distortion in the output waveform. Furthermore, the size of the capacitor C_1 for producing a reasonable output voltage may be large. Variations of the front of the generated impulse were achieved by adding inductances of different magnitudes in series with C_1. Such a method proved tedious and, in most cases, uneconomical.

    A preferable method is to energize the transformer from an AC supply for less than half a cycle. The energization of the transformer is initiated via a thyristor or an ignitron to be triggered at the required phase angle. The transformer remains ener-

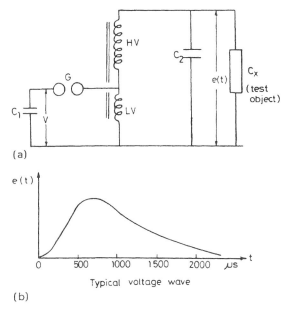

(a)

e(t)

Typical voltage wave

(b)

Figure 16.19 Circuit for producing a switching impulse voltage using a test transformer energized from a charged capacitor.

gized until the supply current passes through a zero value. Then the energy stored in the transformer is discharged through the test object. For generating a much higher output voltage, several transformer units were arranged in cascade (Anis et al., 1975). The advantages inherent in this method are mainly its small high-frequency distortion in the output voltage and the elimination of the energy storage capacitors C_1 (Fig. 16.19).

3. For producing damped oscillations, the source of high voltage is a Tesla coil, which consists of an air-cored transformer. Its LV side is connected to the DC or AC supply circuit through a capacitor C_1 and a series element (R_s, L_s) (Fig. 16.20a). C_2 is the equivalent capacitance of the high-voltage winding and the test object. On closing the switch S_1, an oscillatory current will flow in the LV circuit, inducing high oscillatory voltage in the HV circuit of the Tesla coil. The frequency of the induced oscillations can be made the same by tuning the two circuits (i.e., when $L_1 C_1 = L_2 C_2$). The generator above has been modified by Sandhaus (1976). The switch S_1 was replaced by a triac for better control of the firing instant (Fig. 16.20b).

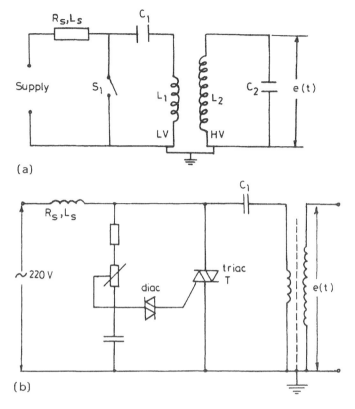

Figure 16.20 (a) Tesla coil equivalent circuit (S_1, switch); (b) switch S_1 is replaced by a triac.

16.6 GENERATION OF IMPULSE CURRENTS

Lightning strokes involve both high-voltage and high-current impulses on transmission systems. Therefore, generation of impulse currents of the order of several hundreds of kiloamperes finds application in testing lightning arresters. They are also used in electric-arc and electric-plasma studies. The waveshapes in common use are the double exponential and rectangular waves (Fig. 16.21).

16.6.1 Double-Exponential Current Impulses

The impulse wave shape is defined according to the IEC recommendations. T_1 is the front duration and T_2 is the time to half peak. According to the IEC (1973), standard waves are 4/10 and 8/20 μs.

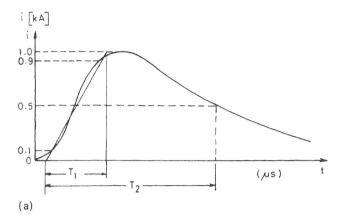

(a)

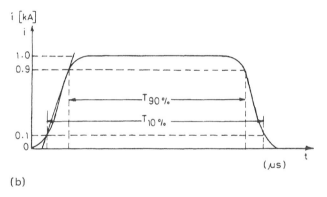

(b)

Figure 16.21 Impulse current double-exponential (a) and rectangular (b) wave-forms.

To produce one of the standard current waveforms, a circuit such as the one shown in Figure 16.22 is normally used. The basic difference between this circuit and those generating impulse voltages (Figs. 16.11 and 16.15) is that here the capacitances are much larger and the resistors are much smaller in magnitude.

If the capacitor C is charged to a voltage V and discharged when the gap G is triggered (Fig. 16.22), the current i can be shown to vary with time according to the relation

$$i(t) = \frac{V}{\omega L} \exp(-\gamma t) \sin \omega t \tag{16.46}$$

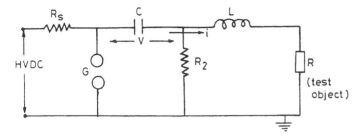

Figure 16.22   Circuit for producing double-exponential current impulses.

where $\gamma = R/2L$ and $\omega = \sqrt{1/LC - R^2/4L^2}$. The equivalent resistance of the test object should be chosen as $R < 2\sqrt{L/C}$ for underdamped oscillation in the current wave. The time taken for the current i to rise from zero to the first peak is

$$T_1 = \frac{1}{\omega}\sin^{-1}\omega\sqrt{LC} \tag{16.47}$$

If the test object is an ideal surge arrester, the following approximate expressions are used for determining the waveforms of the impulse current (Haefely, 1982):

$$\text{Peak current} = \frac{V - V_r}{\sqrt{L/C}} \tag{16.48}$$

where V_r is the residual voltage of the surge arrester.

$$\text{Front time } T_1 \simeq 1.25\sqrt{LC} \tag{16.49}$$

$$\text{Time to half-peak value } T_2 \simeq 2.5\sqrt{LC} \tag{16.50}$$

16.6.2   Rectangular Current Impulses

The rectangular impulse wave as defined by the IEC recommendations is shown in Figure 16.21. The duration $T_{10\%} < 1.5T_{90\%}$ (IEC, 1970). The rectangular current impulses are generated by discharging an artificial transmission line with lumped L and C elements into the test object through a sphere gap. If the line is charged to a DC voltage V and discharged through the test object of resistance R, the current pulse is given by $I = V/(R + Z_0)$, where Z_0 is the surge impedance of the line. A pulse voltage of $RV/(R + Z_0)$ is developed across the test object. The duration $T_{90\%}$ of the current wave is estimated as

$$T_{90\%} \simeq 2\frac{n-1}{n}\sqrt{L/C} \qquad\qquad (16.51)$$

where n is the number of L–C stages (Modrusan, 1977).

16.7. SOLVED EXAMPLES

(1) A Cockcroft–Walton-type voltage doubler has capacitances C_1 and C_2 of $0.01\,\mu\text{F}$ and $0.05\,\mu\text{F}$, respectively. The supply voltage V_s is $100\,\text{kV}$ at a frequency of $50\,\text{Hz}$. If the load current to be supplied is $5\,\text{mA}$, find: (a) ripple voltage; (b) voltage drop; (c) average output voltage; and (d) ripple factor.

Solution:
(a) Ripple voltage $\Delta V = \dfrac{I_1}{C_2 f}$

$$= \frac{5\times 10^{-3}}{0.05\times 10^{-6}\times 50} = 2000\,\text{V} = 2.0\,\text{kV}$$

(b) Voltage drop $= \delta V + \Delta V/2$

$$= \frac{I_1}{f}\left(\frac{1}{C_1}+\frac{1}{2C_2}\right)$$

$$= \frac{5\times 10^{-3}}{50}\left(\frac{10^6}{0.01}+\frac{10^6}{2\times 0.05}\right)$$

$$= \frac{5\times 10^{-3}\times 10^6}{50}\times 110$$

$$= 11\times 10^3\,\text{V} = 11\,\text{kV}$$

(c) $V_{\text{av}} = 2V_s(\text{max}) - (\delta V + \Delta V/2)$
$\qquad = 2\times 100\sqrt{2} - 11 = 271.8\,\text{kV}$

(d) RF $= \dfrac{\delta V + \Delta V/2}{2V_s(\text{max})} = 3.89\%$

(2) A Cockcroft–Walton-type voltage tripler has capacitances C_1, C_2, and C_3 of $0.01\,\mu\text{F}$, $0.05\,\mu\text{F}$, and $0.10\,\mu\text{F}$, respectively. The supply voltage V_s is $100\,\text{kV}$ at a frequency of $50\,\text{Hz}$. If the load current to be supplied is $5\,\text{mA}$, find: (a) ripple voltage; (b) voltage drop; (c) average output voltage; and (d) ripple factor.

Solution:

(a) Ripple voltage $\Delta V = \dfrac{I_1}{f}\left(\dfrac{2}{C_1} + \dfrac{1}{C_3}\right)$

$$= \frac{5 \times 10^{-3}}{50}\left(\frac{2 \times 10^6}{0.01} + \frac{1 \times 10^6}{0.1}\right)$$

$$= \frac{5 \times 10^{-3} \times 10^6}{50} \times 210$$

$$= 21{,}000\,\text{V} = 21\,\text{kV}$$

(b) Voltage drop $= \delta V + \Delta V/2$

$$= \frac{I_1}{f}\left(\frac{1}{C_2} + \frac{1}{C_1} + \frac{1}{2C_3}\right)$$

$$= \frac{5 \times 10^{-3}}{50}\left(\frac{10^6}{0.05} + \frac{10^6}{0.01} + \frac{10^6}{2 \times 0.1}\right)$$

$$= \frac{5 \times 10^{-3} \times 10^6}{50} \times 125$$

$$= 12{,}500\,\text{V} = 12.5\,\text{kV}$$

(c) $V_{\text{av}} = 3V_s(\text{max}) - (\delta V + \Delta V/2)$
$= 2 \times 100\sqrt{2} - 12.5 = 411.76\,\text{kV}$

(d) $\text{RF} = \dfrac{\delta V + \Delta V/2}{3V_s(\text{max})} = 2.95\%$

(3) A Cockcroft–Walton-type voltage multiplier has twelve stages with capacitances all equal to $0.15\,\mu\text{F}$. The supply voltage V_s is $200\,\text{kV}$ at a frequency of $50\,\text{Hz}$. If the load current to be supplied is $5\,\text{mA}$, find: (a) ripple voltage; (b) voltage drop; (c) average output voltage; (d) ripple factor; and (e) optimum number of stages for minimum voltage drop.

Solution:

(a) Ripple voltage $\Delta V = \dfrac{I_1}{fC}\dfrac{n(n+1)}{2}$

$$= \frac{5 \times 10^{-3}}{50 \times 0.15 \times 10^{-6}}\frac{12 \times 13}{2}$$

$$= 52 \times 10^3\,\text{V} = 52\,\text{kV}$$

(b) Voltage drop $= \delta V + \Delta V/2$

$$= \frac{I_1}{fC}\left[\frac{2}{3}n^3 + \frac{n^2}{2} - \frac{n}{6} + \frac{n(n+1)}{4}\right]$$

$$= \frac{5 \times 10^{-3}}{50 \times 0.15 \times 10^{-6}}\left[\frac{2}{3} \times 1728 + \frac{144}{2} - \frac{12}{6} + \frac{12 \times 13}{4}\right]$$

$$= 840 \times 10^3 \, \text{V} = 840 \, \text{kV}$$

(c) $V_{av} = 2nV_s(\text{max}) - $ voltage drop
$= 2 \times 12 \times 200\sqrt{2} - 840 = 5948 \, \text{kV}$

(d) $\text{RF} = \dfrac{\delta V + \Delta V/2}{2nV_s(\text{max})} = \dfrac{840}{2 \times 12 \times 200\sqrt{2}} = 12.4\%$

(e) $n_{\text{optimum}} = \sqrt{\dfrac{V_s(\text{max})fC}{I_1}} = \sqrt{\dfrac{200\sqrt{2} \times 10^3 \times 50 \times 0.15 \times 10^{-6}}{5 \times 10^{-3}}}$

$= 20.59$
$= 20$ stages

(4) A 100 kVA, 220 V/250 kV testing transformer has 8% leakage reactance and 1.5% resistance (based on transformer rating). A long cable has to be tested at 500 kV using this transformer as a resonant transformer operating at 50 Hz. If the charging current of the cable at 500 kV is 4.0 A, find the inductance of the required series inductor. Assume 2% for the resistance of the used inductor and the connecting leads. Neglect dielectric loss of the cable as well as the no-load current of the transformer. What will be the input voltage and power to the transformer?

Solution: Reactance of the cable

$$X_c = V_c/I_c = 500 \times 10^3/4.0 = 125 \, \text{k}\Omega$$

Leakage reactance of the transformer

$$X_L = \%X_l \times \left(\frac{\text{kV}^2}{\text{kVA}}\right)_{\text{base}} \times 10^3$$

$$= \frac{8}{100} \times \frac{(250)^2}{100} \times 10^3 = 5 \times 10^4 \, \Omega$$

$$= 50 \, \text{k}\Omega$$

At resonance, $X_L = X_C$. Hence, the additional series reactance

$$= 125 - 50 = 75\,k\Omega$$

Inductance of the required series inductor

$$= \frac{75 \times 10^3}{2\pi \times 50} = 238.8\,H$$

Total circuit resistance

$$R = \frac{3.5}{100} \times \left(\frac{kV^2}{kVA}\right)_{base} \times 10^3 = \frac{3.5}{100}\left(\frac{250^2}{100}\right) \times 10^3 = 21.875\,k\Omega$$

The maximum current I that can be supplied by the transformer $= 100/250 = 0.4\,A$
The exciting voltage on the transformer secondary

$$= IR = 0.4 \times 21.875 \times 10^3\,V = 8.75\,kV$$

The input voltage to the transformer primary

$$= 8.75 \times 10^3 \times \frac{220}{250 \times 10^3} = 7.7\,V$$

The input power to the transformer

$$= \frac{7.7 \times 100}{220} = 3.5\,kW$$

(5) In a Van de Graaff generator, Fig. 16.23, the belt carries a charge of density $0.5\,\mu C/m^2$ at a velocity of $10\,m/s$. The width of the belt is 10 cm. If the leakage paths between the dome and base have a resistance of $10^{14}\,\Omega$, calculate the charging current and the potential difference between the dome and the base.

Solution: Let ρ_s be the surface charge density on the belt, and u and w respectively the speed and width of the belt; the charging current is

$$I = \rho_s uw$$

$$\rho_s = 0.5 \times 10^{-6}\,C/m^2, u = 10\,m/s, w = 0.1\,m$$

$$I = 0.5 \times 10^{-6} \times 10 \times 0.1 = 0.5 \times 10^{-6}\,A = 0.5\,\mu A$$

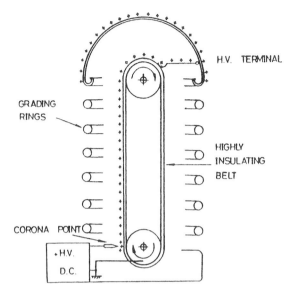

Figure 16.23   Principle of operation of the Van de Graaff generator, shown producing positive high voltage.

At steady state, the excess in the charge collected on the dome is just equal to the charge lost through the leakage paths. Thus, the charging current is equal to the leakage current and the potential difference between the dome and base is

$$V = IR_{\text{leakage}} = 0.5 \times 10^{-6} \times 10^{14} = 0.5 \times 10^{8}\,\text{V} = 50\,\text{MV}$$

(6) The elements of circuit (c) of Figure 16.11 for producing lightning impulse voltages are: $C_1 = 0.125\,\mu\text{F}$, $C_2 = 1\,\text{nF}$, $R_1 = 360\,\Omega$, $R_2 = 544\,\Omega$. What is the wave generated? Derive an expression for the generated wave if the charging voltage V_0 is 100 kV.

Solution: With reference to Table 16.1,

$$\theta = \sqrt{C_1 C_2 R_1 R_2} = \sqrt{0.125 \times 10^{-6} \times 1 \times 10^{-9} \times 360 \times 544} \simeq 4.95\,\mu\text{S}$$

$$\eta = 1/[1 + (1 + R_1/R_2)C_2/C_1] = 1/\left[1 + \left(1 + \frac{360}{544}\right)\frac{10^{-9}}{0.125 \times 10^{-6}}\right]$$

$$= 0.987$$

$$\alpha = \frac{R_2 C_1}{2\theta\eta} = \frac{1 \times 544 \times 0.125 \times 10^{-6}}{2 \times 4.95 \times 0.987 \times 10^{-6}} \simeq 6.9$$

Corresponding to $\alpha = 6.9$, one can obtain, from Figure 16.12,

$$T_2/\theta = 10.1, \qquad \therefore T_2 = 10.1 \times 4.95 = 50\,\mu s$$
$$T_2/T_1 = 45, \qquad \therefore T_1 = 50/45 = 1.02\,\mu s$$

Thus the generated lightning impulse is a $1.02/50\,\mu s$ wave.
From equations (16.41) and (16.42),

$$\alpha_1 = \frac{\alpha - \sqrt{\alpha^2 - 1}}{\theta} = \frac{6.9 - \sqrt{6.9^2 - 1}}{4.95 \times 10^{-6}} = 0.015\,\mu s^{-1}$$

$$\alpha_2 = \frac{\alpha + \sqrt{\alpha^2 - 1}}{\theta} = \frac{6.9 + \sqrt{6.9^2 - 1}}{4.95 \times 10^{-6}} = 2.77\,\mu s^{-1}$$

According to equation (16.40),

$$e(t) = \eta \frac{\alpha V_0}{\sqrt{\alpha^2 - 1}}[\exp(-\alpha_1 t) - \exp(-\alpha_2 t)]$$

$$= 0.987 \times \frac{6.9 \times 100}{\sqrt{6.9^2 - 1}}(e^{-0.015t} - e^{-2.77t})$$

$$= 99.75(e^{-0.015t} - e^{-2.77t}), \qquad t \text{ in } \mu s$$

which is the equation of the waveform of the generated impulse.

(7) What is the wave generated by circuit (b) of Figure 16.11? The circuit elements are the same as those of example (6). The charging V_0 is the same ($= 100\,kV$).

Solution: With reference to Table 16.1,

$$\theta = \sqrt{C_1 C_2 R_1 R_2} \simeq 4.95\,\mu s$$

$$\eta = 1/[1 + C_2/C_1 + R_1/R_2] = 1/\left(1 + \frac{10^{-9}}{0.125 \times 10^{-6}} + \frac{360}{544}\right) = 0.6$$

$$\alpha = \frac{R_2 C_1}{2\eta\theta} \simeq 11.45$$

From Figure 16.12, one obtains:

$$T_2/\theta = 16.25, \qquad \therefore T_2 = 16.25 \times 4.95 = 80.43\,\mu s$$
$$T_2/T_1 = 120, \qquad \therefore T_1 = 80.43/120 = 1.067\,\mu s$$

As calculated in example (6),

$$\alpha_1 = 0.0088\,\mu s^{-1}, \qquad \alpha_2 = 4.62\,\mu s^{-1}$$

$$E(t) = 0.6 \times \frac{11.45 \times 100}{\sqrt{11.45^2 - 1}} (e^{-0.0088t} - e^{-4.62t})$$

$$= 60.2(e^{-0.0088t} - e^{-4.62t}), \qquad t \text{ in } \mu s$$

This confirms that circuit (c) of Figure 16.11 is more efficient than circuit (b).

(8) The capacitive elements of circuit (c) of Figure 16.11 for producing an impulse voltage are: $C_1 = 0.125\,\mu F$ and $C_2 = 1\,nF$. Determine the wave-shaping resistors R_1 and R_2 required to generate a $250/2500\,\mu s$ switching impulse. What is the circuit efficiency η?

Solution:

$$T_2/T_1 = 2500/250 = 10$$

From Figure 16.12, one obtains values of α and T_2/θ corresponding to $T_2/T_1 = 10$:

$$\alpha = 4, \qquad T_2/\theta = 6$$
$$\therefore \theta = 2500/6 = 416.7\,\mu s$$

From Table 16.1, for circuit (c):

$$X = \frac{1}{\alpha^2}\left(1 + \frac{C_2}{C_1}\right) = \frac{1}{4 \times 4}\left(1 + \frac{10^{-9}}{0.125 \times 10^{-6}}\right) = 0.063$$

$$R_1 = \frac{\alpha\theta}{C_2}(1 - \sqrt{1-X}) = \frac{4 \times 416.7 \times 10^{-6}}{10^{-9}}(1 - \sqrt{1-0.063})$$
$$= 53.3 \times 10^3\,\Omega = 53.3\,k\Omega$$

$$R_2 = \frac{\alpha\theta}{C_1+C_2}(1 + \sqrt{1-X}) = \frac{4 \times 416.7 \times 10^{-6}}{0.125 \times 10^{-6} + 10^{-9}}(1 + \sqrt{1-0.063})$$
$$= 26.1 \times 10^3\,\Omega = 26.1\,k\Omega$$

$$\eta = 1/[1 + (1 + R_1/R_2)C_2/C_1]$$
$$= 1/[1 + (1 + 43.3/21.1)(10^{-9}/0.125 \times 10^{-6})]$$
$$= 0.976$$

(9) An impulse generator has eight stages, each stage having a capacitor rated $0.16\,\mu F$ and $125\,kV$. The load capacitor is $1\,nF$. Find the values of the wave-shaping resistors needed to generate a $1.2/50\,\mu s$ lightning impulse wave. What is the maximum output voltage of the generator if

the charging voltage is 120 kV? What is the energy rating of the generator?

Solution:

$$n = 8, \qquad nC_1 = 0.16 = 8C_1, \qquad C_1 = 0.02 \ \mu F$$

$$C_2 = 1 \, \text{nF} = 0.001 \, \mu F$$

$$T_2/T_1 = 50/1.2 = 41.67$$

From Figure 16.12, one obtains values of α and T_2/θ corresponding to $T_2/T_1 = 41.67$:

$$\alpha = 6.4 \qquad T_2/\theta = 9.5$$
$$\theta = 50/9.5 = 5.26 \, \mu s$$

The impulse generator of Figure 16.15 follows the circuit (c) of Table 16.1:

$$X = \frac{1}{\alpha^2}\left(1 + \frac{C_2}{C_1}\right) = \frac{1}{6.4 \times 6.4}\left(1 + \frac{10^{-9}}{0.02 \times 10^{-6}}\right) = 0.026$$

$$R_1 = \frac{\alpha\theta}{C_2}(1 - \sqrt{1-X}) = \frac{6.4 \times 5.26 \times 10^{-6}}{0.001 \times 10^{-6}}(1 - \sqrt{1 - 0.026}) = 440 \, \Omega$$

$$R_2 = \frac{\alpha\theta}{C_1 + C_2}(1 + \sqrt{1-X}) = \frac{6.4 \times 5.26 \times 10^{-6}}{0.021 \times 10^{-6}}(1 + \sqrt{1 - 0.026})$$
$$= 3185 \, \Omega$$

Per stage, the shaping resistors have values of R_1/n and R_2/n;

i.e., $R_1/n = 55 \, \Omega, \qquad R_2/n = 398 \, \Omega$

The DC charging voltage for eight stages is

$$V_0 = 8 \times 120 = 960 \, \text{kV}$$

$$\eta = 1/[1 + (1 + R_1/R_2)C_2/C_1] = 1/\left[1 + \left(1 + \frac{440}{3185}\right)\left(\frac{10^{-9}}{0.02 \times 10^{-6}}\right)\right]$$

$$= 0.95$$

Maximum output voltage

$$= \eta V_0 = 0.95 \times 960 = 912 \, \text{kV}$$

Energy rating

$$= \tfrac{1}{2}C_1 V_0^2$$
$$= \tfrac{1}{2} \times 0.02 \times 10^{-6}(960)^2 = 9216 \text{ J} = 9.22 \text{ kJ}$$

(10) An impulse generator has 12 stages, each having a capacitance of
0.125 µF, front and tail resistances of 70 Ω and 400 Ω, respectively. If
the load capacitance is 1000 pF, find the front and tail times of the
lightning impulse produced.

Solution:

$$n = 12, \qquad nC_1 = 0.125, \qquad \therefore C_1 = 0.125/12 \simeq 0.01 \text{ µF}$$

$$C_2 = 1000 \text{ pF} = 0.001 \text{ µF}$$

$$R_1/n = 70, \qquad \therefore R_1 = 70 \times 12 = 840 \text{ Ω}$$
$$R_2/n = 400, \qquad \therefore R_2 = 400 \times 12 = 4800 \text{ Ω}$$

The generator of Figure 16.15 follows the circuit (b) of Table 16.1.

$$\theta = \sqrt{C_1 C_2 R_1 R_2} = \sqrt{0.01 \times 10^{-6} \times 0.001 \times 10^{-6} \times 840 \times 4800}$$
$$= 6.35 \times 10^{-6} \text{ s} = 6.35 \text{ µs}$$

$$\eta = 1/[1 + C_2/C_1 + R_1/R_2] = 1/\left[1 + \frac{0.001}{0.01} + \frac{840}{4800}\right] = 0.784$$

$$\alpha = \frac{R_2 C_1}{2\eta\theta} = \frac{4800 \times 0.01 \times 10^{-6}}{2 \times 0.784 \times 6.36 \times 10^{-6}} = 4.82$$

Corresponding to $\alpha = 4.82$, one can obtain, from Figure 16.12,

$$T_2/\theta = 7, \qquad \therefore T_2 = 7 \times 6.35 = 44.45 \text{ µs}$$
$$T_2/T_1 = 25, \qquad \therefore T_1 = 44.45/25 = 1.77 \text{ µs}$$

Thus the front and tail times of the generated impulse are 1.77 µs and
44.45 µs, respectively.

(11) A double-exponential impulse-current generator has a total capacitance of 8 µF and a total inductance of 8 µH. The charging voltage is
25 kV. Determine the equation of the generated impulse current if the
time to the first peak is 8 µs.

17

High-Voltage Measurements

H. ANIS *Cairo University, Giza, Egypt*

17.1 INTRODUCTION

This chapter describes various methods for measuring high voltages and high transient currents. Also described are means of measuring electric fields normally associated with high voltages. Table 17.1 lists the most common methods of measuring high voltage, together with their modes of usage. Cathode-ray oscilloscopes are not included in the table, as their versatility with their probes and accessories make them suitable for almost all kinds of measurements.

17.2 SPHERE GAPS

When sphere gaps are used in high-voltage measurement, one of the spheres is earthed and the other is connected to high voltage. The sphere gap arrangement is shown in Figure 17.1. A horizontal sphere gap arrangement is sometimes preferred at lower measured voltages. The procedure is to establish, for a particular test circuit, a relation between the peak voltage, determined by sparkover between the spheres, and the reading of a voltmeter on the primary or input side of the high-voltage source. This relation should be within 3% (IEC, 1973).

Table 17.1 Methods of Measuring High Voltage

Method of measurement	DC		AC				Impulse	
	Mean	Peak	rms	Peak	Waveform		Peak	Waveform
Sphere gaps		x		x			x	
Peak voltmeter				x				
Electrostatic voltmeter	x (rms)		x					
Voltage transformer			x	x	x			
Resistor in series with milliammeter	x		x					
Resistive divider	x	x	x	x	x		x	x
Capacitive divider			x		x			x

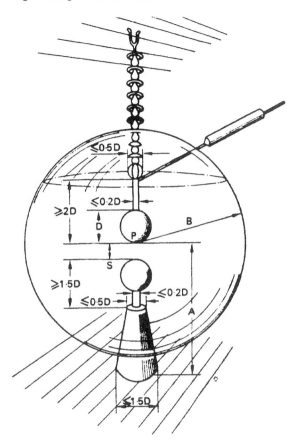

Figure 17.1 Vertical sphere gap.

Standard values of sphere diameter are 6.25, 12.5, 25, 50, 75, 100, 150, and 200 cm. Table 17.2 gives the clearance limits in sphere gap arrangements, stipulated by international standards (IEC, 1960).

The effect of humidity is to increase the breakdown voltage of sphere gaps by up to 3% (Kuffel and Zaengl, 1984). Temperature and pressure, however, have a significant influence on breakdown voltages. The breakdown voltage V_s is related to its value under normal atmospheric conditions by

$$V_s = kV_n \tag{17.1}$$

where k is a factor related to the relative air density (RAD) δ described in Chapter 5. Table 17.3 gives the relation between the RAD and the correction factor k.

Table 17.2 Clearances around the Sphere Gap (Fig. 17.1)

Sphere diameter d (cm)	Minimum value of A	Maximum value of A	Minimum value of B
Up to 6.25	7D	9D	14S
10–15	6D	8D	12S
25	5D	7D	10S
50	4D	6D	8S
75	4D	6D	8S
100	3.5D	5D	7S
150	3D	4D	6S
200	3D	4D	6S

Table 17.3 Correction Factor for Atmospheric Pressure and Temperature

δ	0.70	0.75	0.80	0.85	0.90	0.95	1.0	1.05	1.10
k	0.72	0.76	0.81	0.86	0.90	0.95	1.0	1.05	1.09

Under impulse voltages, the voltage at which there is a 50% break-down probability is recognized as the breakdown level. Table 17.4 lists the standardized disruptive discharge voltages from the results of a large number of international experiments (IEC, 1960). The accuracy of the values of Table 17.4 is generally within 3%, except for the bracketed figures, where the accuracy is about ±5%.

17.3 PEAK VOLTMETERS

The measuring circuit of a peak voltmeter is shown in Figure 17.2 and is made up of two diodes, a properly chosen capacitor, and a milliammeter that can be recalibrated for peak voltage. The arithmetic mean value of the rectified current as indicated by the instrument is given by

$$I = \frac{1}{T} \int_0^{T/2} i_1(t)\mathrm{d}t = \frac{C}{T} \int_{-v}^{+v} \mathrm{d}v$$
$$= \frac{C}{T}\left[V\left(\frac{T}{2}\right) - V(0) \right]$$

(17.2)

If the voltage is symmetric about zero and its peak value is V, then

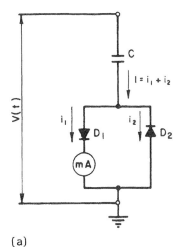

(a)

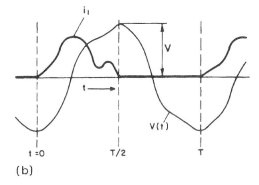

(b)

Figure 17.2   Peak voltmeter circuit.

$$\bar{I} = 2fCV \tag{17.3}$$

Therefore,

$$V = \frac{\bar{I}}{2fC} \tag{17.4}$$

The current waveform should ideally be monitored on an oscilloscope to determine whether there will be more than one zero crossing in a half-cycle. A peak voltmeter may be of either the analog or the digital type (Malewski and Dechamplain, 1980).

Table 17.4 Peak Values of Breakdown Voltages[a] (kV) of Sphere Gaps with One Sphere Grounded at Atmospheric Reference Conditions[b]

Sphere gap (cm)	Sphere diameter[b] (cm)															
	6.25		12.5		25		50		75		100		150		200	
	a	b	a	b	a	b	a	b	a	b	a	b	a	b	a	b
0.5	16.9		16.5													
1	31.4		31.2													
1.5	44.7	45.1	44.7													
2	57.5	58	58.0													
2.5	68.5	70	71.5		71.5											
3	78.0	80.5	84.0		84.5											
3.5	(86.0)[c]	(90.0)	95.5	96.5												
4	(93.5)	(99)	106	108	110											
4.5	(99.0)	(106)	117	120												
5	(105)	(113)	127	132	135	136	136		136							
5.5	(110)	(120)	136	143												
6	(114)	(125)	144	152	158	160										
6.25	(115)	(125)	148	157												
7			(159)	(170)	181	184										
7.5								199	200	199						
8			(171)	(186)	203	207										
9			(182)	(200)	222	229										
10			(192)	(210)	240	250	259		260	261	262		262		262	
11			(200)	(225)	257	268										
12			(208)	(230)	271	286										

Table 17.4 Continued

Sphere gap (cm)	6.25		12.5		25		50		75		100		150		200	
	a	b	a	b	a	b	a	b	a	b	a	b	a	b	a	b
12.5			(210)	(235)	277	294	315	317	321	322						
15					(309)	(331)	367	374	380	381	383	384				
17.5					(336)	(362)	413	425	435	440						
20					(362)	(389)	452	472	483	497	500		500		500	
22.5					(379)	(409)										
25					(393)	(426)	520	545	575	595	605	610				
30							(575)	(610)	655	685	700	715	730	735	735	740
35							(620)	(660)	725	755	785	800				
40							(660)	(705)	(785)	(820)	862	885	940	950	960	965
45							(690)	(730)	(835)	(875)	925	965				
50							(720)	(760)	(880)	(925)	1000	1020	1110	1130	1160	1170
60									(955)	(1005)	(1090)	(1130)	1260	1290	1320	1360
70									(1010)	(1050)	(1180)	(1220)	1370	1410	1460	1520
75									(1025)	(1070)	(1210)	(1260)	1420	1460	1510	1590
80											(1240)	(1290)	1460	1500	1570	1670
90											(1300)	(1350)	(1550)	(1600)	1690	1790
100											(1340)	(1390)	(1630)	(1690)	1810	1900
110													(1700)	(1760)	(1910)	(2000)
120													(1770)	(1830)	(1990)	(2030)
130													(1840)	(1900)	(2070)	(2160)

Sphere diameter[b] (cm)

Table 17.4 Continued

Sphere	Sphere diameter[b] (cm)															
	6.25		12.5		25		50		75		100		150		200	
gap (cm)	a	b	a	b	a	b	a	b	a	b	a	b	a	b	a	b
140													(1890)	(1950)	(2140)	(2240)
150													(1930)	(1990)	(2210)	(2310)
160															(2280)	(2370)
170															(2330)	(2430)
180															(2370)	(2470)
190															(2420)	(2510)
200															(2450)	(2540)

[a]: Values for AC, DC, and negative impulses. b: Values for positive lightning and switching impulses (if different from a).

[b] 25°C and 101.3 kPa.

[c] Values in parentheses are accurate only to ±5% as their corresponding gaps exceed 1/2 sphere diameter.

Referring to equation (17.4), it appears that, in addition to the instrument error, the error in the measured voltage dV is a function of errors in the current di, frequency df, and capacitance dC. It can be proven that

$$dV = di + df + dC \qquad (17.5)$$

Impulse peak voltmeters are different in that they contain active circuits and a memory device and possibly a storage oscilloscope (Hyltén-Cavallius, 1988).

17.4 ELECTROSTATIC VOLTMETERS

An electrostatic voltmeter utilizes the force existing between two opposite plates. The force is created by the process in which a change in stored electrostatic energy is converted into mechanical work. Referring to Figure 17.3, the electrostatic voltmeter is seen to be made up of two parallel plates. One is fixed and the other has a very small movable part that is restrained by a spring. The force of attraction $F(t)$ created by the applied voltage causes the movable part—to which a mirror is attached—to assume

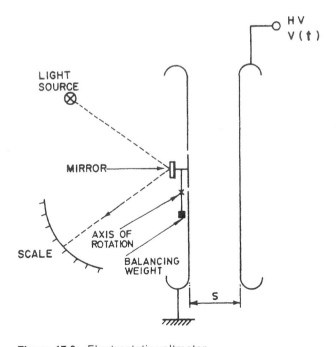

Figure 17.3 Electrostatic voltmeter.

a position at which a balance of forces takes place. An incident light beam will therefore be reflected toward a scale calibrated to read the applied voltage magnitude. Assuming the capacitance between plates to be C, the stored electrostatic energy W in the system will be $W(t) = (1/2)CV^2(t)$. A change $dW(t)$ in stored energy will be faithfully converted into mechanical work. Therefore,

$$dW(t) = -F(t)\, dS \tag{17.6}$$

where dS is the change in S, the separation between the plates (Fig. 17.3). The attraction force as a function of time is thus

$$|F(t)| = \frac{dW(t)}{dS} = \frac{1}{2}V^2(t)\frac{dC}{dS} \tag{17.7}$$

The actual force is the arithmetic mean of expression (17.7), given by

$$\bar{F} = \frac{1}{2}\left(\frac{dC}{dS}\right)\left[\frac{1}{T}\int_0^T V^2(t)\, dt\right] \tag{17.8}$$

where T is the period of variation, if any. Equation (17.8) relates the force to the rms value of the applied voltage V:

$$\bar{F} = \frac{1}{2}\left(\frac{dC}{dS}\right)V^2 \tag{17.9}$$

In the most common attracted disk type of electrostatic voltmeter, the factor dC/dS can be simply evaluated by recalling that

$$C = \frac{A\varepsilon_0}{S} \tag{17.10}$$

where A is the area of the plate and ε_0 is the permittivity of free space. Therefore, the force will be given by

$$|\bar{F}(t)| = \frac{A\varepsilon_0}{2S^2}V^2 \tag{17.11}$$

If V is in volts and A and S^2 are in m^2 the force is in newtons. According to equation (17.11), the factor $A\varepsilon_0/S^2$ is used to control the range of measurement.

17.5 VOLTAGE TRANSFORMERS

Voltage transformers are used to measure high AC voltage accurately. The voltage on the secondary side is closely proportional in amplitude to and almost in phase with the voltage on the primary side. Two alternatives are, however, preferred at very high voltages, namely, capacitive and cascaded voltage transformers (IEEE Committee, 1981).

17.5.1 Capacitive Voltage Transformers

A capacitive voltage transformer consists of a capacitive divider – with C_1 and C_2 as the high and low voltage arms, respectively – used in conjunction with a conventional auxiliary transformer which further steps down the divider output voltage, typically about 10 kV, to the desired secondary value. By adjustment of the inductance, which may consist wholly or partly of leakage inductance of the auxiliary transformer, to equal $1/(2\pi f)^2(C_1 + C_2)$, the voltage drop across C_2 due to the current drained from the divider is largely compensated so that the overall ratio is nearly independent of burden and is the product of the divider and transformer ratios.

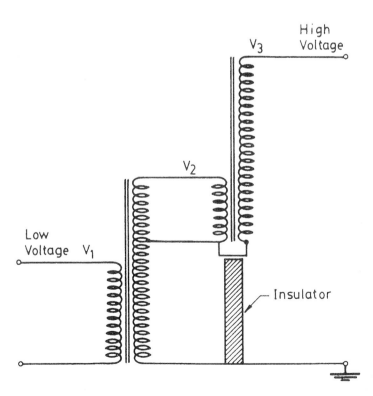

Figure 17.4 Cascaded voltage transformer circuit.

17.5.2   Cascaded Voltage Transformers

In cascaded voltage transformers the primary winding is distributed over a series of coils wound on separate cores which are mounted in oil-filled porcelain containers, stacked in series. The secondary coil is wound on the lowest core. The power-frequency voltage is thus uniformly distributed over a number of coils, each of which need not be insulated from its core for more than 100 kV.

17.6   AMMETER IN SERIES WITH A HIGH IMPEDANCE

An impedance can be used in series with a microammeter or a milliammeter for the measurement of high voltages. Neglecting the impedance of the instrument, the current through the instrument will be proportional to the applied voltage. Evidently, if the impedance is a resistor, the current is in phase with and faithfully represents the voltage. The instrument may be replaced by an oscilloscope if the voltage waveform is to be recorded.

Wire-wound or thin-film resistors may be used. Wire-wound resistors are usually preferred for their superior stability in service and their extremely small temperature coefficient, which reaches $\pm 0.01\%/K$. Freedom from corona discharge and improved cooling may be achieved by immersing the resistor in insulating oil.

17.7   POTENTIAL DIVIDERS FOR AC AND DC

A potential divider consists of two impedances, Z_1 and Z_2, connected in series, to which the voltage to be measured is applied; the components that constitute the impedances are referred to as the high- and low-voltage arms of the divider. Connection between the low-voltage arm and the measuring instrument must be made through a shielded coaxial cable to avoid the adverse effects of stray capacitance between that connection and the high-voltage arm. High-voltage dividers generally consist of either resistors or capacitors, but sometimes a combination of resistors and capacitors, either in series or in parallel, is used, depending on the type of voltage to be measured.

17.7.1   Resistive Potential Dividers

A resistive potential divider is usually employed for the measurement of direct voltages. If, however, the ripples on the DC voltage are to be recorded as well, a resistive/capacitive potential divider will then be more suitable (Kind, 1978).

17.7.2 Resistive Dividers for AC Voltage Measurement

The divider resistive arms should—in the case of alternating voltage—be looked upon as impedances Z_1 and Z_2, which generally possess resistive and reactive components. Therefore,

$$Z_1 = R_1 + jX_1$$

and

$$Z_2 = R_2 + jX_2$$

Under AC conditions the divider ratio, in magnitude, is

$$\left|\frac{Z_1 + Z_2}{Z_2}\right| = \left[\frac{(R_1 + R_2)^2 + (X_1 + X_2)^2}{R_2^2 + X_2^2}\right]^{1/2} \tag{17.12}$$

The output voltage V_2 will be shifted in phase from the measured voltage V_1 by

$$\theta = \tan^{-1}\frac{X_1 + X_2}{R_1 + R_2} - \tan^{-1}\frac{X_2}{R_2} \tag{17.13}$$

Therefore, for the output voltage V_2 to represent V_1 faithfully, this phase shift should vanish. This, according to (17.13), amounts to ensuring that

$$\frac{X_1}{R_1} = \frac{X_2}{R_2} \tag{17.14}$$

The high-voltage arm will consist of many resistor elements stacked in series. As the length of the stack increases, so does its capacitance to surrounding conductors. The influence of earth capacitance C_e on the effective impedance can be derived by considering the general case of a stack earthed at one end and having inherent series resistance $Z = R_1$ and admittance to earth $Y = j\omega C_e$, both assumed uniformly distributed along its length (Fig. 17.5). The currents i_0 and i_1 at the earthed and high-voltage ends of the stack, respectively, due to an applied voltage V are then

$$i_0 = \frac{V}{R_1}\frac{\xi}{\sinh \xi} \tag{17.15}$$

$$i_1 = \frac{V}{R_1}\frac{\xi}{\tanh \xi} \tag{17.16}$$

where

$$\xi = \sqrt{ZY}$$
$$= \sqrt{R_1 C_e \omega}$$

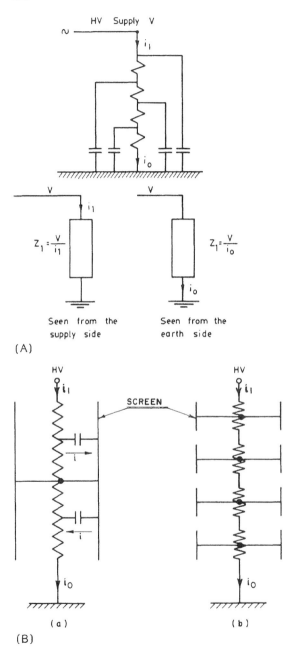

Figure 17.5 (A) Unscreened high-voltage resistance; (B) screened high-voltage resistance, the screen being one piece (a) or subdivided (b).

The effective impedance $Z_1(V/i_0$ or $V/i_1)$ of the stack may be derived by expanding the hyperbolic functions in equations (17.15) and (17.16), giving

$$\left.\begin{array}{l} \text{at the earthed end: } Z_1 \simeq R_1\left(1+\dfrac{j\xi^2}{6}\right) \\[2em] \text{at the high-voltage end: } Z_1 \simeq R_1\left(1-\dfrac{j\xi^2}{3}\right) \end{array}\right\}$$

$$(17.17)$$

When this resistance is used as the high-voltage arm of a resistive divider whose low-voltage-arm resistance is R_2, the effective divider ratio is thus

$$\frac{i_0 R_2}{V} = \frac{R_2}{R_2 + R_1(1 + j\xi^2/6)} \tag{17.18}$$

It appears from equation (17.18) that the divider ratio is sensitive to frequency. Therefore, resistive dividers are not suitable for high-frequency alternating voltages.

17.7.2.1 Unscreened Resistors

Most high-voltage resistors—also, high-voltage arms of resistive dividers—may be considered as a vertical column of length L (m) and diameter $2r$ standing on a grounded plane (Fig. 17.5A). For this configuration the equivalent stray capacitance to earth is given approximately by

$$C_e = \frac{111L}{2\ln(L/r) - 1.1} \qquad \text{pF} \tag{17.19}$$

When $\ln(L/r)$ is greater than about 3, the capacitance depends primarily on L and not on r; it is also not much affected by the shape and disposition of the earthed surface since, if earth is looked upon as a coaxial cylinder of radius L (m), its capacitance is only slightly different, that is,

$$C_e = \frac{111L}{2\ln(L/r)} \qquad \text{pF} \tag{17.20}$$

17.7.2.2 Screened Resistors

Screening is the process of canceling or neutralizing the effect of stray capacitances. It is basically accomplished by surrounding the high-voltage resistor by a conducting screen maintained at the mean potential of the resistor (Fig. 17.5B). Capacitive currents will flow between the screen and the resistor in one direction within the upper portion of the resistor and in the opposite direction within the lower portion. The two currents i_0 and i_1, viewed from either end of the resistor, will see an effective impedance Z_1 of value

$$Z_1 \simeq R_1(1 - j\omega R_1 C_s/12) \tag{17.21}$$

which is intermediate between the two values of Z_1 in the case of unscreened resistors given by equation (17.17). It is preferred, however, to divide the resistor into identical smaller units, each contained in a screen that is maintained at the mean potential of the unit (Fig. 17.5B). This action poses less risk of flashover between the resistor and the screen extremities. At 50 or 60 Hz the practicable voltage limit for a high-precision screened resistor is about 100 kV.

17.7.3 Capacitive Potential Dividers

Resistive potential dividers suffer from two main drawbacks, power losses and stray capacitance to earth. These factors limit their use to voltages below 100 kV at 50 Hz and even lower voltages at higher frequencies. Capacitive potential dividers are therefore more suitable to use with AC voltage, particularly at high voltages and high frequencies. They are limited only by their internal inductances or the dielectric losses of their components (Fig. 17.6a).

The low-voltage capacitor C_2 will normally consist of a fixed unit of low loss angle (air, mica, or polystyrene dielectric). It will, in general, be shunted by a high resistance R_2, which may either be introduced deliberately to avoid the accumulation of a random charge on C_2 or be inherent in the measuring instrument. High-voltage capacitors C_1 can be either screened or unscreened.

17.7.3.1 Unscreened Capacitors

Unscreened capacitors, which may be used as the high-voltage component of a capacitive divider, normally take the form of a stack of identical cylindrical units. In this case, if the series capacitance of the stack is C_1 and the stray capacitance to earth is C_e, then, substituting C_e/C_1 for ξ^2 in equations similar to (17.17) and (17.18), with C_e not greater than C_1, the approximate effect of C_e is to decrease the effective capacitance at the earthed end by $C_e/6$ and to increase it at the other end by $C_e/3$. According to equation (17.19), which also applies here, $C_e \simeq 20\,\mathrm{pF/m}$ in practice, so that the correction due to it in any particular case can be estimated.

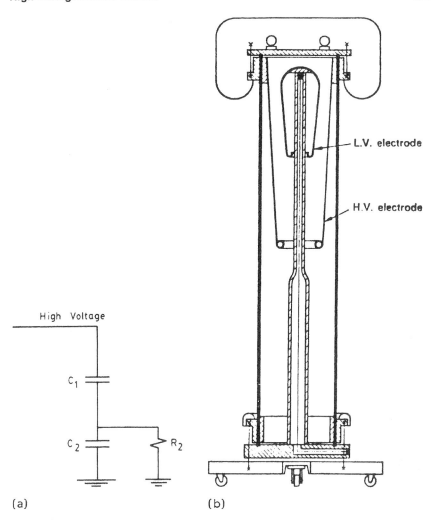

High Voltage

C_1

C_2 R_2

(a)   (b)

Figure 17.6 (a) Capacitive voltage divider; (b) typical screened capacitor.

17.7.3.2 Screened Capacitors

For voltages up to about 30 kV (peak), capacitors with parallel-plate electrodes can be used and screened by being enclosed in a metal case. When filled with carefully dried gas they have almost zero power factor and a highly stable value of capacitance. At higher voltages a capacitor with coaxial cylindrical electrodes, as that shown in Figure 17.6b, may be used as the high-voltage component. The outer cylinder is flared at the ends to avoid discharges. The capacitance C_1 is calculable from the electrode dimensions:

$$C_1 = \frac{111L}{2 \ln(r_1/r_2)} \qquad \text{pF} \tag{17.22}$$

where L is the effective length of the LV electrode (m), r_1 the radius of the outer cylinder (m), and r_2 the radius of the inner cylinder (m).

The maximum stress on the dielectric occurs at the surface of the inner electrode and, for a given voltage and size of outer cylinder, is minimal when the ratio r_1/r_2 is equal to e ($= 2.718$). At atmospheric pressure and with smooth clean electrode surfaces, the capacitor can function with voltage gradients at the inner electrode surface of about 14 kV/cm peak without partial discharges. Higher voltage gradients can be attained by increasing the gas pressure (compressed-gas capacitors).

17.8 DIVIDERS FOR IMPULSE VOLTAGES

With impulse voltages, the complete waveform is to be recorded. Therefore, a potential divider for impulse voltages should have a good impulse fidelity. Both the divider and the connection leads must be considered when this impulse fidelity is to be assessed.

Instead of exploring the frequency characteristics of a divider, it is more common to examine its response to a step voltage applied at its high-voltage terminals. If a unit step voltage that is defined by

$$v_{-1}(t) = \begin{cases} 0 & t < 0 \\ 1 & t > 0 \end{cases} \Bigg\} \tag{17.23}$$

is applied to a divider, its response (output) is $u(t)$. The output with any other input voltage $v_1(t)$ can be found accordingly. If the Laplace transform of the input voltage $v_1(t)$ and of the step response $u(t)$ are, respectively, $V_1(s)$ and $U(s)$, the Laplace transform of the output voltage may be given by

$$V_2(s) = V_1(s)\, U(s)s \tag{17.24}$$

In the time domain the output voltage would be

$$v_2(t) = L^{-1}[V_2(s)] \tag{17.25}$$

Equation (17.25) can be solved analytically whenever $u(t)$ takes an analytical form. Otherwise, numerical methods can be invoked to evaluate $v_2(t)$ on the basis of the superposition theorem:

$$v_2(t) = u(0)v_1(t) + \int_0^t v_1(\lambda)u'(t - \lambda)\, d\lambda \tag{17.26}$$

For faithful reproduction of $v_1(t)$, the wave shape of the response $u(t)$ should be as close as possible to the applied step voltage $v_{-1}(t)$. A reasonable measure of the faithfulness of the response is the total area enclosed between the real and the ideal responses (Fig. 17.7). This area will have the dimensions of time and is thus referred to as the response time of the divider. On

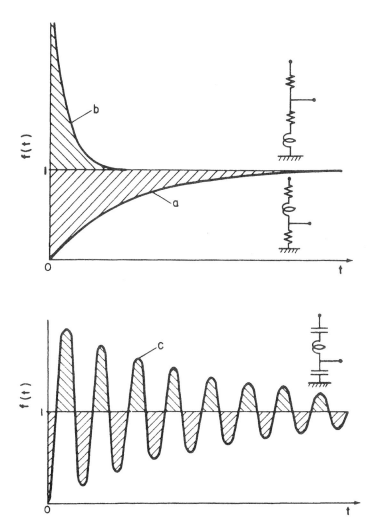

Figure 17.7 Possible responses of impulse dividers: (a) for resistive divider with inductive high-voltage arm; (b) for resistive divider with inductive low-voltage arm; (c) for capacitive divider with inductive high-voltage arm.

the basis of this representation, two criteria are used simultaneously to evaluate the response faithfulness:

1. The total response time (net enclosed area) must be as small as possible.
2. The response should settle down to the correct value in a time much shorter than the rise time of the voltage wave to be measured.

17.8.1 Resistive Dividers

With unscreened resistive dividers, the considerations in Section 17.7.2 relating to the effect of stray capacitance to earth (C_e) of the high-voltage arm of a resistor divider apply where impulses are concerned. With an applied voltage step of amplitude V, equations (17.15) and (17.16) still hold, where $Z = R_1$ and $Y = sC_e$ (s being the operator d/dt). The calculated values of the response times of the resistor are ($C_e R_1/6$) and ($-C_e R_1/3$) at the earthed and high-voltage ends, respectively. These are identical with the time constants derived from equation (17.17). A typical divider response is shown in Figure 17.8 together with its voltage constituents. An upper limit on the value of R_1 should be imposed such that the response time of the high-voltage arm is made much shorter than the rise time of the measured impulse voltage. When a much shorter divider response is desired, the response time of the low-voltage arm should be made equal to that of the high-voltage arm. Since the low-voltage arm is mainly inductive, its time constant is L_2/R_2. Curve c of Figure 17.8 shows the case where $L_2/R_2 = C_e R_1/6$, thus reducing the total response time to zero. Only a small reduction in the final settlement time, however, results.

17.8.2 Capacitive Dividers

In a capacitive divider the high- and low-voltage arms can be assumed to have capacitances C_1 and C_2 in series with lead resistances r_1 and r_2 and inductances L_1 and L_2 in series, and also leakage resistances R_1 and R_2 across C_1 and C_2. The response of the divider to a step voltage will be an oscillation, about unity, of frequency

$$f = \frac{1}{2\pi}\sqrt{\frac{C_1 + C_2}{(L_1 + L_2)C_1 C_2}} \tag{17.27}$$

which is damped out at a time constant of $(L_1 + L_2)/(r_1 + r_2)$. Unless the leakage resistances are ensured to be inversely proportional to the capaci-

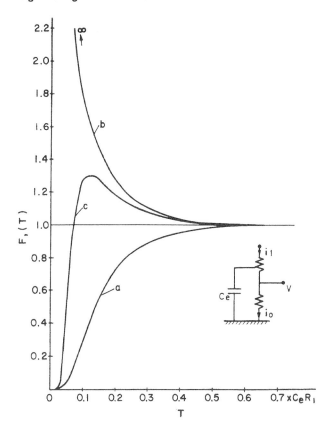

Figure 17.8 Response of resistive divider to a unit step voltage: (a) current at earthed end i_0; (b) current at high-voltage end i_1; (c) divider output v when $L_2/R_2 = R_1 C_e/6$.

tances (i.e., $R_1 C_1 = R_2 C_2$), the eventual settlement output would drop, after some time, to a value of $R_2 C_2/R_1 C_1$.

Additional damping resistances are usually connected in the lead on the high-voltage side of a capacitive potential divider. A total series resistance value is aimed at which critically damps out the oscillation; it should be nearly $2\sqrt{(L_1 + L_2)(C_1 + C_2)/C_1 C_2}$. The resistances should be well distributed along the divider. Such a mixed capacitive/resistive divider acts under high-frequency voltages as a resistive divider, and under low-frequency voltages as a capacitive divider.

Mixed capacitive/resistive potential dividers are also used where the introduced resistors, R_1 and R_2, are connected across the capacitors C_1 and

C_2, respectively. Their response is determined initially by the capacitances and ultimately by the resistors. If $R_1 C_1$ is made equal to $R_2 C_2$, the overall response will be ideal except for an initial effect of the inherent inductance.

17.8.3 Response Time of the Measuring System

The response time may be defined, in view of linearly rising voltages (ramps), as that time beyond which the difference in amplitude between the input and output voltages remains constant. If a linearly rising voltage is applied to the measuring system, equation (17.26) can be used to predict the output. Let the input voltage be

$$v_1(t) = kt$$

where k is a constant slope. The output voltage according to equation (17.26) is

$$v_2(t) = k \int_0^t u(\lambda)\,d\lambda \qquad (17.28)$$

which may be expanded to

$$v_2(t) = k\left[t - \int_0^t [1 - u(\lambda)]\,d\lambda\right]$$

The difference between the ideal and actual normalized responses v_1 and v_2, respectively, is then

$$v_1(t) - v_2(t) = k \int_0^t [1 - u(\lambda)]\,d\lambda$$

For all t above a certain value, the difference above becomes a constant proportional to k. Therefore,

$$k \int_0^\infty [1 - u(\lambda)]\,d\lambda = kT$$

according to which the response time T is found to be (Kuffel and Zaengl, 1984)

$$T = \int_0^\infty [1 - u(\lambda)]\,d\lambda \qquad (17.29)$$

as defined earlier.

17.9 THE MEASURING CABLE

The output of a potential divider is connected to the oscilloscope via a coaxial cable. For a loss-free cable, the velocity v of propagation of the electromagnetic wave is equal to $\sqrt{1/LC}$, where L and C are the inductance and capacitance per unit length of the cable. When the electromagnetic waves are required to travel along an actual cable, this expression gets simplified to the form

$$v \simeq \frac{3 \times 10^8}{\sqrt{\varepsilon_r \mu_r}} \quad \text{m/s}$$

where ε_r is the relative permittivity of the insulation surrounding the conductor and μ_r is the relative permeability of the conductor material. Generally, $\mu_r = 1.0$ for most of the conductors used in cables (IEC, 1962).

17.9.1 Cable Termination

To avoid errors due to reflections at its ends, the cable should be terminated at one end, or preferably both ends, by a resistance equal in value to its characteristic impedance Z_0, which for low-loss cables and very high frequencies is a pure resistance. The cable is thus essentially compatible with a resistor divider, to which it may be connected as shown in Figure 17.9. Matching is achieved by making

$$Z_0 = R_3 + \frac{R_1 R_2}{R_1 + R_2}$$

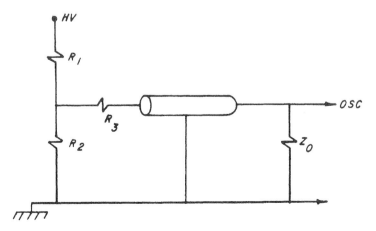

Figure 17.9 Measuring cable connection to a resistive divider.

Normally, R_2 is much smaller than R_1. Therefore,

$$Z_0 \simeq R_3 + R_2 \qquad (17.30)$$

The low-voltage-arm resistance R_2 is now shunted by a resistance R_3 in series with the input impedance to the cable Z_0. Therefore, the divider ratio becomes

$$\frac{V_2}{V_1} = \frac{Z}{Z + R_1} \qquad (17.31)$$

where

$$Z = \frac{R_2(Z_0 + R_3)}{Z_0 + R_2 + R_3}$$

which, according to (17.30), becomes

$$Z = \frac{R_2(2Z_0 - R_2)}{2Z_0} \qquad (17.32)$$

With a capacitive divider, the cable input resistance R_3 is chosen with a value $= Z_0$; its far end is not shunted. Thus, with a unit function voltage applied to the divider, a voltage $C_1/2(C_1 + C_2)$ is initially injected into the cable, is doubled by reflection at the open end, and is absorbed without significant reflection when it returns to the input end. Thus the voltage at the open end jumps to $C_1/(C_1 + C_2)$.

In cases where the divider capacitance is intentionally made small, the cable capacitance C_e is considered and its far end is loaded by Z_0 and C_3 in series. When $C_1 + C_2 = C_3 + C_c$, the initial and final values of the response at the oscilloscope end are equal and a flat-topped overshoot occurs.

With a divider consisting of resistance and capacitance in parallel, the same cable connection is used. This is justified by the fact that the initial portion of the divider's response is controlled by the capacitive components. If the high- and low-voltage capacitances C_1 and C_2 are shunted by resistances R_1 and R_2 and the time constants R_1C_1 and $R_2(C_2 + C_3 + C_c)$ are adjusted to be equal and also large compared with the cable delay time, the response of the divider will be practically the same as if R_1 and R_2 were absent.

It the mixed divider has its resistances and capacitances in series, the cable arrangement can be such that the low-voltage arm is essentially placed at the output of the delay cable (Fig.17.10a). The time constants R_1C_1 and $Z_0(C_2 + C_c)$ should be made equal to ensure equal initial and final response values. The arrangement is generally suitable whenever the cable capacitance is much smaller than C_2. When this condition is not valid, the circuit

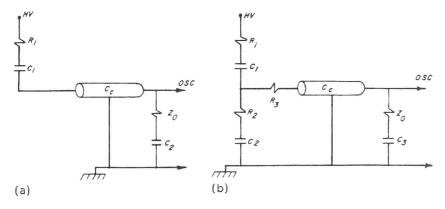

Figure 17.10 Measuring cable connection to a capacitive/resistive divider.

of Figure 17.10b can be used, in which matching is ensured at the cable's sending end by making

$$Z_0 = R_3 + \frac{R_1 R_2}{R_1 + R_2}$$

that is,

$$Z_0 \simeq R_3 + R_2 \qquad (17.33)$$

17.9.2 Digital Recording

Digital instruments have significant advantages over the older analog technology. A digital recorder has an analog-to-digital converter (ADC) where the input signal is sampled at discrete-time intervals as defined by a clock. The samples are then converted into digital codes, which represent the magnitude of the input at these particular instants of time. These codes are stored into memory locations proportional to the time at which the samples were taken. Once recorded, the contents of the memory can be transferred to a computer for analysis or be displayed repetitively on the oscilloscope with the aid of a digital-to-analog converter (DAC). The upper limits of the performance of a digital scope are specified by the sampling rate and number of ADC bits. The number of bits defines the maximum amplitude resolution of the scope. The larger the number of bits, the greater will be the digital scope's resolution or precision; high-speed 8, 16, and 32 bit digital scopes are available. The sampling rate defines how many times per second the input signal amplitude will be converted to a digital code by the ADC. To minimize the sampling error, the sampling rate should be as high as possible. Modern digital scopes have maximum sampling rates up to a few GHz.

The following advantages of using digital—rather than analog—recording can be outlined:

1. A permanent record of the transient, which can be displayed repetitively, is made. This record can be manipulated in a computer to calculate rise times, as well as statistical quantities from repeat measurements. In addition, having the data in a digital file enables digital signal processing applications such as spectral analysis.
2. Most digital scopes contain two or more independent ADCs. Thus, two or more signals (e.g., current and voltage) can be recorded simultaneously without chopping or alternating on repeat events.
3. With a suitable sampling speed and number of bits, digital recording is much more accurate than that obtained with an analog oscilloscope.

17.10 MEASUREMENT OF TRANSIENT CURRENTS

Two methods are generally used to measure transient currents: the use of resistive, low-ohmic shunts, and the measurement of currents by their inductive effects.

17.10.1 Resistive Shunts

A resistive, low-ohmic shunt is designed as a four-terminal device (Fig. 17.11) into which current I is fed via current terminals X and X' and from which a voltage V is tapped via potential terminals Y and Y'. The impedance V/I under AC conditions can be expressed as $R + j\omega L$, where R and L are the effective values of its resistive and inductive components. The shunt should therefore be designed so that, over the frequency band con-

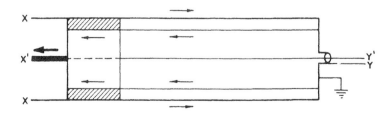

Figure 17.11 Resistive shunt with the shape of a coaxial tube.

cerned in any measurements in which it is involved, R is substantially equal to the DC resistance and $\omega L/R \ll 1$. Both these conditions require the diameter or thickness of the resistive element, depending on whether it is a wire or a ribbon, to be less than the nominal depth to which alternating current at the upper limit of the frequency band will penetrate a slab of the same material. The skin depth d_s is calculable from the frequency f, resistivity ρ, and relative permeability μ_r of the material by the equation $d_s = \sqrt{\rho/\pi\mu_r\mu_0 f}$. It amounts at 1 MHz to 0.35 and 0.33 mm for constantan and manganin, respectively, which are available in wire or sheet from and are nonmagnetic, for which it can be assumed that $\mu_r = 1$.

Two forms of shunt are in common use: the coiled bifilar and the coaxial tube shunt. The former is used in the secondaries of current transformers and other applications where the transient current is relatively limited in magnitude. Coiled bifilar shunts normally have resistances in the order of a few ohms. Coaxial tube shunts may have resistances as small as microohms and are used mainly for measuring large currents (Hebner et al., 1977).

The use of resistive shunts to record transient currents can be represented by the equivalent circuit shown in Figure 17.12a. The impedance Z is the input impedance of the coaxial cable and its matching resistances. To evaluate the adverse effect of the inductive component of the shunt, an incident step current I is assumed. The voltage appearing across the impedance Z is what a recording oscilloscope will show. The time function of the voltage can be shown to take the form

$$v(t) = IZ\left(\frac{R}{R+Z} + \frac{Z}{R+Z}e^{-t/T}\right) \tag{17.34}$$

where $T = L/(R+Z)$ is the time constant. The response of the shunt is sketched in Figure 17.12b.

17.10.2 Current Measurement by Magnetic Coupling

In this method a toroidal coil known as the Rogowski coil surrounds the path of the current to be measured. The time-varying magnetic field induces a voltage at the terminals of the coil which is proportional to the rate of change of current with time:

$$v = M\frac{di}{dt} \tag{17.35}$$

where M is the mutual inductance between the sensing coil and the measured current path. These current sensors have recently been superseded by Hall-effect transducers (Berkebile et al., 1981; Tokayo et al., 1982).

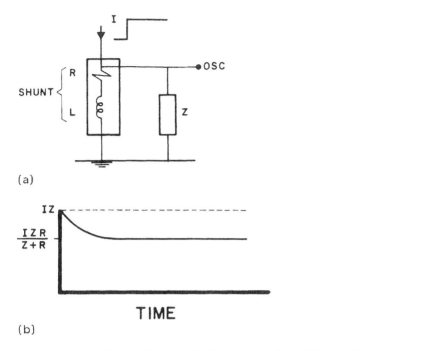

(a)

(b)

TIME

Figure 17.12   (a) Equivalent circuit of a resistive shunt; (b) typical response.

17.10.3   Hall-Effect Current Transducers

If an electric current flows from left to right along a thin strip of metal or semiconductor and if a magnetic field acts vertically upward, the electrons will be deflected toward its back edge. A potential difference is produced betweeen opposite points on the edges, which is known as the Hall voltage. For a constant magnetic field, the Hall voltage is proportional to the current. This property can be used to measure high DC and AC and even transient currents.

In the Hall-effect current transducer the current-carrying conductor is surrounded by an iron-cored magnetic circuit that produces a magnetic field into the air gap. The Hall voltage is amplified and measured or recorded on an oscilloscope. Current probes based on the Hall effect are manufactured to have a bandwidth extending from DC to 50 MHz and more (Holt, 1988).

17.11   MEASUREMENT OF ELECTRIC FIELD

Field measurements play an important role in the construction and problem solving of equipment using high voltages or subjected to high electric fields

(IEEE Working Group, 1983). Most electric field measuring devices are based on the principle that the electric field is proportional to the electric flux density D. When this changes with time due to natural phenomena or by mechanical movement, a displacement current results. The displacement current density is given by

$$\bar{J}_d = \frac{\partial \bar{D}}{\partial t} \tag{17.36}$$

17.11.1   Passive Capacitance Probe

The capacitance probe is essentially a small disk mounted flush with the surface of an electrode or a screen. In either case the device measures the displacement charge induced on the probe surface due to the electric flux density. The electrode version shown in Figure 17.13 can be used to determine the field at the ground plane of a rod–plane gap. For a probe with an effective head area of A and with an electric flux of ψ coulombs normal to the probe surface, the electric field is given by

$$\begin{aligned}
E &= \frac{D}{\varepsilon_0} = \frac{\psi}{\varepsilon_0 A} \\
&= \frac{1}{\varepsilon_0 A} \int i_d \, dt
\end{aligned} \tag{17.37}$$

where ε_0 is the permittivity of free space. By measuring the probe current i_d, the field can be obtained if the measured current is integrated with respect to time. The integration process can be performed automatically by inserting a capacitor C_m between the probe and ground, as seen in Figure 17.13. The

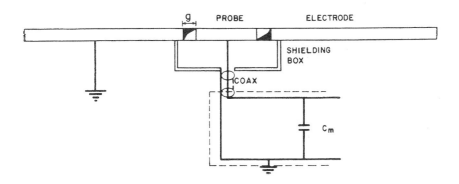

Figure 17.13   Capacitance probe.

voltage V across the measuring capacitor is proportional to the electric field, which can be calculated from

$$E = \frac{C_m V}{\varepsilon_0 A} \qquad (17.38)$$

The effective area A is a function of the width of the annular ring (g), where the effective radius can be approximated by the probe radius and one-half the ring width.

17.11.2   Electric Field Mill

An electric field mill consists of a grounded multivane rotor turning in front of a multisegment sensor plate. The grounded rotor alternately exposes the sensor plates to the ambient electrostatic field and shields them (Fig. 17.14). When each sensor plate is shielded, its charge is expelled (Stark, 1979). The induced charge from the displacement current is given by

$$q = \int \bar{D} \, d\bar{A}$$
$$= \varepsilon_0 E A \qquad (17.39)$$

assuming a static uniform field. The arithmetic mean value of the current for the sensor plates to be exposed and shielded once is then

$$\frac{1}{T/2} \int_0^{T/2} \frac{dq}{dt} \, dt = f(q_{max} - q_{min})$$

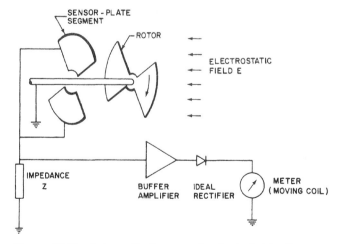

Figure 17.14   Schematic diagram of a single field mill.

where f is the frequency of exposure and shielding. This current is the maximum current and can be expressed in terms of capacitance, where

$$i_{max} = fV(C_{max} - C_{min}) \tag{17.40}$$

The function describing the area exposed is between a triangular and sinusoidal waveform and can be approximated by

$$a = \frac{A(1 + \sin \omega t)}{2} \tag{17.41}$$

ω being the angular velocity. The charge on the sensor is then

$$q = \frac{\varepsilon_0 EA(1 + \sin \omega t)}{2} \tag{17.42}$$

The instantaneous voltage across the impedance Z (see Fig. 17.14) is

$$\begin{aligned} v(t) &= Z\frac{dq}{dt} \\ &= \frac{\varepsilon_0 EA\omega}{2} Z \cos \omega t \end{aligned} \tag{17.43}$$

The impedance Z consists normally of a resistance R and a capacitance C in parallel. If $(\omega RC)^2 \ll 1$, the peak voltage is approximated as

$$\hat{V} = \frac{\varepsilon_0 EA\omega R}{2} \tag{17.44}$$

where the field is a function of frequency. If $(\omega RC)^2 \gg 1$, the peak voltage is approximated as

$$\hat{V} = \frac{\varepsilon_0 EA}{2C} \tag{17.45}$$

where the field is not a function of frequency. If the sinusoidal assumption is relaxed and a pure triangular waveform is used, the results obtained are essentially identical to those of equations (17.44) and (17.45). Therefore, if the impedance Z is a capacitance (e.g., 100 nF) and if the output voltage V is fed into a high-impedance amplifier (e.g., $R = 10\,M\Omega$), the restriction for equation (17.45) is satisfied at power frequency.

17.11.3 Free Body Probes

Free body probes, sometimes called electric dipole sensors, are self-contained and normally have no electrical connection to ground. These types of probe generally consist of two electrodes of a specific geometry separated by a small insulating gap. The principle of operation can be shown by examining the probe with two hemispheres. The charge induced on one hemisphere of radius r is given by

$$Q = \int_{S/2} \bar{D} \, \mathrm{d}\bar{A}$$

$$= 3\pi r^2 \varepsilon_0 E \tag{17.46}$$

from which it follows that the short circuit current between the two hemispheres is

$$i(t) = \frac{\mathrm{d}Q}{\mathrm{d}t} = 3\pi r^2 \varepsilon_0 \frac{\mathrm{d}E}{\mathrm{d}t}$$

$$= 3\pi r^2 \varepsilon_0 j\omega E \tag{17.47}$$

for AC fields. The effective area of the sphere, therefore, becomes

$$A = 3\pi r^2 \tag{17.48}$$

This means that the effective distance between the poles of the dipole, L_e, is related to the capacitance of the hemispheres, C_a, by

$$L_e = \frac{3\pi r^2 \varepsilon_0}{C_a} \tag{17.49}$$

The open-circuit voltage across the probe is equal to the product of the electric field and the effective distance L_e. The stray capacitance C_s acts with C_2 as a capacitive divider.

17.12   OPTICAL MEASUREMENTS IN HIGH VOLTAGE

Electro-optical and magneto-optical devices are developed to measure electric and magnetic fields, voltage, current, and space charge. The advantages of using optical techniques in high-voltage measurements are:

1. The electrical isolation of the measurement system from the system which is being measured.
2. Ground current effects can be minimized because optical coupling eliminates the wires connecting the two systems.
3. The effects of electromagnetic interference on the measurement system can be avoided by physically separating the measuring system from the higher power system which is the source of the interference.
4. Increased precision and accuracy over conventional electrical measurements.

Normally, optical measurement techniques are based on one of three electro-optical effects: Pockels, Kerr, or Faraday. The parameter to be measured changes the optical transmission characteristics of the sensor. The voltage or the current to be measured produces an electric or a magnetic field. This field interacts with the optically active material to change the light beam. An electro-optical system to measure electrical quantities is shown in Figure 17.15.

The Kerr, Faraday, and Pockels effects relate to specific interactions between the field to be sensed and the optically active material according to the general expression

$$\Delta n = \alpha + \beta F + \gamma F^2 + \cdots \tag{17.50}$$

where Δn is a change in the index of refraction of the material, F is an electric or a magnetic field, and α, β, and γ are constants. If F denotes an electric field and α and β are the only non-zero coefficients, then equation (17.50) describes the Pockels effect. If F denotes an electric field and α and γ are the only non-zero coefficients, then equation (17.50) describes the Kerr effect. Finally, if F denotes a magnetic field and α and β are the only non-zero coefficients, then equation (17.50) describes the Faraday effect.

The Kerr effect is the most relevant effect for high-voltage and dielectric measurements. The basic equation describing the electro-optic Kerr effect is (Thompson and Luessen, 1986)

$$n_{\mathrm{p}} - n_{\mathrm{n}} = \lambda B E^2 \tag{17.51}$$

where n_{p} is the component of the index of refraction parallel to the applied field and n_{n} is the component perpendicular to the field, λ is the wavelength, B is the Kerr coefficient, and E is the electric field. The Kerr effect at low frequency arises because the molecules in the optically active material tend to align with the applied electric field. This alignment produces an anisotropy in the index of refraction. Thus, a light beam polarized in the direction of the applied field and a beam polarized perpendicular to that direction will propagate through the material with different velocities. The relative permittivity of nitrobenzene, a commonly used fluid in Kerr effect systems, and hence the Kerr coefficient, is independent of frequency up to about 1 GHz.

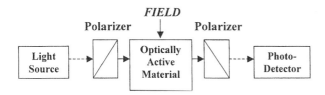

Figure 17.15 Typical electro-optical measurement system.

In measuring electric fields, distinction is drawn between the applied electric field and the sensing electric field. The sensing field is the lower intensity electric field associated with the light beam which is used to detect the difference between the indices of refraction. The applied fields are of higher intensity and may be provided by an external electrode system for the lower frequencies or by a high-power laser for the higher frequencies.

According to equation (17.51), the difference between the indices of refraction, n_p and n_n, is not measured directly. Rather, the phase shift is measured between the electric field component of the light beam polarized in the direction of the orienting field, E_p, and the component polarized in an orthogonal direction, E_n. The phase shift Φ between the perpendicular and parallel components of the light beam can be shown to be given by

$$\Phi = 2\pi BE^2 L \tag{17.52}$$

where L is the effective optical path length that also includes the effect of fringing fields from the electrode edges.

A portable Kerr system (Thompson and Luessen, 1986) for the measurement of high-voltage pulses is shown in Figure 17.16. The complete system is less than 50 cm long and can be used up to 100 kV.

An optical system using linearly polarized incident light is shown in Figure 17.17. The light beam from a laser source is expanded in diameter from 2.0 mm to 50 mm by a beam expander. This expanded light beam is transmitted through the polarizer to become linearly polarized light (polarized plane at an angle $\Phi_p = \pi/4$ from the vertical X axis) with its x and y electric field components having the same amplitude and the same phase. Then it is transmitted through the electric field-stressed liquid dielectric in the test cell in the Z direction. The two field components (e_{1x}, e_{1y}) of incident light transmitting through the dielectric travel at different velocities, resulting in a phase shift between the two electric field components (e_{2x}, e_{21y}) of light after passing through the liquid dielectric. For the parallel electrode system used, the electric field magnitude and direction are constant along the optic path in the Z direction. This optic phase shift $\Delta\theta$ is linearly proportional to the squared electric field magnitude E^2, as given earlier.

17.12.1   Recent Improvements in Kerr Effect Measurements

Measurement sensitivities have been greatly improved by the use of electric field modulation and elliptically polarized incident light modulation. The dynamic behavior of two-dimensional electric field distributions can be observed by the use of a two-dimensional lock-in amplifier, high-quality optical devices, highly sensitive cameras, and computer image processing techniques. Furthermore, a CT (computer tomography) technique has

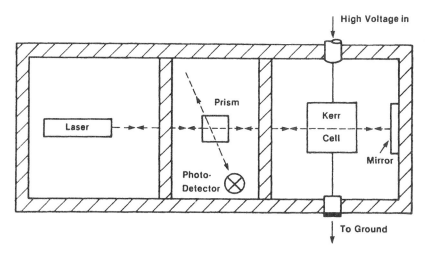

Figure 17.16 Block diagram of a portable Kerr system used to measure high-voltage pulses.

been recently used to measure three-dimensional electric field distributions (Takada et al., 1996).

17.12.2 Kerr Effect for Field Measurements in Oil

Information about the electrical field distribution and the accumulated space charge distribution in oil is vital to both apparatus design and problem diagnosis. Accumulated space charge in oil leads to field concentration and may induce breakdown. Measurements based on the electro-optical Kerr effect are particularly effective in those cases. Based on the electro-optical Kerr effect, attempts were made to develop measurement techniques for the two-dimensional (Zhu and Takada, 1997) and three-dimensional (Shimizu et al., 1996) distributions of the electrical field in dielectric liquids. For rapid measurements, computer-controlled electronic cameras can be

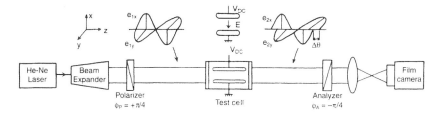

Figure 17.17 Principle of operation of an electro-optic measurement device.

used for optical detection. The influence of system nonuniformity could be eliminated by the use of both optical and electrical modulations. To decrease the effect of the noise component resulting from the electrohydro-dynamic motion of the dielectric liquid, a diagnostic image lock-in technique was proposed.

Also, a Kerr electro-optic measurement system was used to measure the electric field in a transformer oil/solid composite insulation system (Okubo et al., 1997).

Measurement of the space charge field (larger than $80\,kV/cm$) in trans-former oil was undertaken by Mahajan and Sudarshan (1994). Transformer oil was used as a Kerr medium in these measurements. Space charges result-ing from charge injection have been found in transformer oil at room tem-perature under electric stresses in excess of $150\,kV/cm$ with AC excitation and $90\,kV/cm$ under DC excitation. The magnitudes of space charges in transformer oil with AC applied voltages varied from 2 to $50\,nC/cm^3$. The magnitudes of space charges with DC excitation varied from $60\,pC/cm^3$ to $10\,nC/cm^3$. The results illustrated the limitations of the electro-optic tech-nique to measure interfacial electric field (gas–solid) with transformer oil as the Kerr medium.

17.13   PARTIAL DISCHARGE MEASUREMENTS

As seen in many sections of this book, partial discharges within the insula-tion are usually a prelude to insulation failure. Their detection, therefore, is a vital task. When an insulation sample is subjected to high voltage, partial discharges cause patterns of current impulses in the leads of the sample.

17.13.1   Partial Discharge Equivalent Circuit

The behavior of internal discharges under AC voltage can be described using the circuit of Figure 17.18. The cavity is represented by a capacitance c, which is shunted by a breakdown path. The dielectric in series with the cavity is represented by a capacitance b, and the sound part of the dielectric is represented by capacitance a. If the circuit is energized by AC voltage, a recurrent discharge occurs where c is charged. When the breakdown voltage of the cavity, V_1, is reached, the insulation breaks down. As the collapsing voltage reaches its extinction value V_2 the discharge stops and c recharges, and so on. The charge q_1 transferred within the cavity is taken as a measure of the partial discharge. If the sample is large compared with the cavity, as will usually be the case, this charge transfer is equal to

$$q_1 \simeq (b + c)\Delta V \qquad\qquad (17.53)$$

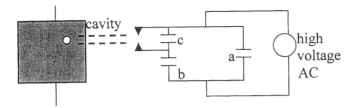

Figure 17.18   Analog circuit of partial discharge.

where $\Delta V = V_1 - V_2$ is the voltage drop within the cavity.

The charge q_1 cannot, in practice, be measured directly with a discharge detector. Instead, the "apparent charge" that flows through the leads of the test object is used to identify the discharge. This charge is given by

$$q = b\Delta V \tag{17.54}$$

It causes a voltage drop $[(b\Delta V)/(a + b)]$ in the test object. Most discharge detectors are capable of determining q. The apparent charge q can be related to the energy P of the discharge as follows:

$$P = \tfrac{1}{2}c(V_1^2 - V_2^2)$$
$$= \tfrac{1}{2}c \times \Delta V \times (V_1 + V_2)$$

If V_2 is neglected, then

$$P \simeq \tfrac{1}{2}c \times \Delta V \times V_1$$

But

$$V_1 = b/(b + c)V_i$$

where V_i is the external inception voltage at which the sample starts to discharge. Therefore,

$$P = \tfrac{1}{2}c\Delta V[b/(b + c)]V_i$$

By neglecting b in comparison with c,

$$P \simeq \tfrac{1}{2}b\Delta V V_i$$
$$\simeq \tfrac{1}{2}qV_i \tag{17.55}$$

17.13.2 Partial Discharge Detection Principle

All circuits used to detect partial discharge impulses can be reduced to the basic diagram of Figure 17.19 which is made up of:

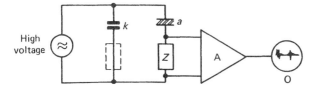

Figure 17.19   Basic diagram for partial discharge detection.

1.   A discharge-free high-voltage source
2.   The tested insulation a
3.   An impedance Z, across which voltage impulses are caused by the discharge impulses in the test object
4.   A coupling capacitance k which facilitates the passage of the high-frequency current impulses
5.   A measurement circuit which amplifies, processes, displays, and stores the detected signal

17.13.2.1   The Detection Impedance

The impedance Z can be connected either in series with the sample, as in the figure, or placed in series with the coupling capacitor. Both methods are equal electrically. However, in practice, if the sample is large, Z is often placed in series with k so that the large charging current of a does not pass through the impedance. Two forms of impedance are commonly used, namely, a noninductive resistor R shunted by a stray capacity C, or an oscillatory RLC circuit.

In the RC circuit the impulse will be unidirectional, of a shape given by

$$V = \frac{q}{a + C\left(1 + \dfrac{a}{k}\right)} \times \exp(-t/Rm) \tag{17.56}$$

where q is the magnitude of the discharge causing the impulse, $q = b\Delta V$, C is the stray capacitance shunting R, and

$$m = \frac{ak}{a + k} + C \tag{17.57}$$

17.13.2.2   Optimal Amplifier Bandwidth

The noise of an amplifier is proportional to the square root of the bandwidth; as the bandwidth is increased, the noise increases. On the other hand, the response factor of the amplifier increases with bandwidth. Optimal sensitivity is obtained when the time constant of the amplifier is made equal to the time constant of the detection circuit. The latter is given by

$$R[ak/(a+k) + C]$$

It is concluded that optimal amplifier sensitivity is obtained when the coupling capacitance k is of the same order of magnitude as the sample a in order to overcome the effect of the stray capacitance C. Also, the sensitivity decreases linearly with the test object capacitance (Kreuger, 1989).

17.13.3 Practical Detection Circuits

The basic detection circuit shown in Figure 17.19 can be used in practice in two different ways, namely, in direct detection and bridge detection circuits.

17.13.3.1 Direct Detection Circuit

Figure 17.20 shows a modern direct partial discharge detection circuit. It has the following features:

1. A low-voltage filter in the power supply is needed to suppress interference from the mains. Such a power-supply filter is a heavy and expensive piece of equipment.
2. A high-voltage filter further suppresses interference or possible small discharges from the transformer.
3. The lead connecting the high-voltage source to the test object must be free of corona.
4. The test object is connected between high voltage and earth to avoid the need for special connections at the object's earth terminal.
5. The coupling capacitor is discharge free and is usually connected to the measurement impedance as shown.
6. The measurement impedance is shunted by an overvoltage protection device. The impedance is either resistive or an attenuated resonance circuit with a relatively wide band. A bandwidth of about 50–100 kHz is sufficient for satisfactory resolution, but wider bandwidths up to 500 kHz are also in use.
7. The amplifier in the PD detector has a bandwidth wider than that of the detection impedance.
8. The impulses are displayed on an elliptical time base.
9. A peak voltmeter serves as a picocoulomb meter and is used to detect the highest impulse.
10. The test circuit is placed in a Faraday cage which is earthed at one single point. The power supply of the detector is filtered.

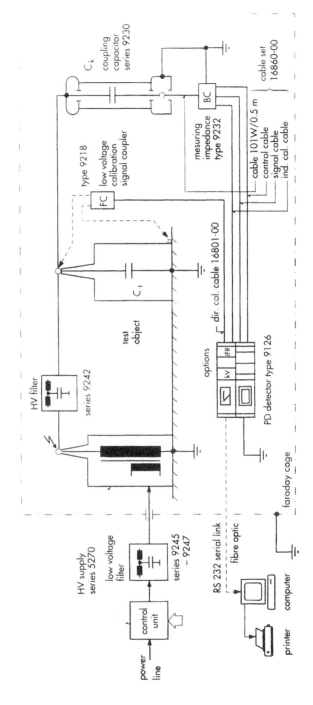

Figure 17.20 Practical direct partial discharge detection circuit. (Courtesy of Haefely-Trench.)

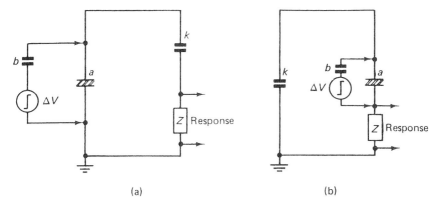

(a)　　　　　　　　　　(b)

Figure 17.21 Principle of calibrating partial discharge detection circuits.

17.13.3.2 Circuit Calibration

To calibrate the detection circuit, a known charge is injected into the test object. This takes the form of a generated step voltage in series with a small capacitor. The calibrator is directly connected to the terminals of the object (Figure 17.21). The relation between the discharges in picocoulombs and the detector output is thus determined.

17.13.3.3 Bridge Detection Circuit

As seen in Figure 17.22, another identical test object (or the calibration capacitor) is used together with the first to form a differential bridge. Discharges in the objects are detected while external discharges are suppressed. Balance is obtained by varying the impedances Z_1 and Z_2.

Although bridge detection is more complicated than straight detection, it has the inherent advantage of suppressing external discharges. Therefore, discharge detection can be performed in the presence of external

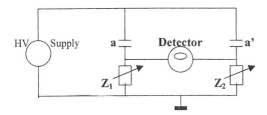

Figure 17.22 Bridge detection circuit.

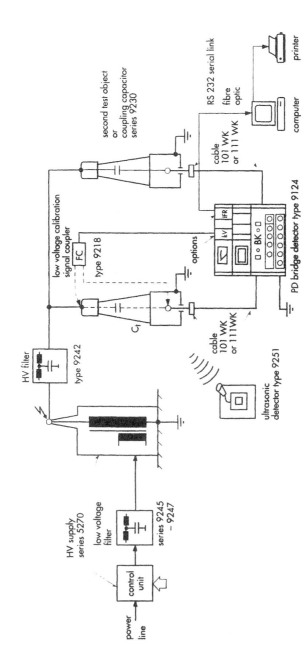

Figure 17.23 Practical bridge partial discharge detection circuit. (Courtesy of Haefely-Trench.)

discharges or other interference. Internal discharges can be easily distinguished from external ones since, unlike the latter, the former are hardly altered by changing the balance. Figure 17.23 displays a complete practical bridge detection arrangement.

17.13.4 Partial Discharge Interpretation

Oscilloscopes are still in general use to observe the impulses of partial discharges. The discharge impulses are usually displayed on an elliptical time-base as shown in Figure 17.24a–d. The ellipse is situated in such a way that top and bottom coincide with the plus and minus crests of the high-voltage sine wave and the ends coincide with the zero crossings. Currently, the display is interfaced with a computer where it is stored, processed, and redisplayed.

The location and shape of the discharge spikes can be used to identify the source of partial discharge. The following typical cases demonstrate this exercise (Nattrass, 1988).

17.13.4.1 Discharges in an Internal Solid Dielectric Cavity

The discharges occur ahead of the voltage peaks on both the positive and negative halves of the waveform, as seen in Figure 17.24a. They are nearly of

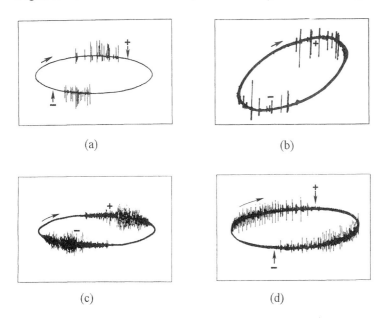

(a) (b)

(c) (d)

Figure 17.24 Partial discharge interpretation displays.

the same amplitude and number on both sides of the ellipse, with some randomness with time. The inception voltage is above the minimum detectable discharge. There is little or no variation in magnitude with increases in voltage, and the discharge extinction voltage is equal to or slightly below the discharge inception voltage.

17.13.4.2   Internal Discharges at Fissures in Solids in the Direction of Field

This phenomenon is typical of some cable problems. Discharges occur in advance of the voltage peaks and are often similar in number and magnitude, although differences of 3:1 are normal, as seen in Figure 17.24b. There is little or no variation in magnitude with a fairly rapid increase and decrease in voltage, and the phenomenon appears similar to the preceding case.

17.13.4.3   Cavity Forming between Metal or Carbon on One Side and Dielectric Material on the Other Side

The discharge phenomenon is in advance of the voltage peaks and is asymmetrical, as seen in Figure 17.24c. Unequal magnitude and number on the two half-cycles are usually in the ratio of 3:1, but approximately 10:1 is needed to clearly distinguish it from the first case above.

17.13.4.4   Voids in Cast Resin Insulation Systems due to Faulty Processing

The formation of electrically conducting products can result from the action of discharge upon the resin at these sites. The discharges occur in advance of the voltage peaks and are often equal on both halves of the waveform, as seen in Figure 17.24d.

17.14   SPACE CHARGE MEASUREMENTS IN INSULATION

The accumulation of space charges in insulation materials has the adverse effect of lowering the breakdown field of the insulation. This is particularly evident in case of high-voltage cables. Several techniques have been developed to directly measure the space charge distribution.

17.14.1   Thermal Pulse Method

The thermal pulse method consists of applying a thermal pulse to one surface of the dielectric by means of a light flash and measuring the electrical response generated by the sample as a function of time while the thermal transient diffuses across the sample. The form of the time dependence of the electrical response carries information about the charge or polarization distribution. Such distribution can be obtained by a deconvolution process. This imposes limitations on the usefulness of the method and necessitates that results should be interpreted appropriately.

17.14.2 Laser Intensity Modulation Method (LIMM)

This method utilizes sinusoidally modulated surface heating of dielectric samples to produce spatially nonuniform temperature distributions through the thickness. In the LIMM technique, each surface of the sample, usually a thin film, is exposed to a laser beam which is intensity modulated in a sinusoidal fashion by an acousto-optic modulator or light chopper. The laser beam is absorbed by the front of the electrode of the sample. The sinusoidal modulation of the laser beam causes a sinusoidal fluctuation in temperature of the front electrode, resulting in propagation of temperature waves into the sample. The temperature waves are attenuated as they progress through the sample and are also retarded in phase. Thus a nonuniformly distributed thermal force acts on the sample. The interaction of the fluctuating temperature and the spatially distributed polarization and space charge produces a sinusoidal pyroelectric current. This current is a unique function of the modulation frequency and the polarization and charge distributions. As with the thermal pulse methods, a mathematical deconvolution technique is required to compute the polarization and space-charge distributions from the current–frequency data.

17.14.3 Pressure Pulse Methods

A number of methods utilizing pressure pulses have been developed to study charge and polarization profiles in the thickness direction of dielectrics (Li and Takada, 1994). In those methods piezoelectric transducers are generally used. When a piezoelectric transducer is exposed to an electric field, acoustic waves are generated in the transducer. The opposite is also true: when pressure is applied, surface charges are induced on the transducer. In Figure 17.25, when an electric field is applied to the first transducer a pulsed acoustic wave $p(t)$ is generated. When this wave is propagated through the second transducer, surface charges on the transducer are induced so that an electric current flows through the outer circuit.

Polymer insulating materials possess neligible piezo strain and stress characteristics. When a space charge is accumulated in a polymeric material, the piezo strain and stress characteristics of the charged material become greatly enhanced (Li et al., 1994; Ahmed and Srinivas, 1997).

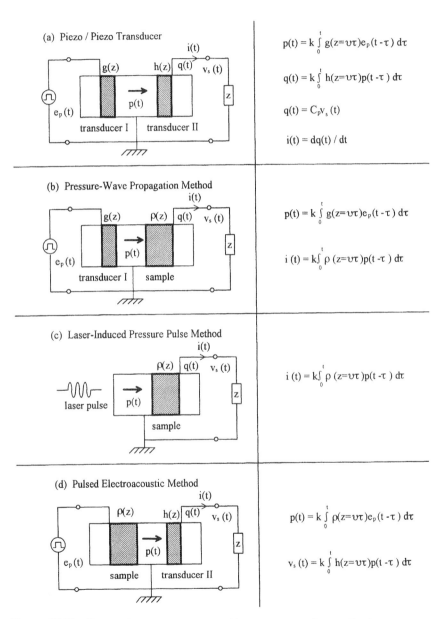

Figure 17.25   Space charge measurement by pressure pulse methods.

17.4.3.1 Pressure Wave Propagation Method

When the second transducer of Figure 17.25a is replaced by a dielectric with space charge accumulation (see Figure 17.25b), a current flows through the outer circuit which depends on the sample's space charge. The current in the outer circuit is proportional to the accumulated space charge if the duration of the acoustic wave $p(t)$ is much narrower than that of the transit time of the acoustic wave across the sample. A space charge distribution of thickness 50–200 μm can be obtained directly. Since the acoustic velocity in a solid polymeric dielectric is about 2000 m/s, if 5% space resolution is desired, then the duration of the applied pulse voltage should be about 1.25–5.0 ns.

17.4.3.2 Laser-Induced Pressure Wave Propagation

As seen in Figure 17.25c, when a laser pulse of about 500 ps is irradiated to the target electrode of the sample, the reaction of the evaporation of the electrode material generates the pulse acoustic wave $p(t)$. The current through the outer circuit is proportional to the accumulated space charge distribution. Therefore, the space charge distribution can be determined. The acoustic wave generated by this method is large, the signal level being about 100 mV. One weak point is that the laser pulse generation setup is large and expensive. This method is used in film samples (thickness of about 50 μm), plate samples (thickness of about 5.0 mm), and coaxial samples.

A thermoelastically generated, laser-induced pressure pulse technique is also used for measuring space charge in dielectric films. It employs a stress wave generated thermoelastically by absorption of a sub-nanosecond laser pulse.

17.4.3.3 Pulsed Electroacoustic Method

When a pulsed electric field is applied to a dielectric with accumulated space charge, as shown in Figure 17.25d, pulsed force is generated from which a pulsed acoustic wave $p(t)$ is launched. The acoustic wave $p(t)$ is proportional to the space charge distribution $p(z)$. This acoustic wave is detected by the second transducer so that the induced charge $q(t)$ on the surfaces of the transducer is proportional to the space charge distribution $p(z)$, with the condition that the duration of the pulsed electric field is the same as in the pressure wave method.

17.14.4   Electro-Optical Methods

17.14.4.1   Photoconductivity Method

This method is based on the absorption of a narrow light beam in a thin photoconductive layer. Very weakly absorbed visible monochromatic light liberates carriers that move under the field of the space charges. By the detection of the photocurrent produced, information can be obtained about the distribution of the space charge. This method is nondestructive only for short illumination times.

17.14.4.2   Electro-Optic Field and Space Charge Mapping

Electro-optical methods consist in the application of the Kerr or Pockel effect. The mechanism of these methods is based on the interaction of polarized light, narrow or expanded, with the field of the space charges. Electro-optical methods have been used mostly on transparent dielectric liquids, but are also used on polymers.

17.14.4.3   Spectroscopic Measurement

The splitting or shift of spectral lines in an electric field is used to determine the magnitude of the field. Variations in the field through a sample are observed by the spatial resolution of the spectroscopic signal in the detection system.

17.15   SOURCES OF ERROR IN HIGH-VOLTAGE MEASUREMENTS

In addition to human and instrument errors, two types of error are known to exist in high-voltage testing: those caused by electromagnetic (EM) interference, and those produced by improper earthing.

17.15.1   Shielding against EM Interference

Electromagnetic noise may infiltrate into the measuring system and be superimposed on the measured signal. This noise is caused by transient potentials and by strong electromagnetic fields associated with the high, rapidly changing voltages and currents. The following measures can be taken to reduce EM noise:

1.   Power-line RF filters may be used with oscilloscopes.
2.   Signals should be transmitted via shielded and matched coaxial cables.

3. With very steep surge measurements, multiple shielding may be used.
4. If necessary, the oscilloscope may be located away from the source of noise in a shielded chamber.

17.15.2 Grounding in High-Voltage Laboratories

High, rapidly changing voltages and currents in a high-voltage laboratory produce transient currents in the ground connections. Following are some of the measures that should be taken to avoid these adverse effects:

1. Multiple grounding of the high-voltage test circuit should be avoided.
2. The ground current loop can be interrupted by winding the cable on a ferrite core, thus raising the impedance of that loop.
3. Impulse generators should be connected to very low-resistance ground composed of a grid with rods driven into the ground close to the generator (Chapter 13).
4. The power supplies to the DC source of impulse generators and to the oscilloscope should be made via isolating transformers.

17.16. PROBLEMS

(1) A high-voltage periodic wave may be approximated by the triangular function shown. Indicate the voltage reading obtained by:

(i) 25 cm sphere gap
(ii) a peak voltmeter
(iii) an electrostatic voltmeter

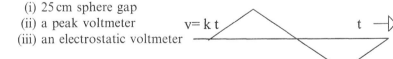

If $k = 20 \times 10^6$ V/s and the wave frequency is 50 Hz, the temperature and pressure during the measurement are respectively $30°C$ and 99 kPa, use the attached table to predict the expected sphere gap separation.

Spacing (cm)	2.0	2.5	3.0	3.5	4.0	4.5
Voltage (kV)	59	73	86	99	112	125

(2) If a peak voltmeter uses a milliammeter whose full-scale deflection is 500 mA, choose a proper value for the internal capacitance of the

voltmeter in order to measure the voltage wave shown, whose magnitude lies between 0 and 5 kV.

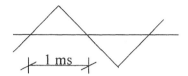

(3) A high-voltage, 60 Hz AC signal of amplitude x kV with a superimposed heavy DC component of y kV is to be measured by a peak voltmeter and an electrostatic voltmeter. Estimate the reading of each instrument.

(4) A capacitive/resistive potential divider is used to measure AC high voltages. The division ratio is to be 1:1000, the high-voltage arm resistance is 600 kΩ, and the high-voltage arm capacitance is 1000 pF. Estimate appropriate values for the low-voltage arm components, justifying your answers. If the tolerated error in phase shift between the divider's input and output is only 5°, estimate the corresponding tolerance in the proposed low-voltage arm resistance.

(5) An unscreened 500 kV resistive divider is used to measure AC voltages at a division ratio of 1:500. If the high-voltage arm resistance is 2.5 MΩ, and the divider is 1.2 m high and 0.15 m in diameter, estimate the error in the maximum measurable voltage.

(6) A Kerr cell is used to measure high votlages of up to 20 kV. The cell uses a dielectric liquid whose Kerr constant is 10^{-15} m/V^2 and uses a uniform field gap of 5 mm. The corresponding maximum detectable phase retardation angle is 40°. It is now required to replace the cell liquid with another whose Kerr constant is 10^{-16} m/V^2 and which has a discharge threshold of 800 kV/cm. What is the maximum phase retardation that can be practically detected?

(7) Estimate the amplitude error in an AC unscreened 2000:1, 100 MΩ resistive divider which is 120 cm high and 10 cm in diameter.

REFERENCES

Ahmed NH, Srinivas NN. IEEE Trans Dielectr Electr Insul 5:644–656, 1997.
Berkebile L, Nilsson S, Sun S. IEEE Trans PAS-100:1498–1504, 1981.
Hebner RE, Malewski RA, Cassidy EC. Proc IEEE 65(11):1524–1548, 1977.
Holt P. Electra 121:69–75, 1988.
Hyltén-Cavallius M. High Voltage Laboratory Planning. Basel, Switzerland: E. Haefely, 1988.

IEC. Recommendations for Voltage Measurement by Means of Sphere-Gaps (One Sphere Earthed). Publication 52. Geneva: International Electrotechnical Commission, 1960.

IEC. Radio Frequency Cables: General Requirements and Measuring Methods. Publication 96-1. Geneva: International Electrotechnical Commission, 1962.

IEC. High Voltage Test Techniques. Publication 62. Geneva: International Electrotechnical Commission, 1973.

IEEE Committee. IEEE Trans PAS-100:4811–4814, 1981.

IEEE Working Group. IEEE Trans PAS-102:3549–3557, 1983.

Kind D. An Introduction to High Voltage Experimental Technique. Braunschweig, Germany: Friedr Vieweg & Sohn, 1978.

Kreuger FH. Partial Discharge Detection in High-Voltage Equipment. London: Butterworths, 1989.

Kuffel E, Zaengl WS. High Voltage Engineering Fundamentals. Oxford: Pergamon Press, 1984.

Li Y, Takada T. IEEE Electr Insul Mag 10:16–28, 1994.

Li Y, Yasuda M, Takada T. IEEE Trans Dielectr Electr Insul 2:188–195, 1994.

Mahajan SM, Sudarshan TS. IEEE Trans Dielectr Electr Insul 2:63–70, 1994.

Malewski R, Dechamplain A. IEEE Trans PAS-99:636–649, 1980.

Nattrass DA. IEEE Electr Insul Mag 4:10–23, 1988.

Okubo H, Shimizu R, Sawada A, Kato K, Hayakawa N. IEEE Trans Dielectr Electr Insul 5:64–70, 1997.

Shimizu R, Matsuoka M, Kato K, Hayakawa N, Hikita M, Okubo H. IEEE Trans Dielectr Electr Insul 4:191–196, 1996.

Stark WB. Proceedings of 3rd International Symposium on High Voltage Engineering, Milan, 1979, Paper 44-08.

Takada T, Zhu Y, Maeno T. IEEE Electr Insul Mag 2:8–20, 1996.

Thompson JE, Luessen LH. Fast Electrical and Optical Measurements. Boston: NATO ASI Series, 1986.

Tokayo, Harumoto Y, Yamamoto H, Yoshida Y, Mukae H, Shimada M, Ida Y. IEEE Trans PAS-101:3967–3976, 1982.

Zhu Y, Takada T. IEEE Trans Dielectr Electr Insul 5:748–757, 1997.

18

Testing Techniques

R. RADWAN and A. EL-MORSHEDY *Cairo University, Giza, Egypt*

18.1 INTRODUCTION

This chapter presents some of the techniques and theory involved in the testing of high-voltage equipment either in the laboratory or in the field. The main concern with electric equipment is that its insulation should withstand its operating voltage and the occasional overvoltage transients expected in the system. A further requirement for the satisfactory performance of the insulation is the minimization of internal discharges in voids present within the dielectric, which may cause its deterioration and eventual breakdown. High-voltage dielectric loss and partial discharge testing can, to some extent, help in the selection and design of insulation for electrical equipment.

Equipment insulation may be classified as self-restoring and non-self-restoring. Self-restoring insulation completely recovers its insulating properties after a disruptive discharge. The method of sampling and test techniques differ from one type of insulation to the other. International and national specifications for testing are outlined to meet the users' and manufacturers' requirements. These specifications describe in detail the tests to be carried out on a specific piece of equipment, their procedure, and acceptable limits of test results. Usually, the specifications are not limited to electrical tests but include other tests, such as mechanical and thermal tests.

A new equipment may pass a high-voltage test but may fail in the same test after some time in service. This is due to contamination of the insulators' surfaces, which can lower their dielectric strength while gradual chemical deterioration due to the reaction of the insulation with air is taking place. Corona discharges in voids in the insulation can gradually cause deterioration (Chapter 8). Voltage transients can initiate tracking or carbonization, and can even puncture the insulation and lead to early failure. This shows the importance of periodic maintenance and in-service tests. The following sections are devoted to various types of tests and voltages to be applied to circuit breakers, cables, transformers, high-voltage insulators, and surge arresters.

18.2   CLASSIFICATION OF TESTS

Some of the tests are performed during the early stages of development and production, others after production and installation. Type tests are performed on each type of equipment before their supply on a general commercial scale so as to demonstrate performance characteristics meeting the intended application. These tests are of such a nature that they need not be repeated unless changes are made in the design of the product. Routine tests are made by the manufacturer on every finished piece of product to make sure that it fulfills the specifications. Acceptance and commissioning tests are made by the purchaser and are self-explanatory. Maintenance tests are usually carried out after maintenance or repair of the equipment.

18.3   TEST VOLTAGES

The conventional forms of test voltages in use can be divided into three main groups: (a) direct voltages, (b) power-frequency alternating voltages, and (c) impulse voltages, which are divided into lightning and switching impulses. Also, in the case of testing machine insulation, alternating voltages of low frequency are sometimes used.

Methods of generating these voltages are described in Chapter 16. In this chapter attention is focused on the levels of test voltages and the techniques of voltage application. Also, the test circuits and precautions to be taken in high-voltage laboratories with regard to earthing, clearances, and interference are discussed. Tables 18.1 and 18.2 list the recommended test voltages adopted for testing equipment for rated AC voltages ranging between 1 and 765 kV (IEC, 1976b).

For equipment with rated voltages of 1 to 300 kV, performance under power-frequency operating voltage, temporary overvoltages, and switching overvoltages is generally checked by a short-duration power-frequency test.

Table 18.1 Recommended Test Voltages for Rated Voltages Less than 300 kV

Rated voltage (rms) (kV)	Rated lightning-impulse withstand voltage (peak) (kV)	Rated power-frequency short-duration withstand voltage (rms) (kV)
3.6	20[b]	10
	40	
4.4[a]	60[c]	19
	75[d]	
7.2	40[b]	20
	60	
12[a]	60	28
	75	
13.2[a]	95[c]	34
13.97[a]	110[d]	
14.52[a]		
17.5	75[b]	38
	95	
24	95[b]	50
	125	
26.4[a]	150	50
36	145[b]	70
	170	
36.5[a]	200	70
52[a]	250	95
72.5[a]	325	140
123[e]	450	185
123	550	230
145[e]	450	185
145[e]	550	230
145	650	275
170[e]	550	230
170[e]	650	275
170	750	325
245[e]	650	275
245[e]	750	325
245[e]	850	360
245[e]	950	395
245	1050	460

[a]Specifications for dielectric tests for the United States and Canada.
[b]For effectively earthed neutral with additional overvoltage protection.
[c]For transformers with rating of 500 kVA and below.
[d]For transformers with rating above 500 kVA.
[e]Reduced insulation permissible only for systems with effectively grounded neutral.

Table 18.2 Recommended Test Voltages for Rated Voltages Above 300 kV

Rated voltage (rms) (kV)	Rated switching-impulse withstand voltage (peak) (kV)	Rated lightning-impulse withstand voltage (kV)
300	750	850
		950
	850	950
		1050
362	850	950
		1050
	950	1050
		1175
420	950	1050
		1175
	1050	1175
		1300
		1425
525	1050	1175
		1300
		1425
	1175	1300
		1425
		1550
765	1300	1425
		1550
		1800
	1425	1550
		1800
		2100
	1550	1800
		1950
		2400

But aging of internal insulation and contamination of external insulation require long-duration power freqency tests. The performance under lightning overvoltage is checked by a lightning-impulse test. For equipment with rated voltages $\geq 300\,\text{kV}$, the performance under switching overvoltages is checked by switching impulse tests. Equipment in systems with effectively earthed neutrals can safely have reduced insulation. Therefore Table 18.2 lists more than one test voltage for each rated AC voltage of the equipment.

18.3.1   Tests with Direct Voltage

Direct voltage is used mainly to test equipment used in high-voltage DC transmission systems. It is additionally used in insulation testing of arrangements with high capacitance, such as capacitors and cables. It is also used in fundamental investigations in discharge physics and dielectric behavior.

The value of the test voltage is defined by its arithmetic mean. The test voltage, as applied to the test object, should not conain AC components corresponding to a ripple factor of more than 5% when normal current is drawn. During the test it is required that the rate of voltage rise above 75% of its estimated final value should be about 2% per second (IEC, 1973). The requirements of the test are generally satisfied if no disruptive discharge occurs on the test object when under the test voltage for the specified duration.

18.3.2   Tests with Alternating Voltage

The voltage used in this test generally has a frequency in the range 40–60 Hz and a sinusoidal shape. The ratio of its peak to its rms value is equal to $\sqrt{2} \pm 5\%$. Partial discharges should not reduce the test voltage. This is usually achieved if the total HV circuit capacitance is within 1000 pF and the circuit current with the test object short circuited is at least 1 A(IEC, 1973).

For dry tests on small samples of solid insulation or insulating liquids, a short-circuit current on the order of 0.1 A rms may suffice. For tests under artificial pollution, the required short-circuit current depends on the ratio of series resistance R_s to the steady-state reactance X_s of the voltage source, including the generator or supply network, at the test frequency. It should be at least 6 A for $R_s/X_s < 0.1$, and at least 1 A for $R_s/X_s = 0.25$ (IEC, 1973).

The value of the test voltage is defined by its peak divided by $\sqrt{2}$. The peak values of voltages can be measured with a sphere gap or a peak voltmeter (Chapter 17). The rated withstand voltage is determined by the same method as that in the direct voltage test.

18.3.3   Tests with Impulse Voltage

A standard lightning impulse voltage has been accepted as an aperiodic impulse that reaches its peak value in 1.2 μs and then decreases slowly in about 50 μs to half its peak value. Switching impulses are characterized by having much longer fronts and total durations (Chapters 14 and 16).

624   **Radwan and El-Morshedy**

18.3.3.1   Rated Impulse Withstand Tests

For tests on non-self-restoring insulation, three impulses are applied at the rated withstand voltage level of the specified polarity. The requirements of the tests are satisfied if no failure occurs. For withstand tests on self-restoring insulation, two procedures are in common use:

1. Fifteen impulses of the rated withstand voltage with the specified shape and polarity are applied. The requirements of the test are satisfied if not more than two disruptive discharges occur (IEC, 1973).
2. The test procedure for determining the 50% disruptive discharge voltage is applied. The test requirements are satisfied if the determined voltage is not les than $1/(1 - 1.3\sigma)$ times the rated impulse withstand voltage, where σ is the per-unit standard deviation of the disruptive discharge voltage (IEC, 1973).

The values of the 50% disruptive discharge voltage V_{50} and its standard deviation σ can be calculated using statistical methods. The two test procedures are the multiple-level (or probit) method and the up-and-down method (IEC, 1973).

18.3.3.2   Multiple-Level Method

At least 10 impulses are applied at each test voltage level. The voltage interval between levels is approximately 3% of the expected 50% disruptive discharge voltage. The value of 50% disruptive discharge voltage is obtained from a curve of disruptive discharge probability versus prospective test voltage. The accuracy of determination increases with the number of voltage applications at each level.

The values of V_{50} and the standard deviation S can be calculated in terms of the set of measured voltages V_i and their relative deviations $Z_j = (V_i - V_e)/\sigma_e$, where V_e and σ_e are the values initially estimated and are to be corrected by reiteration. Thus

$$V_{50} = \frac{\overline{V(\overline{Z^2})} - \overline{Z} \cdot \overline{ZV}}{(\overline{Z^2}) - (\overline{Z})^2}$$
$$S = \frac{\overline{ZV} - \overline{V} \cdot \overline{Z}}{(\overline{Z^2}) - (\overline{Z})^2}$$

(18.1)

in terms of the average values of V and Z. Some weighting coefficients can be applied to the readings at each voltage level and the confidence limits can be calculated, as is well known (ANSI, 1968).

18.3.3.3 Up-and-Down Method

A voltage V is chosen that is approximately equal to the expected 50% disruptive discharge level. ΔV is the voltage interval and is approximately equal to 3% of V. One impulse is applied at the level V. If this does not cause a disruptive discharge, the next impulse should have the level $V + \Delta V$. If a disruptive discharge occurs at the level V, the next impulse should have the level $V - \Delta V$. This procedure is continued until a sufficient number of observations has been recorded.

To calculate V_{50} and σ, the voltage readings corresponding to either the withstands or discharges can be used. If the total number of either type of event is N, with n_i shots (≥ 20) at each voltage level V_i, the lowest level being V_0 and the highest being V_k, then

$$V_{50} = V_0 + \Delta V\left(\frac{A}{N} \pm 0.5\right) \tag{18.2}$$

$$S = 1.62\Delta V\left(\frac{NB - A^2}{N^2} + 0.029\right) \tag{18.3}$$

where

$$N = \sum_{i=0}^{k} n_i, \qquad A = \sum_{i=0}^{k} in_i, \qquad \text{and} \qquad B = \sum_{i=0}^{k} i^2 n_i$$

Evidently, the same statistical manipulation of the test results applies to all similar tests.

18.3.3.4 Tests with Switching Impulse

The standard switching impulse is an impulse having a time to crest T_1 of $250\,\mu s$ and a time to half-value T_2 of $2500\,\mu s$. It is described as a $250/2500$ impulse. The test procedures for switching impulses are, in general, the same as those for lightning impulse testing, and similar statistical considerations apply.

With switching impulses, disruptive discharges may occur at random times before the crest. In presenting the results of disruptive discharge tests, the relationship of discharge probability to voltage is generally expressed in terms of the prospective crest value.

18.4 TESTS WITH IMPULSE CURRENTS

Transient currents of large amplitude are experienced during the discharge of energy-storing devices, lightning strokes, and some short circuits. If these currents have a definite shape, they are referred to as impulse currents. Impulse currents for testing normally take either of two shapes, the double-exponential or the rectangular shape, as defined in Chapter 16. Two standard shapes are in use for impulse currents: the double exponential of $8/20$ or $4/10\,\mu s$, and the rectangular, with a virtual duration of $500\,\mu s$, $1000\,\mu s$, or $2000\,\mu s$.

18.5 SAFETY PRECAUTIONS IN THE LABORATORY

Extreme caution and safety awareness are essential items of high-voltage test procedures. The following safety features are considered for the safety of the operating personnel, installations, and apparatus:

1. It is essential that the high-voltage equipment be properly designed and manufactured to permit testing without unnecessary danger.
2. The actual danger zone of the high-voltage circuit must be clearly marked and protected from unintentional entry by walls or metallic fences.
3. All doors should be interlocked to remove high voltage automatically when opened.
4. Before touching the high-voltage elements after testing, visible metallic connection with earth must be established.
5. All metallic parts of the setup that do not carry potential during normal service must be grounded reliably.
6. It is preferred that the region of the high-voltage apparatus be matted by a closely meshed copper grid. The earth terminals of the apparatus are connected to it noninductively using wide copper bands.
7. All measuring and control cables and earth connections must be laid avoiding large loops.
8. The measuring signal is transferred to the measuring device (e.g., oscilloscope) via coaxial cables.
9. Shielding of a high-voltage setup by a Faraday cage is necessary for complete elimination of external interference, as sensitive measurements are often required in high-voltage experiments.
10. The clearance between test object and extraneous structure should be at least $1.5S$, where S is the flashover distance between the electrodes of the test object. In this case the effect of such structures on the test results will be negligibly small.

A typical test arrangement of a high-voltage laboratory containing an impulse generator, a high-voltage transformer, and a potential divider is shown in Figure 18.1.

18.6 NONDESTRUCTIVE TESTING

Nondestructive electrical tests are usually carried out on the equipment insulation to ensure that its electrical characteristics comply with the specifications without destroying it. These tests include partial discharges, radio interference, dielectric loss angle, and insulation resistance measurements (Kind, 1978).

18.6.1 Partial Discharge Measurements

Corona discharges may occur on the surface of an insulator or in voids within its volume. The presence of corona may be detected by several non-electrical and electrical methods (Nattrass, 1988). The electrical discharge detection methods make use of the current impulses accompanying discharges in the cavity. Before we discuss these methods, it is appropriate to explain what goes on in a void within an insulating material subjected to an alternating voltage stress.

A specimen of an insulating material containing a gas void can be represented as shown in Figure 18.2a and its equivalent circuit in Figure 18.2b. The capacitance C_v represents the void and C_b the dielectric above and below it. C_a represents the rest of the dielectric. When an alternating voltage V_a in excess of that corresponding to the breakdown threshold of the gas in the void is applied to the dielectric a partial discharge (PD) will start in it. The process of partial discharge in the void is illustrated in Figure 18.3. The voltage appearing across the void if there were no discharge is V_v and is given by the expression

$$V_v = \frac{V_a}{1 + (d/d_1 - 1)/\varepsilon_r} \tag{18.4}$$

where d and d_1 are the thicknesses of the insulating specimen and the gas void, respectively, and ε_r is the relative permittivity of the dielectric. Partial discharge in the void will start at a voltage V_i on the positive half-cycle and approximately $-V_i$ on the negative half-cycle. $\pm V_e$ is the voltage at which the discharge stops. The discharge in the void will be accompanied by a sharp current pulse, as indicated in Figure 18.3. This may be repeated several times on the increasing part of the positive half-cycle. At point m the voltage across the void reverses its polarity, since at this instant V_v is decreasing, and discharge will continue with almost regular negative current

Figure 18.1   Layout of a high-voltage laboratory. (Courtesy of IREQ.)

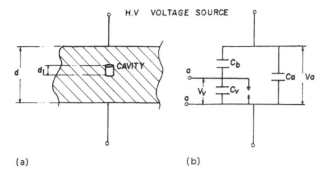

Figure 18.2   Void representation; (b) its equivalent circuit.

pulses. The electrical partial discharge detection methods are classified as straight or balanced methods.

18.6.1.1   Straight Detection Methods

In straight detection methods PD measuring instruments measure the charge released within the discharging sites of the test specimen. The simplest circuit used for such measurement contains a series impedance connected between the test object and ground, as illustrated in Figure 18.4.

Partial discharge current measurement is difficult, and precautions taken to improve the fidelity of display of the current waveshape have been discussed in detail (Nattrass, 1988). Partial discharge measuring instruments available measure the apparent charge (i.e., the charge transfer that takes place at the terminals of the specimen). The amplified discharge pulses are displayed oscillographically, superimposed on a power-frequency elliptic timebase, as shown in Figure 18.5. The discharge shown in Figure 18.5a is

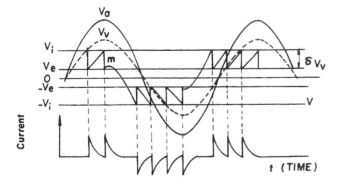

Figure 18.3   Voltage and current traces of partial discharge in a void.

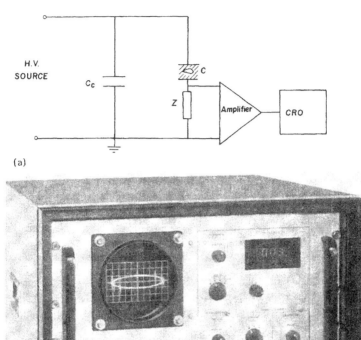

(a)

(b)

Figure 18.4 Partial discharge detection circuit (a) and apparatus (b). (Courtesy of Hypotechnics.)

for a gas void and is characterized by its symmetry around both voltage peaks, while that for a point-to-plane gap is characterized by its regular pattern around the negative voltage peak (Fig. 18.5b).

The dependence of the apparent charge Q_a on the actual charge Q_v in the test specimen can be obtained by considering the discharge model shown in Figure 18.2. When breakdown occurs in the void, charge transfer amounts to

$$Q_v = \left(C_v + \frac{C_a C_b}{C_a + C_b} \right) \delta V_v \tag{18.5}$$

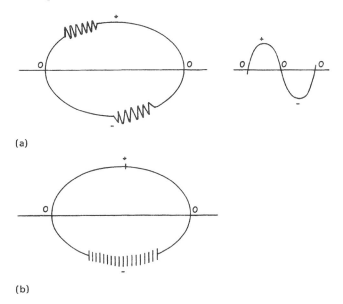

(a)

(b)

Figure 18.5 Partial discharge display: (a) void; (b) point to plane.

where δV_v is the voltage drop in the void at breakdown (Fig. 18.3).

The capacitance C_a is usually much greater than C_b and Q_v can be aproximately expressed as

$$Q_v = (C_v + C_b)\delta V_v \tag{18.6}$$

After the void breakdown the system restores the voltage across the capacitor C_b to its original value, and this requires a charge Q_a to be supplied to C_b. This charge is the apparent charge and is given by

$$Q_a = C_b \delta V_v \tag{18.7}$$

From equations (18.6) and (18.7) the apparent and actual charges are related by the approximate expression

$$Q_a = \frac{C_b}{C_v + C_b} Q_v \tag{18.8}$$

Commercial partial discharge measuring instruments are usually calibrated in terms of apparent charge. Figure 18.6 shows a simplified circuit for PD detection, employing a passive RC differentiator to sense the high-frequency signals generated by PD pulses. The RC network acts as a single-pole high-pass filter whose function is analogous to that of the power-frequency separation filter in conventional PD detection equipment. With

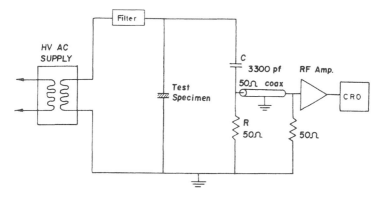

Figure 18.6   A simplified circuit of a PD measuring system that utilizes a passive *RC* differentiator as the sensing element.

low-noise amplifiers, the apparatus sensitivity could reach 0.01 pC (Bilodeau et al., 1987; Kreuger, 1989).

18.6.1.2   Balanced Detection Methods

Balanced detection methods are much more sensitive than straight detection methods. In the straight PD detection methods discharges in any part of the test circuit and not within the test sample may be detected and displayed along with the discharge impulses in the sample. This implies the use of discharge-free high-voltage sources or the provision of filters. Discharges on the high-voltage leads or loose earth connections, although they can be recognized, should be eliminated. In addition, noise may be picked up from a variety of possible sources having nothing to do with the test setup (e.g., nearby thyristor-controlled machines, ultrasonic generators, and arcing contacts).

The detection of PD with bridges similar to that of Schering has been adopted for more than 50 years. The high-voltage arms of the bridge contain the test sample and a separation capacitor. The low-voltage arms contain balancing resistors and capacitors. The output of the bridge is supplied to and displayed on an oscilloscope through a filter and an amplifier. External disturbances are thus considerably reduced. If the separation capacitor and test samples are of equal capacitance, complete rejection of external inter-ference is possible. Otherwise, the bridge would be balanced only at one frequency, resulting in less effective rejection of external interferences. Screening of the low-voltage arms and good earthing will no doubt substan-tially decrease external interference.

18.6.2 Radio-Interference Tests

Corona discharges are known to produce radio noise over a considerable portion of the radio-frequency spectrum (Chapter 5). Measurement of radio interference from HV equipment can be carried out by a circuit similar to that shown in Figure 18.6 except that a radio-noise meter is used instead of the PD amplifier and oscilloscope.

18.6.3 Dielectric Loss Measurements

Losses always occur in dielectrics due to conduction, polarization, and ionization (Chapters 7 and 8). Dielectric losses cause certain electrical effects, which can be utilized for nondestructive high-voltage testing (Kind, 1978).

18.6.3.1 Schering Bridge

The Schering bridge, devised by Schering in 1919, has since been widely used to measure the capacitance and loss angle of high-voltage insulators, capacitors, and cable samples. The bridge comprises two high-voltage (HV) arms and two low-voltage (LV) arms. The HV arms are the test piece and a standard capacitor. This capacitor should have no significant losses over the full working range. A suitable design employs a smooth electrodes with corona shields, insulated with compressed gas. The LV arms are an adjustable precision capacitor and resistors. Both the LV arms and the null detector are shielded from the high-voltage circuit to eliminate any errors in the measurements caused by the effect of stray capacitances. Proper grounding of the circuit is essential for safety and accuracy. Each of the LV arms is shunted by an overvoltage protective device, which operates at a few tens of volts in case either of the high-voltage arms fails. The capacitance C_2 and loss angle δ of the test piece (Fig. 18.7) can easily be evaluated:

$$C_2 = C_1 \frac{R_4}{R_3} \tag{18.9}$$

$$R_2 = \frac{C_4 R_3}{C_1} \tag{18.10}$$

$$\tan \delta = \omega C_4 R_4 \tag{18.11}$$

18.6.3.2 Bridge Incorporating Wagner Earth

The Schering bridge is earthed at the low-voltage end of the high-voltage source. The capacitance of the detector's screened leads as well as the stray capacitances of branches AB and AD affect the balance conditions. To overcome this drawback, auxiliary arms are used for maintaining points B

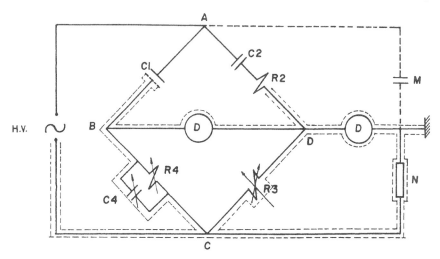

Figure 18.7   Schering bridge incorporating Wagner earth.

and D at earth potential under balance. In Figure 18.7 an additional arm N is connected between the low-voltage terminal and earth (Kuffel and Zaengl, 1984). The stray capacitance of the high-voltage terminal to earth is represented by a capacitance M. The arrangement becomes equivalent to a six-arm bridge. The ratio M/N must be balanced for phase as well as magnitude with the main bridge arms. At balance the terminals of the detector are at earth potential. The capacitances between the leads and screens are in parallel with the impedance N and thus do not affect the balance conditions. An electronic device for automatic balancing of Wagner earth could be developed.

18.6.4   Insulation Resistance Measurement

The magnitude of the resistance of a solid or liquid insulation varies with its temperature, its moisture content, the applied voltage, and whether it is AC or DC (Chapters 7 and 8). It is evident that if the resistance or the loss factor of the insulation in a certain machine, transformer, or cable is measured periodically, under almost identical service conditions, the results will indicate its degree of aging. When the measured insulation resistance starts to decrease considerably below the reading of the previous test, the machine, transformer, or cable should be taken out of service. Its insulation should be more carefully tested and overhauled to avoid sudden failure during operation.

The insulation resistance test can be performed using an adjustable high-voltage DC source (IEEE Committee, 1981). The DC voltage applied during the test should be raised very slowly up to no more than 80% of the acceptance test values. The initial current component corresponds to charging the circuit capacitance. After several seconds, or minutes, the circuit current decreases to its steady component, which corresponds to conduction through the insulation. This last component is the quantity to be measured, recorded, and compared with the previous readings during similar tests.

18.7 CIRCUIT BREAKER TESTING

Testing of circuit breakers requires highly equipped laboratories and sophisticated testing procedures. Standard specifications give and describe in detail the tests to be carried out on circuit breakers and how they are performed (IEC, 1972).

18.7.1 Mechanical Test

Mechanical failures represent more than 80% of the total failures in circuit breakers. The circuit breaker is subjected to 1000 operating cycles, with no voltage or current in its main circuit. During the test, lubrication of the mechanical parts is allowed according to the manufacturer's instructions. However, mechanical adjustment or replacement of any part is not permissible. After the test a complete check on the circuit breaker is performed to make sure that all mechanical linkages and contacts are in good condition.

18.7.2 Temperature Rise Test

During operation the temperature of any part of the circuit breaker should not exceed the specified limits of temperature rise. These limits depend on each individual part and the circuit breaker type. The temperature rise test is carried out with normal rated current flowing in the main circuit and the circuit breaker mounted as under normal service conditions. The maximum observed temperature rise is then compared with the stipulated limits.

18.7.3 Insulation Tests

Dielectric tests are carried out to make sure that the circuit breaker withstands the overvoltage expected within the power system with a reasonable safety factor. The main dielectric tests are impulse and power-frequency voltage tests. The impulse voltage waveshape usually used is the standard wave $1.2/50\,\mu s$, with voltage magnitudes as given in the relevant standard specifications (Tables 18.1 and 18.2). Power-frequency voltage tests are applied for 1 minute to indoor and outdoor circuit breakers when dry or wet (i.e., under simulated rain), respectively. The breakder under test should withstand the specified voltage tests without flashover or puncture.

18.7.4   Short-Circuit Tests

A circuit breaker should be capable of making and breaking the circuit under short-circuit conditions without appreciable deterioration of its components or change in its performance. Short-circuit tests are usually carried out according to test duties that specify the test current, percentage of DC component, and transient and power-frequency recovery voltages. These tests are numerous and it is difficult to discuss them all in detail within the space of this section. However, they are classified as follows:

1. Breaking current tests
2. Making current tests
3. Short-time current tests
4. Operating sequence tests
5. Single-phase short-circuit tests
6. Short-line fault tests
7. Out-of-phase switching tests
8. Capacitor charging-current breaking tests
9. Small inductive current breaking tests

The short-circuit tests need high-power testing plants to meet the ever-increasing circuit breaker ratings, with currents reaching the order of 10 kA or more at voltages ranging up to 750 kV. Such plants are extremely expensive. The synthetic method of circuit breaker testing provides an alternative solution. In the basic form it consists of two independent voltage and current sources (Fig. 18.8). The current source injects the current into the circuit breaker under test at a relatively reduced voltage, while the voltage source injects a high-voltage transient across the circuit breaker contacts at the moment it interrupts the current. The high-voltage source usually contains a large capacitor bank.

Actual synthetic testing circuits are much more elaborate than the simplified one shown in Figure 18.8. They would include elaborate control and instrumentation schemes, in addition to a backup circuit breaker for protecting the sources in case of failure of the breaker under test. The voltage source could be connected either across or in series with the current circuit (Lythall, 1986).

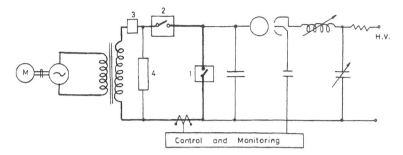

Figure 18.8 Synthetic testing circuit: (1) breaker under test; (2) master breaker; (3) synchronizing switch; (4) overvoltage protector.

18.8 CABLE TESTING

Cables are subjected to electrical and thermal stresses while in service. They also undergo mechanical stresses during their installation and repair. These mechanical stresses may cause insulation cracking, and hence the leakage current increases, leading to thermal breakdown of the cable insulation.

18.8.1 Conductor Resistance

The conductor resistance of a complete length or a specimen of cable is measured and the result is corrected to a temperature of 20°C and 1 km length in accordance with the formula

$$R_{20} = \frac{R_m}{l_c[1 + \alpha_{20}(\theta - 20)]} \qquad \Omega/\text{km} \qquad (18.12)$$

where the temperature coefficient α_{20} is 3.93×10^{-3} for copper and 4.03×10^{-3} for aluminum, and R_m is the measured resistance in ohms, l_c the sample length in kilometers, and θ the ambient temperature in °C.

18.8.2 High-Voltage Test

The high-voltage test is carried out under power-frequency alternating voltage or direct voltage. Because of the much lower power needed under DC, it is used for commissioning and maintenance tests of complete cable networks. According to IEC 502 (IEC, 1978), the power-frequency test voltage is $(2.5U_0 + 2)$ kV for cables of rated voltages (U_0) up to and including 6 kV, and $2.5U_0$ kV for cables of higher-rated voltages.

18.8.3   Partial Discharge and Loss-Angle Tests

Partial discharge tests are carried out to ensure that the level of the discharge does not exceed a specified limit. Single-core and multicore cables are tested by applying a voltage of 1.25 times the rated voltage between the conductor and the metallic screen. The magnitude of the PD impulses should not exceed 20 pC for cables insulated wth butyl rubber, EPR, and XLPE; and 40 pC for PVC cables (IEC, 1987b). Also, the loss factor $\tan \delta$ is measured at different voltages up to 1.25 times the rated voltage, using a Schering bridge.

18.9   TESTING OF POWER TRANSFORMERS

The following tests are required on most power transformers, as recommended by IEC 76-1 and 2 (IEC, 1976c,d), and 76-3-1 (IEC, 1987a).

1. Routine tests
    a. Measurement of winding resistance
    b. Measurement of voltage ratio and check of voltage vector relationship
    c. Measurement of impedance voltage, short-circuit impedance, and load loss
    d. Measurement of no-load loss and current
    e. Tests on tap changers
2. Type tests
    a. Dielectric tests
    b. Temperature rise tests

Some special tests may be required, such as a short-circuit test, measurement of zero-sequence impedance on three-phase transformers, and measurement of acoustic sound level. In this section, only type tests are discussed; other tests are given in the standard specifications.

18.9.1   Insulation Levels and Dielectric Tests

It is essential to specify higher test voltages for the internal insulation of the transformer than for the external insulation of the system. A failure in the non-self-restoring internal insulation would be catastrophic and normally leads to the transformer being withdrawn from service for a long period.

The recommended clearances among the various terminals and windings are referred to the rated withstand voltage of the internal insulation of the transformer (IEC, 1976a, 1987a). For equipment rated voltage less than 300 kV the reference voltage for external clearance is the lightning impulse voltage, whereas for equipment of higher rating the switching impulse voltage is the one to be employed.

18.9.2 Temperature Rise Test

The temperature rise values for the windings, cores, and oil of transformers designed for operation at normal altitudes are listed in tables according to the cooling medium (IEC, 1976d). For oil-immersed transformers, the temperature rise tests include the determination of top oil temperature rise and of winding temperature rise.

18.10 TESTING OF SURGE ARRESTERS

Surge arresters are identified by their rated voltage, rated frequency, nominal discharge current, and class of long-duration discharge (IEC, 1970). Some of the type tests made on nonlinear resistor-type arresters for AC systems are summarized below.

18.10.1 Power-Frequency Sparkover Tests

Dry and wet tests are made on complete arresters. The voltage initially applied to the arrester should be of a low value, to avoid sparkover of its series gaps. The permissible time during which the applied voltage may exceed the rated voltage of the arrester is in the range 2–5 s. After sparkover the test voltage is switched off by automatic tripping within 0.5 s. The power-frequency sparkover voltage is the average of five test results.

18.10.2 Impulse Sparkover Tests

These tests are made on the same samples as those used for power-frequency sparkover tests. Five positive and five negative impulses of the standard shape are applied to the test sample, and the series gaps of the arrester must spark over on every impulse. If, in either series of five impulses, the gaps fail to spark over once only, an additional 10 impulses of that polarity are applied and the gaps must spark over on all these impulses. Arresters with rated voltages exceeding 100 kV are additionally tested under switching impulses.

18.10.3 Impulse Current Withstand Tests

The rated voltage of these tests must be in the range 3–6 kV. Each sample is subjected to two current impulses of the standard shape with peak values depending on the arrester class. Before each test, the samples must be at approximately the ambient temperature. Subsequently, the power-frequency sparkover voltage is determined. It should not change by more than 10% (IEC, 1970). Also, no evidence of puncture or flashover of the nonlinear resistors or significant damage to the series gaps or grading circuit should occur.

18.11 TESTING OF INSULATORS

Insulators used on overhead transmission lines and in substations are subjected to type tests, sample tests, and routine tests.

18.11.1 Type Tests

The main type tests include the following:

1. *Dry lightning impulse withstand test.* The dry insulator is tested under positive and negative 1.2/50 μs impulse voltage waves. Two test procedures are in common use for the lightning impulse withstand test: the withstand procedure with 15 impulses and the 50% flashover voltage procedure (IEC, 1979).

2. *Dry switching impulse withstand test.* The insulator is tested under positive and negative 250/2500 μs impulse voltage waves. The 50% flashover procedure is normally used.

3. *Dry power-frequency withstand test.* This test is applicable only to insulators for indoor use. The applied test voltage is the specified dry power-frequency withstand voltage adjusted for atmospheric conditions.

4. *Wet power-frequency withstand test.* This test is applicable only to insulators for outdoor use. The applied test voltage is the specified wet power-frequency withstand voltage adjusted for atmospheric conditions. The characteristics of artificial rain are given in IEC 168 (IEC, 1979). The wet power-frequency withstand voltage is determined by the same procedure as that described in determining the dry power-frequency withstand voltage.

5. *Artificial pollution tests.* These tests are intended to provide information on the behavior of outdoor insulators under conditions representative of pollution in service. The pollution tests may be made either to determine the maximum degree of pollution withstood by the insulator under a given voltage or to determine its withstand voltage for a specified degree of pollution. The pollution tests involve application of the pollution and simultaneous or subsequent application of voltage. The pollution tests fall into two categories: the saline fog method and the pre-deposited pollution method (IEC, 1975).

18.11.2 Sample Tests

The sample tests on insulators include the following two principal tests:

1. *Power-frequency puncture test.* The insulator, after having been cleaned and dried, is completely immersed in a tank containing a suitable insulating medium (usually oil) to prevent surface discharges. The test voltage is applied on the insulator and then raised rapidly to the minimum puncture voltage specified. No puncture may occur below the specified minimum puncture voltage (IEC, 1979).

2. *Temperature cycle test.* This test is performed on individual insulator units made of ceramic or toughened glass. The unit is cyclically immersed in hot- and cold-water baths maintained with a temperature difference of 50 K. The insulator is left in each bath for $(15 + 0.7m)$ minutes, with a maximum of 30 minutes, m being the mass of the insulator in kilograms. This heating–cooling cycle is repeated three times in succession. The insulator is then examined. It should suffer no cracks or damage and should withstand a subsequent dry flashover test. For insulators made of annealed glass, the test procedure is slightly different (IEC, 1979).

NOTE

From the discussion above, it is evident that each item or type of equipment has its specific tests. Tests on pieces of equipment other than those mentioned above are described in detail in their relevant standard specifications. The ideas behind these tests, and the general procedures, are essentially similar to those described above.

18.12. PROBLEMS

(1) (a) Differentiate between type and routine tests.

(b) Explain briefly why portable voltage sources, instead of AC sources, are used to test cables on site.

(c) What is meant by nondestructive testing of electrical equipment? Mention three of these tests and explain briefly why they are carried out on electrical equipment.

(2) (a) What is the effect of partial discharge on power cables?

(b) Oil-impregnated paper cables are used at voltages less than 33 kV, whereas at higher voltages high-pressure oil-filled cables are used. Explain this statement.

(c) What is meant by the apparent charge in the process of partial discharge in voids in solid insulating materials? Deduce an expression relating this charge to the actual charge.

(3) A specimen of solid insulating material 1 cm thick has a cylindrical air void 0.5 mm thick. Calculate the voltage to be applied to the specimen

to start discharge in the void, given that the dielectric constant of the material is 4. Also, calculate the ratio between the apparent and actual charges.

(4) Determine the series resistance and capacitance and the loss tangent of a cable test piece using the Schering bridge in Figure 18.7. The bridge has the following elements in its branches under the balance condition:

$C_1 = 100 \, \text{pF}$ $R_3 = 620 \, \Omega$
$R_4 = 3180 \, \Omega$ $C_4 = 0.00125 \, \mu\text{F}$

(5) What are the sources of error affecting the balance condition in a Schering bridge? Show how these sources are eliminated.

(6) Explain why, after cable maintenance, the insulation resistance measurement test is performed before the high-voltage test.

(7) Why are synthetic testing circuits now commonly used for short-circuit tests of circuit breakers?

(8) What are the various tests to be carried out on power transformers?

(9) What electrical tests are applicable to indoor and outdoor insulators?

(10) Sketch the two common voltage waveshapes used for high-voltage testing of electrical equipment against lightning and switching surges.

(11) A tube of solid insulating material has inner and outer diameters of 3 and 3.25 cm respectively. The resistivity of the tube material is $1.56 \times 10^{13} \, \Omega \cdot \text{cm}$ and its dielectric is 3.5. Calculate the loss tangent of the tube and its dielectric loss per meter when used in a 20 kV system.

(12) A specimen of 10 m long of a high-voltage cable is tested at 100 kV. The cable has a capacitance of 0.75 F/km. Calculate the charging current and the kVA drawn from the testing source. Neglect the cable inductance and losses.

REFERENCES

ANSI. Standard Techniques for Dielectric Tests. Publication 68.1. New York: American National Standards Institute, 1968.

Bilodeau TM, Shea JJ, Fitzpatrick GJ, Sarjeant WJ. IEEE Electr Insul Mag 3:4, 1987.

IEC. Lightning Arresters, Part 1, Non-linear Resistor Type Arresters for A.C. Systems. Publication 99-1. Geneva: International Electrotechnical Commission, 1970.

IEC. High Voltage Alternating Current Circuit Breakers. Publication 56-4. Geneva: International Electrotechnical Commission, 1972.

IEC. High Voltage Test Techniques. Publication 60-1, 2. Geneva: International Electrotechnical Commission, 1973.

IEC. Artificial Pollution Tests for HV Insulators for AC Systems. Publication 507. Geneva: International Electrotechnical Commission, 1975.

IEC. Insulation Co-ordination, Part 1, Terms, Definitions, Principles and Rules. Publication 71-1. Geneva: International Electrotechnical Commission, 1976a.

IEC. Insulation Co-ordination, Part 2, Application Guide. Publication 71-2. Geneva: International Electrotechnical Commission, 1976b.

IEC. Power Transformers. Publication 76-1. Geneva: International Electrotechnical Commission, 1976c.

IEC. Power Transformers. Publication 76-2. Geneva: International Electrotechnical Commission, 1976d.

IEC. Extruded Solid Dielectric Insulated Power Cables for Rated Voltages from 1 kV up to 30 kV. Publication 502. Geneva: International Electrotechnical Commission, 1978.

IEC. Tests on Indoor and Outdoor Post Insulators of Ceramic Material or Glass for Systems with Nominal Voltages Greater than 1000 V. Publication 168. Geneva: International Electrotechnical Commission, 1979.

IEC. Power Transformers. Publication 76-3-1. Geneva: International Electrotechnical Commission, 1987a.

IEC. Partial Discharge Tests for Cables. Publication 885-I, II. Geneva: International Electrotechnical Commission, 1987b.

IEEE Committee. IEEE Trans PAS-100:3292–3300, 1981.

Kind D. Introduction to High Voltage Experimental Technique. Braunschweig, West Germany: Friedr Vieweg & Sohn, p 52.

Kreuger F. Discharge Detection in High Voltage Equipment. London: Butterworth, 1989.

Kuffel E, Zaengl WS. High-Voltage Engineering Fundamentals. Oxford: Pergamon Press, 1984.

Lythall RT. The J & P Switchgear Book. London: Butterworth, 1986.

Nattrass D. IEEE Electr Insul Mag 4(3):10–23, 1988.

19

Applications of High-Voltage Engineering in Industry

M. ABDEL-SALAM *Assiut University, Assiut, Egypt*

19.1 INTRODUCTION

Electrostatics is currently found at the basis of many major industries related to environment preservation, communications, processing of mineral ore resources, and so on. In the majority of these industries, the unique properties of high-voltage electrostatic fields and forces are utilized to collect, direct, deposit, separate, or select very small or lightweight particles. The principles of application of electrostatics are discussed for the following selected industries: electrostatic precipitation, electrostatic separation, electrostatic painting, electrostatic spraying of pesticides in orchards, electrostatic imaging, electrostatic printing, electret transducers, transport of light materials, paper manufacture, smoke detection, electrostatic spinning, electrostatic pumping, electrostatic propulsion, air cleaning from gaseous pollutants, ozone generation, and biomedical applications.

19.2 ELECTROSTATIC PRECIPITATION

Electrostatic precipitation is essentially the charging of dust particles in a gas and their subsequent separation under the effect of the electric field (Gordon and Peisakhov, 1972; Oglesby and Nichols, 1978). These processes may occur within a single zone or be distributed over two zones, where the first zone—the charging zone—is intended to charge the particles, and the second zone—the collecting zone—is designed to settle the particles.

The stream of ions charging the dust particles is produced by means of a corona discharge (Chapter 5) in an inhomogeneous electric field. These ions interact with the particulates entrained in the gas and impart charge to the dust, which then experiences a force towards the collecting electrode where it is held by electrostatic forces until it is removed by mechanical rapping (Fig. 19.1).

Two systems of electrodes are used in electrostatic precipitators to obtain an inhomogeneous electric field: a wire conductor enclosed in a cylindrical pipe (pipe-type precipitators), or a row of wire conductors located between plates (plate- or duct-type precipitators). The electrodes around which a corona discharge is formed are termed the discharge electrodes, whereas the electrodes receiving the charged dust particles, deposited by the action of the electric field, are the collecting electrodes.

Figure 19.2 displays schematically the charging of dust particles in a precipitator.The particles are charged as a result of bombardment by ions energized in the electric field, for particles larger than 1 µm, and through collision with ions which participate in the continuous thermal motion of molecules without the aid of an electric field, for particles in the sub-micron range.

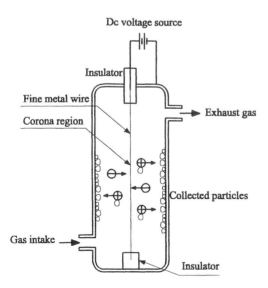

Figure 19.1   Pipe-type electrostatic precipitator.

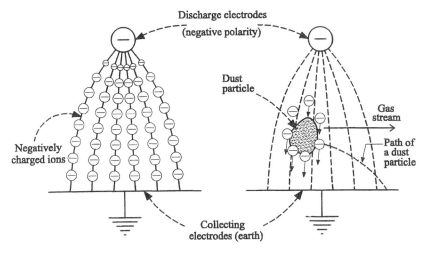

Figure 19.2 Schematic diagram illustrating the charging of dust particles in an electrostatic precipitator.

A dust cake on a collecting electrode consists of only 10–50% dust, depending on the size of the particles, the remaining part being channels (cracks) filled with gas. Because of a difference in the values of relative permittivities of dust and gas, the lines of electric field concentrate inside the channels (Fig. 19.3). When the voltage is high, an electric breakdown takes place across the dust cake, and the gas inside the channels is ionized. The outlet of each channel acts then as a point discharge on the collecting electrode. This phenomenon is termed back corona (Abdel-Salam and Singer, 1991).

High resistivity particles cause back corona to be formed in the precipitated layer and this has a number of deleterious effects. The most significant is the release of ions which move countercurrent with respect to the dust particles and partly neutralize their charges, so that dust collection is adversely affected and current intensity in the precipitator greatly increases. This is in addition to a reduction in the sparkover voltage, which consequently reduces the precipitator's performance.

Pulse charging was proposed as an alternative solution to back discharge. A train of pulses, usually superimposed on a DC level value, provides a high peak electric field for particle charging. However, the mean collector current in this case depends not only on the peak voltage but also on the pulse duty cycle. In this way the voltage and current can be decoupled by changing the mark/space ratio of the applied pulses, and high charging fields can be established without back discharge. Pulse charging can produce a uniform

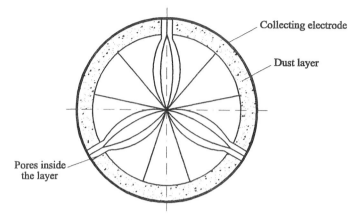

Figure 19.3 Schematic diagram illustrating the arrangement of field lines on formation of back corona.

distribution of ionic current on both the discharge and collecting electrodes (Masuda and Hosokawa, 1988). The disadvantage of pulse charging is the ion shortage that occurs when the current must be lowered extremely to cope with high dust resistivity. This results in a low charging rate which impairs the performance even if back discharge is avoided.

To overcome this shortcoming, prechargers have been proposed. The fundamental aim of all precharger systems is to separate the charging and collecting processes into two separate stages. In the charging stage, different techniques are used to eliminate or reduce the effects of back corona, whereas the collecting stage is similar for all prechargers and comprises only parallel plates so that no back corona can be formed. Different types of precharger have been developed and tested at pilot-scale stage, including the trielectrode charger, high-intensity charger, boxer charger, and cold-pipe charger (Masuda, 1981).

Precipitator efficiency can reach or even exceed 99%, depending on several design parameters, including the dimensions and geometry of the gas duct, gas temperature and velocity, average size and resistivity of the particulates, and corona discharge intensity (Landham et al., 1987).

19.3   ELECTROSTATIC SEPARATION

Electrostatic separation is the selective sorting of solid species by means of utilizing forces acting on these species in an electric field. The main items of the separator are a charging mechanism in the charging zone, an external electric field in the separating zone, and a feeding and product collection system.

The charging of two different species entering the separating zone results in: (a) particles bear electric charges of opposite sign; or (b) only one type of particle bears an electric charge; or (c) particles bear the same sign of charge, but the magnitude of the electric charge is significantly different. Although there are many ways (Lawver and Dyrenforth, 1973; Sadiku, 1994; Iuga et al., 1998; Hoferer et al., 1999) to charge solid particles, the most common mechanisms are as follows.

1. Charging by contact and frictional electrification is the mechanism most frequently used to selectively charge and electrostatically separate two species of different materials such as phosphate and quartz. The ore, composed of small particles of quartz and phosphate, is vibrated on its way from the hopper to the forming chute (Fig. 19.4a). Phosphate particles enter the separating zone with a net positive charge while the quartz particles bear a net negative charge.

A coal beneficiation system has been developed on the basis of electrostatic separation (Masuda, 1981), where the external field in the separating zone deflects the coal-rich and ash-rich particles in opposite directions towards the collecting-plate compartments.

2. Charging by ion or electron bombardment is used, in which solid particles pass through a corona discharge (Chapter 5) from a fine wire or a series of needle points positioned parallel to a grounded rotor of the separator. The particles are charged by bombardment with the corona ions. The charged particles rapidly share their charge with the grounded rotor and are thrown from the rotor in a trajectory determined by centrifugal force, gravity, and air resistance. The dielectric or poorly conducting particles lose their charge slowly and are thus held to the surface of the rotor by the image force associated with their surface charge. The well conducting particles are thrown free of the rotor by a combination of centrifugal force and gravity (Fig. 19.4b).

3. Charging by conductive induction is a charging mechanism suitable for separating well conducting particles from well insulating particles. A grounded rotor is located close to a positive drum (Fig. 19.4c). When conductive particles coming from the hopper pass over the rotor they become negatively charged and attracted toward the positive drum. However, insulating particles fall down by gravity.

19.4 ELECTROSTATIC PAINTING/COATING

This is a kind of electrostatic precipitation of powder or liquid paint on the surface of an object to be coated (or painted) (Inculet, 1977/1978).

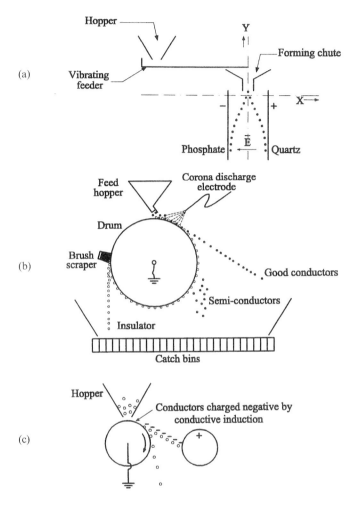

Figure 19.4 (a) Electrostatic separator based on charging by contact and frictional electrification. (b) Electrostatic separator based on charging by ion or electron bombardment. (c) Electrostatic separator based on conductive charging.

In liquid paint, a liquid jet issues from the reservoir and extends along the axis of a concentric charging cylinder where a potential V_p is applied between the jet and the cylinder. The jet charges and breaks up into charged droplets while it is in the cylinder. The electric field plus space charge effects between the jet and the grounded object deposit the paint droplets on the surface of the object to be painted, not only on the front side but also on the back side of the object (Fig. 19.5).

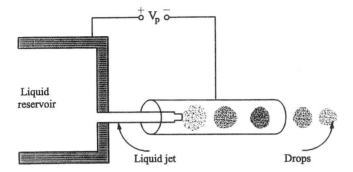

Figure 19.5 Schematic diagram of electrostatic generation of liquid droplets for painting.

In powder painting, the paint particles are charged by bombardment with corona ions (Chapter 5) moving under the influence of the prevailing electric field between the corona electrode and the object being grounded (Fig. 19.6). The paint precipitation stops at a certain thickness called the "limiting thickness" because of back discharge (Abdel-Salam and Singer, 1991) which results in craters that impair the quality of finished coat.

In its simplest form, an electrostatic coating operation is visualized as taking place in the following manner. The object to be coated is grounded and supported so that its surface can be approached without obstruction. The coating material, after being charged, is sprayed into the space above

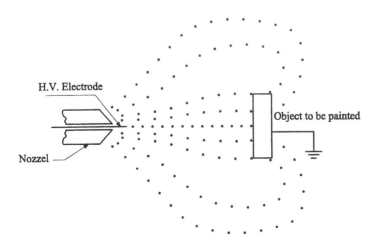

Figure 19.6 Electrostatic powder painting.

the surface in the form of finely divided particles. There is an attraction of the particles to the surface and, as a result, they move toward and accumulate on the surface to form the coating. The various electrostatic coating applications are somewhat sophisticated modifications of this simple situation. They differ from one another in the manner in which the particles are forced, the means by which they are charged, etc.

Electrostatic painting is widely used in the continuous coating lines of automobiles, electric appliances, furniture, and so on.

19.5 ELECTROSTATIC SPRAYING

As has been established, the application of a sufficiently high-voltage field at the surface of a liquid coming from a nozzle will lead to the formation of charged droplets in the form of a spray, as discussed for electrostatic painting.

Conventionally, agricultural pesticides are applied in the form of water-based sprays using a hydraulic-atomizing nozzle. The major problem with this method, however, is that a large portion of the spray is lost and the target deposition efficiency is less than 25% (Abdel-Salam and Singer, 1991; Abdel-Salam et al., 1993 and 1995; Law, 1995). The spray loss and accordingly the deposition efficiency can be attributed to both wind drift out of the target area (exodrift) and deposition to the soil during the process (endodrift).

As with industrial coating techniques, electrostatic pesticide spraying technology offers an attractive alternative, which will increase not only the pesticide deposition efficiency but the biological efficacy as well. The reason behind this is the small droplet sizes that can be achieved in electrostatics-based systems. That is, in addition to being essential for increasing the biological efficacy, the generation of small droplets will increase the droplet's charge-to-mass ratio. This will, in effect, increase the coulomb force acting on the droplet to a value which is high enough to overcome the mass-dependent gravitational and inertial forces (Law, 1989), with a subsequent increase of the deposition efficiency of pesticides.

In pesticide spraying, a nozzle has been developed to generate a concentrated jet which forms an electrified cloud inside the tree foliage. The mutual repulsion of the particles and their coulomb attraction to the leaves (electrically grounded through the tree itself) ensure a superior coverage on both sides of the leaves (Law and Cooper, 1989) (Fig. 19.7).

19.6 ELECTROSTATIC IMAGING

Electrostatic imaging is a process by which an ordered arrangement of electric charges is deposited on a surface as a latent electrostatic image

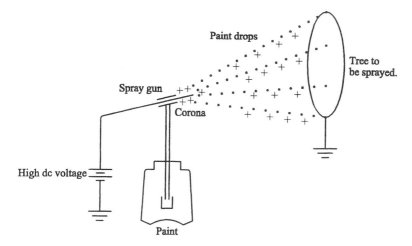

Figure 19.7   Electrostatic spraying of trees.

and subsequently "developed" to convey visual information to an observer (Inculet, 1977/78; Masuda, 1981; Schein, 1992).

Consider a system of a metal plate (aluminum) covered by a very thin layer of insulating material (alumina), which in turn is covered by a layer of photoconducting material (selenium). Initially the entire surface of the photoconductor is charged uniformly by a positive charge, as shown in Figure 19.8a. The charge on the top surface comes from the positive ions produced by a corona discharge (Chapter 5) on a wire traveling above the photoconductor. The negative charge on the lower surface comes from the metal plate below and is drawn by attraction to the upper positive charges. Since the photoconductor is an insulator in the dark, the negative charges can get no farther than the bottom of the photoconductor.

When light from the document to be copied is focused on the photo-conductor it becomes a conductor, and the negative charges on the lower surface can combine with those of positive polarity on the upper surface to neutralize each other. However, the positive surface charge in dark areas is maintained on the photoconductor (Fig. 19.8b). Thus, a "latent" positive-charge electrostatic image is produced on the surface of the photoconductor (Fig. 19.8c).

Development of the electrostatic image is the process of supplying powder (toner) to the charged areas to provide visibility of the image. The toner particles are charged negatively and attracted to the positive image by reason of the external electric field generated by this image (Fig. 19.8d). If a sheet of paper is placed over the development image and sprayed with a

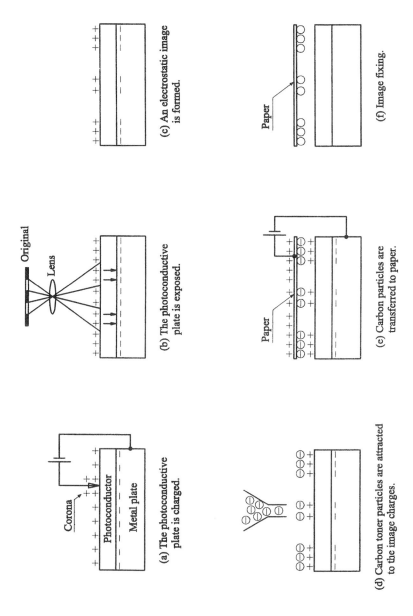

Figure 19.8   Steps in electrostatic imaging.

positive charge, the electric field penetrating the sheet will act on the negative toner and cause it to transfer to the paper (Fig. 19.8e). When the sheet is stripped away from the imaging surface, a well-toned image is still present. A second sheet of paper can then be placed over the image and the transfer process is repeated. Later, the toner is thermally fixed to form a permanent image (Fig. 19.8f).

19.7   ELECTROSTATIC PRINTING

Many typewriters and computer line printers use so-called impact printing, i.e., characters are cast on metal type, which makes an impact on an ink ribbon to produce prints on paper. New requirements for performance and speed exceed the capabilities of most impact printing technologies. Electrostatic printing is an alternative method.

Electrostatic printers are classified into ink-jet and ink-spray printers (Swatik, 1973; Ashley, 1977; Buehner et al., 1977; Carnahan and Hou, 1977). The printing ink is formed into macroscopic droplets which are imaged by electric field, external to the form (print surface). The form serves only as a receptor for the imaged ink droplets and, upon contact, a visible image is instantly produced. This technique is viable on standard forms (sheets of paper), and no image development or fixing is required. Electrostatic printers are usually known as nonimpact printers.

The ink-jet printing technique produces instantly visible images on standard forms by the electrostatic deflection of charged ink droplets into electrically programmable dot matrix patterns. This principle is analogous to the deflection of electrons in a cathode ray tube. The electrically conducting ink is forced through a nozzle to form a thin jet, which then breaks up into droplets under the influence of surface tension and the mechanical vibrations in the nozzle. At the point where droplets are forming, a high-voltage field controlled by a computer is applied to the jet to give the newly formed droplets an electric charge related to the signal (Fig. 19.9). This is why the ink-jet printer is known as the charge-modulation type of ink printer. The drops then move into a deflection region where a steady transverse electric field deflects them by an amount depending on their acquired charge. This deflection causes them to strike the print surface (usually a piece of paper) at different points, creating an image.

In the ink-spray printing technique, a reservoir containing the ink is pressurized at a low level (a few centimeters of water) sufficient to form a convex meniscus of ink at the opening of a vibrating nozzle but not high enough to cause an outflow of ink. The conductive ink, maintained at ground potential, is attracted by the electric field of the gate. When the

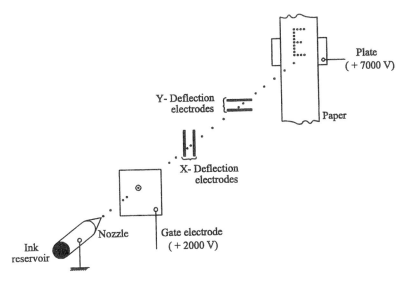

Figure 19.9   Ink-jet printer.

electrostatic attraction force exceeds the surface tension of the meniscus, droplets are produced. Expressed in terms of potential V, the droplets are produced when the potential between the gate electrode and the meniscus is

$$V > 2D(\gamma/\varepsilon d)^{0.5}$$

where γ = ink surface tension, ε = ink permittivity, d = nozzle orifice diameter, and D = distance between orifice and gate (Inculet, 1977/78). The droplets are accelerated and then imaged into the desired dot-matrix format by electrostatic deflection. Since the droplets are produced with a nearly uniform specific charge, programmable deflection is obtained by varying the magnitude of the deflection field. This is accomplished by electronically controlling the voltage applied to the deflection electrodes (Fig. 19.10). This is why the ink-spray printer is known as the field-modulation type of ink printer. Two sets of electrodes are employed to obtain both horizontal and vertical deflection, to allow for printing on stationary forms.

Since the droplets are very small they can be quickly accelerated and are able to produce high-quality hard copy much faster than electronic (impact) printers. There are commercially available electrostatic printers with capabilities of printing in excess of 30,000 lines/minute. With the current concern for noise from impact printers, electrostatic printers have a definite advantage as the level of noise is only that generated by the hardly visible droplets of ink landing on the paper.

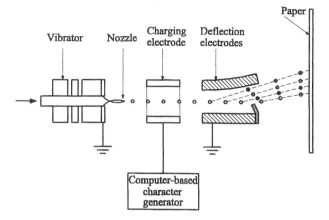

Figure 19.10 Ink-spray printer.

19.8   ELECTRET TRANSDUCERS (MICROPHONES)

While the majority of electrostatic applications in industry involve fine particles, the handling and the phenomena associated with very thin films have also attracted industrial interest. A good example is the electret transducer (microphone) shown in Figure 19.11.

An electret is a dielectric material with a permanent dipole moment or charge separation. The word "electret" indicates that this is the electric counterpart of "magnet." In the electret microphone, an electret film, only a few micrometers thick, is metallized with a thin layer of foil less than 1 μm thick and backed with a porous metal plate (Fig. 19.11). Sound waves strike the electret film, moving it with respect to the metal plate, thus changing the distance b depending on the loudness of the sound. As the surface charge density ρ_s is constant (permanent charge density), the electric field E_b and hence the surface charge on the metal plate changes with the distance b. Such a change of the charge on the plate is used as an output signal after being amplified. The higher the voltage applied to the metal plate, the higher is the electric field E_b (see Example 5) and the better is the performance of the microphone. The excellent frequency response and reliability have resulted in an extension of the use of this device to headsets for telephone operators (Crowley, 1986).

19.9   ELECTROSTATIC TRANSPORT OF LIGHT MATERIALS

The feasibility of electrostatic transport of light materials which triboelectrify themselves on a stationary "conveyer" belt, as shown schematically in

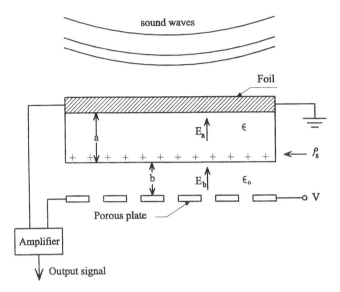

Figure 19.11 Electret microphone.

Figure 19.12, appears very promising (Inculet, 1977/78). The three-phase high-voltage supply *U–V–W* generates a traveling field at the surface of the conveyor. The charged particles will be transported along the field. The electrostatic conveying is independent of the sign of the charge; thus, both positively and negatively charged particles will transport in the same direction.

19.10 SANDPAPER MANUFACTURE

Another example of what electrostatics can do for industry is evidenced in the lining up and gluing of billions of particles in the right position with a uniform distribution, as shown in Figure 19.13. Abrasive particles of quartz, silicon carbide, diamond, etc. are poured onto a continuous conveyor or belt. As they arrive in the strong electric field between two parallel electrodes stressed by a voltage up to 100 kV, they line up along the field lines with sharp points facing the electrodes. The electric charge which they acquire from the semiconducting belt causes them to be propelled and implanted in the adhesive. Any particle which strikes an area already fully covered by other particles falls back and is recovered for reprocessing as surplus abrasive material (Inculet, 1977/78).

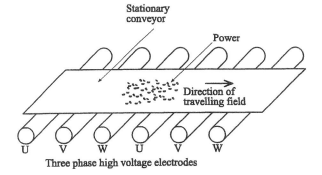

Figure 19.12 Electrostatic conveying of tribo-electrified materials.

19.11 SMOKE PARTICLE DETECTOR

The construction is shown schematically in Figure 19.14. The design of such detectors is based on the fact that electrically charged particles of smoke have substantially lower mobility than ions. During the normal surveillance periods, two continuous and equal ion currents i_1 and i_2 flow through the air of the two chambers made conductive by radiation from a radioactive material (Chapter 3). The current i_1 flows in the sensing chamber whose walls have so many holes that smoke particles flow in easily, in contrast to the low-response chamber (Inculet, 1977/78). In large chambers, the radio-active material may be replaced by a corona source of ions (Chapter 5) for sensitive detection of smoke particles.

A sudden increase in the number of smoke particles in the ambient will be followed by an immediate increase in the particle density in the sensing

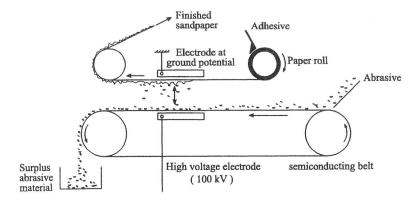

Figure 19.13 Principle of electrostatic manufacture of sandpaper.

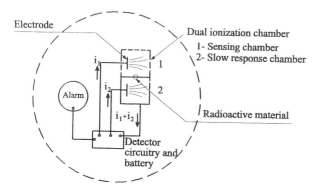

Figure 19.14   Smoke particle detector.

chamber and by a slower increase in the corresponding particle density in the second chamber. The current i_1 will decrease on account of the attachment of ions to the smoke particles. The decrease is more than sufficient to actuate the solid-state circuitry, which sounds the alarm. Nowadays, the low cost of smoke detectors has made them accessible to practically everyone interested in acquiring a warning device, which undoubtedly will save lives.

19.12   ELECTROSTATIC SPINNING

Figure 19.15 shows a schematic of an electrostatic device to make a thread from fibers with minimum use of air to avoid quality degradation due to contamination of particles (Masuda, 1981). The air-conveyed fibers are fed from a feeder into a high-voltage DC field between the gap plate and conical electrodes and are aligned along the field lines. At the same time they are attracted towards the rotating shaft by a gradient force caused by the narrowing gap. A mother thread is inserted from the top of the rotating shaft into its hole to grasp the vertically aligned fibers. When it is pulled upwards, the fibers are pulled into the hole to be spun continuously to produce a thread.

19.13   ELECTROSTATIC PUMPING

In high-voltage power equipment, the current-carrying components must be cooled to carry away the heat caused by ohmic losses. Generally, the fluids are more effective in cooling when they make direct contact with the high-voltage conductors, but this requires that they must be electrical insulators such as oils. The coolant must also circulate to carry away the heat, so some

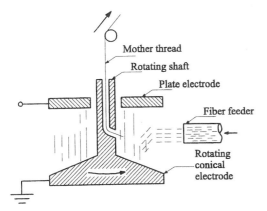

Figure 19.15   Electrostatic spinning.

means of pumping by thermal convection or mechanical pumps is generally included. A more effective method in coolant circulation is based on the force transmitted to the cooling fluid by charges in a high-voltage field (see Example 2). This electrohydrodynamic pumping has been known for some time, but practical implementation in a transformer (Melcher, 1976) and in a cable (Crowley and Chato, 1979) has been reported.

19.14   ELECTROSTATIC PROPULSION

An ordinary rocket engine propels the rocket forward by ejecting fuel backwards. The propulsion force F obtained is expressed as

$$F = \frac{d}{dt} mv = v \frac{dm}{dt}$$

where dm/dt is the rate of ejection of the propellant and v is the ejection velocity, which is constant. For chemical fuel, the ejection velocity may be supersonic, of the order of 10^3 m/s. It is difficult to achieve much higher velocities using chemical fuels.

Electrostatic propulsion uses ionized particles as propellant. The particles are accelerated by electrostatic force to a much higher velocity. Thus, for the same propulsion force, electrostatic propulsion consumes less mass than a chemical engine does (Shen and Kong, 1983).

Figure 19.16 illustrates the basic operating principle of electrostatic propulsion. Propellant enters the ionization chamber, where it is ionized either by heat or by bombardment with electrons (Chapter 3). Positive ions are accelerated and ejected through the screen. Electrons are absorbed

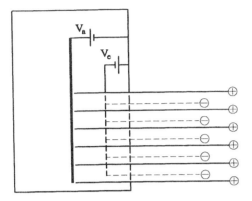

Figure 19.16 Electrostatic propulsion concept.

by the anode. These electrons must also be ejected with the positive ions at the same velocity; otherwise, the rocket will be increasingly negatively charged.

The velocity of the positive ion v_i is given by

$$v_i = \sqrt{\frac{2q_i V_a}{m_i}} \tag{19.1}$$

where q_i is the charge on the positive ion and m_i is its mass. V_a is the accelerating voltage for positive ions, which may reach several tens of kilovolts.

Similarly, the velocity of the electron v_e is given by

$$v_e = \sqrt{\frac{2q_e V_e}{m_e}} \tag{19.2}$$

where $q_e = 1.6 \times 10^{-19}$ C and $m_e = 9.11 \times 10^{-31}$ kg.

To neutralize the ejected ion, v_e must be equal to v_i. Thus, the voltages V_a and V_e must be related by the following equation

$$V_a/m_i = V_e/m_e \tag{19.3}$$

In equation (19.3), it is assumed that the ions carry a single positive charge. Because $m_i \gg m_e$, $V_a \gg V_e$.

To estimate the magnitude of the propulsion force in terms of the ion current I, being related to the density of ions n_i as

$$I = n_i v_i q_i A = dq_i/dt \tag{19.4}$$

where A is the cross-section of the ion beam, the rate of mass ejection is

$$\frac{dm_i}{dt} = n_i v_i m_i A$$

Thus, the propulsion force F is expressed as

$$F = v_i \frac{dm_i}{dt} = \frac{v_i m_i I}{q_i} \tag{19.5}$$

Electrostatic propulsion is used on synchronous satellites to adjust their position and to adjust antenna orientations. It is potentially useful for exploring deep space because of the large propulsion force-to-fuel ratio as compared with chemical engines (Langmuir et al., 1961).

19.15 AIR CLEANING FROM GASEOUS POLLUTANTS

Sulfur oxides (SO_x), oxides of nitrogen (NO_x), and unburned hydrocarbons are considered to be hazardous to health. City air may have unacceptable levels of SO_x and NO_x because of heavy concentrations of automobiles and factories. Room air can be polluted by cigarette smoke, which contains various poisonous gases including ammonia. Animal farms also emit ammonia, which has an unpleasant odor, is unhygienic, and should be removed together with other pollutants like SO_x, NO_x, etc. which contribute to acid rain. Sulfur oxides are the most serious pollutants, followed closely by nitrogen oxides and hydrocarbons.

If a gas containing SO_x, NO_x, water vapor, oxygen and ammonia interacts with electrons of energy in the range 5–20 eV, the SO_x and NO_x are transformed to sulfuric and nitric acids, respectively. This transformation takes place in the presence of O, OH, HO_2 and NH_2 radicals that have been formed due to the interaction of the electrons with water vapor, oxygen, and ammonia present in the gas to be treated. Sulfuric and nitric acids in the presence of ammonia are subsequently transformed into salts in the form of solid particulates. These particulates may be removed by means of an electrostatic precipitator (Civitano and Sani, 1992; Yan et al., 1998).

Production of electrons with energy between 5 eV and 20 eV may be obtained by the use of DC coronas (Chapter 5). However, the process is not energy efficient and the performance is poor (Mizuno et al., 1984; Yang and Yamamoto, 1998). The poor performance is probably due to a small ionization zone in the DC corona (small active treatment volume), and a large amount of energy is expended on ion migration which does not contribute to the production of radicals. The use of pulsed-streamer corona discharge in a nonuniform electrode geometry avoids these difficulties (Masuda, 1993; Mizuno et al., 1998). Streamers propagate across the entire gas-treatment volume between the electrodes, where ionizing molecules and free energetic electrons are available. This results in a larger active volume for gas treatment. The streamers leave positive ions, which do not contribute to the power consumption because no significant movement of the ions occurs within the short pulse period. The result is a large improvement in power efficiency.

19.16 OZONE GENERATION

Ozone is a virtually colorless gas with an acrid odor and very strong oxidizing properties. Ozone is very toxic to humans and animals, causes serious damage to plant life, and produces deterioration in many materials. Ozone is the triatomic form of oxygen that has been formed by recombining highly excited oxygen atoms or molecules with one another. Unlike other agents such as chlorine, oxidation with ozone leaves no toxic residues that have to be removed and disposed of because the process yields only "oxygenated" products and oxygen. In practice, ozone is completely nonselective and immediately starts to oxidize everything it comes into contact with. This property makes it a very powerful bactericide and viricide as well as a strong bleaching agent. The ozone molecule is only moderately stable and, in the absence of oxidizable substances, will decay to form stable diatomic oxygen. The instability of ozone has the consequence that it can be neither stored nor transported over distances requiring long periods of time. This means it must be produced on site when and where it is required, which has the added environmental and safety advantages of not having to store or transport potentially dangerous chemicals in large quantities (Kogelschatz and Eliasson, 1995).

The traditional way of producing ozone is by means of dielectric barrier discharge or so-called silent electrical discharge. Dry feed gas (either oxygen or air) is passed through a narrow discharge gap. One side of the gap is formed by a grounded electrode, the other by a dielectric, normally glass, in contact with a high-voltage electrode (Fig. 19.17). When high voltage at high frequency is applied to the high-voltage electrode, ozone is produced in the feed gas by microdischarges taking place in the discharge gap (Eliasson et al., 1987; Braun et al., 1991; Abdel-Salam et al., 1997).

Ozone has many applications in industry. Among these applications are sterilization processes and control of microbials, purification of drinking water, and treatment of waste water instead of the use of chlorine, control of survival rate of farmed fishes, cleaning, deodorization, and making healthy indoor air, $deNO_x$ and $deSO_x$ processes by eliminating these hazardous pollutants from flue gases, ozone therapy, and protection of the earth's surface from dangerous ultraviolet radiation.

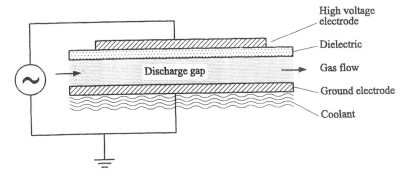

Figure 19.17   Electrode configuration of silent or dielectric-barrier discharge.

19.17   BIOMEDICAL APPLICATIONS

There are needs in biotechnology for the manipulation of small objects, such as cells, chromosomes, biological membranes, and nucleic acid and protein molecules. Biological cells range in size from less than a micrometer to several hundred micrometers, and molecules are even smaller, measured in nanometers. High-voltage electrostatic forces are highly suitable for handling, characterization, and separation of these fine particles. With the use of electrostatic effects, these objects can be manipulated collectively or even individually. In addition, because the electrostatic force is a "surface force" distributed around the surface of an object, it enables gentle manipulation without applying too much stress to the object (Washizu et al., 1992; Mizuno and Washizu, 1995).

Another aspect of electrostatic effects is associated with the breakdown of membranes. When a pulsed electric field of moderate magnitude is applied, a cell membrane breaks down and becomes permeable, but a resealing process follows due to the fluidity of the membrane. This process is called "reversible breakdown" and is used for bringing foreign material into the cell interior. In particular, when this process is conducted in a DNA solution, it can be used to inject foreign genes into the cells (transfection). Alternatively, when a partial breakdown occurs at the contact point of two cells, they may fuse into one, to yield a hybrid between these two cells (electrical cell fusion). On the other hand, if the pulse is too strong, irreversible breakdown takes place and the cell is destroyed. This effect can be used for sterilization.

19.18.   SOLVED EXAMPLES

(1) A Florida phosphate ore, consisting of small particles of quartz and phosphate rock, is separated into its components by applying a uni-

form electric field E in an electric precipitator, as shown in Figure 19.4a. Assuming zero initial velocity and displacement, determine the separation between the particles after falling 1 m. Consider $E = 800\,\text{kV/m}$ and charge-to-mass ratio $q/m = 10\,\mu\text{C/kg}$ for both positively and negatively charged particles. Ignore the coulombic force between particles and the effects of air resistance.

Solution:
The electrostatic force is acting horizontally while the gravitational force (weight) is acting vertically on the particles; thus

$$q\bar{E} = m\frac{d^2x}{dt^2}\bar{a}_x \qquad \text{or} \qquad \frac{d^2x}{dt^2} = \frac{q}{m}E \tag{19.6}$$

Integrating twice gives

$$x = \frac{q}{2m}Et^2 + c_1 t + c_2 \tag{19.7}$$

where c_1 and c_2 are integration constants to be determined from initial conditions. Similarly,

$$-mg = m\frac{d^2y}{dt^2} \qquad \text{or} \qquad \frac{d^2y}{dt^2} = -g \tag{19.8}$$

Integrating twice, one obtains

$$y = -\tfrac{1}{2}gt^2 + c_3 t + c_4 \tag{19.9}$$

Since the initial displacement is zero,

$$x(t = 0) = 0, \qquad \therefore c_2 = 0$$
$$y(t = 0) = 0, \qquad \therefore c_4 = 0$$

Also, since the initial velocity is zero,

$$dx/dt|_{t=0} = 0, \qquad \therefore c_1 = 0$$
$$dy/dt|_{t=0} = 0, \qquad \therefore c_3 = 0$$

Thus

$$x = \frac{qE}{2m}t^2, \qquad y = -\tfrac{1}{2}gt^2 \tag{19.10}$$

When $y = -1$ m,

$$t^2 = \frac{1 \times 2}{9.8} = 0.204$$

and

$$x = \tfrac{1}{2} \times 10 \times 10^{-6} \times 8 \times 10^5 \times 0.204 = 0.816 \text{ m}$$

The separation between the particles is

$$2x = 1.632 \text{ m}$$

(2) An electrohydrodynamic ion-drag pump is shown in Figure 19.18 for pumping oil from left to right. The oil contains a uniform charge density ρ_0 which is generated at the left electrode and collected at the right electrode. The pump electrodes are two parallel plates, each of area A, spaced a distance d. The potential difference between the electrodes is V_0. Derive expressions for the force density (i.e. force acting on a unit volume) and the pumping pressure. For a charge density $\rho_o = 30 \ mC/m^3$, calculate the pumping pressure if the pump electrodes are stressed by a voltage $V_0 = 30 \ \text{kV}$.

Solution:
Poisson's equation states that

$$\nabla^2 \phi = -\rho_0/\varepsilon$$

as ϕ depends only on x; hence

$$\frac{d^2\phi}{dx^2} = -\rho_0/\varepsilon \tag{19.11}$$

Integrating twice gives

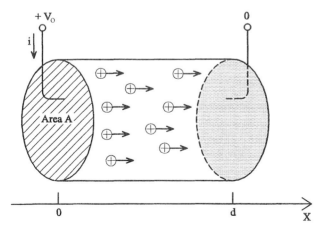

Figure 19.18 Electrostatic ion-drag pump.

$$\phi = \frac{-\rho_0}{2\varepsilon} x^2 + c_1 x + c_2 \tag{19.12}$$

where c_1 and c_2 are integration constants to be determined by applying the boundary conditions

$$\phi(x = 0) = V_0, \qquad \therefore c_2 = V_0$$

$$\phi(x = d) = 0, \qquad \therefore c_1 = \frac{\rho_0 d}{2\varepsilon} - \frac{V_0}{d} \tag{19.13}$$

$$\therefore \phi = \frac{-\rho_0}{2\varepsilon} x^2 + \left(\frac{\rho_0 d}{2\varepsilon} - \frac{V_0}{d} \right) x + V_0$$

The electric field

$$E = -d\phi/dx = \frac{V_0}{d} + \frac{\rho_0}{\varepsilon} \left(x - \frac{d}{2} \right) \tag{19.14}$$

The force density f is

$$f = \rho_0 E = \frac{\rho_0 V_0}{d} + \frac{\rho_0^2}{\varepsilon} \left(x - \frac{d}{2} \right) \tag{19.15}$$

The net force F acting on the oil volume filling the space between the pump electrodes is obtained as

$$F = \int f \, dv = \int_0^d f A \, dx = \rho_0 V_0 A \tag{19.16}$$

where dv is an elemental volume of the oil.
The force per unit area or the pumping pressure P is

$$P = F/A = \rho_0 V_0 \tag{19.17}$$

For the given numerical data,

$$P = 30 \times 10^{-3} \times 30 \times 10^3 = 900 \text{ N/m}^2$$

(3) In a xerographic copying machine, the surface of the photoconductor above the grounded metal plate (Fig. 19.8), is initially charged uniformly with surface-charge density ρ_s, as shown in Figure 19.19. The photoconductor has thickness a and permittivity ε_2. As explained in Section 19.6, light from the document to be copied is focused on the photoconductor so that the charges on the lower surface combine with those on the upper surface to neutralize each other. The image is developed by pouring a charged black powder over the surface of the photoconductor. The electric field attracts the charged powder, which is later transferred to paper and melted to form a permanent image. Determine the electric field inside the photoconductor and in the space above the photoconductor up to an upper grounded plate at distance d from the lower grounded plate (Fig. 19.19).

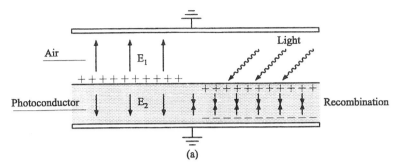

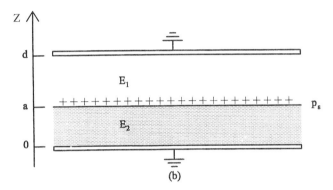

Figure 19.19 (a) Xerographic copying machine. (b) Modeled version of the machine.

Solution:
Consider the modeled version of Figure 19.19a as in Figure 19.19b. In a space-charge-free field, one applies Laplace's equation. As the potential depends only on z,

$$\nabla^2 \phi = \frac{d^2 \phi}{dz^2} = 0$$

Integrating twice gives

$$\phi = Az + B \tag{19.18}$$

Let the potentials above and below the photoconductor surface be ϕ_1 and ϕ_2, respectively,

$$\phi_1 = A_1 z + B_1, \qquad z > a \tag{19.19}$$

$$\phi_2 = A_2 z + B_2, \qquad z < a \tag{19.20}$$

The boundary conditions at the grounded plates are

$$\phi_2(z = 0) = 0 \tag{19.21}$$

$$\phi_1(z = d) = 0 \tag{19.22}$$

Also, the boundary conditions at the surface of the photoconductor are

$$\phi_1(z = a) = \phi_2(z = a) \tag{19.23}$$

$$D_{1n} - D_{2n} = \rho_s|_{z=a} \tag{19.24}$$

i.e.,
D_n is the normal electric flux density at the photoconductor surface.

$$\rho_s = \varepsilon_0 E_{1n} - \varepsilon_2 E_{2n} = -\varepsilon_0 \frac{d\phi_1}{dz} + \varepsilon_2 \frac{d\phi_2}{dz} = -\varepsilon_0 A_1 + \varepsilon_2 A_2$$

With the use of the four boundary conditions in equations (19.21)–(19.24), one can determine the unknown constants A_1, A_2, B_1, and B_2 and hence the potentials ϕ_1 and ϕ_2 above and below the photoconductor surface

$$E_1 = -\frac{d\phi_1}{dz} = -A_1 = \frac{\rho_s}{\varepsilon_0 \left[1 + \dfrac{\varepsilon_2}{\varepsilon_0} \dfrac{d}{a} - \dfrac{\varepsilon_2}{\varepsilon_0} \right]} \tag{19.25}$$

$$E_2 = -\frac{d\phi_2}{dz} = -A_2 = \frac{-\rho_s \left(\dfrac{d}{a} - 1 \right)}{\varepsilon_0 \left[1 + \dfrac{\varepsilon_2}{\varepsilon_0} \dfrac{d}{a} - \dfrac{\varepsilon_2}{\varepsilon_0} \right]} \tag{19.26}$$

(4) In an ink-jet printer, assume that an ink drop of diameter 0.03 mm is charged negatively with 100×10^{-15} C. The density of the ink is $2000 \, \text{kg/m}^3$. The deflection plate is charged to a voltage $V_0 = 3500 \, \text{V}$ with a spacing d of 2 mm (Fig. 19.20). The length of the deflection plate L_1 is 15 mm and the distance L_2 from the exit end of the deflection plate to the print surface (print sheet) is 12 mm. The velocity v_z of the ink drop along the deflection plate is assumed 25 m/s. Find the vertical displacement of the drop on the print surface.

Solution:
For the coordinate system of Figure 19.20, the following equations are applicable inside the deflection plates:

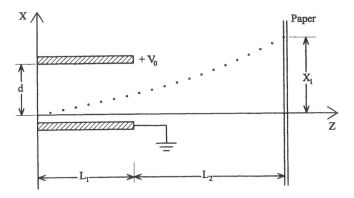

Figure 19.20   Coordinate system for calculating displacement of ink droplets.

$$m\frac{\mathrm{d}^2x}{\mathrm{d}t^2} = qE = q\frac{V_0}{d} \tag{19.27}$$

$$z = v_z t \qquad (v_z = \text{constant}) \tag{19.28}$$

where m and q are the mass and charge of the ink drop. E is the electric field between the deflection plates. Integrating equation (19.27) twice gives

$$\frac{\mathrm{d}x}{\mathrm{d}t} = v_x = \frac{qV_0}{md}t + c_1 \tag{19.29}$$

$$x = \frac{qV_0}{2md}t^2 + c_1 t + c_2 \tag{19.30}$$

The initial conditions at the entrance end of the deflection plates are

$$V_x(t = 0) = 0 \qquad \text{and} \qquad x(t = 0) = 0$$

Therefore, $c_1 = 0$, $c_2 = 0$ and

$$v_x = \frac{qV_0}{md}t \tag{19.31}$$

$$x = \frac{qV_0}{2md}t^2 = \tfrac{1}{2}v_x t \tag{19.32}$$

Outside the deflection plates, $x > L_1$

$$x = x_0 + v_{x0}(t - t_0) \tag{19.33}$$

$$z = v_z t \tag{19.34}$$

where t is the travel time of the drop completing its journey through the deflection plates. x_0 and v_{x0} are its position and velocity at the exit end of the plates (corresponding to travel time t_0).

To obtain a numerical answer, let us first calculate the mass of the drop

$$m = \tfrac{4}{3}\pi(\tfrac{1}{2} \times 0.03 \times 10^{-3})^3 \times 2000 = 2.82 \times 10^{-11} \text{ kg}$$

The velocity in the z direction is not affected by the deflection voltage. Therefore

$$t_0 = L_1/v_z = \frac{15 \times 10^{-3}}{25} = 6 \times 10^{-4} \text{ s}$$

$$v_{x0} = \frac{qV_0}{md}t_0 = \frac{100 \times 10^{-15} \times 3500}{2.82 \times 10^{-11} \times 2 \times 10^{-3}} \times 6 \times 10^{-4} = 3.72 \text{ m/s}$$

$$x_0 = \tfrac{1}{2}v_{x0}t_0 = \tfrac{1}{2} \times 3.72 \times 6 \times 10^{-4} = 11.16 \times 10^{-4} \text{ m}$$

The time t_1 required for the drop to reach the print surface is

$$t_1 = (L_1 + L_2)/v_z = 27 \times 10^{-3}/25 = 1.08 \times 10^{-3} \text{ s}$$
$$x_1 = x_0 + v_{x0}(t_1 - t_0)$$
$$= 11.16 \times 10^{-4} + 3.72(1.08 \times 10^{-3} - 0.6 \times 10^{-3}) = 29 \times 10^{-4} \text{ m}$$

which is the vertical displacement of the drop on the print surface.

(5) An electret loudspeaker is made from a layer of plastic of thickness $a = 25\,\mu\text{m}$ and relative permittivity $\varepsilon_r = 2.8$, backed with a metal foil and stretched in front of a porous electrode with the plastic facing the electrode at a distance $b = 75\,\mu\text{m}$ (Fig. 19.11). If the microphone electrode and the metal foil are held at ground potential ($V = 0$), find the electric stress E_a in the foil/plastic laminate and the charge density on the microphone electrode. Neglect the motion of the laminate, which maintains a surface charge of $\rho_s = 25\,\mu\text{C/m}^3$.

Solution:
The fields in the regions above and below the charged surface have constant values E_a and E_b and satisfy the boundary conditions

$$E_a a + E_b b = V \tag{19.35}$$

$$\varepsilon E_a - \varepsilon_0 E_b = \rho_s \tag{19.36}$$

Solving equations (19.35) and (19.36) simultaneously, one obtains

$$E_a = \frac{\varepsilon_0 V + b\rho_s}{a\varepsilon_0 + b\varepsilon} \tag{19.37}$$

$$E_b = \frac{\varepsilon V - a\rho_s}{a\varepsilon_0 + b\varepsilon} \tag{19.38}$$

As $V = 0$, the fields E_a and E_b are expressed as

$$E_a = \frac{b\rho_s}{a\varepsilon_0 + b\varepsilon} \tag{19.37a}$$

$$E_b = \frac{-a\rho_s}{a\varepsilon_0 + b\varepsilon} \tag{19.38a}$$

The charge density ρ_{se} on the microphone electrode is

$$\rho_{se} = \varepsilon_0 E_b \tag{19.39}$$

Generally, the distance b is much greater than the electret (laminate) thickness a, so that the field is mainly inside the electret. This field is $E_b \approx \rho_s/\varepsilon$.

Substituting in equations (19.37a) and (19.38a), and with the given numerical values,

$$E_a = \frac{75 \times 10^{-6} \times 25 \times 10^{-6}}{25 \times 10^{-6} \times 8.84 \times 10^{-12} + 75 \times 10^{-6} \times 2.8 \times 8.84 \times 10^{-12}}$$
$$= 0.9 \times 10^6 \text{ V/m}$$

which is the stress in the foil/plastic laminate.

$$E_b = \frac{-25 \times 10^{-6} \times 25 \times 10^{-6}}{25 \times 10^{-6} \times 8.84 \times 10^{-12} + 75 \times 10^{-6} \times 2.8 \times 8.84 \times 10^{-12}}$$
$$= 0.3 \times 10^6 \text{ V/m}$$

Thus the charge density ρ_{se} on the microphone electrode is

$$\rho_{se} = \varepsilon_0 E_b = 8.84 \times 10^{-12} \times 0.3 \times 10^6 = 2.652 \times 10^{-6} \text{ C/m}^2$$
$$= 2.652 \text{ }\mu\text{C/m}^2$$

(6) A unipolar smoke detector consists of two parallel electrodes, separated by a distance $d = 0.01$ m and stressed by a steady applied voltage $V = 100$ V. One electrode contains a radioactive material which ionizes the air nearby, forming ions of both polarities (Fig. 19.14). One polarity is immediately absorbed by the electrode for collection with a mobility $\mu_i (= 1.5 \times 10^{-4} \text{ m}^2/\text{s·V})$. The charges are supplied in such abundance that the current is limited only by the space charge that builds up in the air between the electrodes.

(a) Find the space-charge-limited current density.
(b) When smoke particles enter the space, the ions stick to them so that
    the charge now moves with the mobility of charged smoke particles
    μ_s which is equal to 0.001 μ_i. What effect will this have on the
    current density?

Solution:

The space between the detector electrodes is filled with ions and the
interelectrode electric field is a one-dimensional field obtained by solving Poisson's equation

$$\frac{d^2\phi}{dx^2} = \frac{-\rho}{\varepsilon_0}$$

where ρ is the ion charge density, which is assumed constant over the
space between the detector electrodes. Integrating the above equation
gives

$$\frac{d\phi}{dx} = \frac{-\rho}{\varepsilon_0}x + c_1 \tag{19.40}$$

where c_1 is a constant of integration to be determined from the perti-
nent boundary condition

$$E(x = 0) = 0, \qquad \therefore c_1 = 0$$

Therefore,

$$E = \frac{\rho}{\varepsilon_0}x \tag{19.41}$$

$$-\int_0^d E dx = -V$$

$$-\int_0^d \frac{\rho}{\varepsilon_0} x dx = -V = \frac{-\rho}{\varepsilon_0}\frac{d^2}{2} \tag{19.42}$$

$$\therefore \rho = 2\varepsilon_0 V / d^2$$

Therefore, the electric field at the collecting electrode is

$$E(x = d) = \frac{\rho}{\varepsilon_0}d = \frac{2\varepsilon_0 V d}{\varepsilon_0 d^2} = \frac{2V}{d} \tag{19.43}$$

which is higher than the geometric electric field ($= V/d$) because of the
space-charge effect.

(a) The current density at the collecting electrode is

$$J(x = d) = \mu_i \rho E(x = d) = \mu_i \rho \frac{2V}{d} = 4\varepsilon_0 \mu_i V^2/d^3 \qquad (19.44)$$

(b) With smoke particles in the interelectrode spacing of the detector, the current density at the collecting electrode

$$J(x = d) = 4\varepsilon_0 \mu_s V^2/d^3$$
$$= (4 \times 8.84 \times 10^{-12} \times 10^{-3} \times 1.5 \times 10^{-4} \times (100)^2)/10^{-6}$$
$$= 53.04 \times 10^{-9} \text{ A/m}^2$$

(7) In an electrostatic precipitator, charged dust builds up on the collecting wall in a layer of thickness a (Fig. 19.21). As the dust layer gets thicker, breakdown may occur, causing a spark (back discharge) which blows the dust back into the exhaust air. Find the thickness of the layer when breakdown occurs at $E_{bd} = 3\,\text{MV/m}$ in terms of the charge density ρ_0 ($= 15 \times 10^{-3}\ C/m^3$) inside the layer. Assume that the discharge electrode is very far away.

Solution:
The electric field is perpendicular to the conducting wall and is uniform in x and y. It is not uniform in z because of the volume charge within the dust layer.

Applying Gauss' law to the surface s of Figure 19.21,

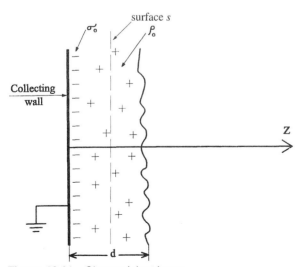

Figure 19.21 Charged dust layer.

$$\oiint_s \overline{E}.\overline{ds} = AE_z = \frac{q_{enclosed}}{\varepsilon_0} = \frac{\sigma_0 A}{\varepsilon_0} + \frac{A}{\varepsilon_0} \int_0^z \rho_0 dz$$

$$= \frac{\sigma_0 A}{\varepsilon_0} + \frac{A\rho_0}{\varepsilon_0} z$$

where σ_0 is the surface charge density on the collecting wall.

$$\therefore E_z = \frac{\sigma_0}{\varepsilon_0} + \frac{\rho_0}{\varepsilon_0} z \tag{19.45}$$

As the collecting wall is grounded, one would expect the total charge to be zero (Adams, 1971), i.e.,

$$\sigma_0 A + \rho_0 A d = 0$$
$$\therefore \sigma_0 = -\rho_0 d \tag{19.46}$$

where d is the thickness of the dust layer.

The electric field within the dust layer is

$$E = E_z = \frac{-\rho_0 d}{\varepsilon_0} + \frac{\rho_0}{\varepsilon_0} z = \frac{\rho_0}{\varepsilon_0}(z - d) \tag{19.47}$$

Equation (19.47) dictates that E_z is negative for $z < d$. This is true because the electric field within the dust layer is directed from the space charge to the collecting wall, i.e., in the negative z direction.

From equation (19.47), the maximum electric field occurs at the collecting walls, i.e., at $z = 0$

$$|E|_{max} = \frac{\rho_0 d}{\varepsilon_0}$$

The thickness d_{bd} of the dust layer at breakdown is obtained as

$$E_{bd} = \frac{\rho_0 d_{bd}}{\varepsilon_0} \tag{19.48}$$

i.e., $d_{bd} = E_{bd}\varepsilon_0/\rho_0$

$$= (8.84 \times 10^{-12} \times 3 \times 10^6)/(15 \times 10^{-3})$$

$$= 1.77 \times 10^{-3} \text{ m} = 1.77 \text{ mm}$$

(8) In an electrostatic propulsion scheme, the propellant is cesium ions whose mass $m_i = 133 \times 1.67 \times 10^{-27}$ kg (corresponding to the mass of cesium, which has an atomic weight of 133) and charge $q_i = 1.6 \times 10^{-19}$ C. The accelerating voltage V_a is 3500 V and the ion current is 0.2 A. Calculate the velocity of the ejected ions and the propulsion force.

Solution:
With reference to equation (19.1), the ion velocity v_i is

$$v_i = \sqrt{\frac{2q_i}{m_i}} V_a = \sqrt{\frac{2 \times 1.6 \times 10^{-19} \times 3500}{133 \times 1.67 \times 10^{-27}}} \simeq 7.1 \times 10^4 \text{ m/s}$$

A chemically fueled rocket engine cannot achieve this velocity. According to equation (19.5), the propulsion force F is

$$F = v_i m_i I / q_i = (7.1 \times 10^4 \times 133 \times 1.67 \times 10^{-27} \times 0.2)/(1.6 \times 10^{-19})$$
$$= 197.1 \times 10^{-4} \text{ N}$$

(9) Assume, in Figure 19.4a, that the voltage applied between the collecting parallel plates is 120 kV, the spacing between plates is 0.6 m, and the vertical dimension of the plates is 1.2 m. The charge-to-mass ratio of a phosphate particle is 10×10^{-6} C/kg. The particle starts free fall at the middle of the top edge of the parallel plates. Find the position of the particle at the exit end of the plate.

Solution:
Set up the coordinate system as shown in Figure 19.4a. The velocity in the y direction is governed by the gravitational force, i.e.,

$$d^2y/dt^2 = -g = -9.8 \text{ m/s}^2$$

Integrating the above equation with respect to t and using the initial condition $dy/dt = 0$ at $t = 0$, one obtains

$$dy/dt = -9.8t \tag{19.49}$$

Performing one more integration with the initial condition $y = 0$ at $t = 0$, one obtains

$$y = -4.9t^2 \tag{19.50}$$

Thus, the phosphate particles will exit the plate at $y = -1.2$ m and $t = t_0$;

$$t_0 = \sqrt{\frac{1.2}{4.9}} \simeq 0.5 \text{ s}$$

On the other hand, the velocity of the phosphate particle in the x direction is governed by the electrostatic force, i.e.,

$$m\frac{d^2x}{dt^2} = qE \tag{19.51}$$

where m and q are the particle mass and charge. E is the drifting field between the plates $(= V/d)$, V is the voltage across the plates, and d is the spacing between plates.

Thus,

$$\frac{d^2 x}{dt^2} = \frac{q}{m}\frac{V}{d}$$
$$= 10 \times 10^{-6} \times 120 \times 10^3 / 0.6 = 2 \text{ m/s}^2$$

Integrating twice and substituting the initial conditions $dx/dt = 0$ and $x = 0$ at $t = 0$, one obtains

$$dx/dt = 2t \qquad \text{and} \qquad x = t^2 \qquad\qquad (19.52)$$

At $t = t_0 = 0.5$ s, the particle exits the plate at

$$x = (0.5)^2 = 0.25 \text{ m}$$

Note that the variable t may be eliminated from the expressions for y and x to obtain an equation

$$y = -4.9x$$

Therefore, inside the parallel plates the trajectory of the phosphate particle is a straight line. The E-field is zero outside the plates, but gravitational force still exists; thus the trajectory of the phosphate particle after its exit from the parallel plates will be a parabola.

(10) In an ink-jet printer, the drops are charged by surrounding the jet of radius 25 μm with a concentric cylinder of radius 750 μm, as shown in Figure 19.5. Calculate the minimum voltage V_p required to generate a charge 50 pC on the drop if the length of the jet inside the cylinder is 120 μm.

Solution:
Assume the liquid of the jet to have an electrical conductivity of about that of water (10^{-6} S/m or greater). With such a high conductivity, the jet will have a negative charge per unit length λ;

$$\lambda = \frac{CV_p}{L} \qquad\qquad (19.53)$$

where C is the capacitance of the cylinder–jet combination, L is the unbroken length of the jet inside the cylinder, and V_p is the applied voltage. The capacitance C is given approximately by

$$C = \frac{2\pi\varepsilon_0 L}{\ln(b/a)} \qquad\qquad (19.54)$$

where b is the inner radius of the cylinder and a is the radius of the jet.

If the drops into which the jet is breaking have radii r, they must come from a length x of the jet which is

$$x = \tfrac{4}{3}\pi r^3 / \pi a^2 = \frac{4r^3}{3a^2} \tag{19.55}$$

Since the jet breaks up into drops while it is in the cylinder and hence has a charge, each drop will carry away a charge q given by

$$q = x\lambda = \frac{4r^3 C V_p}{3a^2 L} \tag{19.56}$$

Upon substitution of the equation for C in equation (19.56), one obtains the drop charge as

$$q = \frac{8\pi V_p \varepsilon_0 r^3}{3a^2 \ln(b/a)}$$

For a jet radius a of 25 μm, the drop radius r will be about 50 μm. For a drop charge of 50 pC,

$$50 \times 10^{-12} = \frac{8\pi V_p 8.84 \times 10^{-12} (50 \times 10^{-6})^3}{3(25 \times 10^{-6})^2 \ln(750/25)}$$

$$\therefore V_p = 11.48 \text{ kV}$$

which is the minimum voltage required for generating drops with a charge of 50 pC per drop.

REFERENCES

Abdel-Salam M, Singer H. J Phys D: Appl Phys 24:2000–2007, 1991.
Abdel-Salam M, Soliman FA, Megahed AA. J Phys D: Appl Phys 26:2082–2091, 1993.
Abdel-Salam M, Soliman FA, Megahed AA. IEEE Ind Appl Mag Nov/Dec:33–41, 1995.
Abdel-Salam M, Mizuno A, Shimizu K. J Phys D: Appl Phys 30:864–870, 1997.
Adams AT. Electromagnetics for Engineers. New York: USA. Ronald Press Company, 1971.
Ashley CT, Edds KE, Elbert DL. IBM Res Dev 21:69–74, 1977.
Braun D, Kuchler U, Pietsch G. J Phys D: Appl Phys 24:564–568, 1991.
Buehner WL, Hill JD, Williams TH, Woods JW. IBM Res Dev 21:2–9, 1977.
Carnahan RD, Hou SL. IEEE Trans IA-13:95–105, 1977.
Civitano L, Sani E. In: Capitelli M, Gorse C, eds. Plasma Technology. New York: Plenum Press, 1992.
Crowley JM. Fundamentals of Applied Electrostatics. New York: John Wiley, 1986.
Crowley JM, Chato JC. Record of IEEE/IAS Annual Meeting, Cleveland, OH, 1979, pp 226–229.

Eliasson B, Hirth M, Kogelschatz U. J Phys D: Appl Phys 20:1421–1425, 1987.

Gordon G, Peisakhov I. Dust Collection and Gas Cleaning, Moscow: Mir, 1972.

Hoferer B, Weinlein A, Schwab AJ. International Symposium on High Voltage Engineering, London, 1999, paper # 5.402.S281.

Inculet I. J Electrostat 4:175–192, 1977/78.

Iuga A, Morar R, Samuila A, Dascalescu L. Conference Record, IEEE/IAS, St Louis, MO, 1998, pp 1953–1960.

Kogelschatz U, Eliasson B. Ozone generation and applications. In: Chang C, Kelly E, Crowley J, eds. Handbook of Electrical Processes. New York: Marcel Dekker, 1995.

Landham E, Dubad J, Brien M, Lindsey C, Plulle W. Proceedings of IEEE Annual Meeting on Electrostatic Processes, Atlanta, GA, 1987, paper # 8-2.

Langmuir, DB, Stuhlinger, Sellen JM. Electrostatic Propulsion. New York: Academic Press, 1961.

Law SE. J Electrostat 23:145–156, 1989.

Law SE. Electrostatic atomization and spraying. In: Chang C, Kelly F, Crowley T, eds. Handbook of Electrical Processes. New York: Marcel Dekker, 1995.

Law SE, Cooper SC. Trans ASAE 32:1169–1172, 1989.

Lawver JE, Dyrenforth WP. Electrostatic separation. In; Moore AD: ed. Electrostatics and its Application. New York: Wiley-Interscience, 1973.

Masuda S. J Electrostat 10:1–15, 1981.

Masuda S. Proceedings of EPRI/NSF Symposium on Environmental Applications of Advanced Oxidation Technologies, San Francisco, CA, 1993.

Masuda S, Hosokawa S. IEEE Trans IA-24:708–713, 1988.

Melcher JR. J Electrostat 2:121–132, 1976.

Mizuno A, Washizu M. Biomedical engineering. In: Chang C, Kelly E, Crowley J, eds. Handbook of Electrical Processes. New York: Marcel Dekker, 1995.

Mizuno A, Clements JS, Davis RH. Conference Record, IEEE/IAS Annual Meeting, 1984, pp 1015–1020.

Mizuno A, Kisanuki Y, Noguchi M, Katsura S, Lee S-H, Hong YK, Kang KO, Shin SY. Conference Records, IEEE/IAS, St Louis, MO, 1998, pp 1790–1793.

Oglesby S, Nichols GB. Electrostatic Precipitation. New York: Marcel Dekker, 1978.

Sadiku MNO. Elements of Electromagnetics. Philadelphia: Saunders College Publishing, 1994.

Schein LB. Electrophotography and Development Physics. 2nd ed. New York: Springer Verlag, 1992.

Shen LC, Kong JA. Applied Electromagnetism. Monterey, CA: Brooks/Cole Engineering Division, 1983.

Swatik DS. Nonimpact Printing. In: Moore AD, ed. Electrostatics and its Application. New York: Wiley-Interscience, 1973.

Washizu M, Shikida M, Aizawa S, Hotani H. IEEE Trans IA-28:1194–1202, 1992.

Yan K, Yamamoto T, Kanazawa S, Ohkubo T, Nomoto Y, Chang J-S. Conference Record, IEEE/IAS, St Louis, MO, 1998, pp 1841–1844.

Yang C-L, Yamamoto T. Conference Record, IEEE/IAS, St Louis, MO, 1998, pp 1767–1772.

20

Safety and Electrostatic Hazards

R. RADWAN and A. EL-MORSHEDY *Cairo University, Giza, Egypt*

20.1 INTRODUCTION

Static electricity is a well known phenomenon. Every one has experienced static discharge at one time or another. The early Greeks, 600 BC, noticed that amber, when rubbed, could attract light objects. They were responsible for the term "electricity" derived from their word for amber. They spent many hours rubbing a small piece of amber on their sleeves and observing how it would then attract pieces of fluff and stuff (Hayt, 1987).

Static electricity affects many industries and diverse environments. The results of static charge buildup are quite noticeable. These results include potentially dangerous electrical shocks which can cause decreases in productivity, machinery jams, fires, and explosions. Static electricity can also cause severe damage to sensitive electronic components, requiring costly repair.

20.2 NATURE OF STATIC ELECTRICITY

Static electricity is an electrical charge at rest, it is generated by unbalancing the molecular construction of relatively nonconductive insulators such as plastics and paper. To understand what static electricity is, we have to learn a little bit about the nature of matter.

681

Everything is made up of atoms. Atoms are made of protons, electrons, and neutrons. Protons have a positive charge, electrons have a negative charge, and neutrons have no charge. Usually, atoms have the same number of electrons and protons. Then, the atom has no net charge, and is said to be neutral. When two different materials are rubbed together, electrons move from one atom to another. Some atoms get extra electrons and become negatively charged, while others lose electrons and become positively charged. When charges are separated like this, it is called static electricity (Fig. 20.1). If two bodies have different charges they attract each other, and if they have the same charge they repel each other.

The majority of electrostatic charges are caused by two different materials being rubbed together. Static electricity can also be generted by friction, pressure, and separation (Fig. 20.2a,b). Friction causes heat which excites the molecular particles of the material. When two materials are then separated, a transfer of electrons from one material to the other may take place. As electrons transfer, the absence or surplus of electrons creates an electric field.

Static electricity can also be generated by rupture of the molecular structure caused by cutting, slitting, and tearing. It can also be generated by rapid temperature change, radioactive emission, and chemical changes in a material. These processes give rise to imbalance between positive and negative ions on the surface of the material.

Lightning is a manifestation of static electricity. Masses of air rise and fall during thunderstorms. They rub against rain clouds, which causes them to become electrically charged. Lightning occurs when two objects of different charges get close to each other (Anderson, 1979).

The amount of static electricity generated depends upon the materials subjected to friction or separation, the amount of friction or separation, and

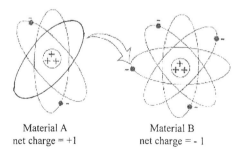

Material A
net charge = +1

Material B
net charge = - 1

Figure 20.1   Electron transfer from material A to material B by separation.

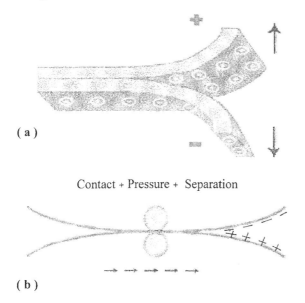

(a)

Contact + Pressure + Separation

(b)

Figure 20.2 (a) Two materials are in contact and are then separated. (b) Friction, pressure, and separation are the major causes of static electricity.

the relative humidity of the environment. Common plastic generally will create the greatest static charge. Low-humidity conditions, such as those created when air is heated during the winter, will also promote the generation of significant static electrical charge. During summer, the air is more humid. The water in the air helps electrons to leak more quickly and so the buildup of charge is not quick.

Many of the common activities man performs daily may generate charges on his body. For example, when he gets out of the car, he gets a shock on closing the door. The source is usually static charges, which build up between his body and the car seat while he is in the seat but remain harmless until he gets up. At that time, he takes considerable electric charge with him as he gets out of the car. If the charge has no discharge path, then a very high potential of several thousand volts can build up very quickly. When he reaches for the door, the high voltage causes a spark, which discharges him quickly to the car. One solution is to hold onto the metal door frame as he gets out of the seat, allowing himself to harmlessly discharge slowly as he gets up.

When a person walks across a carpeted floor on a dry day, his body builds up an electrical potential of several thousand volts, which discharges through the air when his fingers get close to the door handle. An arc of

Table 20.1   Man's Common Activities and the Voltages Generated

Means of generation	Electrostatic voltage (kV)		
	Relative humidity		
	10%	40%	55%
Person walking across carpet	35	15	7.5
Person walking across vinyl tile	12	5	0.3
Worker at bench	6	0.5	0.4
Ceramic dips in plastic tube	2	0.7	0.4
Ceramic dips in styrofoam	14.5	5	3.5
Circuit packs as bubble plastic cover is removed	26	20	7

electric current is produced to overcome the normally insulative qualities of air. Table 20.1 gives man's common activities and the corresponding generated electrostatic voltages for different values of relative humidity (Unger, 1997).

20.3   TRIBOELECTRIC SERIES

When two different materials are rubbed together, one becomes positively charged and the other becomes negatively charged. Scientists have ranked materials in order of their ability to hold or give up electrons. This ranking is called the triboelectric series. Table 20.2 gives a short triboelectric series that provides an indication of the ordering of some common materials (Adams, 1986). The way to use a triboelectric series is to note the relative positions of the two materials of interest. The material that is charged positive will be the one that is closer to the positive end of the series and the material closer to the negative end will be charged negatively. In fact, the work function of the material determines its position in the series. Materials with higher work function tend to appropriate electrons from materials with lower work function.

There are several problems in using this triboelectric series. Real materials are seldom very pure and often have surface finishes and/or contamination that strongly influence triboelectrification. Besides, the spacing between materials on the series does not give the magnitude of the charge separated with a high confidence level. Many factors strongly influence the results. Among these factors are the difference in the electronic surface energy, surface finish, electrical conductivity, and mechanical properties.

Table 20.2   Triboelectric Series Chart

Positive end of series
- Human hands
- Asbestos
- Glass
- Mica
- Human hair
- Nylon
- Wool
- Fur
- Lead
- Silk
- Aluminum
- Paper
- Cotton
- Steel
- Wood
- Amber
- Hard rubber
- Nickel & copper
- Brass & silver
- Gold & platinum
- Synthetic rubber
- Orlon
- Saran
- Polyethylene
- Teflon
- Silicone rubber

Negative end of series

20.4   BASIC LAWS OF ELECTROSTATIC ELECTRICITY

Static electricity has been studied for hundreds of years. Many scientists have studied static electricity; Ampère, Priestley, Franklin, Faraday, Volta, and Coulomb are among them. Their laws that describe electrical behavior remain definitive.

Charles Coulomb performed an elaborate series of experiments in 1785. He used a delicate torsion balance, invented by himself, to determine quantitatively the force exerted between two objects each having a static charge of electricity. Coulomb stated that the force between two very small charged objects separated in vacuum or free space by a distance which is large compared to their size is proportional to the charge on each and inversely proportional to the square of the distance between them. This can be written as

$$F = k\frac{q_1 q_2}{r^2} \tag{20.1}$$

where q_1 and q_2 are the positive or negative quantities of charge, r is the separation between charges, and k is a proportionality constant. If the International System of Units (SI) is used, q is measured in coulombs (C), r in meters (m), and the force should be in newtons (N). This will be achieved if the constant of proportionality k is written as

$$k = 1/4\pi\varepsilon_0$$

The constant ε_0 is called the permittivity of free space and has the magnitude 8.854×10^{-12} F/m. The force acts along the line joining the two charges and is repulsive if the charges are alike in sign and attractive if they are of opposite sign (Fig. 20.3).

The force F is a vector, having both magnitude and direction. Rewriting equation (20.1) as a vector equation and substituting the value of k, we have

$$\underline{F} = k\frac{q_1 q_2}{r^2}\underline{u}_r \tag{20.2}$$

This is the complete vector expression for Coulomb's law in rationalized SI units, where $\underline{u}_r$ is the unit vector in the direction of $\underline{F}$.

20.5   MATERIALS AND STATIC ELECTRICITY

Some metals, like silver and copper, have one elecron in the outermost occupied shell of the atom. This electron is so loosely held that it migrates easily from atom to atom when an electrostatic field is applied. Materials that permit such electron motion are called conductors. Silver and copper are examples of good conductors, their resistance to electronic motion being relatively slight. Some good conductors have two electrons in the outermost occupied shell. A few, such as aluminum, have three. In all conductors these

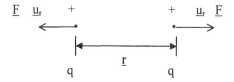

Figure 20.3   Two positive point charges and the resulting repulsive force.

electrons are loosely bound and can migrate rapidly from atom to atom. Such electrons are called free charges. When an ungrounded conductor becomes charged, the entire volume of the conductive body assumes a charge of the same voltage and polarity. A charged conductor can be neutralized by connecting it to ground (Fig. 20.4).

In other substances, the electrons may be so firmly held near their normal position that they cannot be liberated by the application of ordinary electrostatic fields. These materials are called dielectrics or insulators. Although a field applied to an insulator may produce no migration of charge, it can produce polarization of the insulator or dielectric, i.e., a displacement of the electrons with respect to their equilibrium positions. The charges of an insulator are often called bound or polarization charges in contrast to the free charges of a conducting material. Because of this, an insulator may retain several static charges of different polarities and potentials at various areas on its surface. This accounts for why certain areas of a material may stick together and others may repel each other. Connecting the insulator to ground will not result in an exchange of electrons, as in the case with conductive materials (Fig. 20.4); therefore, other means must be used for neutralizing static charges on insulators.

Certain other materials, which have properties intermediate between conductors and insulators, are called semiconductors. Under some conditions such substances may act like insulators but, with the application of heat or sufficient field, may become a fair conductor.

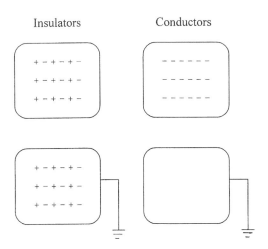

Figure 20.4 Insulators and conductors have different charge and grounding capabilities.

In many cases, the choice of material and its electrical properties is a key factor in controlling the generation and safe dissipation of static electricity. For example, conductive polymers have properties suitable for a wide range of applications. The electrical properties of materials, such as resistivity and charge decay properties, are very important in the specification of materials for electrostatic uses. These properties must be measured according to national or international standards.

20.6 ELECTROSTATIC DISCHARGES (ESD)

ESD is defined as the transfer of charge between bodies at different elecrical potentials. It arises when a static charge accumulates on a nonconducting object and then finds a path to ground. The buildup of charge is caused by man-made fibers such as nylon brushing against an insulator. This takes place every day in offices and homes, particularly if these have carpets woven from man-made fibers (Moore, 1997). A person walking across a carpeted room will build up a charge as ions transfer from the fibers to the person who becomes charged. Once a person is charged, and depending on the relative humidity, a discharge will occur if the person touches a surface at a lower potential. If, for example, this discharge is via a personal computer keyboard, then many thousands of volts could instantaneously appear across voltage-sensitive components in the computer. ESD can cause direct and indirect damage to semiconductor devices and electronic systems. ESD impacts productivity and product reliability of the electronic systems. Despite the great effort during the past 20 years, ESD still affects production yields, manufacturing costs, product quality, product reliability, and profitability.

There are many industries and commercial operations where electrostatic phenomena are either exploited to achieve useful ends or dealt with as a nuisance or hazard to be avoided. Examples of the beneficial exploitation of ESD include xerographic copy machines and electrostatic paint sprays. On the other hand, there are electrostatic nuisances which can slow production, affect product quality or damage equipment, and electrical discharges which can cause fires or explosions leading to injury or death. The estimated average loss in component manufacture ranges from 16 to 22%, in contractors and subcontractors 8 to 15%, and in users 27 to 33% (Halperin, 1990).

20.7 STATIC ELECTRICITY PROBLEMS

Static electricity is acquiring increasing importance as a source of industrial hazards. It is a phenomenon that creates problems which cost industry millions of dollars per year. In Europe and USA an accident is caused every day by electrostatic discharges. The problems caused by static electricity discharge are described below.

20.7.1 Production Problems and Slowdowns

Static charge on an automatic sheet feeder, and in some cases with hand-fed operations, may result from the separation or tearing away of one sheet from the next in the sheet pile. A static charge generates an electric field which acts like a magnet. This magnet repels similar charges and attracts opposite or neutral charges. This accounts for the attraction between charged materials and machine frames or rollers causing machine jams. In many cases the machine operator will be forced to run at a slow speed to avoid the problems caused by static electricity (Cross, 1987).

20.7.2 Dust Attraction

Many products, such as plastic molding materials and film, develop high static charges on their surfaces. These highly charged surfaces will attract airborne dust, sometimes from over one meter away, and hold it tightly to the surface. Subsequent operations, such as molding or printing, and their end products can be seriously affected by that contamination.

20.7.3 Shocks to Personnel

When handling highly charged materials, people receive unpleasant static electricity shocks either directly from the material or indirectly. Some people are more susceptible to holding charges than others, usually because of the grounding properties of the shoes they are wearing. An example of direct shock is when a person gets out of a car seat and then touches the door handle. Workers may receive shocks at the delivery end of a dryer's conveyor. The constant heat change of the mesh belt allows it to become highly charged, and operators are often in contact with that part of the machine. Operators are also often shocked when they pull off barrier film or delaminate credit cards.

20.7.4 Fires and Explosions

Static electricity is usually found on nonconductive materials where high resistivity prevents the movement of the charge. There are two situations where the static charge can move quickly and be dangerous in a combustible atmosphere. The first one is where a grounded object intensifies the static field until it overcomes the dielectric strength of the air and allows current to flow in the form of a spark. The second one is where the charge is on a floating conductor such as an isolated metal plate. The charge in this case is very mobile and will flash to a proximity ground at the first opportunity (Bodurtha, 1980).

20.7.5   Image Spiders and Webbing

Static electricity is responsible for ink spidering and webbing. After the squeegee is rubbed against the screen repeatedly, the friction causes a charge to build up in the polyester mesh. The screen, normally negatively charged, is forced onto a plastic substrate, also negatively charged, in the printing cycle. The two negative charges repel each other, causing the ink to misbehave and creating a mark on the image.

20.8   HAZARDS OF ELECTROSTATIC ELECTRICITY IN INDUSTRY

Electrostatic electricity is encountered in most industrial processes and other daily life activities. Accumulation of static charges on workpieces during their processing may result in voltages as high as several tens of kilovolts. As a result of these high voltages electrostatic discharge (ESD) may occur, leading to either total or partial damage of these pieces. The cost of damage and fires from ESD amounts to several hundred million dollars a year.

Some industries are very sensitive to ESD even at a few hundred volts, such as the petroleum and chemical industries. Other industries and activities such as textile, fertilizer, paper, rubber, wheat flour, grain transportation, hospitals, etc. suffer from ESD and strict measures should be taken to eliminate, or reduce to an acceptable level, the accumulation of static charges.

Daily activities of personnel produce voltages on their bodies as high as 30 kV when walking across the floor. This voltage depends on the flooring material, whether it is a carpet or vinyl, and the shoes and socks they wear. For vinyl flooring the voltage is in the order of a few kilovolts.

The following sections will discuss ESD hazards in some industries and daily activities.

20.8.1   Electronic Industry

As electronic technology advances, electronic components tend to become smaller and smaller. As the size of the components is reduced, the microscopic spacing of insulators and circuits within them is also reduced, thus increasing their sensitivity to ESD. ESD damage can occur in electronic components from as little as 20 V. Although this minute discharge cannot be felt, the effect on electronic devices can be devastating (Boxleitner, 1989).

Table 20.3 Static Voltage Tolerances of Various Types of Component

Device type	Range of ESD susceptibility (V)
MOSFET	100–200
JFET	140–10,000
CMOS	250–2000
Schottky diode, TTL	300–2500
Bipolar transistor	380–10,000
SCR	680–1000

Objects used in the workplace, such as styrofoam coffee cups, flooring materials, storage bins, desktops, and even ordinary clothing, are all sources of static generating materials. Without an effective ESD program in place, workers who handle electronic parts may damage them without any outside indication.

Static damage of components by operating personnel is fast becoming one of the most significant problems facing the electronics industry. Technological advances in chip design, that make possible higher circuit densities and higher performance, quite often result in higher static susceptibility. Table 20.3 gives the static voltage tolerances of various types of component (Boxleitner, 1989).

If static potentials can be kept below 100 V, the static problems should not exist. The basic concept of complete static protection for electronic components is the prevention of static buildup where possible, and the quick reliable removal of already existing charges.

Static damage to components can take the form of upset failures or catastrophic failures. Upset failures occur in two forms, direct and latent. Direct catastrophic failures occur when a component is damaged and will never function again. This is the easiest type of ESD damage to find since it usually can be detected during testing.

Latent failures occur when ESD weakens the component to the point where it will still function properly during testing, but over time the wounded component will give poor system performance and eventually lead to complete system failure. The cost of repair is very high because latent failures occur after final inspection or in the hands of a customer. This type of damage is hard to find and it severely affects the reputation of the company's product.

An upset failure occurs when an ESD has caused a current flow that is not significant enough to cause total failure, but in use may intermittently result in loss of software or incorrect storage of information.

20.8.2 Grain Transportation

Grain such as wheat is transported by either pipes or conveyor belts. When transported by pipe, the relative motion of the grain to the pipe walls causes friction and hence electrostatic charges are generated on the grain. As the grain continues to flow along the pipe it acquires more and more charges. The amount of charge acquired by the grain particles depends on their flow speed, surface area, and charge leakage. If the charge builds up on the grain particles such that the electric field between them and the pipe walls exceeds the withstand flashover voltage of the ambient, a spark will develop. This spark may lead to explosion of the transportation system and injuries to nearby personnel.

Estimation of the electric field in such a system is rather difficult unless the charge density is known. A surface charge density (ρ_s) of approximately $10 \ \mu C/m^2$ is taken, based on the fact that the limiting surface charge in air is $25 \ \mu C/m^2$. However, the potential and electric field along the pipe can be given by the following equations (Crowley, 1986).

$$V = \frac{\rho_0 a^2}{4\varepsilon} \left(1 - \frac{x^2}{r^2} \right) \tag{20.3}$$

$$E = \frac{\rho_0 x}{2\varepsilon} \tag{20.4}$$

where V and E are the radial potential and electric field at distance x, ρ_0 is the volume charge density in the pipe, x the radial distance from the pipe centerline, r the pipe radius, and ε the permittivity of the medium in the pipe (grain). Crowley (1986) showed that voltages of the order of few megavolts may be generated during grain transportation. Of course, these values are very pessimistic since charge leakage from grain particles to earth through the pipe reduces these voltages to much smaller values.

Conveyor belts are also used in grain industries such as grain elevators, cereals, flour mills, and like materials. In these industries the atmospheric air is contaminated with fine dust particles which are highly explosive when subjected to ESD of sufficient energy. Electrostatic discharge can start from two main sources. The first is from the conveyor belt itself as a result of friction between the belt and grain particles. The second is from the workers walking around and generating electrostatic charges by friction with the floor. Under some favorable conditions the charge accumulated on a worker is enough to initiate a spark when touching a metallic casing or frame of a machine. Another negative effect of electrostatic charges in these industries is the ability of statically charged bodies to attract fine dust particles which accumulate on them, thus necessitating frequent cleaning.

20.8.3 Petroleum Industry

Electrostatic charge generation in the petroleum industry is a vital concern, since the system is susceptible to sparks which may lead to fires and costly damage to property and personnel. Electrostatic discharge may occur when charge accumulates on the surfaces of two materials in contact with each other and then suddenly separated. The energy accompanying the ESD depends on the charge accumulated and hence on the type of the two materials, humidity, and surface area of the materials in contact.

In the petroleum industry the ambient atmosphere usually contains flammable vapors and gases. If the energy in the discharge exceeds the minimum ignition energy of the vapor or gases, ignition will occur. The energy of discharge is given by

$$U = \tfrac{1}{2}CV^2 \tag{20.5}$$

where U is the energy in joules, C is the capacitance between the discharging bodies in farads, and V is the voltage in volts.

The minimum ignition energy of flammable substances depends on their percentage in the ambient air, length of spark, and geometry of the sparking electrodes (Huang et al., 1997). The minimum ignition energy of benzene and natural gases ranges between a fraction of a millijoule and a few millijoules.

Pure gasoline and other products do not generate static charges when flowing in pipes unless they are contaminated with moisture and minute particles.

20.8.3.1 Filling Petrol Tanks

Petrol tanks are filled by pumping gasoline or other products through a pipe and feeding it into the tank at its top or bottom. During pumping the liquid in the pipe gains static charges by friction with the pipe wall. If the pipe and tank are not properly earthed, static charges accumulate in the liquid and on the pipe. If the tank is filled at high velocity, liquid droplets are formed and this will add extra static charges, in the same way as cloud charging, to the liquid and hence to the tank. The voltage of a large insulated tank may reach several hundreds of kV relative to ground during the filling process and long high-energy sparks are produced to ground or to nearby grounded metal structures. Sparks as long as about 3 m may take place between the surface of the gasoline and the tank wall, leading to fire or explosion.

To reduce static charge accumulation, good earthing and bonding of pipes and tanks is the usual protective measure. Tanks used for storing petroleum products should be effectively earthed. The resistance to earth should not exceed 7 Ω. If this value cannot be achieved by direct contact of the tank base with earth when the tank is completely isolated from all pipelines and other connections, external earthing electrodes should be used. For tanks of diameter up to 30 m a minimum of two earthing electrodes is used, and for tanks larger than 30 m a minimum of three electrodes is recommended. The electrodes should be symmetrically spaced around the tank and each electrode should have a separate connection to the tank.

To avoid static electrification by liquid atomizing during the filling process, so-called floating roof tanks are commonly used. In these tanks filling is carried out at the bottom.

When filling fuel tanks, sparks may develop from their walls and gauging rods or from any projecting parts inside the tank to the fuel surface. Also, any floating object may develop a spark to the tank walls (Fig. 20.5).

When loading or unloading fuel tankers, flexible rubber hoses are used. Again, static charges are generated by friction between the liquid fuel and the hose walls. When tankers are offshore the total length of the loading line is rather long and the static charge generated may be considerably large. To avoid the risk of fires and costly losses the hose is constructed with embedded spiral metallic wires so that the surface and volume resistivity of the hose is decreased to a value of the order of 1 MΩ. The embedded wires are properly connected to the metallic end connectors. The hose ends should be earthed to avoid charge accumulation. The smaller the grounding resistance, the more efficient the charge dissipation to earth. An earthing resistance of a few ohms is normally adopted.

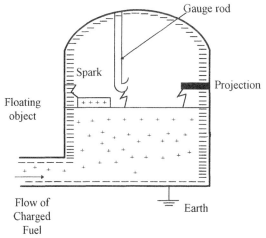

Figure 20.5   Sparks in fuel tanks during the filling process.

In gas stations the hose for fueling cars is earthed so that no sparks occur during filling of car tanks. Fuel truck tanks, when arriving at gas stations after a long drive, will have gained considerable static charge and if they immediately start unloading sparks may occur between the hose connector and the intake of the truck tank. In the presence of fuel vapor, fire may occur. To avoid this problem the truck and tank should be connected to a grounded point in the gas station before unloading for enough time to dissipate their charges to earth. A long time ago it was a common practice to connect a metallic chain to the truck tank which touched the ground when running on roads and streets. With concrete roads a spark may occur when the chain strikes the road surface. In the presence of fuel vapor this may be more dangerous than if the chain did not exist. When loading a tank truck in a loading station, earthing and bonding of the steel rack, pipes, and downspout (tank fill pipe) are essential (Fig. 20.6).

The velocity of filling of tank trucks has a considerable effect on charge accumulation. A simple formula relating the velocity and filling pipe diameter has been recommended by the API (1991) as follows:

$$vd < 0.5 \tag{20.6}$$

where v is the velocity in meters per second and d is the inside diameter of the filling pipe in meters. In addition to this restriction, the linear flow velocity should never exceed $7 \, \mathrm{m/s}$. However, if the product is contaminated with water drops and/or foreign particles, the flow velocity should be restricted to $1 \, \mathrm{m/s}$.

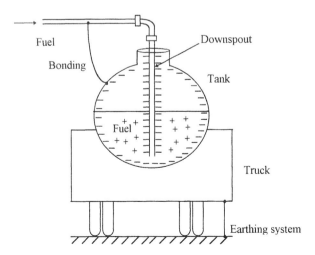

Figure 20.6 Bonding of a tank truck.

20.8.3.2   Fuel Flow in Pipelines

Crude liquid oil is pumped from the production fields to refineries through metallic pipelines over long distances. The flow of oil, with a considerable amount of contamination, water, and foreign particulate, generates electrostatic charges depending on the oil flow speed, amount of contaminant, and material and surface condition of the pipeline. Like grain transportation by pipes, unless the pipeline is grounded, large quantities of static charges are generated which may lead to ESD between the oil and the pipe wall or at the exit end of the pipe. Long pipelines are constructed from several sections tightly joined together to prevent oil or vapor leakage. The pipeline sections are bonded and each section is earthed to avoid accumulation of static charge.

20.8.4   Chemical Industries

Chemical industry facilities such as fertilizer factories suffer from fires and explosions due to static charges unless they are properly protected. The final chemical product usually passes through several industrial processes: crushing, grinding, pulverizing, mixing, and belt or pipe transportation. All these processes are accompanied by electrostatic charge generation. Many of the elements of the chemical products are volatile and flammable. Also, tiny dust particles possess static charges and their accumulation on metallic parts may cause ESD. In some cases, in a dry atmosphere, minute sparks may occur between the particles themselves, leading to explosion or fire. In some industries humidification of the ambient may be a solution to reduction of static charges, but in most chemical processes increased humidity has an adverse effect on the product.

In chemical industries strict measures should be taken to avoid ESD. Among these measures are bonding and earthing of all metal parts, cleaning the accumulated dust by vacuum cleaner rather than blowing as this aggravates the problem. Blowing air at high speed at the dust particles increases the static charges on the particles and metal surfaces. Covering the floor with antistatic material is an additional protective measure.

20.8.5   Textile Industry

The textile industry suffers from electrostatic charges generated during the manufacturing processes. When leaving rollers, textiles generate static charges by contact difference of potential which in some cases amounts to several kilovolts. Textiles have large resistivities and that leads to charge accumulation higher than with other products of low resistivity due to the long charge decay time.

The process of rolling textile fabrics at high speed generates high charges and may result in sparks that damage their surface and in some cases cause fires. When printing, textiles are more susceptible to fires due to the flammable volatile materials used in the printing process. In the wool and cotton industries static charges affect the fibers during carding and spinning by weakening and causing irregularities of the yarns. Controlled ambient humidity, in the range of 50%, is very important in the textile industry since it affects the tensile strength of the yarns and at the same time reduces the ESD.

Hazards from electrostatic charges in the paper industry are similar to those of the textile industry since the rolling and printing are more or less the same.

The rubber industry is liable to ESD during the various manufacturing stages. With the increase of use of synthetic rubber, the problem of ESD is more serious than with natural rubber. Synthetic rubber is made mainly from petroleum and contains volatile flammable elements. To reduce static charge accumulation, earthing of all metallic parts in the manufacturing system is a usual practice. Nowadays static charge eliminators, such as charge neutralizers (Section 20.10.2), are used.

20.8.6 Hospitals

Hospitals, especially operation theaters, suffer from static electricity hazards. Personnel scuffing along corridors with high resistance floors accumulate large static charges which may cause ESD. In operation theaters the presence of anesthetics and probable leakage during daily work may cause serious accidents to the patient, doctors, and nursing staff. These accidents may be fires, burns, or explosions. Also, during dry winter days, changing synthetic bed sheets and similar clothes may lead to ESD. To eliminate ESD hazards, flooring should be of an antistatic material with relatively low resistivity. Accumulation of static charges on moving personnel can be appreciably decreased by wearing heel grounders to offer a continuous path between them and the antistatic flooring. If the decay time of charge on personnel is a few tens of milliseconds, there will be no chance of charge accumulation since the time to take one step is longer than that. Also, great care should be taken to reduce anesthetic vapor in the ambient and to use a good ventilation system. Bed sheets and clothes of surgeons and nursing staff should be of conducting material.

20.8.7 Belt Drives

Mechanical rotational energy is transmitted by pulleys and belts in many industrial applications. If the belt material is a good insulator and in tight contact with the pulley surface, a static charge is generated on both surfaces. Due to the low conductivity of the belt material, accumulation of static charges will occur. The potential of the belt may reach several kilovolts and that may lead to sparks at some points in the system. These sparks are not only fatal to workers but also may lead to fires and explosions in some industries. Also, motors' rotors rotate through bearings with a thin film of lubricating oil having reasonable insulating properties, giving a chance of charge accumulation on the rotor. A potential difference of a few kilovolts is developed between the rotor and the stator. Induction motors usually have an air gap of less than one millimeter. Under operating conditions the air insulation in the gap, between the rotor and stator, is hot and a spark may develop therein, leading to damage in the motor windings. Also, sparks may develop at the bearings in the thin film of oil. Accidents due to electrostatic electrification such as hair pulling of female workers have been reported in many places.

To eliminate these hazards special precautions should be taken. The belt and pulley material should be made of a material having a relatively high conductivity to reduce charge accumulation to a reasonable value. If the belt material is not conductive, oil dressing is applied to it at intervals which may be several months with some oils. Also, the rotor can be connected permanently to earth through carbon brushes.

20.9 HAZARDS FROM ELECTRICAL EQUIPMENT AND INSTALLATIONS

Power system plants contain several elements installed either indoor or outdoor, many of which are sources of static charge generation. These charges may affect other installations which are not parts of the power plant, or affect the performance of some elements in the system.

In the following sections two examples will be discussed, namely, static charges in power transformers and from electric power transmission lines.

20.9.1 Static Charges in Power Transformers

Transformers are major elements in power plants and the possibility of their damage and outage is of paramount concern for continuity of supply and avoidance of long repair and maintenance times. Large power transformers are cooled by pumping away hot oil at the top of the tank, through pipes to radiators, and then back to the tank. During oil flow, the oil generates static charges by friction with the iron core, pipes, and radiator. If these are not properly earthed, static charges accumulate on the core and piping system, leading to voltages of a few kV appearing at some points in the transformer.

Under some conditions sparks may develop inside or outside the transformer. If these sparks occur frequently, damage and degradation of the winding and oil will occur, leading to transformer outage. Static electrification due to liquid dielectric flow has been found to be the cause of at least a dozen failures of large forced-oil-cooled power transformers in compact HVDC substations (Goto et al., 1988; Gasworth et al., 1988).

Static charge accumulation in transformers depends on the condition of the oil (fresh or aged), speed of flow, water content, material of pipes and its surface roughness, and temperature of oil. Experimental work reported by Radwan et al. (1992) showed that oil, when in friction with steel, aluminum, and brass, showed static electrification in descending order. Friction with pressboard resulted in static charge eight times that with steel.

To reduce static charge accumulation in transformers and the cooling system, good earthing of core, tank, and piping system is a major factor. Mixing anti-static additives (ASAs) with oil helps to a great extent in reducing static charge generation. There are several ASAs such as chlorothiazide, hexamine, and theophylline. It was found (Radwan et al., 1992) that the addition of 10 ppm of theophylline to oil reduces static electrification by 77%, at the expense of reducing the breakdown strength of oil by only 5.6%.

20.9.2 Hazards from HV Power Lines

Electric energy is transmitted by transmission lines for long distances at high voltages up to 765 kV and soon to reach 1000 kV. These lines are installed in areas with nearby normal daily activities. However, a safe distance is usually reserved for these lines, named "Right-of-Way" (ROW), where buildings and other permanent structures are forbidden (Table 20.4). The values shown in this table may differ from one country to another. However, the difference is usually in the order of $\pm 20\%$.

Farmers, animals, plants, pipelines, fences, and vehicles are subjected to various effects in the vicinity of power lines. These effects could result from capacitive, inductive, and conductive coupling between the power lines and other objects. This section will be devoted to capacitive coupling since it is voltage dependent.

The basic principle of capacitive coupling is the induced charges from voltage sources in system capacitances and hence the division of the voltage between them. The voltage induced in a vehicle parked under a transmission line can be calculated from the expression (Fig. 20.7)

$$V_{oe} = V_l \frac{C_{lo}}{C_{lo} + C_{oe}} \tag{20.7}$$

Table 20.4   Right-of-Way of HV Lines

Voltage level (kV)	Right-of-way (m)
500	29
220	24
132	15
66	14
33	12
20	5
11	5

where V_l is the HV line voltage, V_{oe} is the voltage between object or vehicle to earth, C_{lo} is the capacitance between HV line and object, and C_{oe} is the capacitance of the object to earth.

Pipelines and fences running parallel to power lines acquire charges and hence voltages, by induction, of several kilovolts when insulated from ground. The values of these voltages depend on the power line voltage, distance from the HV line, and phase arrangement of the power line. The open circuit voltage V_{op} on an insulated pipeline is given by (CIGRE, 1995)

$$V_{op} = 0.25 U h_p \sqrt{\frac{h_1^2}{d_{1p}^4} + \frac{h_2^2}{d_{2p}^4} + \frac{h_3^2}{d_{3p}^4} - \frac{h_1 h_2}{d_{1p}^2 d_{2p}^2} - \frac{h_2 h_3}{d_{2p}^2 d_{3p}^2} - \frac{h_3 h_1}{d_{3p}^2 d_{1p}^2}} \quad (20.8)$$

where U is the line voltage between phases in kV, h_p is the height of the pipeline above the ground in meters, h_j is the mean height of phase conductor j in meters ($j = 1, 2, 3$), and d_{jp} is the distance between phase conductor j and the pipeline in meters (Fig. 20.8). If the pipeline is earthed at

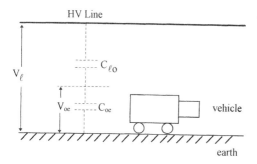

Figure 20.7   Induced voltages on vehicles parked under HV lines.

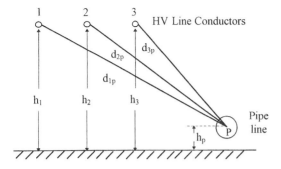

Figure 20.8 Induced voltages from HV conductors in pipelines.

one point, a short circuit current will flow to ground, given by

$$I_{sc} = j\omega cl V_{op} \tag{20.9}$$

where c is the capacitance of the pipeline to earth per unit length and l is its length.

If persons or workers touch the pipeline during or after installation they may experience an electric shock. The severity of the electric shock depends on the voltage acquired by the pipeline. The body current of a person touching the pipeline is approximately given by the short circuit current at one point of the pipeline. This assumption can be verified by the circuit shown in Figure 20.9 (CIGRE, 1995), as follows:

$$
\begin{aligned}
I_p &= \frac{V_{op}}{\dfrac{1}{j\omega C_p} + Z_b} \\
&= V_{op} \frac{j\omega C_p}{1 + j\omega C_p Z_b}
\end{aligned}
\tag{20.10}
$$

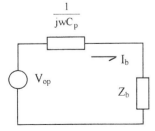

Figure 20.9 Equivalent electric circuit representing a person touching a pipeline.

where C_p is the total capacitance of the pipeline to ground and Z_b is the body contact resistance to ground. The capacitive impedance $1/\omega C_p$ is usually much higher than the body contact impedance to earth; thus the body current is approximately equal to the short-circuit current I_{sc} in equation (20.9). The body current is a function of the length of parallelism between power line and pipeline and is relatively independent of the contact impedance.

The capacitance per unit length between pipeline and earth varies between about 20 pF/m and 40 pF/m, depending on the ratio of height to radius, about 20 for $h/r > 5$ and 40 for $h/r = 2$, where h is the height of the pipeline from ground level and r is the radius of the pipeline (CIGRE, 1995).

Metallic fences are treated like pipelines in estimating the induced voltages and short-circuit currents.

The above discussions show that precautions should be taken to protect personnel and workers from lethal shocks when touching unearthed pipelines and fences running parallel to HV lines for a considerable distance. Earthing of pipes and fences is a safeguard against such shocks. An earthing resistance of about 5 Ω is adequate for such installations.

Fuel tank trucks unloading under or near overhead transmission lines may be subjected to fires. The voltage acquired by these trucks may reach several kilovolts and, with the truck almost insulated from earth, a spark may develop to earth. In the presence of fuel vapor or leakage during unloading, ignition may occur.

Animals suffer from injuries and lethal shocks when they come into contact with metallic structures in the vicinity of HV lines. Tall trees near HV lines suffer from corona discharge on sharp edges of their leaves, which affects their growth.

20.10 STATIC ELIMINATORS AND CHARGE NEUTRALIZERS

Static eliminators are used for industrial applications as a countermeasure against electrostatic problems. They are used to neutralize static charges during the handling and processing of sheets or webs of material, mainly in textiles, plastics, converting, printing, and similar processes. They play a vital role in maximizing production, reducing waste, and improving safety. Static eliminators include such as antistatic spray and brushes which stop buildup of static charges on lenses, disc drives, photographic materials, and metal rollers.

20.10.1 Grounding

In a static-safe work station, static charge removal from conductors is accomplished by the appropriate use of grounded table tops, wrist straps, and floor mats. The rate of charge removal is restricted by the capacitance of the conductor and the resistance of the path through which it flows to ground.

Grounding of bodies is considered most effective in the case of a charged conductive object. On the other hand, when the charged object is dielectric, grounding is not effective. Instead, neutralization with atmospheric ions, humidification, and use of antistatic agents are effective means in such cases. By grounding and bonding, the hazards of building up high charges of static electricity can be greatly reduced.

Grounding is the process of connecting one or more conductive objects to the ground. Grounding of all the plant machinery, metal parts, and enclosures is a must because it will continuously discharge static electricity.

Wrist straps and other connections used to ground employees should be solidly grounded where static-safe workstations are used for semiconductor, electronic, and explosive work. Grounding prevents the wrist straps from becoming a shock hazard in the event of a short circuit from a voltage to the wrist strap conductor.

Grounding is only an aid to reducing static electricity problems; it is not a complete solution. For example, grounding the operators will not drain off static electricity from their clothes, nor will it drain off static electricity from a plastic container one may be holding.

Bonding is the process of connecting two or more conductive objects together by means of a conductor. Bonding will equalize the potential between two adjacent non-current-carrying metal parts on enclosures. Approved grounding clamps are acceptable only for static bonding.

20.10.2 Charge Neutralization

Charge neutralizers are of four main types: passive units, active units, air ionizers, and radioactive units. The four types produce air ionization and achieve charge neutralization by migration of air ion countercharge to the charged surface (Chubb, 1985).

20.10.2.1 Passive Neutralizers

Passive neutralizers operate by using the charge on the surface to be neutralized to generate an electric field at nearby grounded surfaces and, by so concentrating this electric field at a number of points of small radii of curvature, to cause local electrical breakdown of air close to these points. The electrical breakdown is localized as a corona by the small radius of curvature and is unable to form a spark channel to bridge the gap to the surface. The breakdown processes produce numerous air ions and these move in the electric field between the corona discharge region and the surface. Ions of opposite polarity to the charge on the surface will move to the surface and tend to neutralize it. This process will operate so long as the electric field at the points is locally above the breakdown strength of air.

Practical passive neutralizers often use an array of fine wires mounted from a support bar. The use of a carbon fiber brush, in which the fiber diameter is around 10 μm, is an attractive alternative to wires.

The disadvantages of passive neutralizers are their inability to operate at very low levels of surface charge density and their tendency to overcompensate. This overcompensation is associated with the aerodynamic displacement of the ion current flow to the surface by the air motion induced by the motion of the web surface. This effect may be minimized by use of a second neutralizer, preferably one with a heated discharge wire or points (Secker, 1984).

20.10.2.2   Active Neutralizers

These units have a similarity to passive neutralizers in that the concentration of electric field at fine points or a wire is used to provide a source of air ionization. The main difference is that a separate AC high-voltage power source is used to generate this electric field. The generation of ions is thus independent of the level of charge on the web surface. The field generated by this charge on the web is involved only as the means of transporting ions of appropriate polarity and quantity to achieve neutralization. Voltages of 6 to 12 kV are commonly used at normal mains frequency. For webs moving at speeds above 2 m/s, a higher frequency energization is needed to achieve uniform neutralization (Secker, 1984).

The risks of shock to personnel and ignition of flammable atmospheres can be avoided with active neutralizers by low-value capacitive coupling of the power supply to each discharge point. Grounding the discharge points and having the nearby high-voltage electrodes embedded within suitably robust encapsulation may also reduce the risk of shock.

20.10.2.3   Air Ionizer Units

These are similar to active neutralizer units but use an air flow or air blast to project air ions of both polarities to a region in which there are surfaces which may have charges needing to be neutralized (Larigaldie and Giboni, 1981). Ideally, the units produce a neutral mixture of positive and negative air ions of appropriate quantity and sign which may be extracted by the electric fields generated by any surface charges. A balancing quantity of ionization will flow to other nearby grounded surfaces to maintain neutrality. In this way the air ions make the air weakly conducting in the region treated. This approach is mainly used in semiconductor manufacturing plants.

The main problems with this approach are the difficulty of ensuring good neutrality of the air ion mixture, the risk of introducing ozone into the working environment, and the danger that electric fields from shielded charges will cause charge to be deposited on other nearby neutral insulating surfaces.

20.10.2.4 Radioactive Neutralizers

Radioactive units use an alpha-particle-emitting source as a means of generating air ionization, from which the charges on the surface to be neutralized may extract an appropriate quantity and polarity of air ions to achieve neutralization. Po 210 is the usual radioactive isotope used because it produces high-energy alpha particles with little gamma particle emission and has a convenient half-lifetime of 138 days. Neutralizing sources having currents around $9\,\mu A/m$ are available (Hadden, 1983).

Radioactive neutralizers do not provide as high neutralizing currents as those of passive and active units. Their performance falls off over a period of several months. They are simple and compact, require no external power source, and should be able to be used in flammable atmospheres with precautions to prevent risk of sparking if highly charged webs can move to touch grounded surfaces.

20.11   LIGHTNING PROTECTION

Lightning is a well known natural phenomenon observed all over the world. The buildup of large quantities of charges in clouds and the development of lightning strokes have been described in Chapter 14.

This section is devoted to protection against lightning strokes to buildings, metallic structures, industrial installations, and overhead power transmission lines.

The main protection schemes against lightning are rods or masts (Franklin's rod), earth wires, and surge diverters and arresters. The function and operation of surge diverters and arresters are covered in Chapter 14.

20.11.1   Lightning Rods

Protection of buildings and similar structures against direct lightning strokes is usually in the form of rods installed on or near them and projecting higher than the protected object.

A lightning protection system consists of the rod (lightning receiver), earthing system, and the current-carrying conductor between the rod and earth. The rod can be made of copper, steel, or any other similar material. The tip of the rod is usually sharp, to create a high electric field to help in directing and attracting the lightning stroke to it. Lightning rods as high as 60 m are sometimes used. In this case a short length of rod fastened to a metallic lattice structure or wooden poles is used.

Lightning rods installed on the ground surface or on the roof of a building have what is called a "protective zone" (Fig. 20.10). There are several methods and empirical formulas to determine the protective zones of single and multiple rods. For a single rod, a cone drawn from its tip with a radius r on the ground surface, or with a solid angle α to the vertical, determines the protective zone of the rod.

According to IEC 1024-1 (IEC, 1993), there are four lightning protection levels having efficiencies of 0.98, 0.95, 0.9, and 0.8. The efficiency of a lightning protection system is given by

$$E = 1 - \frac{N_c}{N_d} \tag{20.11}$$

where N_c is the value of the annual accepted frequency of flashes for the considered structure, and N_d is the actual annual value of frequency of lightning flashes to the structure.

The design of a lightning protection system depends on several factors such as the activity of lightning in the area, configuration of the structure to be protected, importance of the structure, and safety of personnel and property.

The protective zone of a rod of height less than 30 m is given by the following empirical formulas (Razevig, 1972):

for $h_x < 2/3h$

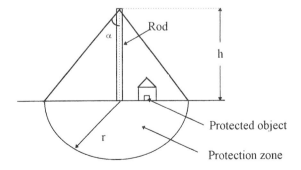

Figure 20.10   Protective zone of a Franklin's rod.

$$r_x = 1.5h\left(1 - \frac{h_x}{0.8h}\right) \tag{20.12}$$

and for $h_x > 2/3h$

$$r_x = 0.75h\left(1 - \frac{h_x}{h}\right) \tag{20.13}$$

where h is the height of the rod and h_x is the height of the protected object. It is clear from these formulas that as the height of the protected object approaches that of the rod the protection zone is decreased until it is zero for equal heights.

For lightning rods of height greater than 30 m the values obtained from equations (20.12) and (20.13) are multiplied by the factor k which is equal to $5.5/\sqrt{h}$.

Other methods for determining the protective zone are the protective angle and the rolling sphere (IEC, 1993). The idea of the protection angle and rolling sphere is illustrated in Figure 20.11. The value of the protection angle and radius of the rolling sphere depend on the protection level required and the height of the rod. In many cases more than one protective rod is required, to increase the area of the protective zone and the degree of protection.

To increase the protective zone and degree of protection and to decrease the height of the rod, the electric field at the tip of the protecting rod should be increased to help direct the lightning stroke to it. The field at the tip of the rod can be increased by applying to it a repetitive impulse voltage of a few tens of kilovolts from an external source (Webster, 1999).

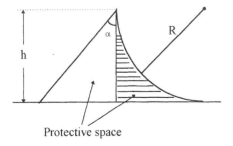

Figure 20.11 Protective zone of a rod determined by protective angle α and rolling sphere of radius R.

20.11.2 Safe Distance From a Lightning Rod

An object within the protective zone of a rod is considered safe from lightning strokes with a certain degree of reliability. This reliability can be as
high as 99.9%, depending on the design of the lightning protection system
(LPS). As the protected object moves near the boundaries of the protection
zone, the reliability of protection decreases, and vice versa. However, the
object should not be too close to the rod otherwise flashover from the rod to
the object, in air or in the earth soil, may occur, depending respectively on
whether the object is insulated from earth or earthed. Figure 20.12 shows a
protective mast and an object within the protection zone. The distances d_a
and d_e are the safe distances between the mast and the object in air and on
earth respectively. The values of these distances can be obtained from the
voltages which appear at the points on the mast facing the object during a
lightning stroke.

   Assuming that the mast has an inductance L in henries/m and the
earthing resistance is R ohms, the voltage at points a and e are given by
the following equations

$$V_a = I_m R + h_1 L \frac{di}{dt} \tag{20.14}$$

and

$$V_e = I_m R \tag{20.15}$$

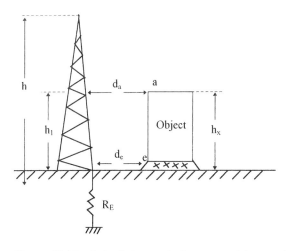

Figure 20.12 Safe distances between an object and a protective mast.

where h_1 is the height of point a facing the mast, di/dt is the rate of rise of the lightning current, and I_m is the maximum value of the lightning current. Knowing these voltages and the flashover voltages per meter of air and earth soil, both safe distances can be calculated.

20.11.3 Lightning Rod Dimensions

The height of the rod h is determined by the protective zone required, the level of protection, and the height of the object to be protected.

The cross-section of the rod should be capable of carrying the lightning current expected in the local area. The rod temperature during discharge of a lightning stroke should not exceed a certain limit, depending on its material, to avoid oxidization. The cross-sectional area can be calculated by considering the energy involved in raising the rod temperature. Assume the resistance per unit length of the rod is r, the lightning current is $i = I_m e^{-t/T}$, and the wave-tail time is T_2. The wave front time is relatively small and hence the heating effect during this period can be neglected. From this approximation the wave-tail time T_2 corresponding to 50% of I_m is approximately 70% of the time constant T of the lightning current wave (Fig. 20.13). The energy in the current wave is given by

$$W = r \int_0^\infty i^2 dt \tag{20.16}$$

$$= r \int_0^\infty I_m \left(e^{-\frac{0.7t}{T_2}} \right)^2 dt$$

$$= r I_m^2 \frac{T_2}{1.4} \tag{20.17}$$

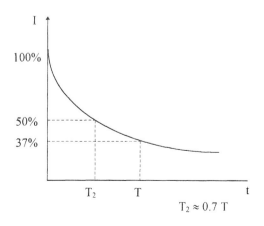

Figure 20.13 Approximate representation of a lightning current wave.

The temperature rise of the rod can be obtained from the expression

$$t = \frac{W}{gac}$$

$$= \frac{\rho I_m^2 T_2}{1.4gca^2}$$

(10.18)

where ρ is the specific resistance of rod material (ohm·cm), a is the cross-section (cm^2), g is the specific gravity (g/cm^3), and c is the specific heat (cal/g · °C).

20.11.4   Lightning Aerial Wires

Aerial wires are used to protect overhead transmission line conductors from direct lightning strokes. They may be single or double bare wires suspended between towers above the phase conductors. Aerial wires have protective zones similar to those of lightning rods. However, since they are used mainly in conjunction with transmission lines conductor, the protective zone is replaced by the protection angle α (Fig. 20.14). The value of α in most existing power lines is in the range of 20–30°.

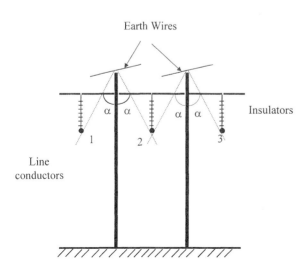

Figure 20.14   Aerial earth wires and protective angle α.

20.12. PROBLEMS

(1) A person having a capacitance of $100\,\text{pF}$ and whose resistance to ground, including his footgear, is $10^9\,\Omega$ is charged to a voltage of $20\,\text{kV}$ during walking on a carpeted floor. Calculate the time constant of the man's body and the times needed to discharge to $10\,\text{kV}$ and $100\,\text{V}$.

(2) The same person as in problem (1) is walking rapidly at the rate of $6.5\,\text{km/h}$ and his step is $0.9\,\text{m}$. Does this person accumulate static charges during his walk?

(3) A person with a capacitance of $120\,\text{pF}$ and whose resistance to ground is relatively large is walking on a carpeted floor. He gains a charge of $12 \times 10^{-9}\,\text{C}$ per step. How many steps does he walk before his body attains $10\,\text{kV}$?

(4) An insulated fuel tank is filled through a pipeline. The tank capacity is 2.4 million gallons and it can be filled in 10 days. If the rate of flow of electric charges in the fuel is $4 \times 10^{-10}\,\text{A}$ per gallon per minute and the tank capacitance is $10\,\text{nF}$, calculate:

  (a) the voltage developed between the tank and ground after 10 minutes of filling;

  (b) if a spark were to develop between the tank and ground, what is the energy in the spark? Is this energy capable of igniting the fuel and air mixture, given that the ignition threshold energy is $10^{-3}\,\text{J}$?

  (c) the tank voltage to earth at the end of the filling period. Is the answer practical? Comment on the answer.

(5) A tank truck is travelling at a speed of $80\,\text{km/h}$ on a concrete road. A charge is accumulated on the truck at a rate of $10^{-11}\,\text{C/km}$. If the truck reaches the end station after $4\,\text{h}$ and its capacitance and resistance to ground are $400 \times 10^{12}\,\text{F}$ and $10^{10}\,\Omega$ respectively, calculate:

  (a) the charge accumulated on the tank truck after its trip;

  (b) the potential of the tank truck to ground just after its arrival;

  (c) the time needed for complete discharge of the truck;

  (d) the time needed for the truck to discharge to $100\,\text{V}$.

(6) An earthed aerial pipeline having a diameter of $0.5\,\text{m}$ runs parallel to a $500\,\text{kV}$ HV transmission line for $500\,\text{m}$. The dimensions and distances (Fig. 20.8) are as follows:

$$h_1 = h_2 = h_3 = 18\ \text{m}$$

$$d_{12} = d_{23} = 12\ \text{m}$$

The horizontal distance between the nearest line conductor and the pipeline is 20 m. The pipeline is 1 m above earth level. Calculate the induced voltage on the pipeline and hence the body current of a person touching it. The capacitance of the pipeline to earth is 30 pF/m.

(7) A 220 kV tower is struck by a lightning stroke at a height of 20 m. The tower footing resistance is 5 Ω and its inductance is 1.5μH/m. If the lightning current is 30 kA and its rate of rise is 25 kA/μs, calculate the tower voltage with respect to earth. If the BIL of the line is 1050 kV, does back flashover occur between the tower body and the phase conductors?

(8) A lightning mast of 20 m height is installed to protect a structure of 8 m height. Calculate the radius of the protective zone of this mast. If the structure height is increased to 15 m, calculate the reduction in the protective zone radius.

(9) A lightning mast, as shown in Figure 20.12, is installed to protect a metallic tank. The mast earthing resistance is 5 Ω and its inductance is 2 μH/m. The tank, with its support, is 7.5 m high. If a direct lightning stroke of 40 kA and a rate of rise of 25 kA/μs strikes the mast, calculate the safe distances d_a and d_e when the tank is insulated from earth and when resting directly on the ground. Assume the permissible electric field in air and ground soil to be 500 and 300 kV/m respectively.

(10) A filling hose with an internal diameter of 128.2 mm is used to fill a petrol tank. Calculate the optimum velocity of fuel flow and the flow rate in gallons per minute.

(11) Calculate the cross-sectional area of a copper lightning rod to protect against a lightning current of 200 kA having a wave shape of 0/125 μs, given the following constants for copper:

Density	8.9 g/cm^3
Specific heat	0.093 cal/g·K
Resistivity at 20°C	1.7×10^{-6} Ω·cm
Resistivity temp. coefficient	0.004
Allowed temperature rise	100K

(12) Discuss and give reasons for the following:

(a) Cotton clothes and bed sheets are used in operation theaters rather than synthetic fabrics.

(b) In grain grinding factories it is not recommended to clean objects from accumulated dust particles by air blowing.

(c) Conductive oil dressings are used with belt conveyors.

(d) Sparks between the rotor and stator may develop in induction motors used for belt drives.

(e) Rubber hoses for filling petrol tanks have embedded metal wires.

(f) Filling petrol tanks at the top at high speed is dangerous.

(g) In the textile industry it is recommended that the humidity should be within 50%.

(h) Loose hair of women's workers in some industries may lead to accidents.

REFERENCES

Adams CK. Nature's Electricity. London: McGraw-Hill, 1986.

Anderson JG. Transmission Line Reference Book 345 kV and Above, EPRI, Palo Alto, CA, 1979. ch. 12.

API. Protection against Ignitions arising out of Static, Lightning, and Stray Currents. Washington: American Petroleum Institute, 1981.

Bodurtha FT. Industrial Explosion Prevention and Protection. New York: McGraw-Hill, 1980.

Boxleitner W. Electrostatic Discharge and Electronic Equipment: A Practical Guide for Designing to Prevent ESD Problems, New York, IEEE Press, 1988.

Chubb JN. The control of static electricity. Electrostatics Summer School 85, University College of North Wales, Bangor, UK, 1985.

CIGRE. Working Group 36.02. Guide on the Influence of High Voltage AC Power Systems on Metallic Pipe Lines. Paris, 1995.

Cross J. Electrostatics: Principles, Problems, and Applications, Adam Hilger, IOP Publishing Limited, Great Britain, 1987.

Crowley JM. Fundamentals of Applied Electrostatics. New York: John Wiley & Sons, 1986.

Gasworth M, Melcher J, Zahn M. IEEE Trans Electr Insul 23:103–115, 1988.

Goto K, Okubo H, Miyamoto T, Tsukioka H, Kohno Y. IEEE Trans Electr Insul 23:153–157, 1988.

Hadden RJB. Industrial static eliminators. International Workshop on Electrostatics, Llandudno, Oyez, 1983.

Halperin S. Guidelines for Static Control Management. China: Eurostat, 1990.

Hayt WH Jr. Engineering Electromagnetics. London: McGraw-Hill, 1987.

Huang J, Liu S, Li C. IEEE Industry Applications Society Annual Meeting, New Orleans, LA., 1997.

IEC. IEC 1024.1: Protection of Structures. Geneva: International Electrotechnical Commission, 1993.

Larigaldie S, Giboni N. J Electrostat, Morgan Hill, CA., Volume 10, page 57, 1981.

Moore AD. Electrostatics: Exploring, Controlling, and Using Static Electricity. Laplacian Press, 1997.

Radwan RM, El-Dewieny RM, Metwally IA. IEEE Trans Electr Insul 27:278–286, 1992.

Razevig DV. High Voltage Engineering. Delhi: Khanna Publishers, 1972, p 430.

Secker PE. Getting rid of electrostatic charge. International Symposium on Electrostatics—Application and Hazards, Southampton, Oyez, 1984.

Unger B. Basic ESD Seminar, China, 1997.

Webster JG, ed. Wiley Encyclopedia of Electrical and Electronics Engineering, vol 9. New York: John Wiley and Sons, 1999, p 714.

Index